Lecture Notes in Computer Science   6637

Commenced Publication in 1973
Founding and Former Series Editors:
Gerhard Goos, Juris Hartmanis, and Jan van Leeuwen

W0246094

Jianliang Xu   Ge Yu   Shuigeng Zhou
Rainer Unland (Eds.)

Database Systems for Adanced Applications

16th International Conference, DASFAA 2011
International Workshops: GDB, SIM3, FlashDB,
SNSMW, DaMEN, DQIS
Hong Kong, China, April 22-25, 2011
Proceedings

 Springer

Volume Editors

Jianliang Xu
Hong Kong Baptist University, Department of Computer Science
Kowloon Tong, KLN, Hong Kong, China
E-mail: xujl@comp.hkbu.edu.hk

Ge Yu
Northeastern University, School of Information Science and Engineering
Shenyang, Liaoning 110004, China
E-mail: yuge@ise.neu.edu.cn

Shuigeng Zhou
Fudan University, School of Computer Science
220 Handan Road, Shanghai 200433, China
E-mail: sgzhou@fudan.edu.cn

Rainer Unland
University of Duisburg-Essen
Institute for Computer Science and Business Information Systems (ICB)
Schützenbahn 70, 45117 Essen, Germany
E-mail: unlandr@informatik.uni-essen.de

ISSN 0302-9743   e-ISSN 1611-3349
ISBN 978-3-642-20243-8   ISBN 978-3-642-20244-5 (eBook)
DOI 10.1007/978-3-642-20244-5
Springer Heidelberg Dordrecht London New York

Library of Congress Control Number: 2011924108

CR Subject Classification (1998): H.3, H.4, I.2, C.2, H.2, H.5

LNCS Sublibrary: SL 3 – Information Systems and Application, incl. Internet/Web
and HCI

Typesetting: Camera-ready by author, data conversion by Scientific Publishing Services, Chennai, India

Printed on acid-free paper

Springer is part of Springer Science+Business Media (www.springer.com)

Preface

Database Systems for Advanced Applications (DASFAA) is an annual international database conference, located in the Asia-Pacific region, which showcases state-of-the-art R&D activities in database systems and their applications. It provides a forum for technical presentations and discussions among database researchers, developers and users from academia, business and industry. DASFAA 2011, the 16th in the series, was held in Hong Kong during April 22–25, 2011.

Among the proposals submitted in response to the call-for-workshops, we carefully selected six workshops, each focusing on a specific area that contributes to the main themes of the DASFAA conference. This volume contains the papers accepted for these six workshops that were held in conjunction with DASFAA 2011. They are:

- First International Workshop on Graph-Structured Data Bases (GDB 2011)
- First International Workshop on Spatial Information Modeling, Management and Mining (SIM3 2011)
- International Workshop on Flash-Based Database Systems (FlashDB 2011)
- Second International Workshop on Social Networks and Social Media Mining on the Web (SNSMW 2011)
- First International Workshop on Data Management for Emerging Network Infrastructures (DaMEN 2011)
- 4th International Workshop on Data Quality in Integration Systems (DQIS 2011)

We are very grateful to the workshop organizers for their tremendous effort in soliciting papers, selecting papers by peer review, and preparing attractive programs. We asked all workshops to follow a rigid paper selection process, including the procedure to ensure that any Program Committee members were excluded from the paper review process of any paper they were involved in. A requirement about the overall paper acceptance rate was also imposed on all the workshops.

We would like to express our appreciation to Dik Lun Lee, Wang-Chien Lee, Kamal Karlapalem, Jeffrey Xu Yu, Myoung Ho Kim, Kam-Fai Wong, Samuel Tam, and many other people for their support in the workshop organization. Our thanks also go to Rainer Unland and Veronika Muntoni for their hard work in compiling this proceedings volume.

April 2011

Jianliang Xu
Ge Yu
Shuigeng Zhou

GDB 2011 Workshop Organizers' Message

Recent applications on graph-structured data, for example, biological and chemical databases, social networks, business process models, ontologies, the Semantic Web and XML, have sparked a renewed interest in graph-structured databases. The database research community has been actively contributing to pressing issues on graph-structured databases including query models and languages, performance optimization, pattern matching and mining, keyword search, query processing on modern architectures and infrastructure, graph visualization and user interfaces, among many others.

The First International Workshop on Graph-Structured Data Bases (GDB) was held on April 22, 2011 at The Chinese University of Hong Kong, in conjunction with the 16th International Conference on Database Systems for Advanced Applications (DASFAA 2011). The GDB workshop targets at bringing academic and industrial researchers together to share their knowledge and opinions on visions, challenges and solutions for graph-structured databases.

The workshop attracted ten submissions in addition to an invited keynote. The submissions are highly diversified, coming from China, France, Germany, Hong Kong, Italy, Singapore, South Korea and Spain. The Program Committee consisted of 15 members. All submissions were peer reviewed by at least two Program Committee members for originality, impact and technicalities. The Program Committee selected five papers to be included in the workshop proceedings (acceptance rate 50%). The accepted papers covered research areas on query models and languages, graph-structured data mining, performance optimization, querying in modern architectures, cloud computing and ontologies.

The workshop would not be possible without the help from many people and organizations. Firstly, we would like to thank the Program Committee members for evaluating the submissions in a prompt and professional manner. Secondly, we are thankful to Lei Chen for accepting our invitation to give a keynote talk. Thirdly, we would like to express our gratitude to the Workshop Co-chairs of DASFAA 2011 for their help in preparing the workshop. Finally, we thank all authors and participants of the workshop for their contributions.

We believe that the first GDB workshop provided an exciting platform for exchanging interesting ideas and insights on novel research problems. We trust that the workshop will become a traditional annual meeting opportunity for researchers on graph-structure databases.

April 2011

Sourav S. Bhowmick
Byron Choi
Wei Wang

SIM3 2011 Workshop Organizers' Message

Nowadays, spatial data exist pervasively in various information systems and applications. The unprecedented amount of spatial data that has been amassed and that is being produced in an increasing speed calls for extensive research on spatial information modeling, management, and mining. The First International Workshop on Spatial Information Modeling, Management and Mining (SIM3 2011) was a half-day workshop held in conjunction with DASFAA 2011. The workshop provided a forum for original research contributions and practical experiences of spatial information modeling, management, and mining.

The workshop received 17 submissions from Asian, North America, and Europe. Through careful reviewing by the Program Committee, eight full papers and two short papers were selected for the presentation and inclusion in the proceedings. We grouped the ten accepted papers into two sessions.

The papers in the first session discussed spatial data compression, storage, and query. Yu et al. presented a stream compressor GDScomp for GML documents. Wei proposed a compression algorithm for GML documents that supports direct querying over the compressed GML documents. Zhu et al. proposed an approach to store and query GML documents based on model-mapping. Wang presented a framework and a prototype of GML data management. Li et al. introduced a multi-layer grid method for skyline queries in distributed environments. In the second session, five papers covered spatial planning, visualization, mining, and information systems. Yuan and Schneider presented an interesting approach for route finding in a gridded environment. Tahir et al. presented a Web-based visualization tool to support map personalization by analyzing users' mouse movement data. Viswanathan and Schneider carefully reviewed the existing modeling strategies for spatial data warehouses and SOLAP from conceptual, logical, and implementation levels. Jin et al. proposed a method to find out the optimal parameter value of the bandwidth for clustering with Bayesian posterior density estimation and MCMC method. Jin presented an extension to Oracle DBMS to support the development of spatial–temporal information management.

A successful workshop requires a lot of efforted from many people. First, we would like to thank the authors for their contributions, and the Program Committee members for reviewing and selecting papers. In addition, we appreciate DASFAA 2011 workshop Co-chairs Jianliang Xu, Ge Yu, and Shuigeng Zhou for the excellent coordination. Finally, we would like to thank the local Organizing Committee for its wonderful arrangements.

April 2011

Xin Wang
Jihong Guan

FlashDB 2011 Workshop Organizers' Message

Recently, new storage media such as flash memory have been developed very quickly, creating big challenges to the architecture of computer systems as well as the design of system software. In particular, NAND flash in the form of solid state disks (SSDs) has been an alternative to traditional magnetic disks, both in the home-user environment and in the enterprise computing environment, due to its shock-resistance, low power consumption, non-volatile, and high I/O speed. The special features of flash memory and other new storage media impose new challenges to traditional data management technologies. As a result, exploiting the characteristics of flash memory and other new storage media has become an important topic of database systems research. The data management community needs to rethink traditional underlying storage architecture, query processing algorithms, indexing mechanism, buffer management schemes as well as many traditional issues in magnetic-disk-oriented database systems in order to adapt to the advances in the underlying storage infrastructure.

The First International Workshop on Flash-Based Database Systems (FlashDB 2011) was held on April 22, 2011 in Hong Kong in conjunction with DASFAA 2011. The overall goal of the workshop was to bring together researchers who are interested in optimizing database performance on flash memory or other new storage media-based infrastructure by designing new data management techniques and tools.

The workshop attracted 12 submissions from Germany, Poland, France, Korea and China. All submissions were peer reviewed by at least three Program Committee members to ensure that high-quality papers were selected. On the basis of the reviews, the Program Committee selected seven papers for inclusion in the workshop proceedings (acceptance rate 58%). The final program of the workshop also consisted of four invited talks. One of them was from IBM Almaden Research, USA, and the other three were from academia, presented by Theo Haerder (University of Kaiserslautern, Germany), Sang-won Lee (Sungkyunkwan University, South Korea), and Jianliang Xu (Hong Kong Baptist University, Hong Kong).

The Program Committee of the workshop consisted of 14 experienced researchers and experts. We would like to thank the valuable contribution of all the Program Committee members during the peer-review process. Also, we would like to acknowledge the DASFAA 2011 workshop Chairs for their great support in ensuring the success of FlashDB 2011, and the support from the Natural Science Foundation of China (No. 60833005) and Tin Ka Ping Foundation.

April 2011

Xiaofeng Meng
Lihua Yue
Peiquan Jin
Bin Cui
Zhiyong Shan

SNSMW 2011 Workshop Organizers' Message

The Second International Workshop on Social Networks and Social Media Mining on the Web (SNSMW 2011) was held in Hong Kong, China on 22 April, 2011, in conjunction with the DASFAA 2011 conference. The aim of the workshop is to provide a premium forum for researchers and industrial practitioners to disseminate their latest research progress and advances in social networks and social media mining. The Web has evolved since its birth. Currently, the role of the Web is not only a medium for information transmission but also a medium for people's collaboration. The Social Web is emerging as an active and non-negligible research area in Web computing. The topics of interest include computational models for social media, social network analysis/mining, community detection and evolution, blog search and retrieval, group interaction, collaboration, and recommendation, trust and privacy techniques for social media, and so on.

This year's workshop attracted 20 submissions from China (including Hong Kong and Taiwan), India, Japan, Korea, Iran, Poland, Greece, Denmark, France and the USA, which cover a broad range of interesting topics in social Web computing. All submissions were rigorously peer reviewed by three Program Committee members. The Program Committee selected 11 papers for inclusion in the proceeding (acceptance ratio is 55%).

The workshop would not have been successful without the help of many organizations and individuals. First, we would like to thank the Program Committee Co-chairs of SNSMW. Guandong Xu coordinated the workshop organizational affairs and prepared the CFP; Lin Li set up a submission system for the workshop. Hong Cheng built the workshop website; Botao Wang helped to form the Program Committee. We also want to extend our great gratitude to Yanhui Gu, who helped to manage the review process for the workshop. Second, we would like to thank the hard work of all Program Committee members for evaluating the assigned papers in a timely and professional manner. In addition, we appreciate the guidance and communication of the DASFAA 2011 workshop Committee, who ensured that the workshop can smoothly. Last but not least, we would like to thank all the authors who submitted very interesting and impressive papers from their recent work. We hope we can continue this workshop in the coming years.

April 2011

Guandong Xu
Lin Li
Hong Cheng
Botao Wang

DaMEN 2011 Workshop Organizers' Message

The emerging network infrastructures such as P2P, mobile and sensor networks, and cloud computing were once lab toys. Nonetheless, they show a strong potential of becoming mainstream in the foreseeable future. While most network-side issues have been addressed or resolved, the data management issues that arise from the real deployment of these infrastructures are ever increasing. In particular, challenges associated with acquiring, storing, processing, and analyzing large-scale data from these heterogeneous networks call for novel data management techniques. The inherently dynamic nature of these networks further poses new research issues, such as privacy and security. The DaMEN workshop aims to facilitate the collaboration between researchers in database and networking areas by presenting cutting-edge research topics and methodologies.

The Program Committee (PC) of the workshop consisted of 18 researchers and specialists from 16 universities and institutions across China, Japan, Singapore and the USA. In a rigorous review process, each submission was reviewed by three experts for its technical merit, originality, significance and relevance to the workshop. Finally, the PC Chairs decided to accept 60% of the submitted papers. The accepted papers span exciting topics from cloud computing to VANETs, and investigate issues such as storage, search and query processing.

We are very grateful for the efforts of all authors who presented their frontier work in related areas. Finally, we appreciate the impeccable support of all members of the PC and the external reviewers, who provided excellent feedback and valuable directions for the authors to improve their work.

Last but not least, the final program of the workshop consisted of an invited keynote talk. The invitation was kindly accepted by Yunhao Liu from the Hong Kong University of Science and Technology. The title was "GreenOrbs: Lessons Learned from Extremely Large-Scale Sensor Network Deployment." We would also like to thank Aoying Zhou and Yoshiharu Ishikawa for serving as the workshop General Co-chairs.

April 2011   Haibo Hu
                                                             Weining Qian

DQIS 2011 Workshop Organizers' Message

Integration systems have been the subject of intense research and development for over three decades. A fundamental aspect of user satisfaction from integration systems is the quality of data they produce. Industry reports indicate that expensive data integration initiatives stemming from migrations, mergers, legacy upgrades etc. succeed in achieving a common technology platform, but are rejected by the user communities due to the presence (or revelation) of poor data quality. Poor data quality is known to compromise the credibility and efficiency of commercial as well as public endeavors. Several developments from industry as well as academia have contributed significantly toward addressing the problem. These typically include analysts and practitioners who have contributed to the design of strategies and methodologies for data governance; solution architects including software vendors who have contributed toward appropriate system architectures that promote data integration and data experts who have contributed to data quality problems such as duplicate detection, identification of outliers, consistency checking and many more through the use of computational techniques.

The DQIS workshop provided a forum for diverse researchers and made a consolidated contribution to new and extended methods for addressing the challenges of data quality in integrating systems. Topics covered by the workshop include data integration, linkage and fusion; entity resolution, duplicate detection, and consistency checking; data profiling and measurement; use of data mining for data quality assessment; methods for data transformation, reconciliation, consolidation; algorithms for data cleansing; data quality and cleansing in information extraction; dealing with uncertain or noisy data; data lineage and provenance; etc. Following the success of MCIS 2008 in New Delhi, India, MCIS 2009 in Brisbane, Australia, and MCIS 2010 in Tsukuba, Japan, the 4th workshop (renamed DQIS) was held on April 22, 2011 at the Chinese University of Hong Kong in conjunction with the 16th International Conference on Database Systems for Advanced Applications (DASFAA 2011). This year, the DQIS workshop attracted eight submissions from Australia, China, Italy, Austria, and the USA. All submissions were peer reviewed by at least three Program Committee members to ensure that high-quality papers were selected. On the basis of technical merit, originality, significance, and relevance to the workshop, the Program Committee decided on four papers to be included in the workshop proceedings (acceptance rate 50%). The workshop also invited two papers and one keynote.

The workshop Program Committee consisted of 15 experienced researchers and experts in the area of data analysis and management. We would like to acknowledge the valuable contribution of all the Program Committee members during the peer-review process. Also, we would like to express our gratitude to the DASFAA 2011 workshop Chairs for their great support in ensuring the success of DQIS 2011.

April 2011

Xiaochun Yang
Shazia Sadiq
Xiaofang Zhou
Ke Deng

DASFAA 2011 Workshop Organization

Workshop Co-chairs

Jianliang Xu	Hong Kong Baptist University, China
Ge Yu	Northeastern University, China
Shuigeng Zhou	Fudan University, China

Publication Chair

Rainer Unland	University of Duisburg-Essen, Germany

First International Workshop on Graph-Structured Databases (GDB 2011)

Workshop Co-organizers

Sourav S. Bhowmick	Nanyang Technological University, Singapore
Byron Choi	Hong Kong Baptist University, China
Wei Wang	The Univesity of New South Wales, Australia

Program Committee

Yasuhito Asano	Kyoto University, Japan
Stephane Bressan	National University of Singapore, Singapore
Lei Chen	Hong Kong University of Science and Technology, Hong Kong, SAR China
James Cheng	Nanyang Technological University, Singapore
Bingsheng He	Nanyang Technological University, Singapore
Yiping Ke	The Chinese University of Hong Kong, Hong Kong, SAR China
Xuemin Lin	University of New South Wales, Australia
Chengfei Liu	Swinburne University of Technology, Australia
Eric Lo	Hong Kong Polytechnic University, Hong Kong, SAR China
Wilfred Ng	Hong Kong University of Science and Technology, Hong Kong, SAR China
Lu Qin	The Chinese University of Hong Kong, Hong Kong, SAR China

Sherif Sakr	University of New South Wales, Australia
Nan Tang	University of Edinburgh, UK
Jeffrey Xu Yu	The Chinese University of Hong Kong, Hong Kong, SAR China
Shuigeng Zhou	Fudan University, China

First International Workshop on Spatial Information Modeling, Management and Mining (SIM3 2011)

Workshop Co-chairs

| Xin Wang | University of Calgary, Canada |
| Jihong Guan | Tongji University, China |

Program Committee

Lars Bernard	Technical University of Dresden, Germany
Michela Bertolotto	University College Dublin, Ireland
Elena Camossi	JRC, ISPRA, Italy
Han Cao	Shannxi Normal University, China
Christophe Claramunt	Naval Academy Research Institute, France
Liqiang Geng	National Research Council, Canada
Bo Huang	The Chinese University of Hong Kong, SAR China
Yan Huang	University of North Texas, USA
Yoshiharu Ishikawa	Nagoya University, Japan
Bin Jiang	University of Gavle, Sweden
Ning Jing	National University of Defence Technology, China
Ki-Joune Li	Pusan National University, Korea
Songnian Li	Ryerson University, Canada
Xiang Li	East China Normal University, China
Eleni Mangina	University College Dublin, Ireland
Gavin McArdle	NCG, Ireland
Zhiyong Peng	Wuhan University, China
Wolfgang Reinhardt	UniBw München, Germany
Angela Schwering	University of Osnabrück, Germany
Wenzhong Shi	Hong Kong Polytechnic University, China
Xiaohua Tong	Tongji University, China
Monica Wachowicz	Wageningen University, The Netherlands
Shuliang Wang	Wuhan University, China
Shuigeng Zhou	Fudan University, China
Fubao Zhu	Fudan University, China

International Workshop on Flash-Based Database Systems (FlashDB 2011)

Workshop General Co-chairs

Xiaofeng Meng   Renmin University of China, China
Lihua Yue   University of Science and Technology of China, China

Program Committee Co-chairs

Peiquan Jin   University of Science and Technology of China, China
Bin Cui   Peking University, China
Zhiyong Shan   Renmin University of China, China

Program Committee

Jianhua Feng   Tsinghua University, China
Theo Haerder   University of Kaiserslautern, Germany
Bin He   IBM Almaden Research, USA
Ioannis Koltsidas   IBM Zurich Research Lab, Switzerland
Sang-Won Lee   Sungkyunkwan University, South Korea
Qiong Luo   Hong Kong University of Science and Technology,
                             Hong Kong, China
Vijayan Prabhakaran   Microsoft Research, USA
Luc Bouganim   INRIA, France
Sivan Toledo   Tel Aviv University, Israel
Jianliang Xu   Hong Kong Baptist University, Hong Kong,
                             SAR China
Da Zhou   China Mobile Research Institute, China

External Reviewers

Guoliang Li   Tsinghua University, China
Yanfei Lv   Peking University, China

Supported by

National Natural Science Foundation of China
Tin Ka Ping Foundation

Second International Workshop on Social Networks and Social Media Mining on the Web (SNSMW 2011)

Workshop Co-chairs

Guandong Xu	Aalborg University, Denmark
Lin Li	Wuhan University of Technology, China
Hong Cheng	Chinese University of Hong Kong, China
Botao Wang	Northeastern University, China

Program Committee

James Bailey	University of Melbourne, Australia
Li Chen	Hong Kong Baptist University, China
Hong Cheng	Chinese University of Hong Kong, SAR China
Peter Dolog	Aalborg University, Denmark
Irene Ggarrigos	University of Alicante, Spain
Yanan Hao	Victoria University, Australia
Yoshinori Hijikata	Osaka University, Japan
Hideyuki Kawashima	Tsukuba University, Japan
Lin Li	Wuhan University of Technology, China
Yuefeng Li	Queensland University of Technology, Australia
Wenxin Liang	Dalian University of Technology, China
Tieyun Qian	Wuhan University, China
Wenyu Qu	Dalian Maritime Univeristy, China
Munehiko Sasajima	Osaka University, Japan
Xiaohui Tao	Queensland University of Technology, Australia
Kenji Tateishi	NEC Corporation, Japan
Athina Vakali	Aristotle University, Greece
Botao Wang	Northeastern University, China
Daling Wang	Northeastern University, China
Yitong WANG	Fudan University, China
Zongda Wu	Wenzhou University, China
Guandong Xu	Aalborg University, Denmark
Zhenglu Yang	University of Tokyo, Japan
Junjie Yao	Peking University, China
Jianwei Zhang	Kyoto Sangyo University, Japan
Yu Zong	West Anhui University, China

First International Workshop on Data Management for Emerging Network Infrastructures (DaMEN 2011)

Workshop General Co-chairs

Aoying Zhou   East China Normal University, China
Yoshiharu Ishikawa   Nagoya University, Japan

Program Committee Co-chairs

Haibo Hu   Hong Kong Baptist University, Hong Kong,
                         SAR China
Weining Qian   East China Normal University, China

Program Committee

Jidong Chen   EMC Research China Lab, China
Yueguo Chen   Renmin University of China, China
Reynold C. K. Cheng   University of Hong Kong, Hong Kong,
                         SAR China
Byron Koon-Kau Choi   Hong Kong Baptist University, Hong Kong,
                         SAR China
Chi-Yin Chow   City University of Hong Kong, Hong Kong,
                         SAR China
Haibo Hu   Hong Kong Baptist University, Hong Kong,
                         SAR China
Hideyuki Kawashima   University of Tsukuba, Japan
Ken C. K. Lee   University of Massachusetts Dartmouth, USA
Weining Qian   East China Normal University, China
Lidan Shou   Zhejiang University, China
Hongzhi Wang   Harbin Institute of Technology, China
Linhao Xu   IBM China Research Lab, China
Wenwei Xue   Nokia Research, China
Ying Yan   Microsoft Search Technology Center, China
Man Lung Yiu   Hong Kong Polytechnic University, Hong Kong,
                         SAR China
Tomoki Yoshihisa   Osaka University, Japan
Rong Zhang   East China Normal University, China
Baihua Zheng   Singapore Management University, Singapore

Fourth International Workshop on Data Quality in Integration Systems (DQIS 2011)

Workshop Co-organizers

Xiaochun Yang	Northeastern University, China
Shazia Sadiq	University of Queensland, Australia
Xiaofang Zhou	University of Queensland, Australia
Ke Deng	University of Queensland, Australia

Program Committee

Lei Chen	Hong Kong University of Science and Technology, SAR China
Jun Gao	Peking University, China
Adam Jatowt	Kyoto University, Japan
Cheqing Jin	East China Normal University, China
Marek Kowalkiewicz	SAP Research, Australia
Qing Liu	CSIRO, Australia
Chaoyi Pang	CSIRO, Australia
Wanita Sherchan	CSIRO, Australia
Yanfeng Shu	CSIRO, Australia
Laurianne Sitbon	Queensland University of Technology, Australia
Bin Wang	Northeastern University, China
John (Junhu) Wang	Griffith Univeristy, Australia
Kai Xu	Middlesex University, UK
Ji Zhang	The University of Southern Queensland, Australia
Ying Zhang	The University of New South Wales, Australia

Table of Contents

Spatial Planning, Visualization, Mining and System

The First International Workshop on Flash-Based Database Systems (FlashDB)

Storage Management for SSD

Invited Talk I

Regular Papers

Energy Efficiency & Hybrid Storage

Invited Talk II

Invited Talk III

Regular Papers

The 2nd International Workshop on Social Networks and Social Media Mining on the Web (SNSMW)

Social Networking and Community Structure

Social Media and Data Mining

The First International Workshop on Data Management for Emerging Network Infrastructures (DaMEN)

Invited Talk

Query and Stream Processing

Storage and Scheduling

Fourth International Workshop on Data Quality in Integration Systems (DQIS)

Invited Talk

Session I

Invited Paper

Regular Papers

Session II

Invited Paper

Regular Papers

Privacy-Preserved Network Data Publishing

Lei Chen

Hong Kong University of Science and Technology
`leichen@cse.ust.hk`

Nowadays, more and more people join multiple social networks on the Web, such as Facebook, Linkedin, and Livespace, to share their own information and at the same time to monitor or participate in different activities. Meanwhile, the information stored in the social networks are under high risk of attack by various malicious users, in other words, peoples privacy could be easily breached via some domain knowledge. Thus, as a service provider, such as Facebook and Linkedin, it is essential to protect users privacy and at the same time provide useful data. Simply removing all identifiable personal information (such as names and social security number) before releasing the data is insufficient. It is easy for an attacker to identify the target by performing different structural queries.

In this talk, I will briefly review the current work on protecting the privacy of published social networks including clustering-based approaches and graph editing methods. Then, I will present a recent work, called k-automorphism, to protect against multiple structural attacks, following by a framework which provides privacy preserving services based on the users personal privacy requests. In the end, I would like to highlight some future work related to this topic.

J. Xu et al. (Eds.): DASFAA Workshops 2011, LNCS 6637, p. 1, 2011.

Towards Efficient Subgraph Search in Cloud Computing Environments*

Yifeng Luo[1], Jihong Guan[2], and Shuigeng Zhou[1]

[1] School of Computer Science, and Shanghai Key Lab of Intelligent Information
Processing, Fudan University, Shanghai, China
{luoyf,sgzhou}@fudan.edu.cn
[2] Dept. of Computer Science & Technology, Tongji University, Shanghai, China
jhguan@tongji.edu.cn

Abstract. This paper proposes an efficient approach to subgraph search
over a large graph database under the MapReduce framework. The main
idea is first to build inverted edge indexes for graphs in the database,
and then to retrieve data only related to the query subgraph by using
the built indexes to answer the query. Experimental results show that
the proposed approach has good performance and scalability.

Keywords: Graph database; Subgraph search; Cloud computing;
MapReduce; Inverted index.

1 Introduction

Graph is now an important data structure, which models objects as vertices
and the pairwise relationships between objects as edges. Graph-based modeling is employed in more and more applications [1], including pattern recognition [5–7], social networks, chem-informatics [3], graph-structured XML query
processing [4] and so on. When a database is used to manage the data of objects
that are represented by graphs, this database is referred to as a graph database.
Usually, a graph database falls into two categories [2]: graph-transaction setting
where a graph database consists of a large number of relatively small graphs,
and single-graph setting where a graph database contains only one large graph.

This paper deals with subgraph search in a large graph database containing
many graphs. Subgraph search is one of the fundamental problems in many practical graph-related applications such as chemical compound search, community
detection in social networks, motif finding in biological networks, and graph-
structured XML query processing. The problem of subgraph search or query can
be described formally as follows: given a graph database $D = \{g_1, g_2, \cdots, g_n\}$
and a graph query q, to answer the query is to find all graphs that contain q in
D. These resulting graphs are supergraphs of q, and q is one of their subgraphs.

---

* This work was supported by National Natural Science Foundation of China under
grants No. 60873040 and No. 60873070. Jihong Guan was also supported by the
Shuguang Scholar Program of Shanghai Education Development Foundation under
grant No. 09SG23.

J. Xu et al. (Eds.): DASFAA Workshops 2011, LNCS 6637, pp. 2–13, 2011.

Efficiently answering subgraph queries in a large graph database is absolutely not a trivial problem. Obviously, to scan the graph database sequentially and check whether a graph in the database is a supergraph of the query graph is prohibitively time-consuming. Existing subgraph search algorithms usually explore graph indexes to boost the processing of subgraph queries, and various indexing strategies have been proposed [8–12]. However, existing approaches are mainly based on centralized computing systems and evaluated on relatively small databases, each of which contains tens of thousands of small graphs with dozen of nodes and edges. Due to the combinatory issue, these approaches will face the scalability problem when dealing with large-scale graph databases, which consist of tens of millions of relatively large graphs with hundreds of nodes and edges. What is more, storage also faces challenge as the data amount of the graph database and its indexes is very huge. So how to scale up subgraph search algorithms to massive graph databases is an urgent and significant research issue, as the graph data amount is expanding drastically.

Cloud computing [17] is emerging as a new computing paradigm that has many merits. A cloud usually contains a cluster of nodes, each of which has computing and storage resources of its own. These resources are shared across the cluster at the cloud's disposal. Fault-tolerance is an intrinsic property of cloud computing. When a job is issued to a cloud, it will automatically split the relatively large job into small tasks, which will be scheduled to different nodes for execution. If some nodes on which some tasks are executing fail, these tasks will be re-assigned to some other running nodes for re-execution. Another important property of cloud computing is load-balancing. A cloud will monitor each node in the cluster and ensure that each node will get almost equal number of tasks to execute. Overall, cloud computing is efficient in handling both CPU intensive and I/O intensive jobs. Considering that subgraph search is both CPU intensive and I/O intensive, so implementing subgraph search on cloud computing platforms to exploit their advantages of scalability and elasticity is a natural choice.

This paper studies subgraph search on large-scale graph databases in cloud computing environments. Concretely, we propose and implement an efficient subgraph search approach under the MapReduce [18] framework that is a typical cloud-oriented parallel programming model. We also conduct experiments to validate the proposed approach. The rest of the paper is organized as follows: Section 2 reviews the related work. Section 3 presents an overview of our subgraph search approach. Section 4 introduces the implementation details of our approach. Section 5 gives the experimental evaluation on the proposed approach. Section 6 concludes the paper and pinpoints some future works.

2   Related Work

Due to space limit, here we give a brief review on the related work, including graph indexing/search and MapReduce-based computing.

2.1   Graph Search and Indexing

In the past years, graph search has been extensively studied as a centralized computing issue in database area. For improving processing efficiency, indexes are widely used. Up to now, various indexing strategies for graph search have been proposed [8–12]. To reduce space overhead, usually only significant graph elements are indexed. So this is a feature-filtering based indexing scheme. The process is like this: indexes are built on the graph database using a set of selected feature $F = \{f_1, f_2, \cdots, f_m\}$, each feature may be an edge, a node, or a keyword appearing in the labels of edges and/or nodes. When a query graph q is issued, the graph database will be checked by the following rule (maybe with a subsequent verification phase): for each graph $g \in G$, if $\exists f \in F$ such that $f \subseteq q$ and $f \not\subseteq g$, then $q \not\subseteq g$, g is filtered out.

Graphgrep [12] builds the indexes by enumerating all paths with length up to L, of all graphs in the database, and filters graphs by paths when doing search. Although it is fast to index paths with length limit and the index size can be kept small, structural information of graphs is lost and the filtering power of paths is limited, which will lead to a large candidate set and subsequently high verification cost. [8] uses subgraphs to keep structural information and to improve filtering power. The feature set consists of discriminative frequent subgraphs mined from the graph database. Though better filtering power is achieved, the verification cost is still high. TreePi [13] builds feature set by mining discriminative frequent subtrees from the graph database, and uses the subtree feature set to filter database graphs. Still, the filtering power is limited and verification cost is relatively high. There are other indexing schemes, either closure-based [11] or coding-based [15, 16]. We will not go into any further detail about graph indexing, interested readers can refer to [1].

In this paper, we also use indexes to enhance graph search efficiency. We build indexes directly using graph edges. Considering the elasticity of resources in cloud platforms, we do not try to filter edges for indexing.

2.2   MapReduce-Based Computing

In this paper, we propose a cloud-based subgraph query approach to decomposing the relatively huge subgraph search job into multiple relatively small tasks, and then run these small tasks on a Hadoop [22] cluster for parallel execution. The Hadoop cluster consists of HDFS [23] and MapReduce [18]. HDFS is a scalable distributed file system that is capable of storing massive data, and it is the open-source implementation of Google's GFS [19]. GFS provides fault tolerance while running on inexpensive commodity hardware, and is capable of delivering high aggregate performance to a large number of clients.

MapReduce is a parallel programming model, it has now become a typical and popular cloud computing framework for data-intensive parallel computation in shared-nothing clusters. A MapReduce job consists of a Map phase and a Reduce phase. In Map phase, workers (computing nodes) called Mappers invoke user-defined map function(s) to process key/value pairs to generate a set of intermediate key/value pairs. In Reduce phase, workers called Reducers invoke user-defined reduce function(s) to merge all intermediate key/value pairs associated with the same intermediate key value. A large input file is first partitioned into several splits, each of which is fed to a Mapper as input. Input splits of different Mappers can be processed in parallel, the results are forwarded to the Reducers for merging.

A number of high-level applications have been developed on MapReduce because of its ease of use for parallel execution. In addition to data management applications [20, 21], there are also some graph-related works on cloud platforms. Kang et al. [25] created a software library using Hadoop that performs typical graph mining tasks in big graphs, including degree distributions and PageRank, diameter estimation [26], connected components and triangle counting. Recently, Gu et al. [24] implemented a breath first search (BFS) algorithm in graph analysis using Sector/Sphere, a MapReduce-like cloud computing programming model. However, none of them deals with subgraph search in large graph databases.

3   Cloud-Based Subgraph Search: An Overview

In this section, we give an overview of our subgraph search approach implemented in the MapReduce framework. The main idea is to build inverted edge indexes for the graphs in the database, and when processing queries, only the data related to the queries is checked, from which the final results are obtained. Fig. 1 shows the overview of the cloud-based subgraph search approach.

Our approach consists of two phases: the off-line index building phase and the online subgraph query processing phase. In the first phase, we build inverted indexes by two MapReduce jobs: one is responsible for building inverted indexes for each unique edge in the graph database, the other is responsible for building indexes over the inverted indexes for each unique edge built in the first phase, which is to construct the mappings between edges and their offsets in the inverted index files. In the second phase, subgraph queries are processed. When a

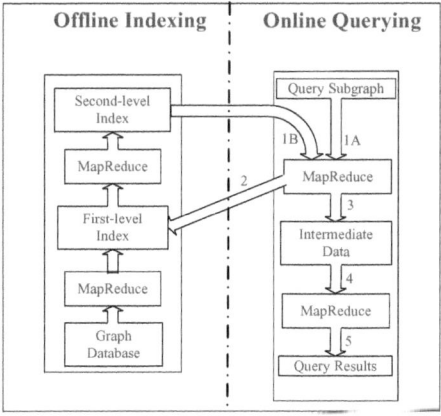

Fig. 1. An overview of cloud-based subgraph search

subgraph query is issued, two MapReduce jobs are launched. The first MapReduce job is to retrieve the candidate results by using indexing information, the second MapReduce job is to evaluate the final query results by employing set intersection operations.

4   Implementation Techniques

In this section we describe the details of the implementation of the cloud-based subgraph search approach.

4.1   Index Building

Two MapReduce jobs are used to build the indexes over the edge set of the graph database. The logics of building index are presented in Algorithm 1. The first MapReduce job takes the graph database file(s) as input and builds inverted indexes over the edge set, as the first-level index. In the graph database, each edge of a graph is represented as a *GraphNo / EdgeLabel* pair where *GraphNo* represents the graph containing the indexed edge, *EdgeLabel* consists of the labels of the edge's two end vertices. If a graph consists of ten edges, then ten *GraphNo / EdgeLabel* pairs will be stored in the graph database, with identical *GraphNo*. The data of graph database is divided into a number of splits, which are distributed over different nodes of the cloud platform.

Algorithm 1. Offline-Index-Building

    INPUT: the graph database D
    OUTPUT: the bi-level index
1: *First-level Indexing Job*
2:    *Map Task*
3:    **for each** graph labeled with G_i **in** graph database D
4:        **for each** edge labeled with $EdgeLabel_i$
5:           **output** $(EdgeLabel_i, G_i)$
6:    *Reduce Task*
7:    **for each** edge labeled with $EdgeLabel_i$
8:        **concat** all graph labels to generate a GraphSet string GS_i: $G_{i1}, G_{i2}, \cdots, G_{in}$
9:        **output** an entry: $(EdgeLabel_i, GS_i)$ **in** a first-level index file t_j
10:
11: *Second-level Indexing Job*
12:    *Map Task*
13:    **for each** entry: $(EdgeLabel_i, GS_i)$ **in** file t
14:        **get** file name of t : $fileName_i$, and offset of this entry in t : $offSet_i$
15:        **output** an entry: $(EdgeLabel_i, fileName_i, offSet_i)$ **in** a second-level index file t'_j

---

The logics in the Map and Reduce functions are very simple. Each record of the input splits is reversed by the Map function to output a *EdgeLabel / GraphNo* pair. A Hash function is used to repartition all outputs of the Mappers, by hashing *EdgeLabel* of each Mapper output pair. Those pairs with equal hash value share the same partition. A partition is a part of the input of a Reducer, which invokes the Reduce function. In the Reduce function, *EdgeLabel / GraphNo* pairs

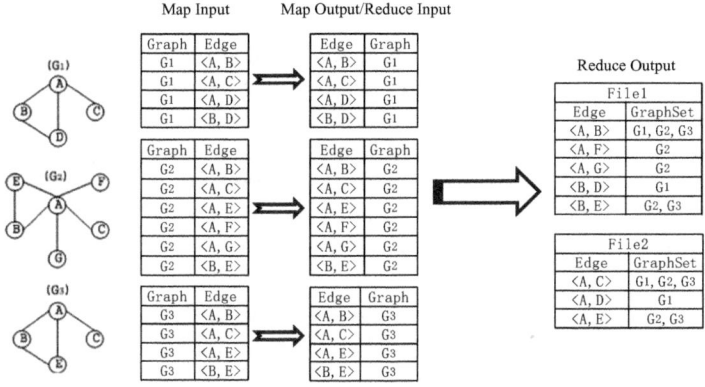

Fig. 2. Building the first-level index

with identical *EdgeLabel* are aggregated together, and the Reduce function generates an *EdgeLabel / GraphSet* pair where the *GraphSet* consists of *GraphNos* with similar *EdgeLabel*. When all Reducers finish execution, the first-level index is built, and multiple first-level index files are created. Each unique edge of the graph database appears only once in all these first-level index files. Which of the first-level index files a certain inverted edge index entry belongs to is determined by the Hash function. The process of the above MapReduce job is illustrated in Fig. 2.

When the first MapReduce job finishes, a second MapReduce job is launched to build the second-level index. The second-level index constructs the mappings between edges and their offsets in the first-level index files. Each of the first-level index files is treated as a whole input split for a Mapper of the second MapReduce job. No Reducers are initiated here. The logic in this Map function is also simple. When a *EdgeLabel / GraphSet* pair in any of the first-level index files is processed by the Map function, the file name and the offset of the current inverted index entry in the first-level index files are recorded. Then the Map function outputs an *EdgeLabel / ⟨FileName, Offset⟩* pair, which corresponds to an entry in the second-level index. The whole process is presented in Fig. 3.

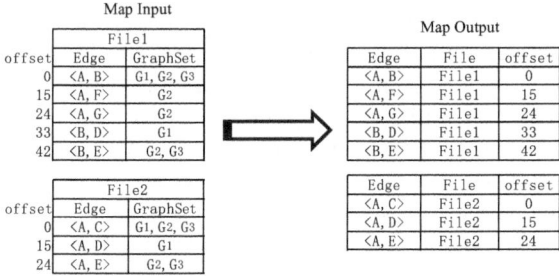

Fig. 3. Building the second-level index

4.2   Subgraph Search

When a query is issued, two MapReduce jobs are initiated to process the subgraph query. The logics of executing graph query are presented in Algorithm 2. The first MapReduce job is used to retrieve the inverted edge index entries whose corresponding edges are contained in the query graph. The second MapReduce job performs a series of set intersection operations to generate the final query results. The query graph is represented as a set of *EdgeLabels*, each of which consists of the labels of the two end vertices.

Algorithm 2. Online-Graph-Querying

    INPUT: second-level index
    PARAMETER: query graph q with l edges
    OUTPUT: final query result
1: *First-level Index Entry Retrieval Job*
2:    *Map Task*
3:    **for each** entry: $(EdgeLabel_i, fileName_i, offSet_i)$ **in** a second-level index file t'_j
4:      **if** edge labeled with $EdgeLabel_i \in q$
5:        $fstream\ in = \mathbf{open}(fileName_i)$
6:        $in.seek(offSet_i)$
7:        $in.read(EdgeLabel_i, GS_i)$
8:        **output** an entry: $(EdgeLabel_i, GS_i)$ **in** an intermediate file f
9:
10: *Graph Set Intersection Job*
11:    *Map Task*
12:    **declare** a set variable: lqr_k to store the local intersection result
13:    **for each** entry: $(EdgeLabel_{ij}, GS_{ij})$ **in** file f
14:      **if** $j == 1$
15:        $lqr_k = GS_{ij}$
16:      **else**
17:        **perform** $lqr_k = lqr_k$ intersects with GS_{ij}
18:    **output** $(null, lqr_k)$ in file f'
19: *Reduce Task*
20:    **declare** a set variable: gqr to store the final query result
21:    **for each** entry: $(null, lqr_{ki})$ **in** file f'
22:      **if** $i == 1$
23:        $gqr = lqr_{ki}$
24:      **else**
25:        **perform** $gqr = gqr$ intersects with lqr_{ki}
26:    **output** $(null, gqr)$

The first MapReduce job takes as input the second-level index files and the query graph. Each of the second-level index files is processed by a Mapper as a whole split. Each Mapper sequentially processes the records in the input splits assigned to it, and all Mappers do their works in parallel. When an *EdgeLabel* / ⟨ *FileName, Offset*⟩ pair is input to a Mapper, the Map function will check whether the *EdgeLabel* is contained in the query graph. If the *EdgeLabel* is contained in the query graph, the Mapper will open the *FileName* file and seek to *Offset* location, and then read a series of *GraphNos* from the *FileName* file. The outputs of Mappers are *EdgeLabel* / *GraphSet* pairs, here *EdgeLabels* are query edges. This process is shown in Fig. 4. No Reducer is initiated in this MapReduce job. The outputs of Mappers are merged into one file for the second MapReduce job to further process.

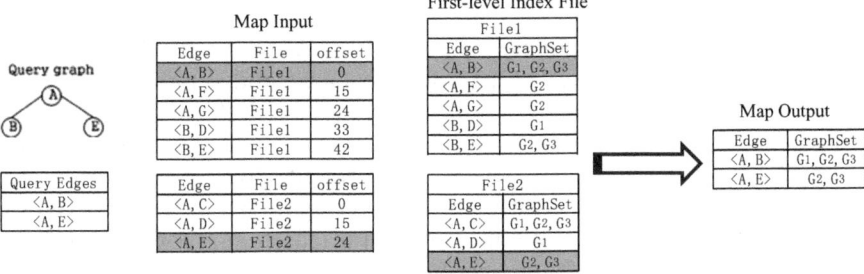

Fig. 4. Retrieving the first-level index entries

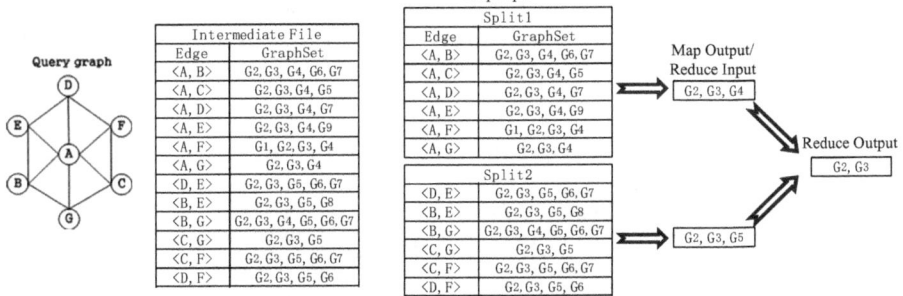

Fig. 5. Evaluating graph sets intersection

The second MapReduce job performs a series of set intersection operations to generate the final query results. Initially, we have tried to do intersection operations on the graph sets output by the first MapReduce job on the coordinator of the MapReduce cluster, but we found that it took too long time for the coordinator to finish the intersection operations by itself. So, we finally decided to start a second MapReduce job to do graph sets intersection operations. The merged output file of the first MapReduce job is taken as input of the second MapReduce job. The input file is partitioned into several splits, each of which is processed by a Mapper. Each Mapper performs intersection operations on the graph sets contained in its input split. If a Mappers finishes its work, it outputs its local intersection results to a Reducer to process further. The Reducer performs intersection operations on the intersection results from different Mappers to generate the final query results. Only one of such Reducer is initiated. The execution process is illustrated in Fig. 5.

5   Experimental Evaluation

In this section, we present the experimental results of evaluating the efficiency of the cloud-based subgraph search approach.

5.1   Experimental Settings

We deploy a cluster with 10 nodes, each of which is a commodity PC. Each node has an Intel Duo Core2 2.93GHz CPU, 3GB memory, and Windows XP OS. We use Hadoop 0.19.2, and compile the source codes under JDK 1.6 in Eclipse 3.3.2. One of the ten nodes is used as coordinator, and the rest nine nodes are used as computing and storage nodes. As we have not found any large-scale graph database suitable for our experiment, we use synthetic datasets to evaluate the efficiency of the proposed cloud-based subgraph search approach. Experiments with real-life datasets are left for our future work. The synthetic datasets are generated according to the Erdos-Renyi random graph model. Three datasets are generated. Major statistics of the three datasets is presented in Table 1. Query graphs are randomly generated with 20 edges, 30 edges, 50 edges and 100 edges respectively.

   We measure the query time to evaluate the efficiency of the cloud-based subgraph search approach. One point is worthy of being mentioned: in a distributed environment, the performance of a distributed algorithm or system is influenced by many factors, including network, computing model, resource scheduling, etc.

Table 1. Statistics of synthetic datasets

Statistics	DataSet1	DataSet2	DataSet3
#graphs	50,000	70,000	100,000
#vertices per graph	30-40	40-70	50-100
#edges per graph	300-400	600-800	800-1000

5.2   Experimental Results

We first measure the query time for querying four randomly-generated graphs over the three datasets. The results are presented in Fig. 6, Fig. 7 and Fig. 8. The results seem not what we have expected. As we initially imagined, if the number of query edges increases, the query time will increase, since more time will be spent on checking whether the *EdgeLabels* of an entry in the second-level index is contained in the query graph. However, as we can see from these three figures, the query time sometimes decreases as the number of query edges increases. The reason behind this is that the computing cost spent on checking is minor, compared with that of data distribution and the subsequent graph set intersection operations. As for the MapReduce job that performs graph set intersection operations, the majority of the execution time is spent on the Reducer. The Reducer performs intersection operations on all the outputs of the preceding Mappers. So the input of Reducer may be relatively of large size, and thus it takes the Reducer much time to finish the intersection operation on the graph sets.

   When the query graph contains more edges, more Mappers will be initiated to perform local graph set intersection operations, and each Mapper gets more graph sets as inputs. So the outputs of the Mappers and thus the input of the Reducer may become smaller, compared with that of query graphs with fewer

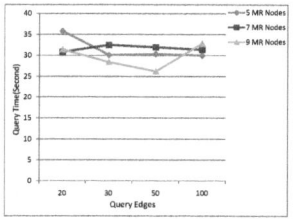 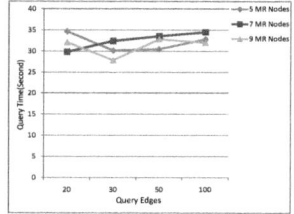 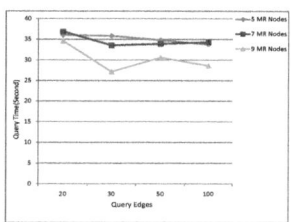

Fig. 6. Results over Data-
Set1

Fig. 7. Results over Data-
Set2

Fig. 8. Results over Data-
Set3

edges. In our experiments, the query edges affect query time little, and the query
time must not decrease as the number of query edges increases. This may be not
a general conclusion. For a general conclusion, more experiments with larger
databases on larger-size clusters are requested, which is also left for our future
work.

We also measure the average query time on each database graph. For DataSet1,
the average query time spent on each database graph is 0.5 ms to 0.7 ms, for
DataSet2, the average query time spent on each database graph is 0.4 ms to 0.5
ms, and for DataSet3, the average query time spent on each database graph is
0.25 ms to 0.4 ms. It seems that the average time spent on a database graph decreases as the size of the graph database increases. The reason is that the cost of
network, data distribution and MapReduce job startup is shared by all database
graphs. The more database graphs are, the less the shared cost by each graph is.
This phenomenon validates the scalability of our subgraph searching approach,
that is, our approach can scales up to massive large-scale graph databases.

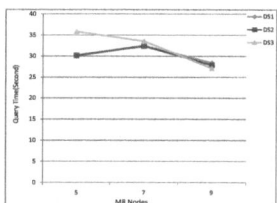 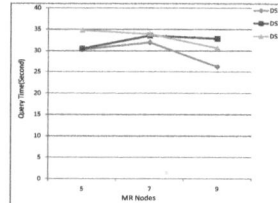 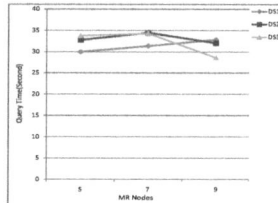

Fig. 9. Results for 30-edge
queries

Fig. 10. Results for 50-edge
queries

Fig. 11. Results for 100-
edge queries

We then measure the query time when changing the size (the number of
MapReduce nodes) of the cloud cluster. The experimental results are presented
in Fig. 9, Fig. 10 and Fig. 11. As we can see from these three figures, the query
time of a given database does not follow a monotone trend, as the number of
MapReduce (MR) nodes increases. When the number of MR nodes increase to
seven, the query time turns longer, compared with that of five-node cluster. But
the query time decreases when the number of MR nodes increase to nine. Our
explanation on this phenomenon is as follows:

For a small-scale graph database, a centralized system may be adequate to process the queries efficiently. In such cases, processing queries in a distributed system must not be advantageous over in a centralized system. For a given graph database, the query time is not determined only by the number of nodes. Adding more nodes must not yield more performance gain, because data distribution and synchronization cost may counteract the benefit of using more nodes. For all the three synthetic datasets we use, a cluster of five computing and storage nodes is adequate. When the number of nodes increases to seven, the query time increases, compared with that of query with a five-node cluster. However, when nine nodes are employed to execute queries, the query time decreases. This is because the benefit of using more nodes surpasses the overhead of data distribution and synchronization as well as the initialization of MapReduce jobs. It is not difficult to predict that the query time will increase again as the number of MR nodes continues to increase. So our conclusion from the experimental results is: it is suitable to employ a large-scale cluster to implement subgraph query for large-scale databases, and a middle-scale cluster should be employed for middle-scale graph databases.

6   Conclusion and Future Work

This paper studies subgraph search over large-scale graph databases in cloud computing environments. A cloud-based subgraph search approach is presented, which uses a bi-level indexing structure to boost the search efficiency. Inverted index over edge set of the graph database is built as the first-level index, and then a second-level index is built to construct the mappings between edges and offsets of their corresponding inverted index entries in the first-level index files. Experiments on synthetic datasets show that cloud-based subgraph search is efficient for large-scale graph databases.

Some optimization and improvements can be done on the proposed approach. On one hand, merging multiple queries into one can reduce I/Os and averaged startup cost of MapReduce jobs per query, and thus can improve query throughput. On the other hand, exploring more efficient lookup strategies over the second-level index will speed up the locating of query edges, as currently the second-level index is sequentially scanned.

References

1. Aggarwal, C.C., Wang, H. (eds.): Managing and mining graph data. Kluwer Academic Publishers, Dordrecht (2010)
2. Kuramochi, M., Karypis, G.: Finding frequent patterns in a large sparse graph. In: Proceedings of SDM (2004)
3. Willett, P.: Chemical similarity searching. J. Chem. Inf. Comput. Sci. 38, 983–996 (1998)
4. Polyzotis, N., Garofalakis, M.: Statistical Synopses for Graph-Structured XML Databases. In: Proceedings of SIGMOD (2002)

5. Beretti, S., Bimbo, A., Vicario, E.: Efficient Matching and Indexing of Graph Models in Content Based Retrieval. IEEE Trans. on Pattern Analysis and Machine Intelligence 23, 1089–1105 (2001)
6. Messmer, B., Bunke, H.: A new algorithm for error-tolerant subgraph isomorphism detection. IEEE Trans. on Pattern Analysis and Machine Intelligence 20, 493–504 (1998)
7. Petrakis, E., Faloutsos, C.: Similarity searching in medical image databases. IEEE Trans. on Knowledge and Data Engineering 9(3), 435–447 (1997)
8. Yan, X., Yu, P., Han, J.: Graph Indexing Based on Discriminative Frequent Structure Analysis. ACM Transactions on Database Systems 30(4), 960–993 (2005)
9. Cheng, J., Ke, Y., Ng, W., Lu, A.: Fg-index: towards verification-free query processing on graph databases. In: Proceedings of SIGMOD (2007)
10. Williams, D.W., Huan, J., Wang, W.: Graph Database Indexing Using Structured Graph Decomposition. In: Proceedings of ICDE (2007)
11. He, H., Singh, A.K.: Closure-Tree.: An Index Structure for Graph Queries. In: Proceedings of ICDE (2006)
12. Giugno, R., Shasha, D.: Graphgrep: A fast and universal method for querying graphs. Proceedings of ICPR 2, 112–115 (2002)
13. Zhang, S., Hu, M., Yang, J.: TreePi: A Novel Graph Indexing Method. In: Proceedings of ICDE, pp. 181–192 (2007)
14. Ferro, A., Giugno, R., Mongiovi, M., et al.: GraphFind: enhancing graph searching by low support data mining techniques. BMC Bioinformatics 9 (2008)
15. Jiang, H., Wang, H., Yu, P., Zhou, S.: GString: A Novel Approach for Efficient Search in Graph Databases. In: Proceedings of ICDE (2007)
16. Zou, L., Chen, L., Jeffrey, Y.L.: A novel spectral coding in a large graph database. In: Proceedings of EDBT (2006)
17. Armbrust, M., Fox, A., Griffith, R., Joseph, A.D., Katz, R., Konwinski, A., Lee, G., Patterson, D., Rabkin, A., Stoica, I., Zaharia, M.: Above the Clouds: A Berkeley View of Cloud Computing. Technical Report, UC Berkeley Reliable Adaptive Distributed Systems Laboratory (February 2009)
18. Dean, J., Ghemawat, S.: MapReduce: Simplified data processing on large cluster. In: Proceedings of OSDI, pp. 137–150 (2004)
19. Ghemawat, S., Gobioff, H., Leung, S.T.: The Google file system. In: Proceedings of SOSP, pp. 29–43 (2003)
20. Olston, C., Reed, B., Srivastava, U., et al.: Pig latin: a not-so-foreign language for data processing. In: Proceedings of SIGMOD, pp. 285–296 (2008)
21. Abouzeid, A., Pawlikowski, K.B., Abadi, D.J., et al.: HadoopDB: An Architectural Hybrid of MapReduce and DBMS Technologies for Analytical Workloads. In: Proceedings of VLDB, pp. 285–296 (2009)
22. http://hadoop.apache.org
23. http://hadoop.apache.org/hdfs/
24. Gu, Y., Lu, L., Grossman, R., Yoo, A.: Processing massive sized graphs using Sector/Sphere. In: Proceedings of the Workshop on Many-task Computing on Grids and Supercomputers (MTAGS 2010), co-located with SC 2010, New Orleans, LA (November 2010)
25. Kang, U., Tsourakakis, C.E., Faloutsos, C.: PEGASUS: A Peta-Scale Graph Mining System - Implementation and Observations, In: Proceedings of ICDM 2009 (2009)
26. Kang, U., Tsourakakis, C.E., Appel, A., Faloutsos, C., Leskovec, J.: HADI: Fast diameter estimation and mining in massive graphs with Hadoop, CMU ML Tech Report CMU-ML-08-117 (2008)

Latency-Optimal Walks in Replicated and Partitioned Graphs

Stefan Plantikow and Maik Jorra

Zuse Institute Berlin, Takustrasse 7, 14195 Berlin, Germany
{plantikow,jorra}@zib.de

Abstract. Executing walks in partitioned, distributed graphs with minimal latency requires reducing the number of network hops taken. This is especially important for graph databases that specialize on executing fast graph traversals. We present fast-forward-search, an algorithm that uses overlapping graph partitionings, i.e. replication, and parallel speculative execution to minimize the number of required network hops. We proof optimality of the algorithm, analyze storage, message, and computational overhead caused by the parallelism of fast-forward, and introduce escapicity, a metric for replica selection that helps reducing that parallelism at the price of lost optimality. Experiments for a set of smaller graphs indicate that fast-forward-search saves between $20 - 90\,\%$ of network hops depending on graph and replication factor and that escapicity outperforms classic measures of network centrality as a metric for replica selection in our scheme.

1 Introduction

Partitioning graph databases in a way that network traffic is minimized is still an open problem. For many application domains either the size of graphs or the number of requests to be handled exceed the capacity of a single machine. Examples include geo information, social networks, bibliographical relationships, information systems (linked data), physical networks (transport, electricity), and biological networks (protein structures) [11].

These applications execute traversal queries [14], i.e. navigate the structure of the graph. The most basic traversals are walks which visit vertices sequentially by following edges. Walks are the building blocks of more complex queries and therefore fast edge traversal is the defining property of a graph database [1][1].

Graphs can be distributed by grouping vertices into partitions and by assigning each partition to a different machine. However, this introduces remote hops between machines at partition boundaries and thus may cause query processing to be dominated by network latencies. This can be disruptive for applications that rely on fast, interactive access to data [6].

---

[1] This separates graph databases from classic RDBMS which require a B-Tree-look-up for each edge traversal [14].

J. Xu et al. (Eds.): DASFAA Workshops 2011, LNCS 6637, pp. 14–27, 2011.

1.1 Problem

Graphs need to be structured for parallel or distributed processing. Graph partitioners [5] commonly reduce inter-partition network traffic by minimizing the cut-size, i.e. the sum of inter-partition edge weight sums. This aims to avoid CPU underutilization due to network congestion. Reduced network usage can be as high as 90 % [2].

However graph partitioners only optimize for the average case. Even when assuming an optimal partitioning, there always exist queries whose execution causes many network hops. Therefore using a single partitioning may not sufficiently limit network usage of partitioned graph databases.

To address this, we show how using overlapping graph partitions may reduce the number of required partition changes for walks in distributed directed read-only graphs. A first evaluation indicates a reduction of $20 - 90\%$, depending on graph, replication factor, and walk length.

We present *Fast-Forward-Search* (FFS), an algorithm for optimal replica selection for walk queries that is based on parallel speculative execution. We show optimality of FFS and discuss its storage, computation, and communication overhead. Additionally, we introduce *Escapicity*, a replica selection metric that reduces the message overhead of FFS at the price of giving up optimality.

We conclude with an evaluation of FFS and escapicity for a set of smaller graphs.

2 Definitions

Let $G = (V, E)$ be a weighted, directed, and connected graph where V is the set of vertices and $E \subseteq V \times V$ is the set of edges. For a vertex $v \in V$, $E_v^{\rightarrow}$ is the set $\{(\{v\} \times V) \cap E\}$ of all edges which start at v. We assume that for every edge $e \in E$ there exists a function $w(e) : E \rightarrow \mathbb{N}_0$ that assigns a positive integer weight to e. We'll also use source vertex normalized weights $w_n((v, u)) : E \rightarrow [0; 1]$ which normalizes an edges $e = (v, u)$ weight $w(e)$ against the sum of all weights of its source vertex' outgoing edges $E_v^{\rightarrow}$, i.e. $w_n((v, u)) = \frac{w((v,u))}{\Sigma_{e \in E_v^{\rightarrow}} w(e)}$.

$\Pi \subseteq 2^{2^V}$ (where 2^V is the power set of V) is a *possibly non-disjoint* partitioning of G iff for each $v \in V$ there exists at least one partition $\pi \in \Pi$ that contains v. We define $\Pi(v)$ to be the set of partitions of $v \in V$, i.e. $\Pi(v)$ is the set $\{\pi \in \Pi \mid v \in \pi\}$. For practical reasons, we limit this work to using a fixed vertex replication factor k and a fixed number of partitions $|\Pi|$.

A walk r in G of length $|r|$ is a sequence of vertices $v_1, v_2, \ldots, v_n$ with $(v_i, v_{i+1}) \in E$ for $i = 1, \ldots, n$. We write $r(i)$ for the ith waypoint of r. In a random walk, all edges (v_i, v_{i+1}) are selected with probability $w_n((v_i, v_{i+1}))$. A partition walk r' of a fixed walk r and partitioning Π completely maps r to a sequence of pairs $r'(i)$ from $\{r(i)\} \times \Pi(r(i))$. To shorten, we write $r'(i)_V$ and $r'(i)_\Pi$ for the vertex and partition components of $r'(i)$. For each walk r, there exists at least one partition walk r' since each vertex is included in at least one partition by definition.

Every partition walk r' is a walk on $G_\Pi = (V_\Pi, E_\Pi)$, the exploded partition graph of G where each $v \in V$ has been replaced with replicated vertices $\{(v, \pi_i) \mid \pi_i \in \Pi(v)\}$ which represent partition membership of v in $\Pi(v)$ and that consists of edges between all (v_i, π_i) and (v_j, π_j) for which $(v_i, v_j) \in E$.

We define the cost $c(r', i)$ of a partition walk r' to be the number of partition changes up to the ith waypoint, i.e. $c(r', i) = |\{j < i \mid r'(j)_\Pi \neq r'(j+1)_\Pi\}|$ and thus $0 \leq c(r', j) \leq j$. To shorten, we write $c(r', |r'|)$ as $c(r')$.

Latency Optimality. A partition walk r' of a walk r and partitioning Π is latency optimal up to i if its costs $c(r', i)$ are minimal, i.e. there exists no other partition walk $r'' \neq r'$ of r and Π with $c(r'', i) < c(r', i)$. We write $c_{min}(r)$ for the minimal cost of any partition walk of r.

3   Optimal Partition Walks

Using above definitions, we show how an optimal partition walk r' may be constructed for a given walk r over G_Π. We defer showing optimality of the construction until the distributed algorithm has been presented (cf. Sect. 5).

Stay-Local Principle. The key insight is that it is sufficient for finding optimal walks to prefer partition-local edges over remote ones, taking maximum local benefit in the process. This may delay jumping to another partition but costs are not increased as the partition change has to occur at some point anyways unless the remaining walk may be completed locally.

Construction. The construction extends this stay-local principle with length-limited lookahead at partition boundaries. At the end of a local walk, the remote follow-up partition is chosen such that the length of the remaining walk that is contained locally in that partition is maximized. Formally, we define $\mathrm{lw}(r, i, \pi_j)$ to be the *longest local partition walk* of r that starts at vertex $(r(i), \pi_j) \in G_\Pi$ and stays in π_j as long as possible. We construct a partition walk r' in G_Π for a walk r in G piecewise by sequentially applying one of two possible operations for a follow-up vertex $r(i)$:

stay-local Given the current partition π_i, append $\mathrm{lw}(r, i, \pi_i)$ to r', starting at $r'(i)$. It may not always be possible to perform this operation (e.g. initially) leading to:
select-partition Select $\pi_j \in \Pi(r(i))$ such that

$$\mathrm{lw}(r, i, \pi_j) \geq \mathbf{max}(\{l \mid l = |\mathrm{lw}(r, i, \pi_x)|, \pi_x \in \Pi(r(i)) \setminus \pi_j\}) \qquad (1)$$

In other words, π_j contains the longest local partition walk that is reachable in step i. As a result of select-partition, $(r(i), \pi_j)$ is chosen as $r'(i)$.

Using these operations, r' is constructed as follows: Initially, we choose the start partition π_s using select-partition for $r(1)$ and add $\mathrm{lw}(r, 1, \pi_s)$ to r' (stay-local). If consequently $|r| = |r'|$, the walk is complete (by definition).

Otherwise, for walks with $c_{min}(r) > 0$, there are at least $c_{min}(r)$ cases in *every* partition walk of r where the partition is changed. For such walks, select-partition and stay-local are executed alternatively until $|r'| = |r|$ and thus r' is complete.

4   Fast-Forward-Search

In this section, we show how the optimal partition walks constructed in the previous section can be found efficiently in a distributed system. To this end, we introduce fast-forward-search (FFS), a distributed partition walk routing algorithm, analyze it's costs, and discuss a possible way to reduce them.

4.1   System Model

We assume a simplified model of a distributed system that stores a replicated and partitioned graph. In this system, each partition is stored on a different physical machine. Machines can communicate with each other via an underlying network. We assume the existence of a routing layer for entering the graph, i.e. a way for finding all partitions that contain the start vertex of a walk, and a routing layer for sending messages to the machine that stores a partition. To focus on the effects of replication on distributed graphs, we deliberately choose not to discuss failures and assume that machines (processes, partitions) do not crash and messages are never lost. We also do not consider updates to the graph.

4.2   Query Model

In this system, we wish to perform some read-only computation (e.g. a fold-left) by traversing the graph along some walk r. We call such computations *queries*. Queries consist of a sequence of *query steps*. A query step is a serializable function that takes the current waypoint (vertex), it's outgoing edges, and optionally application data as input, and either produces a subsequent-vertex and query step or a final query result. Query steps are required to be deterministic (i.e. do not rely on input from the environment). Note that no a priori knowledge of the complete walk is required. Instead the walk is generated as the computation proceeds, starting from the initial query step and start vertex.

4.3   Fast-Forward-Search

After these considerations, we are now ready to define FFS. FFS is a latency-optimal distributed partition walk routing algorithm inspired by the *stay-local* and *select-partition* rules.

Optimal partition walks may trivially be found using global search. For example, equation (1) could be computed easily in parallel by broadcasting a follow-up query step to all vertex replicas of the corresponding follow-up vertex at a partition boundary. However this apporach would lead to tree-like, excessive branching since *every* partition boundary encountered would cause the creation of k (replication factor) additional messages.

The central idea for avoiding this branching is to ensure that after broadcasting a query step to the k vertex replica partitions, only one of those partition will perform the next broadcast step and to select that partition such that it contains a maximum local walk of the remainder of the query.

In a distributed system, this requires all involved partitions to reach agreement on selecting this single winner partition. FFS achieves this by using shared information and speculative execution. More precisely, FFS exploits that each vertex replica always knows all replicas of its follow-up vertices, i.e. the partitions they are contained in. Together with knowledge about the partitions currently searched by FFS, this is sufficient to simulate which other replicas cannot execute the next step locally. Inversely a partition which cannot continue locally knows if other partitions can. In this way, partitions execute the query speculatively for as long as they can. Finally, when none can continue, all remaining partitions contain a maximum local walk and a follow-up broadcast step cannot be delayed any longer. In that case, it suffices to uniquely select a single winning partition deterministically. This winner then continues by executing the broadcast for the next query step.

Figure 1 shows the pseudo-code that is run in parallel by a set of partitions $replicas_{||}$ that contain replicas of vertex v. If local query execution returns QUERYRESULT($result$), this $result$ is delivered to the application by $\pi = $ selUnique($replicas_{||}$. Alternatively a follow-up QUERYSTEP with $qstep_w$ and vertex w is returned. If this step can be processed locally, the set of replicas executing in parallel ($replicas_{||}$) is recomputed as $replicas_{\cap w}$ by cutting the current $replicas_{||}$ with $\Pi(w)$, the vertex replicas of w, and the query continues locally with $qstep_w$. Otherwise the winner partition selUnique($replicas_{||}$) broadcasts $qstep_w$ to a set of partitions selPartitionsFrom($\Pi(w)$) $\subseteq \Pi(w)$ that hold vertex replicas of w for executing the subsequent query step instead.

```
 1: on msg FASTFWD(v, qstep_v, replicas_|| ⊆ Π(v))
 2:  ▷ Executed at each partition π ∈ replicas_|| in parallel
 3:     ▷ Run follow-up query step locally
 4:     next ← qstep_v(v, E⃗_v, payload_π(v))
 5:     if next = QUERYRESULT(qresult) then
 6:        ▷ Deliver result if query has finished
 7:        if selUnique(replicas_||) = π then
 8:           [deliver qresult]
 9:        end if
10:     else if next = QUERYSTEP(w, qstep_w) then
11:        ▷ Search for best partition (agreement by mutual simulation)
12:        replicas_∩w ← replicas_|| ∩ Π(w)
13:        if w ∈ π then
14:           ▷ stay-local
15:           msg ← FASTFWD(w, qstep_w, replicas_∩w)
16:           [send msg toPartitions {π}]
17:        else if replicas_∩w = ∅ ∧ selUnique(replicas_||) = π then
18:           ▷ select-partition i.e. broadcast at boundary of a best partition
19:           replicas_⊆w ← selPartitionsFrom(Π(w))
20:           msg ← FASTFWD(w, qstep_w, replicas_⊆w)
21:           [send msg toPartitions replicas_⊆w]
22:        end if
23:     end if
24: end on msg
```

Fig. 1. Fast-Forward-Search (FFS): Latency optimal execution of walks in distributed, partitioned graphs using replica selection based on parallel speculative execution

4.4 Cost Analysis

We next examine the storage, computational, and message overhead caused by FFS for a constant message fan-out $1 < f = |\text{selPartitionsFrom}(\Pi(v))| \leq k$.

Storage Costs. To examine the storage requirements, we assume a simple, adjacency list-like scheme that is roughly following the storage layout of the neo4j graph database [14]. In this scheme, each of the $k \cdot n$ replicated vertices stores its id, $k-1$ partition ids for its replicas, and the id of its first outgoing edge. Each of the $k \cdot m$ replicated edges stores ids of source and destination vertices, and its own edge id. Edges are grouped into double-linked lists according to their source vertex. Therefore they have to store previous and next edge id pointer for their source vertex' edge list. External (out-of-partition) edges additionally store $k - 1$ extra partition ids for destination vertex replicas.

With this scheme, required storage grows roughly linear with the number of graph elements for fixed k. The overhead per replica increases linearly with increased replication degree and is independent from overall graph size. It is caused by the need to store $k - 1$ partition pointers in external edges. Although $O(k^2)$ partition ids may need to be stored per unreplicated edge, it turns out that this does not dominate storage requirements when partition ids are small compared to node/edge ids and payload. For $k \leq 6$, the overhead is $\leq k + 1$ for a graph with 10^{12} elements and a constant degree in the range 20-110.

Number of Messages. FFS eliminates $\alpha \cdot h$ remote hops from a walk that requires h hops when no replication is used. We assume that partition selection sends f messages initially when broadcasting to find a suitable start partition. Therefore FFS requires $f + \lceil (1 - \alpha) h \rceil f$ messages, while unreplicated partitioning needs $1 + h$ messages for a walk with h hops.

Figure 2 shows the ratio of messages required by FFS to messages required with plain, disjoint partitioning for $f = 1, \ldots, 5$ and $\alpha \in \{0.1, 0.3, 0.5, 0.7\}$. For long enough walks, the overhead converges to a constant factor $\leq f$.

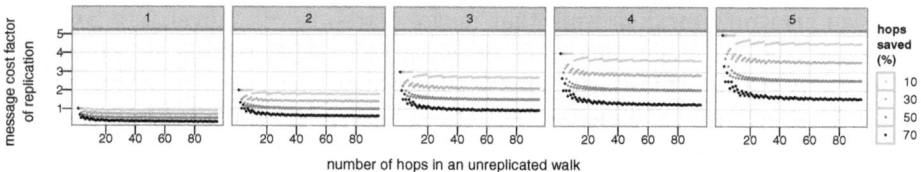

Fig. 2. Ratio of number of messages required by replication to number of messages required without replication in relation to percentage of hops saved. (Columns $\hat{=}$ message fan out)

Computational Cost. Branching out to f destinations at partition boundaries causes duplication of work, i.e. the whole query is executed f times in the worst case. However subqueries that started in parallel will stop as soon as the f selected partitions start to differ with respect to the walk generated by the query. Therefore subquery duplication may be used as an indicator for partition overlap and latency gain potential. This is further discussed in sect. 6.

4.5   Lowering the Fan-Out

To reduce the message, and by extension, the computational overhead of FFS, we need to use small f. This in turn requires finding a suitable metric for selecting f vertex replicas (selPartitionsFrom in alg. 1).

To give an intuition about the metric we're looking for, we view a single partition π_i as an edge-weighted flow network, where source vertices of incoming inter-partition edges are replaced with a single source vertex v_{source} and destination vertices of outgoing inter-partition edges with a single sink vertex v_{sink}. Resulting identical edges are coalesced and assigned the sum of their weights.

By viewing partitions as flow networks, it becomes easy to see that the target vertex replicas that are selected at a partition boundaries by FFS should have a high distance to the v_{sink} of their respective partition's flow networks.

There are several well-known metrics for rating vertices in a graph. Classical centrality metrics [10] give a clue about the position of a node in a network, w.r.t. information propagation from or to the node, like betweenness, closeness, and flow-betweenness. Other metrics like PageRank [12] rate node significance by taking the amount and relevance of connected neighbors into account.

These metrics may be calculated for partition flow networks, but the result values will only indicate if a vertex is positioned close to the network's border. However they do not express if vertices are closer to the source or to the sink. The only exception to this is flow-betweenness which indicates how easy it is to reach the partition's sink with a random walk. In any case, computing these metrics is costly due to the need to look at the complete flow graph and thus it might be advisable to try more simple, ad-hoc approaches instead.

Ratio-Metric. Normalized edge weights $w_n((v_i, v_j))$ express the probability that an arbitrary random walk that moves across v_i will traverse (v_i, v_j) in the next step. For an edge that leads to v_{sink}, this translates into the probability that the partition will be left. This is sufficient to compute the probability that any edge taken at v_i will lead outside the current partition in the next step. This in turn leads to the simple *ratio-metric*. Using this metric selPartitionsFrom can discard vertex replicas with high next-step exit probability.

Escapicity. This ratio-metric can be refined further by recursively considering the neighborhood of vertices. We call this refined metric *escapicity* (Fig. 3). Escapicity is an estimate for the probability that an arbitrary random walk that starts at a replicated vertex v will exit v's partition π_{local} in i steps, i.e. escapicity is a measure of v's closeness to its partition's sink. The parameter i determines the radius of the neighborhood of v that is considered in computing the escapicity and for breaking cycles.

```
1: function ESCAPICITY(v ∈ V, i ≥ 0)  ▷ at π_local
2:   ▷ π_local ⊆ V are the local partition's vertices
3:      if v ∉ π_local then
4:         return 1.0
5:      else if outdeg(v) = 0 then
6:         return 0.0
7:      else if i = 0 then
8:         ▷ Per partition constant
9:         return 1.0 − |(π_local × π_local)∩E| / |(π_local × V)∩E|
10:     else
11:        return    ∑       w_n((v,u))·ESCAPICITY(u,i−1)
                  (v,u)⊂E⃗_v
12:     end if
13: end function
```

Fig. 3. Escapicity is a measure of escape likelihood (closeness to border) of a replicated vertex that considers neighbors at a distance of at most i hops

More precisely, escapicity is defined by recursion over the neighbors of v and decreasing i. External vertices are always scored 1.0. Vertices with an $outdeg(v) = 0$ are always scored 0.0 For remaining vertices, for $i = 0$, an estimate of the average escapicity score for any vertex from V_{local} is used. For $i > 0$ the edge weighted average of the $i − 1$ escapicity scores of v's direct neighbors is used, i.e. cycles are broken by using a score with lower i (less information) when revisiting a vertex.

5   Proof of Optimality

We now proof that partition walks r' that have been constructed by FFS as described in Sect. 3 are always minimal in the number of required partition changes and thus latency optimal, i.e. there exists no r'' such that $c(r'') < c(r')$. To do this, we compare r'' and r' element-wise and show that for all $r(x)$, r' is optimal up to x.

Let's examine a step from $r(x − 1)$ to $r(x)$ in both r' and r''. This step may either be performed locally in the current partition or may require a partition change. Thus given $r'(x − 1)_\Pi = \pi_i$ and $r''(x − 1)_\Pi = \pi_j$ we may encounter one of the following possible steps only:

(1) r' and r'' don't change their partition, i.e.
      $c(r', x) = c(r', x − 1)$ and $c(r'', x) = c(r'', x − 1)$,
(2) r'' exits π_j but r' can stay in π_i and hence
      $c(r'', x) = c(r'', x − 1) + 1$ yet $c(r', x) = c(r', x − 1)$,
(3) the inverse of (2), r' exits while r'' stays and hence
      $c(r', x) = c(r', x − 1) + 1$ yet $c(r'', x) = c(r'', x − 1)$,
(4) r' and r'' both switch and $c(r', x) = c(r', x − 1) + 1$ and $c(r'', x) = c(r'', x − 1) + 1$,
(5) or v_x is the last element of the complete walk r.

Using this list of possible steps, we define a final-state-transducer which looks at local walks of r' and r'' at waypoint x. Depending on *available knowledge* about the lengths of these walks, the state-transition function δ considers two states:

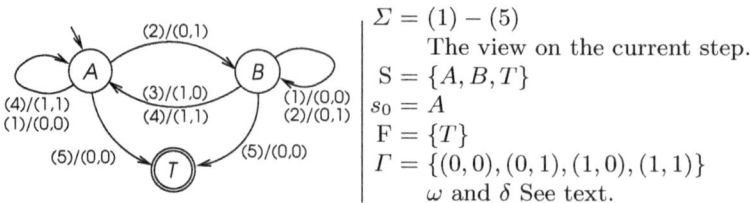

$$\Sigma = (1) - (5)$$
The view on the current step.
$$S = \{A, B, T\}$$
$$s_0 = A$$
$$F = \{T\}$$
$$\Gamma = \{(0,0), (0,1), (1,0), (1,1)\}$$
ω and δ See text.

Fig. 4. Transducer that iterates over walks and delivers the cost increment for every step

In *state A*, it is known that $|\mathrm{lw}(r', x, \pi_i)| \geq |\mathrm{lw}(r'', x, \pi_j)|$. By definition of select-partition this is true after r' has chosen a new partition and thus makes A the initial state. In the second *state B*, no information about the relation of $|\mathrm{lw}(r')|$ and $|\mathrm{lw}(r'')|$ is available. The result of the output function ω is the cost change for r' and r'' for the step, it is added to the pair $(c(r', x-1), c(r'', x-1))$ yielding $(c(r', x), c(r'', x))$. The resulting transducer is defined in Fig.4. The transducer is straightforward but note that in A, step (3) is excluded by the definition of select-partition, and that in B, (3) leads back to A for the same reason.

For all state changes between A and B at $r(x)$, $c(r', x) \leq c(r'', x)$ is satisfied and $c(r')$ can never grow larger than $c(r'')$. This implies that whenever the final state T is reached, there is no r'' with lower cost than r' and it follows that $c(r')$ is minimal. This concludes that the constructed r' is an optimal partition walk of r.   (Optimality)∎

Remark: $c(r', x) \leq c(r'', x)$ indicates that the *Stay-Local Principle* is sufficient but not necessary for an optimal walk r'.

6   Evaluation

For the evaluation we conducted experiments in order to address two topics: First, we wanted to show how much the number of network hops of a random walk in a distributed, partitioned graph G can be reduced when the walk is executed in the replicated graph G_Π using FFS.

Second, we lowered the message fan-out $f \leq k$ of select-partition-steps and with it the subquery duplication overhead by using various vertex replica selection metrics and investigated the impact on the fraction α of saved network hops.

Additionally, we examined how all this depends on the replication factor k and the length of used walks.

6.1   Generating Graph Partitionings

To increase the probability that an edge may be traversed locally, we want to ensure that the partitions of each replica of each vertex differ. However standard graph partitioners (like [8]) do not replicate vertices and create no overlapping partitions.

To generate non-disjoint partitionings, we partition a graph multiple (k) times, using METIS [8] as state-of-the-art graph-partitioner. To ensure that the resulting partitionings differ, all edge weights are set to uniform random values in each iteration. The result partitioning Π is the union of all partitions found in all iterations. Consequently each vertex will be assigned to a different partition in each iteration and $|\Pi(v)| = k$ for all vertices $v \in V$.

6.2 Experimental Setup

An *evaluation sequence* for a graph consists of k iterations. For every iteration, we generated a complete but different partitioning with $m = 12$ partitions.

An *experiment*, for an evaluation sequence and a set of walks, records the number of hops needed for every walk in every iteration of G_Π for a set of partition walk routing algorithms (i.e. FFS, or FFS using some metric). We used fixed sets of graphs and walks described below.

Graphs. We used the following graphs: scaladocs – the hierarchical link graph of the HTML documentation of the scala programming language, lattice_2d – a 2D-Lattice where each vertex has 4 edges, kleinberg – a scale free graph generated with the Kleinberg model, barabasi_sfg – another scale free graph that was generated with the Barabási-Albert model, and random – an Erdős-Rényi random graph. Characteristic measures of these graphs are given in table 1.

Walk Generation. For each graph, we generated 3000 walks of different length, starting at 5, followed by all multiples of 10, up to 140. We used the unpartitioned graph with its initial (random) weights for walk generation. The generation of a walk starts by selecting a start vertex uniformly at random. The successor v_j of each other vertex v_i in the walk is chosen with probability $p_j = w_n \left((v_i, v_j) \right)$.

We decided to focus on queries that are mostly cycle free (i.e. contain very few reuses of the same edge) by trying to avoid picking the same edge twice. This was achieved by setting the weight of a previously encountered edge to 0. Under this scheme, repeated vertex visits may cause $\forall e \in E_v^{\rightarrow} : w(e) = 0$ but then edges are chosen completely random with probability $p_j = \frac{1}{|E_v^{\rightarrow}|}$. Figure 5 shows the average hop-count for the generated walks.

6.3 Experimental Results

Figure 6 depicts the relative impact of FFS on the number of remote hops. Results show that this impact is highly graph-dependent. However, in all graphs, $20-30\,\%$ hops are saved for $k \in \{2, 3\}$ latency-optimal partition routing, i.e every

Table 1. Graphs used in experiments

Name	Vertices	Edges	Avg.Degree	Indegree Min.	Max.	Std.dev.	Outdegree Min.	Max.	Std.dev.
scaladocs	1548	25697	16.60	0	1528	64.21	0	947	29.79
kleinberg	2304	11520	5.00	4	10	0.98	5	5	0
barabasi_sfg	2003	3989	1.99	1	10	1 27	0	13	1.52
lattice_2d	2025	8100	4.00	4	4	0	4	4	0
random	2000	24905	12.45	2	26	3.58	2	27	3.46

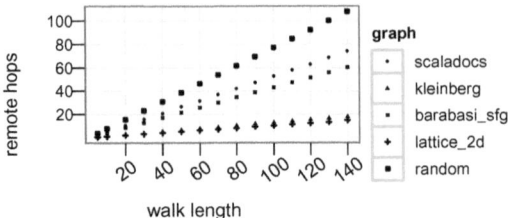

Fig. 5. Average number of remote hops of generated walks

3rd to 5th hop is eliminated. Three of the analyzed graphs achieve substantially higher gains from using FFS at $k = 3$: kleinberg (60 %), barabasi_sfg (40 %), and the 2D-Lattice (≈ 80 %!). Higher savings are achievable by increasing k, however returns diminish as k increases and choosing $k > 4$ seems unreasonable when replicating for performance (latency) only.

For all graphs and k, the gain appears to converge against a constant value (*target gain*). Short walks of length < 20 achieve higher gain. This happens because short walks which have few hops in the unreplicated scenario are converted into hop-less walks by the used replication scheme. This "short walk bonus" gets higher for graphs with a larger target gain.

Subquery duplication. To asses the degree of repeated computations, i.e. subquery duplication d, we measured the actual number of query steps (visits of replicated vertices). Fig. 7 plots them as factor d of vertex visits at $k = 1$, i.e. $1 \leq d \leq k$. In the experiments d always is smaller than k. This is an indicator that our partitioning scheme based on random weights actually generated differing partitionings for the walks. Additionally, $\frac{k-d}{k}$ goes up, as k is increased, i.e. replication reduces relative subquery duplication.

Impact of Metrics. Finally, we analyzed various metrics (Fig. 9) as scores for selecting only a small number of vertex replicas at partition boundaries in order to reduce message fan-out and subquery duplication.

Initial attempts with classic centrality measures were not sufficiently successful. Best results were achieved with betweenness and coreness but both did only work for some graphs. Surprisingly, flow centrality appeared not to be a suitable predictor, too. This is shown in Fig. 8 exemplarily for the Barabási-Albert graph with $3 \leq k \leq 6$, for the other graphs the results are similar.

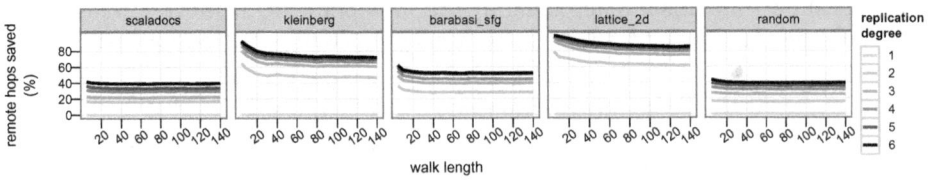

Fig. 6. Percentage of hops saved by latency-optimal routing with FFS when k overlapping partitionings (graph replicas) are used to execute random walks of varying length

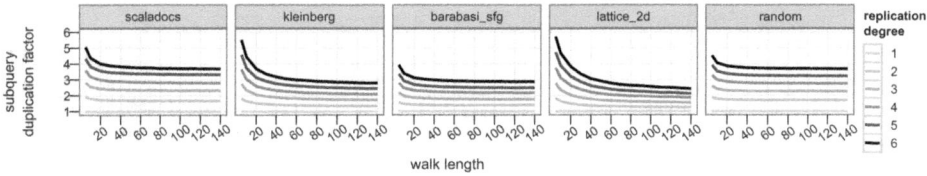

Fig. 7. Subquery duplication factor of partition walks, i.e. the number of additional vertex replicas visited as an effect of speculative execution in relation to walk length, replication degree, and graph

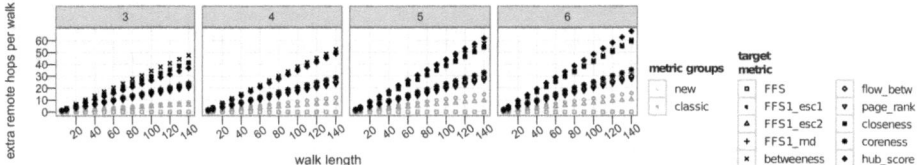

Fig. 8. Overhead in absolute number of hops against optimal FFS on the example of Barabási-Albert. Fan-out was restricted to 1 for all metrics.

After initially good results with the ratio-metric we developed escapicity as it's generalization. Fig. 9 shows the loss in optimality caused by selected metrics for all graphs and $3 \leq k \leq 6$ in terms of required additional hops per walk length. The two variants of FFS2 always achieve minimal overhead, i.e. are closest to finding optimal walks.

In general, an increased fan-out predictably boosts each metric but is not sufficient to heal a bad metric, e.g. for barabasi_sfg escapicity with a fan-out of 1 is better than the random metric with a fan-out of 2.

Using the ratio-metric is always worse than using escapicity with $i = 1$. Using escapicity with more information by increasing the radius i seems to give slight gains with the exception of barabasi_sfg and scaladocs. More experiments with higher i and larger partitions are needed to further analyze this.

7   Related Work

Partitioning. Graph partitioning is NP-complete in general, but there exists a multitude of heuristic partitioning algorithms [5, 15]. Partitioning graphs in a distributed scenario needs to deal with the lack of a global view. Common approaches are growth-based [4] or diffusion-based [2] partitioners.

Replication. Overlapping partitions (i.e. replicated vertices) have been used in VLSI for delay minimization and to optimize chip layout [7,9]. These approaches are static, limited to current-flow-networks, and do not need to address replica selection. Pujol et al. [13] use one-hop replication to optimize network usage for partitioned social graphs. Schism [3] uses graph partitioning and replication of tuples to distribute relational DBMS' across different resource nodes. Therefore

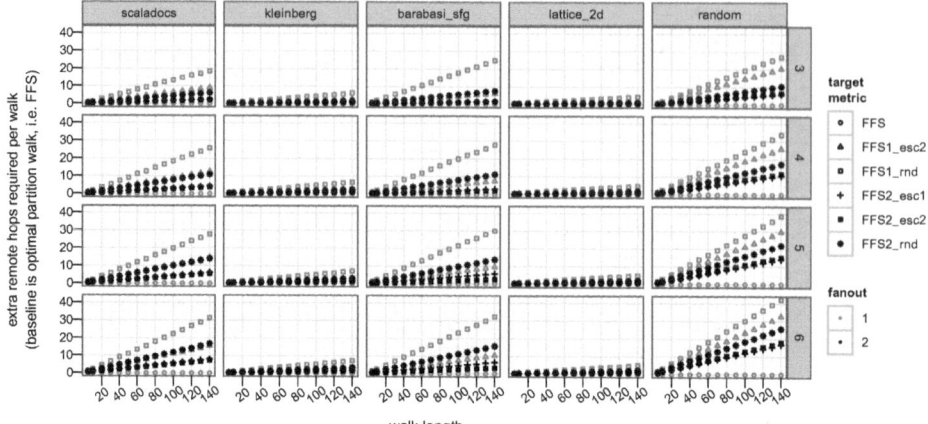

Fig. 9. Overhead in absolute number of hops against optimal FFS caused by lowering the fan-out at partition boundaries and selecting vertex replicas according to some precomputed metric, measured for various walk lengths, $3 \leq k \leq 6$ replicas, and graphs. *Target metrics* are coded according to the following scheme: FFS<fan-out>_<metric> where metric is rnd (random) or esc$i$ (escapicity with i-steps).

Schism generates a graph from SQL traces, partitions the graph and replicates tuples with the goal to minimize distributed transactions. However the graph itself isn't used for query processing, it is transformed to an SQL script and fed back into the original DBMS.

Replica Selection. In large networks, like the WWW, it is common practice to replicate objects for the sake of availability, latency and reliability [16]. Replicated objects require a client to be able to select the replica which suits him best. Since clients are interested in fetching the object, not the environment where it is located, the selection-metrics are mostly distance or latency based.

8   Conclusion

We introduced fast-forward-search (FFS), a new approach for minimizing the latency of walks in distributed graphs that rests on finding hop-minimal paths in multiple, overlapping partitionings by employing parallel speculative execution. We presented a proof of FFS's optimality, described how to generate overlapping partitions with standard graph partitioners by using randomized edge weights and experimentally measured that FFS may save between $20 - 80\%$ of network hops depending on graph type and replication factor k.

We showed that the storage, message, and computational costs of our approach are bounded by the replication factor k and discussed how to reduce them by lowering the message fan-out. Finally, to select fan-out vertices intelligently, we introduced escapicity, a vertex replica selection metric, and evaluated experimentally that escapicity has a very low overhead compared to classic measures of network centrality as a predictor for vertex replica selection.

For the future, we intend to repeat the experiments for larger graphs, and look at the impact of graph size, number of partitions, and alternative replication and graph partitioning schemes on the performance of FFS.

References

1. Angles, R., Gutiérrez, C.: Survey of graph database models. ACM Comput. Surv. 40(1) (2008)
2. Averbuch, A., Neumann, M.: Partitioning Graph Databases. Master's thesis, KTH Stockholm (2010)
3. Curino, C., Jones, E., Zhang, Y., Madden, S.: Schism: a workload-driven approach to database replication and partitioning. Proceedings of the VLDB Endowment 3(1) (2010)
4. Derbel, B., Mosbah, M., Zemmari, A.: Fast distributed graph partition and application. In: 20th International Parallel and Distributed Processing Symposium, IPDPS 2006, p. 10. IEEE, Los Alamitos (2006)
5. Elsner, U.: Static and dynamic graph partitioning: A comparative study of existing algorithms. Ph.D. thesis, Technische Universität Chemnitz (2002)
6. Hastorun, D., Jampani, M., Kakulapati, G., Pilchin, A., Sivasubramanian, S., Vosshall, P., Vogels, W.: Dynamo: Amazon's highly available key-value store. In: Proc. SOSP, pp. 205–220 (2007)
7. Hwang, L., El Gamal, A.: Min-cut replication in partitioned networks. IEEE Transactions on Computer-Aided Design of Integrated Circuits and Systems 14(1), 96–106 (2002)
8. Karypis, G., Kumar, V.: A fast and high quality multilevel scheme for partitioning irregular graphs. SIAM J. Sci. Comput. 20, 359–392 (1998)
9. Kring, C., Newton, A.: A cell-replicating approach to minicut-based circuit partitioning. In: 1991 IEEE International Conference on Computer-Aided Design, ICCAD 1991, Digest of Technical Papers, pp. 2–5. IEEE, Los Alamitos (2002)
10. Lehmann, K., Kaufmann, M.: Decentralized algorithms for evaluating centrality in complex networks, p. 9 (2002)
11. Martínez-Bazan, N., Muntés-Mulero, V., Gómez-Villamor, S., Nin, J., Sánchez-Martínez, M.A., Larriba-Pey, J.L.: Dex: high-performance exploration on large graphs for information retrieval. In: Proceedings of the Sixteenth ACM Conference on Conference on Information and Knowledge Management, CIKM 2007, pp. 573–582. ACM, New York (2007)
12. Page, L., Brin, S., Motwani, R., Winograd, T.: The pagerank citation ranking: Bringing order to the web (1999)
13. Pujol, J.M., Siganos, G., Erramilli, V., Rodriguez, P.: Scaling online social networks without pains. In: 5th International Workshop on Networking Meets Databases, NetDB 2009, co-located with SOSP (October 2009)
14. Rodriguez, M.A., Neubauer, P.: The graph traversal pattern. Computing Research Repository (CoRR) abs/1004.1001 (2010)
15. Schaeffer, S.E.: Graph clustering. Computer Science Review 1(1), 27–64 (2007)
16. Vingralek, R., Breitbart, Y., Sayal, M., Scheuermann, P.: Web++: A system for fast and reliable web service. In: Proceedings of the Annual Conference on USENIX Annual Technical Conference, p. 13. USENIX Association (1999)

Graph-Based Matching of Composite OWL-S Services

Alfredo Cuzzocrea[1], Juri Luca De Coi[2],
Marco Fisichella[2], and Dimitrios Skoutas[2]

[1] ICAR-CNR and University of Calabria, Italy
cuzzocrea@si.deis.unical.it
[2] Forschungszentrum L3S, Hannover 30167, Germany
{decoi,fisichella,skoutas}@L3S.de

Abstract. Existing techniques for Web service discovery focus mainly
on matching functional parameters of atomic services, such as inputs
and outputs. However, one of the main advantages of Web services is
that they are often composed into more complex processes to achieve a
given goal. Applying such techniques in these cases, ignores the workflow
structure of the composite process, and therefore may produce matches
that are not very accurate. To overcome this limitation, we propose in
this paper a graph-based method for matching composite services, that
are semantically described as OWL-S processes. We propose a graph representation of composite OWL-S processes and we introduce a matching
algorithm that performs comparisons not only at the level of individual
components but also at the structural level, taking into consideration the
control flow among the atomic components. We also report our preliminary results of our experimental evaluation.

1 Introduction

Web services are a key technology for enabling interoperability and software reuse.
Service discovery is the process of matching a service request with a service advertisement, and it is based on comparing their descriptions, such as their input and
output parameters. Service composition deals with composing services to create
complex processes that achieve a desired goal given an initial state. This is an important feature, since it allows atomic services to be combined in a flexible way to
complete complex tasks. In the Semantic Web, service descriptions are semantically annotated using concepts from domain ontologies in order to facilitate and
improve the precision of their automatic discovery and composition.

Service composition is a very challenging task, either when performed at design time or, especially, online. Given also that reusability is a key concern in
service-oriented architectures, this makes the discovery of existing composite services an important problem. When an application needs to create a composite
service to fulfill a given goal, it is more effective and efficient to first search a
repository of existing compositions to find similar ones. Then, the best matches
identified can be modified, extended or combined, to produce the desired composite service instead of composing one from scratch. Moreover, when browsing

J. Xu et al. (Eds.): DASFAA Workshops 2011, LNCS 6637, pp. 28–39, 2011.
© Springer-Verlag Berlin Heidelberg 2011

a repository of composite processes, the user may find some interesting process and then issue a "more like this" query to retrieve additional results.

In this paper, we focus on composite Semantic Web services described in OWL-S [3], since OWL-S provides different parts for explicitly describing the *profile* and the *model* of a service. The service profile is mainly aimed at supporting service discovery, and it includes the functional parameters of the service, which are the ones typically used for the matchmaking between service descriptions. The service model is primarily aimed at the specification of composite services. In particular, it describes the internal components and the control flow of the composite process. In a typical service discovery scenario, a query is formulated as the description of a desired service and the result is a ranked list of advertised services, the description of which matches the request, according to a matchmaking algorithm that employs one or more matching criteria. Hence, existing discovery methods do not differentiate between atomic and composite services. The service profile is also used for the matchmaking of composite services, which means that a complex process is treated as a "black box"; its most integral part, the process model, is not taken into account. This severely reduces the accuracy of the results, introducing both false positives and false negatives.

To address this problem, we propose a graph-based method for matching composite services. Matchmaking is performed on the service model rather than the service profile, which includes the structural part of the composite service. In OWL-S, composite services can be composed from atomic ones or from other simpler composite services, allowing several levels of nesting. To specify how component services are combined together, a set of *control constructs* is provided, similar to the typical control structures found in programming languages. These allow services to be executed sequentially, in parallel, conditionally or in a loop.

The proposed method performs the matching on two levels. It matches both the atomic services of the composite process, using their service profiles, as well as the way these services have been composed to create the composite process. To avoid the details and specificities of the OWL-S process model, the composite process is first transformed to a graph representation, containing its component services and their interactions. Matchmaking is then performed considering node similarities and finding common (sub)structures between the two graphs that represent the requested and the available composite process. To increase the efficiency of the search, a two step approach is followed. Initially, a set of candidate graphs is identified considering mappings between pairs of nodes. Then, the best candidates are selected and their structural similarity to the query graph is taken into account in order to filter out false positives and determine the final ranking of the results.

The rest of the paper is structured as follows. The next section describes our graph-based representation of composite OWL-S processes. Section 3 introduces the matchmaking algorithm. A preliminary experimental evaluation is presented in Section 4. Section 5 discusses related work, and Section 6 concludes the paper with directions for future work.

2   Graph Representation of OWL-S Processes

The OWL-S description of a Web service comprises three main parts. The *Service Profile* specifies the functional parameters of the service, namely inputs, outputs, pre-conditions and effects; it may also contain information about the provider and the category of the service, as well as plain text description. The *Service Model* describes the components and the structure of a composite process. The *Service Grounding* specifies the details required by an agent to invoke the service, such as communication protocol, message formats and port numbers. In contrast to typical service discovery approaches that rely on the service profile for matching atomic services, our matchmaking algorithm utilizes the information provided by the service model to perform graph-based matchmaking between complex processes.

In our approach, we represent composite processes as graphs in order to facilitate their matching. Given a composite OWL-S process P, we describe below how the corresponding graph representation, denoted as G_P, is derived. Let $\mathcal{C}$ be the set of control constructs supported in OWL-S. Each occurrence of a control construct $C \in \mathcal{C}$ is represented by a pair of nodes, C_b and C_e, that denote its begin and its end part, respectively. We denote the sets of such nodes as $\mathcal{C}_b$ and $\mathcal{C}_e$, respectively. These nodes allow us to represent the part of the process that is enclosed by this control construct and, hence, to represent the nesting of processes. Moreover, each atomic service s in a composite process is represented by a graph node s. For the sake of simplicity, we use the same symbol to refer both to the node and to the service it represents, since the distinction is typically clear from the context. The set of all the atomic services that are contained in the composite process P is denoted as $\mathcal{S}_P$. Thus, a composite process P is represented by a graph $G_P = (V, E)$, with node set $V = \mathcal{C}_b \cup \mathcal{C}_e \cup \mathcal{S}_P$ and edge set $E \subseteq (\mathcal{C}_b \times \mathcal{C}_b) \cup (\mathcal{C}_b \times \mathcal{S}_P) \cup (\mathcal{S}_P \times \mathcal{S}_P) \cup (\mathcal{S}_P \times \mathcal{C}_e) \cup (\mathcal{C}_e \times \mathcal{C}_e)$. The edges in the graph denote the control flow, as it will be explained below. In the following, we list the control constructs provided by OWL-S.

- *Sequence (SQ)*. It encloses a list of components to be executed in the specified order.
- *AnyOrder (AO)*. It encloses a bag (according to the OWL-S definition) of components to be executed sequentially, but without imposing any restriction on the ordering.

Table 1. Construction of the process graph

Control construct	Added edges	Control construct	Added edges
Sequence($s_1, s_2, \ldots s_{n-1}, s_n$)	$(SQ_b, s_1), (s_1, s_2), \ldots,$ $\ldots, (s_{n-1}, s_n), (s_n, SQ_e)$	AnyOrder($s_1, s_2, \ldots s_{n-1}, s_n$)	$(AO_b, s_1), (s_1, s_2), \ldots,$ $\ldots, (s_{n-1}, s_n), (s_n, AO_e)$
Split($s_1, s_2, \ldots s_n$)	$(SP_b, s_1), (SP_b, s_2), \ldots, (SP_b, s_n),$ $(s_1, SP_e), (s_2, SP_e), \ldots, (s_n, SP_e)$	SplitJoin($s_1, s_2, \ldots s_n$)	$(SJ_b, s_1), (SJ_b, s_2), \ldots, (SJ_b, s_n),$ $(s_1, SJ_e), (s_2, SJ_e), \ldots, (s_n, SJ_e)$
Repeat (s) While (*cond*)	$(RW_b, s), (s, RW_e)$	Repeat (s) Until (*cond*)	$(RU_b, s), (s, RU_e)$
Choice($s_1, s_2, \ldots s_n$)	$(CH_b, s_1), (CH_b, s_2), \ldots, (CH_b, s_n),$ $(s_1, CH_e), (s_2, CH_e), \ldots, (s_n, CH_e)$	If (*cond*) Then (s_1) Else (s_2)	$(IF_b, s_1), (IF_b, s_2),$ $(s_1, IF_e), (s_2, IF_e)$

- *Split (SP)*. It encloses a bag of components to be executed in parallel.
- *SplitJoin (SJ)*. It encloses a bag of components to be executed in parallel. The difference between SP and SJ is that the latter specifies barrier synchronization, i.e., all the included components need to finish their execution before the control construct is considered to be finished.
- *Choice (CH)*. It encloses a bag of components, one of which can be chosen for execution.
- *IfThenElse (IF)*. It encloses two components, one of which is executed based on whether a specified condition is true or false.
- *RepeatWhile (RW)*. It encloses a component that is executed in a loop, as long as a specified condition is true.
- *RepeatUntil (RU)*. It encloses a component that is executed in a loop, until a specified condition becomes true.

Note that we do not include conditions in our graph representation and in our matching algorithm. This is out of scope of this paper, given that OWL-S does not dictate any specific language for expressing such logical conditions. A possible extension to address the issue of conditions is to include them as labels on the nodes that correspond to control constructs having a condition, or on the outgoing edges of these nodes. Then, during the matching, an appropriate reasoner for the language used to express these conditions can be invoked to determine the degree of similarity between the condition in the requested service and the one in the advertised service, e.g., by inferring whether one condition implies the other (or its negation).

Table 1 specifies how the graph edges are constructed for each of the OWL-S control constructs. For simplicity, the table assumes only atomic services as components inside a control construct. If instead of an atomic service s there exists a nested composite process P' enclosed by a control construct C, then: (a) an edge (v, C_b) is added instead of each incoming edge (v, s); (b) an edge (C_e, v) is added instead of each outgoing edge (s, v); (c) the representation of the subprocess P' is computed recursively and added to the graph. Note that for each new occurrence of a control construct, a new pair of corresponding begin and end nodes is introduced. Some examples are shown in Figure 1.

3   Matching OWL-S Processes

In this section, we present our matching algorithm for composite OWL-S processes. First, we discuss how the degree of match *dom* is computed between atomic components and then how the structural similarity is taken into account.

3.1   Matching Atomic Components

Let G_R and G_P be the graph representations of a requested and a candidate composite services R and P, respectively. In the following, we show how to compute the degree of match between two nodes $r \in V(G_R)$ and $s \in V(G_P)$. Recall from Section 2 that each node in the graph corresponds either to an atomic service or to the begin or end part of a control construct. We compute the degree of match only between nodes of the same type, i.e., only for the cases that: (a) $r \in \mathcal{S}_R$ and $s \in \mathcal{S}_P$; or (b) $r \in \mathcal{C}_b$ and $s \in \mathcal{C}_b$; or (c) $r \in \mathcal{C}_e$ and $s \in \mathcal{C}_e$. For all other combinations, the degree of match is zero.

First, we define the degree of match between nodes that correspond to atomic services (also denoted by r and s). The degree of match is computed based on the input and output parameters of these services. For atomic services, an offer s matches a request r if: (a) the outputs offered by s match the outputs requested by r, and (b) the inputs provided by r match the inputs required by s. Consequently, to compute the degree of match for the inputs of r and s, we find the best match for each input of s and we normalize based on the number of input parameters of s:

$$dom_{IN}(r,s) = \frac{\sum_{u \in IN_s} \max_{v \in IN_r} \{sim(u,v)\}}{|IN_s|} \tag{1}$$

Similarly, for the outputs of r and s, we find the best match for each output of r and we normalize based on the number of output parameters of r:

$$dom_{OUT}(r,s) = \frac{\sum_{u \in OUT_r} \max_{v \in OUT_s} \{sim(u,v)\}}{|OUT_r|} \tag{2}$$

In Equations 1 and 2, IN_p and OUT_p denote, respectively, the set of input and output parameters of an atomic process p. The function sim computes the similarity between two individual input or output parameters. There are two basic alternatives for defining this function. The first one is to compare the corresponding parameter classes u' and v' in the ontology $\mathcal{O}$, as defined in the OWL-S service descriptions. In this case, we compute the similarity based on the number of common ancestors of these two classes:

$$sim(u,v) = \frac{|\{w \in \mathcal{O} \mid u' \sqsubseteq w \wedge v' \sqsubseteq w\}|}{|\{w \in \mathcal{O} \mid u' \sqsubseteq w\} \cup \{w \in \mathcal{O} \mid v' \sqsubseteq w\}|} \tag{3}$$

The second alternative is to compare the parameter names using some common string similarity measure, such as cosine similarity or Jaccard similarity. A hybrid similarity measure, combining both alternatives, can also be used [11].

The overall degree of match between r and s is then computed by aggregating the partial degrees of match for the inputs and the outputs, e.g., as the (weighted) average. This can be extended to include scores derived from additional matching criteria.

Next, we discuss how to define the similarity between nodes corresponding to control constructs. As shown by the description of the OWL-S control constructs in Section 2, some of them have similar functionality, and therefore, replacing one with another when searching for similar composite processes should incur a lower penalty. We examine the following cases.

SEQUENCE *and* ANYORDER. Both of these control constructs specify the execution of components in sequence; the difference between the two is that the latter does not specify the exact order but instead it allows any possible ordering (as long as there are no overlaps in the execution of two different components). Hence, a node SQ in the graph of the requested process R can be matched with a node AO in the graph of a candidate process P with a low effect on the similarity of the two processes. This case, however, is not symmetric. If AO appears in R and SQ in P, then the offered process is more restrictive than the requested one; hence, the match should be allowed, but with a lower score.

SPLIT *and* SPLITJOIN. Both of these control constructs specify the execution of components in parallel; the difference is that the latter specifies also barrier synchronization. Hence, it is permitted to match one of them with the other one with low effect on the similarity of the two processes.

REPEATWHILE *and* REPEATUNTIL. Both of these control constructs specify the execution of the enclosed component in a loop; their difference consists in whether the condition is checked at the beginning or at the end of each iteration. Therefore, matching these two control constructs should have a low effect on the process similarity.

Based on these observations, we define the similarity between two nodes denoting control constructs as follows:

$$dom_C(r, s) = \begin{cases} 0.9 & \text{for the pair (SQ, AO)} \\ 0.5 & \text{for the pair (AO, SQ)} \\ 0.8 & \text{for the pairs (SP, SJ), (SJ, SP),} \\ & \text{(RW, RU) and (RU, RW)} \\ 0 & \text{otherwise} \end{cases} \qquad (4)$$

These values hold when both of the nodes correspond either to the begin or to the end part of a control construct; otherwise, the degree of match is zero.

3.2 Matching Process Structure

To take into account the workflow structure of composite services, we measure the "overlap" in their graph representations. For this purpose, we compute their maximum common subgraph. This technique is often used in other applications, such as searching and mining databases of chemical structures, pattern recognition or computer vision [2].

Given a graph $G = (V, E)$, a subgraph of G is a graph $G' = (V', E')$ such that $V' \subseteq V$ and $E' = E \cap (V' \times V')$. Moreover, a graph $G = (V, E)$ is isomorphic to another graph $G' = (V', E')$ if there exists a bijective function $f : V \to V'$ such that for any edge $e = (v_1, v_2) \in E$ there exists an edge $e' = (f(v_1), f(v_2)) \in E'$. Then, the maximum common subgraph MCS of two graphs G_1 and G_2 is defined as the largest subgraph of G_1 that is isomorphic to a subgraph of G_2. Given the above, we can define the degree of match between a request graph G_R and a candidate graph G_P.

Definition 1. *Let G be the maximum common subgraph between a request graph G_R and a candidate graph G_P and f be the corresponding bijective function that maps G to a subgraph of G_P. The degree of match between G_R and G_P is defined as:*

$$dom_{\mathcal{G}}(G_R, G_P) = \frac{\sum\limits_{v \in V_G} dom_{\mathcal{V}}(v, f(v))}{|G_R|} \qquad (5)$$

where V_G denotes the set of nodes of graph G, $|G_R|$ is the number of nodes in the query graph and $dom_{\mathcal{V}}$ is a function that computes the degree of match between pairs of graph nodes based on their type, as described in Section 3.1. Notice that the MCS of two graphs is not necessarily unique; in that case, the one with the highest degree of match is considered.

A problem that arises from this approach has to do with the computational complexity, since the maximum common subgraph isomorphism problem is known to be NP-hard. Therefore, reducing the number of maximum common subgraphs to be computed becomes a critical issue. We address this problem based on the following observation. From Equation 5, it can be seen that, in order for two graphs to have a large degree of match, they should have a large number of node pairs with high degree of match. Indeed, if there are only a few nodes in G_R that can be mapped with high similarity to nodes of G_P, then the sum in the numerator of Equation 5 can not be large. However, this is a necessary but not sufficient condition, since the sum is computed over the nodes of the identified maximum common subgraph. Hence, if the two graphs have many similar nodes but a small maximum common subgraph, then the sum would be again small. This is desired in order to prevent matches between composite services that are different from a structural point of view.

To make the search process more efficient, we identify first those candidate graphs that have nodes that can be mapped with high similarity to the nodes of the query graph, and we select the top-k' ones. This provides a list with candidate matches for the query which contains also false positives due to the reason explained previously. Then, we apply Equation 5 on this subset in order to compute the actual degree of match and to obtain the final list of top-k matches, after filtering out the false positives. The value of k' has to be larger than k to account for the presence of false positives, but it can still be significantly smaller than the total number of candidate graphs to be examined.

To obtain the top-k' list of candidate graph matches, we apply a process based on the Hungarian algorithm (also referred to as the Kuhn-Munkres algorithm) [13], which has also been applied in a similar way to provide an approximation for the graph edit distance [15]. The algorithm can be used to solve the assignment problem in polynomial time and relies on a square cost matrix $\{c_j^i\}$, where each element c_j^i represents the cost of assigning the job j to the worker i. The output of the algorithm is the assignment minimizing the overall cost. We use this to compute the optimal assignment between the nodes of the query graph G_R and those of the candidate graph G_P. Based on the similarity between graph nodes, computed as described in Section 3.1, we construct a $|G_R| \times |G_P|$ cost matrix, where the cost for each pair of nodes is calculated as $c_v^u = 1 - dom_V(u, v)$. In the general case, the number of nodes of the two graphs is not the same, which means that not every node of the one graph can be mapped to a node of the other. To deal with this case, we introduce the concept of ε-node. The assignment of a node v to an ε-node (resp. of an ε-node to a node v) denotes that there is no mapping from (resp. to) node v. In other words, this corresponds to removing (resp. adding) a node in the graph. To make the cost matrix a square matrix, we introduce $||G_R| - |G_P||$ ε-nodes and we add the corresponding rows or columns in the matrix, as needed. We also set the cost for an assignment involving ε-nodes to 1. The optimal assignment between nodes is provided by the output of the algorithm. The overall cost of the assignment is computed as the sum of the costs of the pairwise mappings. The results are sorted in increasing order of cost and the top-k' ones are selected. For each one of these results, the degree of match to the query is then computed according to Equation 5, as explained previously, in order to obtain the final ranking.

4   Experimental Evaluation

Since existing approaches to service discovery focus on atomic services, we are not aware of an appropriate benchmark for evaluating the task of composite service matchmaking. To overcome this limitation, we have conducted experiments on a synthetically generated dataset of composite OWL-S processes, which were composed randomly from a set of publicly available real-world atomic OWL-S services. We describe first our experimental setup and methodology and then we present our results.

We implemented a synthetic generator for composite OWL-S processes. For each process, the generator first selects randomly one control construct, and then chooses how many atomic services or control constructs will be nested in it. This number is bound by a minimum and maximum value specified in a configuration file, which also defines the probability for selecting an atomic service or a given control construct. The number of maximum nested levels for control constructs is also specified. The atomic OWL-S services are selected from the OWLS-TC v2 collection[1], which is a publicly available collection of OWL-S services used to evaluate and compare different matchmaking algorithms. It comprises 1007 services from 7 different domains. All these are atomic services, hence we could not use the provided queries and their corresponding relevance sets to evaluate our matchmaking algorithm for composite processes. Using the generator, we created a dataset comprising the graph representations of 100 composite OWL-S processes and 10 queries.

We have implemented the matchmaking algorithm described in Section 3 in Java, reusing existing libraries whenever possible. In particular, we used the OWL-S API[2] for parsing the descriptions of OWL-S services in order to match

---

[1] http://projects.semwebcentral.org/projects/owls-tc/
[2] http://www.mindswap.org/2004/owl-s/api/

their inputs and outputs, the SimPack[3] library for the computation of the maximum common subgraph and a Java implementation of the Hungarian algorithm[4].

For each one of the 10 queries, we first ran the matchmaking process based on the Hungarian algorithm to obtain a candidate list of matches, and we selected the 20 graphs with the lowest assignment cost. These are graphs that contain nodes with high similarity to the nodes of the query graph, but do not necessarily match the structure of the requested process. Then, for each one of these top-20 candidate matches, we computed the degree of match to the query based on the maximum common subgraph and we retrieved the top-10 results.

We compared the lists of top-10 graphs before ($\mathcal{L}_H$) and after performing the last step ($\mathcal{L}_{MCS}$). Our purpose was to examine how much the former ranking differs from the latter one, i.e., for how many graphs and how much the ranking changes once the structural similarity between the query and candidate processes is taken into account. For this purpose, we used Spearman's footrule distance [10], a commonly used distance measure for comparing different rankings. In particular, we used the extended version proposed by Fagin et al. [6], denoted in the following by F^*, which handles also the case where the compared rankings do not refer to the same set of items. This measure can be applied to our case as follows:

$$F^*(\mathcal{L}_H, \mathcal{L}_{MCS}) = \frac{\sum_{i \in \mathcal{G}} |pos(i, \mathcal{L}_H) - pos(i, \mathcal{L}_{MCS})|}{maxF^*} \qquad (6)$$

where $\mathcal{G}$ denotes the set of graphs in the two ranked lists and the function $pos(i, \mathcal{L})$ returns the position of i in the list $\mathcal{L}$ if $i \in \mathcal{L}$ and $|\mathcal{L}| + 1$ otherwise. $maxF^*$ denotes the maximum possible value that the *numerator* can take, which equals to $n(n+1)$, assuming that the lists to be compared consist of n elements. Higher values of this measure indicate higher difference between the rankings. For two identical rankings the value is zero, whereas the maximum value 1 is obtained when the two lists do not have any elements in common.

Fig. 2 shows the Spearman's distance between the $\mathcal{L}_H$ and $\mathcal{L}_{MCS}$ rankings for our 10 queries. As shown, in all cases the set of results and/or their ranking is affected after the structural similarity is taken into account. This is because some of the initial matches are identified as false positives and they are removed or ranked lower.

We examine in more detail how the results change for queries 1 and 3, which are the ones with the lowest and highest Spearman's distance, respectively. Since the query graphs are also contained in the dataset, the top-1 result in all cases is an exact match with the query graph itself. Table 3 shows the graph IDs in the $\mathcal{L}_H$ and $\mathcal{L}_{MCS}$ rankings for these two queries. Notice, for example, how graph 73, which does not appear in the $\mathcal{L}_H$ list of query 1, is ranked 4th in the $\mathcal{L}_{MCS}$ for the same query, whereas graph 42, initially at the 5th rank, does not appear

---

[3] http://www.ifi.uzh.ch/ddis/simpack.html
[4] http://sites.google.com/site/garybaker/hungarian-algorithm/assignment

Table 2. Spearman's distance for each one of the 10 queries

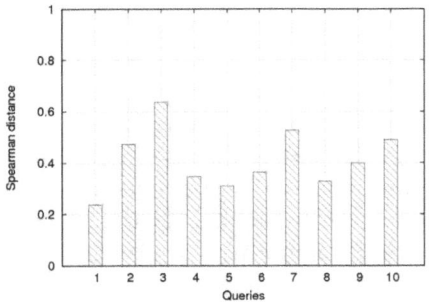

Table 3. Top-10 graphs for queries 1 and 3

Query 1		Query 3	
$\mathcal{L}_H$	$\mathcal{L}_{MCS}$	$\mathcal{L}_H$	$\mathcal{L}_{MCS}$
9	9	27	27
10	62	3	62
62	10	83	73
6	73	71	96
42	6	96	57
22	22	28	99
89	35	95	9
35	46	73	35
30	89	76	53
3	83	53	10

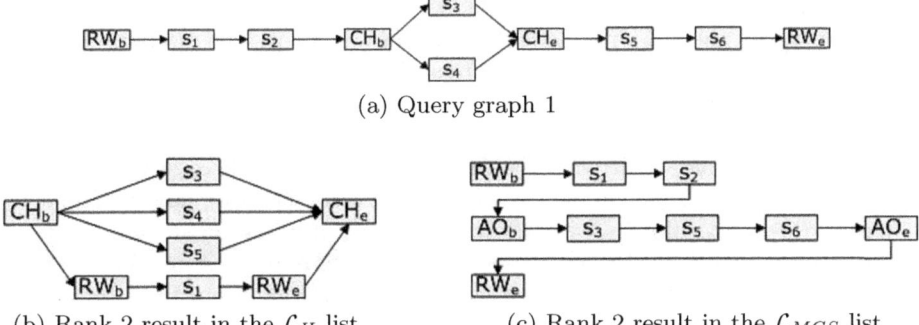

(a) Query graph 1

(b) Rank 2 result in the $\mathcal{L}_H$ list

(c) Rank 2 result in the $\mathcal{L}_{MCS}$ list

Fig. 1. A sample of query results

in the final list of the top-10 results. The differences are even more apparent for the two result lists of query 3.

As an illustrative example, Fig. 1 shows the results at rank 2 returned by the algorithm taking (Fig. 1c) and not taking (Fig. 1b) structural similarity into account for query 1 (Fig. 1a). The result returned at rank 2 in the $\mathcal{L}_{MCS}$ list is clearly more similar to the query than the one returned in the $\mathcal{L}_H$ list. Structural similarity plays a major role in realizing that (c) is more similar to (a) than (b), despite the fact that (a) and (b) share a higher number of nodes, and for this reason result to a lower assignment cost for the Hungarian algorithm.

5   Related Work

Traditional service discovery approaches employ IR-based techniques, such as keyword search on the textual descriptions of the services or matching of parameter names using common string similarity measures. A clustering algorithm is used in [5] to group parameter names into semantically meaningful concepts, which are used to identify similar services. An online search engine for Web services is *seekda*[5], which crawls and indexes service descriptions from the Web. Users can search for services using keywords, tag cloud navigation or faceted browsing, e.g., by country or service provider.

For services on the Semantic Web, logic-based matching is applied to increase the accuracy of the discovery process [14,12]. A reasoner is used to infer equivalence, subsumption or disjointness between the ontology classes describing the compared service parameters and the type of match is characterized accordingly as *exact, plug-in, subsumes, subsumed-by* or *disjoint.* In [1], the problem of matching requested and offered parameters in Semantic Web service descriptions is modeled as the one of matching bipartite graphs. Furthermore, the degree of match can be computed as a continuous, normalized value in the $[0, 1]$ interval, by defining some similarity measure between classes in the ontology [4,18]. Hybrid solutions have also been proposed for combining IR and logic-based techniques [11,9]. Ranking match results combining multiple matching criteria has been proposed in [17].

However, all the aforementioned approaches deal with the discovery of atomic services, ignoring the internal structure and components of a composite process. Our approach addresses this limitation by proposing a graph-based matchmaking algorithm for composite services.

Further work has dealt with the problem of workflow discovery. A search engine for workflows has been presented in [16], which allows for keyword queries to be issued over workflows. A workflow is retrieved if it contains components that match the keywords in the query. *myExperiment*[6] is another search engine for scientific workflows. Again, search is based on keyword queries or tags. In [7], workflow descriptions are augmented with constraints derived from properties about the workflow components used to process data, as well as the data itself. However, structural similarity is also not taken into account during matchmaking. Finally, [8] presents an approach and a tool to discover workflows that employs matching at the workflow structure level. However, they consider generic workflows and therefore they do not deal with how to match individual components or how to handle control constructs. To the best of our knowledge, our method is the first one to address the problem of discovering composite OWL-S services.

6   Conclusions and Future Work

We have proposed a graph-based method for matching composite OWL-S services. In contrast to existing approaches, which deal with atomic services, we focus on the internal components and structure of composite services and we perform the matching based on their process model. We employ a graph representation of composite OWL-S services, where the nodes represent atomic services and OWL-S control constructs. Based on this, the matching algorithm computes the degree of

---

[5] http://seekda.com/
[6] http://www.myexperiment.org/

match between two composite processes based on both node similarity and structural similarity.

As future work, we plan to conduct a more thorough experimental evaluation, and to extend our algorithm to consider conditions on graph nodes, as well as graph indexing to increase search efficiency.

References

1. Bellur, U., Kulkarni, R.: Improved matchmaking algorithm for semantic web services based on bipartite graph matching. In: ICWS, pp. 86–93 (2007)
2. Bunke, H., Shearer, K.: A graph distance metric based on the maximal common subgraph. Pattern Recognition Letters 19(3-4), 255–259 (1998)
3. Burstein, M., et al.: OWL-S: Semantic markup for web services. In: W3C Member Submission (November 2004)
4. Cardoso, J.: Discovering semantic web services with and without a common ontology commitment. In: IEEE SCW, pp. 183–190 (2006)
5. Dong, X., Halevy, A.Y., Madhavan, J., Nemes, E., Zhang, J.: Similarity search for web services. In: VLDB, pp. 372–383 (2004)
6. Fagin, R., Kumar, R., Sivakumar, D.: Comparing top k lists. In: SODA, pp. 28–36 (2003)
7. Gil, Y., Kim, J., Puga, G.F., Ratnakar, V., González-Calero, P.A.: Workflow matching using semantic metadata. In: K-CAP, pp. 121–128 (2009)
8. Goderis, A., Li, P., Goble, C.A.: Workflow discovery: the problem, a case study from e-science and a graph-based solution. In: ICWS, pp. 312–319 (2006)
9. Kaufer, F., Klusch, M.: WSMO-MX: A logic programming based hybrid service matchmaker. In: ECOWS, pp. 161–170 (2006)
10. Kendall, M., Gibbons, J.D.: Rank Correlation Methods. Edward Arnold, London (1990)
11. Klusch, M., Fries, B., Sycara, K.P.: Automated semantic web service discovery with OWLS-MX. In: AAMAS, pp. 915–922 (2006)
12. Li, L., Horrocks, I.: A software framework for matchmaking based on semantic web technology. In: WWW, pp. 331–339 (2003)
13. Munkres, J.: Algorithms for the assignment and transportation problems. Journal of the Society for Industrial and Applied Mathematics 5(1), 32–38 (1957)
14. Paolucci, M., Kawamura, T., Payne, T.R., Sycara, K.P.: Semantic matching of web services capabilities. In: Horrocks, I., Hendler, J. (eds.) ISWC 2002. LNCS, vol. 2342, pp. 333–347. Springer, Heidelberg (2002)
15. Riesen, K., Bunke, H.: Approximate graph edit distance computation by means of bipartite graph matching. Image Vision Comput. 27(7), 950–959 (2009)
16. Shao, Q., Sun, P., Chen, Y.: WISE: A workflow information search engine. In: ICDE, pp. 1491–1494 (2009)
17. Skoutas, D., Sacharidis, D., Simitsis, A., Sellis, T.: Ranking and clustering web services using multicriteria dominance relationships. IEEE T. Services Computing 3(3), 163–177 (2010)
18. Skoutas, D., Simitsis, A., Sellis, T.: A ranking mechanism for semantic web service discovery. In: IEEE SCW, pp. 41–48 (2007)

Design Non-recursive and Redundant-Free XML Conceptual Schema with Hypergraph

(Extended Abstract)

Joseph Fong, Wai Yin Mok, and Haizhou Li

Department of Computer Science, City University of Hong Kong, Hong Kong
csjfong@cityu.edu.hk

Abstract. Data Type Definition(DTD) and XML Schema Definition(XSD) are the logical schema of an XML model, but there is no standard format for the conceptual schema of an XML model. Conceptual modeling is a very important first step for constructing a database application. A conceptual model describes a system that is being built. Abstract ideas are made concrete as the ideas are represented in a formal notation. A formal conceptual model has a number of advantages. First, it helps designers understand and document the application under construction. Second, it facilitates development of algorithms that derive the underlying database schemas. In this paper, a real world of interest is described in a conceptual-model Hypergraph, which is a generic conceptual model. It is a Hypergraph because its hyperedges, or simply edges, are not necessarily binary. Its vertices represent sets of objects and its edges represent relationships among the vertices. Edges in a Hypergraph can be directed or undirected, depending on whether the underlying relationships are functional or non-functional. As opposed to relational databases, in this paper we are interested in constructing XML database applications with "good" properties. Two properties are particularly outstanding. First, the database should not have redundant data because redundant data lead to multiple-update problem once a single copy is modified. Second, since joins are expensive, the number of generated scheme trees, which are a generic hierarchical storage structure, should be as few as possible in order to reduce the number of joins required to answer a query. Users can draw a Hypergraph as XML conceptual schema with data relationships among elements as a result of specified functional dependency and multivalued dependency.

Keywords: Hypergraph, XML conceptual schema, XML Schema Definition, Data Type Definition, Data Modeling Technique .

1 Introduction

Generating schemas for data storage from conceptual models has a long-standing tradition. Its advantages are clear: (1) understandability, allowing both customer and developer to communicate effectively about the data concepts to be included and the

J. Xu et al. (Eds.): DASFAA Workshops 2011, LNCS 6637, pp. 40–52, 2011.

constraints to be enforced and (2) formality, allowing algorithmic derivation of schemas with good properties regarding the space needed to store the data and the time needed to query and update the data.

Following this conceptual-model tradition, we seek for algorithms to derive good XML schemas when the intended usage is for data storage in native XML databases [2]. We consider XML data schemas to be "good" when they prevent redundancy and are stored in as few schemas as possible. Preventing redundancy reduces storage and allows for simple constraint-satisfying update checks, thus reducing both space and time. Storing data in as few schemas as possible reduces query processing time when joins across populated schema instances are necessary.

By using a new definition of Hypergraph acyclicity, we will show that a polynomial-time algorithm exists that guarantees the generation of the fewest number of redundancy-free scheme trees. Further, the algorithm needs neither the URSA nor the BCNF assumption. Under URSA, there is a unique relationship among any set of attributes.

We provide a new definition of acyclicity for Hypergraphs that leads to a quadratic-time algorithm that generates the fewest redundancy-free XML scheme trees from an acyclic Hypergraph. At the same time, there is no loss of information in the sense that every relation of any edge and every object of any vertex in the Hypergraph is accounted for.

This research introduces Hypergraphs, which can be easily integrated into a database design tool. If a Hypergraph is acyclic, the proposed algorithms will be able to generate the fewest redundancy-free XML scheme trees that collectively cover the Hypergraph. The resulting XML scheme trees can then be straightforwardly translated into XML schemas, which define the underlying XML database. Hence, the proposed algorithms have the potential to be integrated with a commercial computer-aided XML database design tool.

Technically, a Hypergraph is a set of multisets of vertices where each vertex represents a set of objects. An edge in an acyclic conceptual-model Hypergraph is a set rather than a multiset. Figure 1 is an example of a Hypergraph based on the specified functional dependency and multivalued dependency from the users requirements:

In figure 1, FD (functional dependency) of Club determines Mascot and vice versa, and MVD (multivalued dependency) of Club determines many Students, and a Club determines many Activities. A circle means optional dependency. For example, a Student and a Course may determine a Grade upon successful completion of the course.

We first present several examples that help clarify our intentions. Example 1 gives an illustrative acyclic Hypergraph along with some valid instance data. Examples 2 and 3 illustrate poor designs: respectively, a design with data redundancy and a fragmented design with more scheme trees than necessary. Example 4 illustrates a good design, given the constraints of the Hypergraph in Figure 1.

Example 1: Figure 2 shows an acyclic Hypergraph H and a valid population of data for H. The data for H states that club b1, whose members are students s2 and s3 and whose mascot is m1, has activities a1 and a2. The data also states that professor p1 teaches courses c1 and c2, but professor p2 does not teach any course. And it states that student s1 earned an A in both courses c1 and c2, student s2 earned a B in course c2, but students s3 and s4 are new students who have not yet earned a grade for any course, although s3, but not s4, has already joined a club.

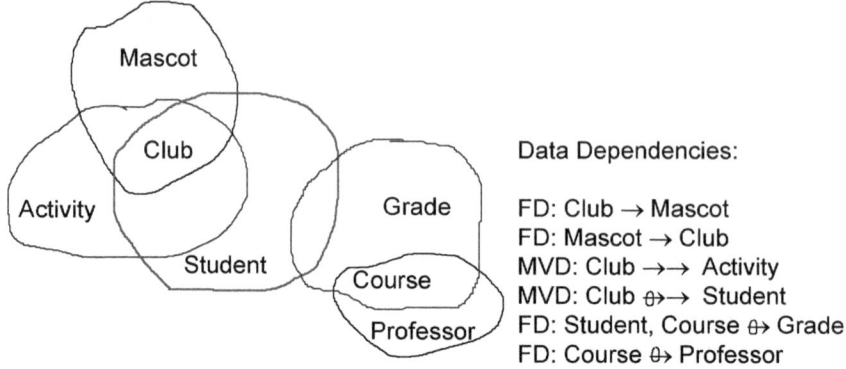

Data Dependencies:

FD: Club → Mascot
FD: Mascot → Club
MVD: Club →→ Activity
MVD: Club θ→→ Student
FD: Student, Course θ→ Grade
FD: Course θ→ Professor

Fig. 1. A Hypergrph with data dependencies

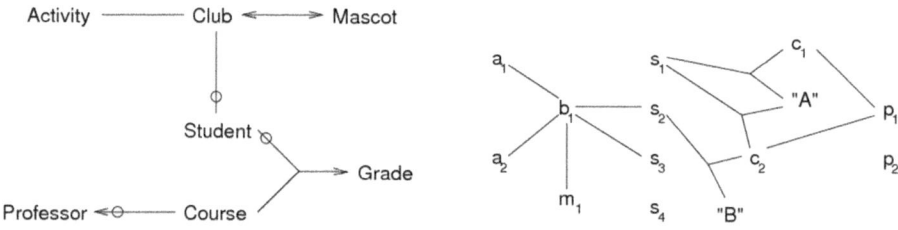

Fig. 2. An acyclic Hypergraph and a valid population of data

Example 2: Figure 3 shows a scheme tree for the Hypergraph H in Figure 2 along with the result of populating it with the instance data in Figure 2. However, the scheme tree and its populated instance in Figure 3 are problematic. The FDs Course → Professor, Club → Mascot, Mascot → Club and the MVD Club →→ Activity are constraints implied by the acyclic Hypergraph in Figure 2 that must hold. As a result, the populated scheme tree in Figure 3 has redundant data. Since course $c2$ appears twice and Course → Professor holds, both appearances of $c2$ must relate to professor $p1$. Similarly, since club $b1$ appears twice and Club → Mascot and Club →→ Activity hold, $b1$'s mascot and activities must appear twice, and since mascot $m1$ appears twice and Mascot → Club holds, $m1$'s club $b1$ must appear twice. In addition, professor $p2$, who does not teach any course, cannot even be included in the populated scheme tree in Figure 3, which results in a loss of data.

Example 3: Figure 4 shows a collection of scheme trees for the Hypergraph H in Figure 1 along with the results of populating them with the data from Figure 2. While the scheme trees and their populated instances in Figure 5 do not have redundancy, they unnecessarily fragment the data. This means we have to combine the data from two or more populated scheme trees to answer some queries. For example, we have to combine the data from the first and last populated scheme trees in Figure 4 to find the students taking courses taught by professor $p1$.

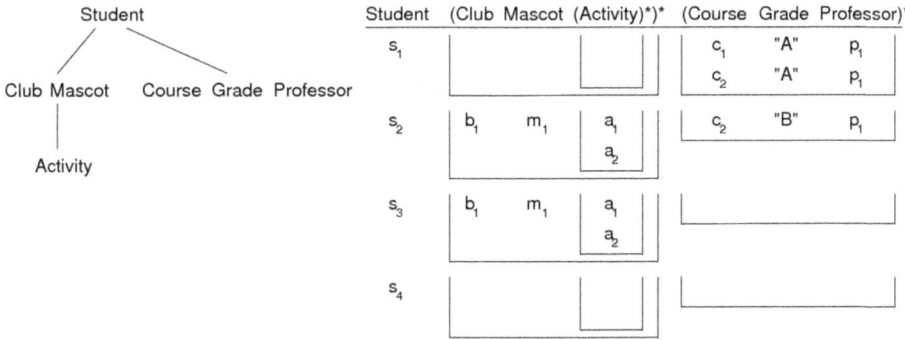

Fig. 3. An incorrect design that leads to data redundancy

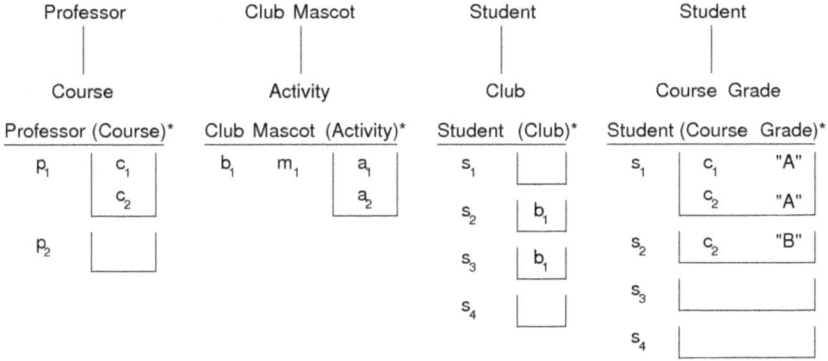

Fig. 4. An incorrect design that leads to fragmentation

Example 4: Figure 5 gives another alternative. It also has no redundancy in its three populated scheme trees. With the constraints given in the Hypergraph H in Figure 1, it is impossible to have fewer than three scheme trees and at the same time accommodate all valid data instances for H and store them without introducing redundancy with respect to the H-given FDs and MVDs and the H-given optional participation constraints.

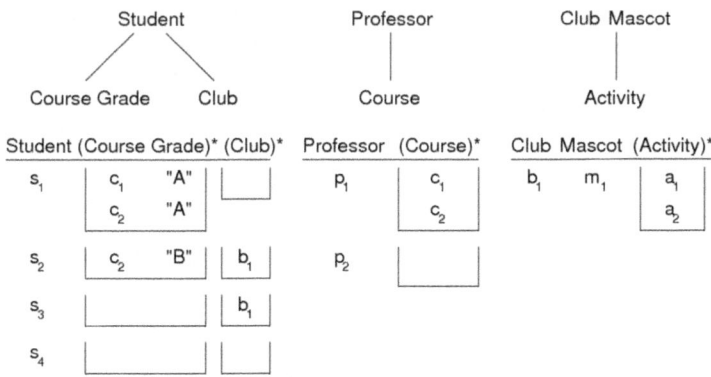

Fig. 5. A correct design that avoids data redundancy and fragmentation

2   Related Work

By way of comparison with the XML normalization work of others [1, 7, 8, 9, 10, 11, 12], we point out that our approach differs significantly. Not only have these other researchers defined their FDs, and thus their normal forms, differently, the basis of our approach is also different from theirs. As opposed to constraints specialized for XML, which are defined in these papers, we rely on standard FD and Hypergraph-generated MVD definitions–both of which can be straightforwardly derived from conceptual-model Hypergraphs. Furthermore, the basis of our approach is conceptual models, which have not been considered at all in other XML normalization work. We believe our approach is more common in practice and in line with the tradition followed by information system developers, who first create conceptual-model instances and then generate database storage structures.

In [1], we prove that generating the fewest redundancy-free XML scheme trees from Hypergraphs is NP-hard. An NP-hard result is clearly undesirable. As a result, we consider placing restrictions on Hypergraphs that may lead to a polynomial-time solution. We observe that if the universal-relation-scheme assumption [3] holds for a Hypergraph H, if H is Graham-reduction acyclic [4], and if each hyperedge in H is in BCNF [5], then we are able to show in [6] that finding the largest redundancy-free XML scheme tree can be done in polynomial time. Continuing this work, we will show that the problem has a quadratic-time solution if the Hypergraph is acyclic—an $O(n^2)$ solution where n is the number of edges in the Hypergraph. To motivate and clarify this result, we provide examples and preliminaries about Hypergraphs, Hypergraph cycles, scheme trees, and Nested Normal Form. We then present an overview of our algorithm; and in the future we will prove its complexity properties.

FIXT[15] used a Structural Graph to extract index and sub index information for efficient XML transformation. Graph is one of the graphical views of the conceptual schema of the XML model. However, the graph cannot easily be mapped to an XML document. XTABLES[16] used the DTD Graph that mirrored the structure of DTD. The graph consisted of elements, attributes, and operators. The function of DTD Graph is to construct the desired relational schema. Lu et al[14] provided a sufficient and necessary condition for the consistency between a DTD Schema and an XML document. A syntactically correct DTD schema might not be consistent with its corresponding XML document. They also proposed a linear algorithm to check the consistency of DTDs in terms of the cyclic DTD graphs, since their algorithm is based on the notion of DTD graphs.

3   Methodology

Two major components of this research are acyclic Hypergraphs and Nested Normal Form. Acyclic Hypergraphs lead to a quadratic-time, algorithmic solution for generating the fewest possible redundancy-free XML scheme trees whereas Nested Normal Form is a necessary and sufficient condition for preventing data redundancy in scheme trees. Their definitions are presented as follows.

Definition 1. A path in a Hypergraph H is a sequence of the form V1, C1, E1, C2, V2, . . . , Vi, C2i−1, Ei, C2i, Vi+1, . . . , Vn, C2n−1, En, C2n, Vn+1, where (1) n ≥ 1, (2) V1, . . . , Vn are vertices of H, (3) E1, . . . , En are edges of H, (4) C1, . . . , C2n are edge-vertex connections of H, and (5) each Ci in the sequence where i is odd conjoins its preceding vertex with its succeeding edge and where i is even conjoins its preceding edge with its succeeding vertex. A path is simple if its vertices are all distinct.

Definition 2. A cycle in a Hypergraph H is a path in H with V1 = Vn+1, where V1 is the first vertex of the path and Vn+1 is the last, and with every other vertex, edge, and edge-vertex connection unique.

Definition 3. A Hypergraph is acyclic if it does not have a cycle.

Example 5: The Hypergraph in Figure 2 is acyclic.

Nested Normal Form (NNF) [13] is a necessary and sufficient condition for removing data redundancy with respect to MVDs and FDs that hold for scheme trees. Since one of our goals is to generate redundancy-free scheme trees, they must all be in NNF. We now proceed to define NNF.

Let T be a scheme tree. Aset(T) denotes the set of attributes of T. Let N be a node in T. Notationally, Ancestor (N) denotes the union of attributes in all ancestors of N, including N. Similarly, Descendent(N) denotes the union of attributes in all descendants of N, including N. Each edge (V,W) in T, where V is the parent of W, denotes an MVD Ancestor (V) →→ Descendent(W). We use MVD(T) to denote the set of all MVDs represented by the edges in T. By construction, each MVD in MVD(T) is satisfied in the total non_nesting of any populated instance of T.

Definition 4. Let U be a set of attributes. Let M be a set of MVDs over U and F be a set of FDs over U. Let T be a scheme tree such that Aset(T) ⊆ U. Let D1 be the set of MVDs that hold for Aset(T) with respect to M ∪ F, and let D2 be the set of FDs that hold for Aset(T) with respect to M ∪ F. T is in NNF with respect to M ∪ F if the following conditions are satisfied.

1. MVD(T) ∪ D2 is equivalent to D1 ∪ D2 on Aset(T).
2. For each nontrivial FD X → A ∈ D2, X → Ancestor (NA) also holds with respect to M ∪ F, where NA is Stethe node in T that contains A.

When NNF's Condition 1 is violated, there is a populated scheme tree that has redundancy with respect to an MVD that holds. When NNF's Condition 2 is violated, there is a populated scheme tree that has redundancy with respect to an FD that holds.

Example 6: The scheme trees in Figures 4 and 5 are all in NNF whereas the scheme tree in Figure 3 is not. To see the violations, let T be the scheme tree S(BM(A)*)*(CGP)* in Figure 3 where S stands for Student, B for Club, and so on. This means that Aset(T) = SBMACGP and MVD(T) = {S →→ BMA, S →→ CGP, SBM →→ A}. The MVDs and FDs that hold for T are {B →→ A | M | SCGP, S →→ BMA | CGP, C →→ P | SGBMA} ∪ {B → M, M → B, SC → G, C → P}. Because B →→ A does not follow from MVD(T) and the FDs that hold for T, NNF's Condition 1 is violated. Consequently, there is a populated instance of T that has redundant data with respect to B $\rightarrow\rightarrow$ A. The populated scheme tree in Figure 3 is one such instance demonstrating the redundancy caused by B $\rightarrow\rightarrow$ A. In addition, C $\rightarrow$ P is an FD that holds for T but C $/\rightarrow$ SCGP. As a result, NNF's Condition 2 is also violated, and therefore there is a populated instance of T that has redundant data with respect to C $\rightarrow$ P. The populated scheme tree in Figure 3 also demonstrates the redundancy caused by C $\rightarrow$ P.

Step 1: Create Hypergraph
In this step, we create Hypergraph in Figure 1 according to user requirements in terms of functional dependencies and multivalued dependency. Many algorithms derive database schemas from various conceptual models. A common goal of these algorithms is to prevent data redundancy in the generated database schemas, and also to keep the number of schemas to a minimum. In addition to satisfy academic curiosity, many of these algorithms are mature enough to be integrated into commercial database design tools. This paper is no different. Beginning with a Hypergraph, which is a generic conceptual model, we develop algorithmic solutions that are able to generate the fewest redundancy-free XML scheme trees if the Hypergraph is acyclic. Further, the proposed algorithms have a quadratic-time complexity. Coupled with our previous results, this research completes the picture that if the Hypergraph is cyclic, the problem is NP-hard; and if the Hypergraph is acyclic, then quadratic-time algorithms exist. Because of its many advantages, the proposed algorithms would be a great benefit to XML database design tools. The following acyclic Hypergraph serves as a running example:

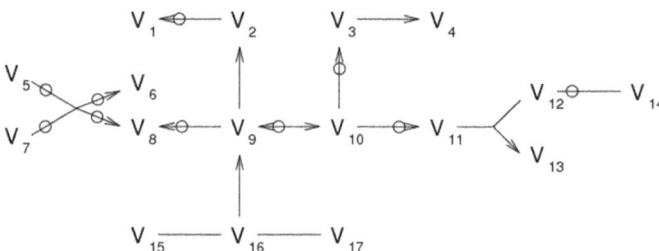

Fig. 6. A sample acyclic Hypergraph

Step 2: Form equivalence classes
Given an acyclic Hypergraph H, we begin by forming equivalence classes of edges based on functional equivalence of FDs given in H. We then observe that the set of vertices in each of the equivalence classes is in 4NF so that they are all redundancy-free with respect to the FDs and MVDs of H. We further observe that each equivalence-class vertex set is a largest possible set of vertices that does not violate 4NF. The set of vertices in each equivalence class constitutes a "degenerate" scheme tree ("degenerate" in the sense that it has only a single root node consisting of the vertices of the edges in the equivalence class). We then observe that an upper bound for the number of fewest redundancy-free scheme trees that collectively cover H is less than or equal to the number of equivalence classes of edges of H. This forest of degenerate scheme trees could be minimal, but normally is not. Our goal is to

combine the equivalence classes of H together into as few scheme trees as possible while retaining the redundancy-free property. In the same example, table 1 shows its equivalent classes.

Table 1. The equivalence classes of the Hypergraph in Figure 6

Edge	Equivalence Class	Equivalence-Class Vertex Set
V5V7 → V6V8	{V5V7 → V6V8}	V5V6V7V8
V2 → V1	{V2 → V1}	V1V2
V3 → V4	{V3 → V4}	V3V4
V12V14	{V12V14}	V12V14
V11V12 → V13	{V11V12 → V13}	V11V12V13
V9 ↔ V10	{V9 → V8, V9 → V2, V9 ↔ V10, V10 → V3, V10 → V11}	V2V3V8V9V10V11
V16 → V9	{V16 → V9}	V9V16
V15V16	{V15V16}	V15V16
V16V17	{V16V17}	V16V17

Step 3: Generate partial ordering
After generating equivalence classes, we organize these degenerate equivalence-class scheme trees into a partial ordering <EqC where Ci <EqC Cj for two distinct equivalence classes Ci and Cj if Ei ∈Ci and Ej ∈Cj such that Ei → Ej. This partial ordering indicates which equivalence classes can potentially be combined in parent-child relationships to form larger redundancy-free scheme trees. It is easy to observe that creating the Hasse diagram for the partial ordering <EqC takes at most quadratic-time in terms of the number of edges in the Hypergraph. In the same example, we generate partial ordering as shown below:

Example 7: Based on functional equivalence of the FDs in the Hypergraph in Figure 5, Table 1 lists its equivalence classes of edges. The Hasse diagram for the <EqC partial ordering on the equivalence classes is shown in Figure 7.

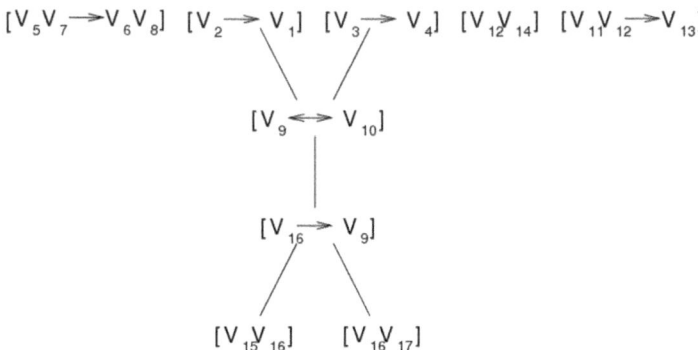

Fig. 7. The Hasse diagram for the <EqC partial ordering for the Hypergraph in Figure 6

Step 4: Build Transitional Tree

The next step is to construct transitional trees from the Hasse diagram for the <EqC partial ordering. However, this is more difficult than one might at first think. Several requirements must be met. (1) Eventually the transitional trees will be transformed into scheme trees. Hence, we must ensure that the resulting scheme trees are in NNF. (2) There cannot be any loss of information. This means that every relation of any edge and every object of any vertex in the given acyclic Hypergraph must be accounted for. (3) The construction process must be done in polynomial-time.

We now give an outline on constructing transitional trees.

(1) Form a set G of maximal equivalence classes with respect to <EqC.
(2) Initialize a set F of transitional trees to empty. Then, for each element C of G, if C has zero non-key connecting vertex, enter C into F. If C has exactly one non-key connecting vertex V, test if V can become a root node. If yes, enter V into F; otherwise, enter C into F.
(3) Test if a remaining maximal equivalence class in the Hasse diagram for <EqC can become a child node of a node in a transitional tree in F. If the answer is yes, make it a child node. Keep testing and adding the remaining maximal equivalence classes to the transitional trees in F until the answer is no for all remaining maximal equivalence classes. When an equivalence class is added as a child node to a transitional tree, remove it from the Hasse diagram.
(4) Go back to (1) until there is no more equivalence class left in the Hasse diagram.

These four steps of (1) (2) (3) and (4) outline a very high-level overview for constructing transitional trees. Much detail has been omitted. For example, testing if an equivalence class can become a child node is fairly complicated and therefore we cannot present its entirety in this proposal. It suffices, however, to say that we need to test if every edge in the equivalence class can be joined completely starting with the designated connecting vertex and every vertex is covered in either the equivalence class being tested or in some other equivalence classes.

Example 8: Four transitional trees are constructed according to our algorithm and they are shown in Figure 8. They will be transformed into scheme trees.

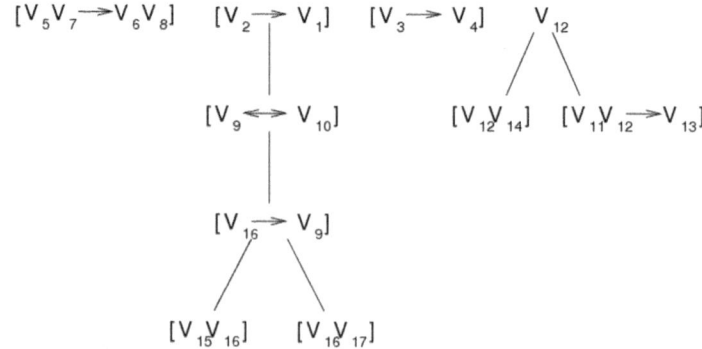

Fig. 8. The transitional trees constructed from the Hasse diagram in Figure 7

This step is fairly straightforward. In the same example, it simply turns a transitional tree into a scheme tree. In the future, we will prove that the resulting scheme trees are in NNF and thus are redundancy-free, and the number of generated scheme trees is also minimal.

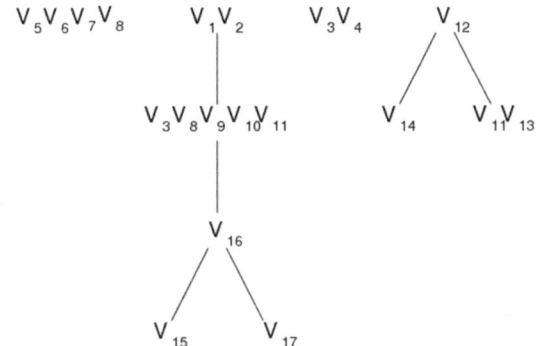

Fig. 9. The NNF scheme trees transformed from the transitional trees in Figure 8

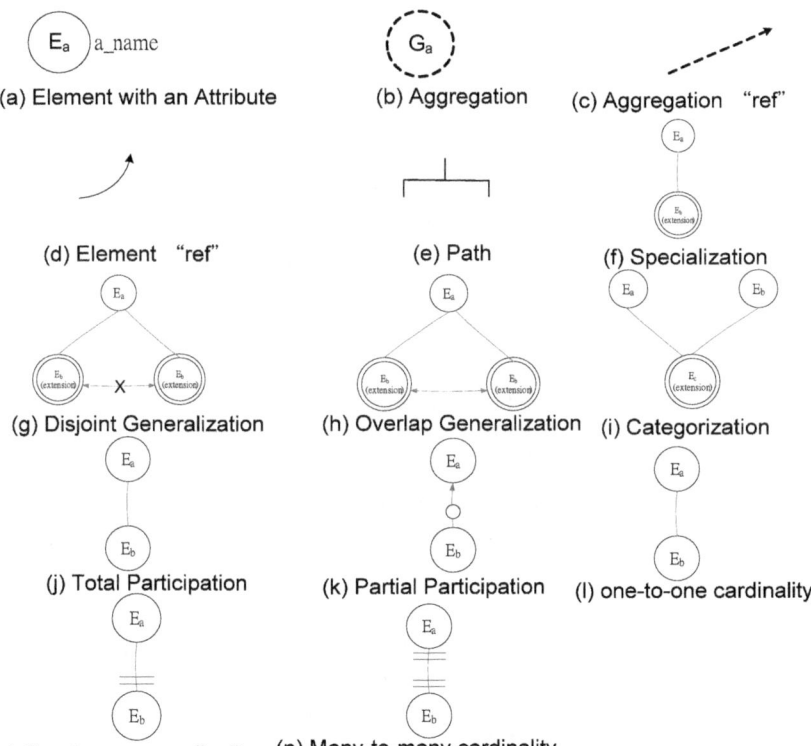

Fig. 10. Data semantics of XML Conceptual Schema in Hypergraph

Example 9: The resulting NNF scheme trees are shown in Figure 9. With the constraints given in the Hypergraph H in Figure 6, it is impossible to have fewer than four redundancy-free scheme trees that collectively cover H.

Step 5: Draw Hypergraph for XML Conceptual Schema
Many researchers have elaborated their views on conceptual schemas of an XML model such as XML trees[17] and DTD Graphs[16]. Most papers used the XML tree to show not only elements, but also attributes and data. The advantage is that a reader can analyze all components of an XML schema in a tree diagram. The disadvantage is that the tree is not easily generated and cannot represent many data semantics. For example, the DTD Graph proposed by IBM's XTABLE[16] is a conceptual schema for an XML model, but it cannot show data semantics. We propose using Hypergraph for representing a conceptual schema of an XML model because it can show data semantics, and it is simple enough for end-user computing. Figure 10 the symbols of a Hypergraph, with various data semantics such as cardinality, generalization, participation, aggregation and categorization.

By adding an artifact root element on top of the derived trees in Hypergraph, we can form a Hypergraph for XML conceptual schema as shown in Figure 11.

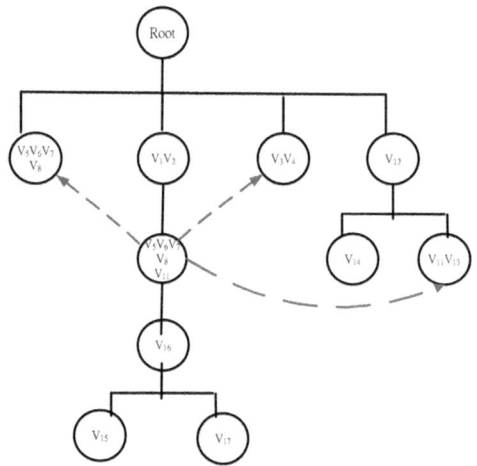

Fig. 11. The derived Hypergraph for XML conceptual schema

4   Conclusion

A log of work have been proposed for the graphical representation of XML. For example, Salim et al[18][19] used UML to represent the expressiveness of XML schema in XSD. However, their approach is reverse engineering XML logical schema XSD into XML conceptual schema UML. Their method lacks of designing an XML conceptual schema from the beginning, and then map into an XML logical schema. Furthermore, UML is not in tree structure as XML database. The contribution of this paper is to introduce Hypergraph in tree structure to design an XML conceptual schema, which can map into an XML logical schema in XSD or DTD without recursive and redundancy, and above all, with many data semantics. The feature of "ref" in Hypergraph can be implemented by ID and IDREF in DTD, as shown in the dash lines in Figure 11, which is much more dynamic than UML. The future research is to derive all data semantics that are possible under an Hypergraph.

References

[1] Mok, W.Y., Embley, D.W.: Generating compact redundancy-free XML documents from conceptual-model Hypergraphs. IEEE Transactions on Knowledge and Data Engineering 18(8), 1082–1096 (2006)
[2] Bourret, R.: XML database products (March 2007),
http://www.rpbourret.com/xml/XMLDatabaseProds.htm
[3] Sagiv, Y.: A characterization of globally consistent databases and their correct access paths. ACM Transactions on Database Systems 8(2), 266–286 (1983)
[4] Maier, D.: The Theory of Relational Databases. Computer Science Press, Rockville (1983)
[5] Codd, E.F.: Recent investigations in relational data base systems. In: IFIP Congress, pp. 1017–1021 (1974)
[6] Mok, W.Y., Fong, J., Embley, D.W.: Extracting a largest redundancy-free xml storage structure from an acyclic Hypergraph in polynomial time. Information Systems 35(7), 804–824 (2010)
[7] Chen, Y., Davidson, S.B., Hara, C.S., Zheng, Y.: RRXF: Redundancy reducing XML storage in relations. In: Proceedings of 29th International Conference on Very Large Data Bases, Berlin, Germany, September 9-12, pp. 189–200 (2003)
[8] Libkin, L.: Normalization theory for XML. In: Proceedings of the 5th International XML Database Symposium, Vienna, Austria, September 23-24, pp. 1–13 (2007)
[9] Schewe, K.-D.: Redundancy, dependencies and normal forms for XML databases. In: Proceedings of the Sixteenth Australasian Database Conference, Newcastle, Australia, January 31-February 3, pp. 7–16 (2005)
[10] Vincent, M.W., Liu, J., Liu, C.: Strong functional dependencies and their application to normal forms in XML. ACM Transactions on Database Systems 29(3), 445–462 (2004)
[11] Wang, J., Topor, R.W.: Removing XML data redundancies using functional and equality-generating dependencies. In: Proceedings of the Sixteenth Australasian Database Conference, Newcastle, Australia, January 31-February 3, pp. 65–74 (2005)
[12] Yu, C., Jagadish, H.V.: XML schema refinement through redundancy detection and normalization. The VLDB Journal 17(2), 203–223 (2008)
[13] Mok, W.Y., Ng, Y.-K., Embley, D.W.: A normal form for precisely characterizing redundancy in nested relations. ACM Transactions on Database Systems 21(1), 77–106 (1996)
[14] Lu, S., et al.: On the Consistency of XML DTDs. Data of Knowledge Engineering 52, 231–247 (2005)
[15] Xiao, J., et al.: FIXT: A Flexible Index for XML Transformation. In: Zhou, X., Zhang, Y., Orlowska, M.E. (eds.) APWeb 2003. LNCS, vol. 2642, pp. 144–149. Springer, Heidelberg (2003)
[16] Funderburk, J.E., et al.: XTABLES: Bridging Relational Technology and XML. IBM Systems Journal 41(4), 616–641 (2002)

[17] Fong, J., Cheung, S.K., Shiu, H.: The XML Tree Model - Toward an XML Conceptual Schema Reversed from XML Schema Definition. Data and Knowledge Engineering 64(3), 624–661 (2008)
[18] Salim, F.D., Price, R., Indrawan, M., Krishnaswamy, S.: Graphical representation of XML schema. In: Yu, J.X., Lin, X., Lu, H., Zhang, Y. (eds.) APWeb 2004. LNCS, vol. 3007, pp. 234–245. Springer, Heidelberg (2004)
[19] Salim, F.D., Price, R., Krishnaswamy, S., Indrawan, M.: UML Documentation Support for XML Schema. In: Proceedings of the 2004 Australian Software Engineering Conference, p. 211. IEEE Computer Society, Los Alamitos (2004)

Classifying Graphs Using Theoretical Metrics: A Study of Feasibility

Linhong Zhu[1], Wee Keong Ng[2], and Shuguo Han[1]

[1] Institute for Infocomm Research, Singapore
{LZHU,SHAN}@i2r.a-star.edu.sg
[2] Nanyang Technological University, Singapore
{AWKNG}@ntu.edu.sg

Abstract. Graph classification has become an increasingly important research topic in recent years due to its wide applications. However, one interesting problem about how to classify graphs based on the implicit properties of graphs has not been studied yet. To address it, this paper first conducts an extensive study on existing graph theoretical metrics and also propose various novel metrics to discover implicit graph properties. We then apply feature selection techniques to discover a subset of discriminative metrics by considering domain knowledge. Two classifiers are proposed to classify the graphs based on the subset of features. The feasibility of graph classification based on the proposed graph metrics and techniques has been experimentally studied.

1 Introduction

The graph representations with a collection of nodes and edges support all aspects of the relational data analysis process. For instance, graphs can be used to represent chemical compound structures, program dependencies, protein-protein interactions, relationships in online social media and so on. In despite of the flourish of graph data, there is lack of efficient approaches for various classical graph theoretical problems. Among those, how to build an efficient automated graph classification models and identify discriminative graph features that separate different graph classes has valuable applications. For example, biologists are interested in studying whether a set of protein (a subgraph in protein-protein interaction network) is functional or not [11].

One conventional solution to classify graphs is first to compute the graph similarity using various measures such as graph-edit distance [3], maximum common subgraph [9] and canonical sequence similarity [2], and then decide whether they belong to the same cluster or not based on a pre-defined similarity threshold. Another solution is that one finds subgraph patterns as features [11,15,16,23] and then represents each graph as a feature vector. In this way, the graph classification problem is converted into the classification of traditional high dimensional data points.

We refer the above two approaches as the isomorphism-based approach. They generally assume a hypothesis to verify the similarity of two graphs: If two

J. Xu et al. (Eds.): DASFAA Workshops 2011, LNCS 6637, pp. 53–64, 2011.
© Springer-Verlag Berlin Heidelberg 2011

graphs are similar in size/shape or they share a set of subgraph patterns, then they belong to the same cluster. As a result, graphs with the same cluster only differ from each other by a small number of nodes or edges. This hypothesis guarantees high accuracy for classification tasks for some particular domains, such as in the chemical data where size/shape differences of chemical component are small. However, this hypothesis is too restrictive to obtain outputs that are not similar in structure but highly related to each other, which can advance some domain analysis to gain significant benefits. As an example, in social network analysis, one interesting task is to predict a user's gender by analyzing his/her connections to others. To address it, one first extracts a subgraph for each user induced by the user and his/her friends. Then the gender prediction problem is converted to graph classification problem. In this application, graphs for the same gender extremely vary in both size and the structure. However, they do share some implicit graph properties. For example, neighborhood graph of a female better fits the small-world phenomenon (popularly known as six degrees of separation [14]) than that of a man.

In lieu of the above issue, this paper studies graph classification problem according to a more general hypothesis, i.e., if two graphs are similar in a set of diverse properties, then they belong to the same class. We call it "property-based graph classification". More specifically, we define the preliminary concept of property-based graph classification as follows.

Problem 1. (Property-based Graph Classification) Given a graph $G = (V, E)$ where V is a (finite) set of vertices and E is a (finite) set of edges, we assume that its properties could be measured by a set of graph theoretical-based metrics (More details about graph metrics will be given in Section 2.1). Then the property-based graph classification problem is to induce a mapping $f(x): \chi \rightarrow \{-1, 0, 1, \ldots, \gamma\}$, from giving training samples $T = \{< x_i, y_i >\}_{i=1}^{L}$, where $x_i \in \chi$ represents properties of a graph G and $y_i \in \{-1, 0, 1, \ldots, \gamma\}$ is a class label associated with training data, and L is the size of training samples. If $y_i \in \{-1, 1\}$, we denote it as a binary property-based graph classification problem.

The objectives of our work consist of two parts: (1) Investigating a set of graph theoretical metrics to facilitate graph classification; (2) Proposing an efficient and effective classification method for the property-based graph classification problem. The challenge lies not only in enumerating graph metrics, but also in selecting suitable metrics on the basis of domain knowledge. We address the challenges as follows:

- We give an extensive survey about the most important and diverse graph metrics that have been proposed and used in a wide range of studies in multiple disciplines. We also propose five novel metrics, non-tree density, cut vertex density, connectivity, vertex compression ratio and edge compression ratio to further discover implicit properties.
- Based on domain knowledge, we present an objective of feature selection combining with a greedy algorithm to perform discriminative feature subset selection.

- We explore an application of graph classification using graph metrics. We design two classifiers based on k-nearest neighbor classifier and reverse k-nearest neighbors classifier to verify the feasibility of graph classification using graph theoretical metrics.

The organization of this paper is as follows: In Section 2, we enumerate a number of graph theoretical metrics and use them to transform graphs into feature vectors. We discuss how to build up graph classifiers in Section 3. In Section 4, we perform experiments to verify the feasibility of our approach. Finally, We review related work in Section 5 and conclude this work in Section 6.

2 Representing Graphs with Theoretical Metrics

2.1 Graph-Theoretical Metrics

In the following definitions of metrics for graph properties, the definitions apply both to undirected and directed graph unless otherwise specified. In addition, we also provide the related references if they are not first proposed by this work.

For easy representation, we use G to denote a single graph and $\mathcal{G}$ to denote a set of graphs. We simply use $|V|$ to represent number of nodes and $|E|$ to represent number of edges. Given two nodes $u, v \in V$, $\sigma(u, v)$ denotes the distance (the length of shortest paths) from u to v. We use $N(v, G)$ to denote a set of neighbors of v in graph G.

Definition 1. (Global Clustering Coefficient [13]) *The global clustering coefficient is defined as:* $GCC(G) = \frac{3\Delta}{|(u,v,w):u,v,w \in V, u,v,w\ is\ connected|}$, *where Δ is number of triangles.*

Global clustering coefficient is based on triplets of nodes. A triplet is three nodes that are connected by edges. A triangle consists of three closed triplets, one centrad on each of the nodes. The global clustering coefficient is $3 \times$ triangles over the total number of triplets.

Definition 2. (Characteristic Path Length [5]) *Characteristic path length is defined as* $CPL(G) = \frac{\sum_{u,v} \sigma(u,v)}{|V| * (|V|-1)}$.

While characteristic path length is a measure of the average distance between any two nodes in the graph, global efficiency is a measure of the rate at which nodes can be reached from each other in the graph. It is approximately inversely related to characteristic path length.

Definition 3. (Global Efficiency [20]) *Global efficiency of a graph is defined as* $GE(G) = \frac{\sum_{u,v} \frac{1}{\sigma(u,v)}}{|V| * (|V|-1)}$.

Definition 4. (Degree Distribution [8]) *The degree distribution of a graph is a function $P: [0, \ldots, k_{max}] \to [0, 1]$, where $P(k)$ is the fraction of the vertices in G that have degree k for $0 < k < k_{max}$, and k_{max} is the largest degree in G.*

Definition 5. (Coreness [17]) *The k-core of G is the largest subgraph G_k of G such that every vertex in G_k has at least k neighbors in G_k, i.e., $\forall v \in V(G_k)$, $|N(v, G_k)| \geq k$.*

Definition 6. (Density) *Given a graph G, we define its graph density as:*

$$d_g = \begin{cases} |E|/|V|(|V|-1) & \text{if } G \text{ is directed,} \\[2mm] 2|E|/|V|(|V|-1) & \text{if } G \text{ is undirected.} \end{cases}$$

In addition, we propose two more density evaluation metrics, non-tree density and cut vertex density. Non-tree density d_{nt}, is defined as $|T|/|E|$, where T is the set of non-tree edges, i.e., remaining edges after removing a spanning tree of a graph. Cut vertex density $d_c = |V_c|/|V|$ (where V_c is the set of cut vertices; i.e., articulation points) is the proportion of cut vertices.

In the following metrics are restricted to directed graphs only.

Definition 7. (Girth [7]) *The girth of a graph is the length of the shortest cycle (of length >3) in the graph.*

Definition 8. (Circumference [7]) *The circumference of a graph is the length of the longest cycle (of length >3) in the graph.*

Definition 9. (Connectivity) *Connectivity $|C|$ is the number of strongly connected component(s).*

Definition 10. (Vertex Compression Ratio) *Vertex compression ratio $r_v = |V^*|/|V|$ (where $|V^*|$ is number of vertices in condensed graph[1]) is the proportion of condensed graph vertices.*

Definition 11. (Edge Compression Ratio) *Similar to Def. 10, edge compression ratio $r_e = |E^*|/|E|$ (where $|E^*|$ is number of edges in the condensed graph) is the proportion of condensed graph edges.*

Definition 12. (Acyclicity) *$d_e = |E_c|/|E|$ (E_c is the set of distinct edges that appear in a cycle) is a measure of the number of edges participating in one or more cycles. If there is no cycle, then $|E_c| = 0$.*

2.2 Feature Selection and Graph Transformation

A graph can be viewed as a collection of features on the basis of above metrics. We may simply normalize the values of each features into the range [0, 1] and transform each graph into a feature vector. In the above setting, we just assume that each metric is equally important. In reality, however, it can not be true especially when considering applications in a specific domain. For example, in

---

[1] We say $G^* = (V^*, E^*)$ is a condensed graph of G if each vertex $v_i^* \in V^*$ corresponds to a strongly connected component C_i in G, and each edge $(v_i, v_j) \in E^*$ if and only if there is at least one edge $u, v \in E$ such that $u \in C_i$ and $v \in C_j$.

social network analysis, a small-world network is mainly characterized by two metrics, namely, a shorter characteristic path length and a a higher clustering coefficient when compared to random graphs. Then a problem comes: given those aforementioned metrics, how can one select a subset of them to reduce the computation cost and achieve good classification performance? We refer it as a discriminative feature selection problem.

In the following, we adopt similar technology from GSSC [12], to formulate the feature selection as an optimization problem. Next we discuss possible selection criteria and propose a greedy algorithm to solve the optimization problem. For simplicity, both term "feature" and "metric" are used to mention any of the aforementioned metrics. In addition, we use f to denote a single feature and $\mathcal{F}$ to denote a feature set. The numeric value of any feature over a graph G is denoted as $f(G)$, and the combination of values of a set of features $\mathcal{F}$ over a graph G is denoted as a vector $\overrightarrow{\mathcal{F}(G)}$.

Mathematically, discriminative feature selection problem could be formulated as:

$$\mathcal{F}_d = \arg \max_{\mathcal{T} \subseteq \mathcal{F}} \{R(\mathcal{T})\} \tag{1}$$

where $R(\mathcal{F})$ is a measurement of discriminative over a set of features $\mathcal{F}$.

We investigate two principles that optimal features should have: representativeness and separability. This guides us to design a suitable discriminative measurement R. Separability means that graphs should be able to be separated from each other based on values of features; while representativeness is important to avoid selecting two highly correlated features. Both separability and representability are desirable in feature selection. Hence, we define R as a multi-criteria measurement:

$$R(\mathcal{F}) = \omega_1 \rho(\mathcal{F}) - \omega_2 \frac{1}{|\mathcal{F}|} \sum_{f_i, f_j \in \mathcal{F}} \text{corr}(f_i, f_j) \tag{2}$$

where $\rho(\mathcal{F})$ denotes separability ratio of features set $\mathcal{F}$, ω_1 and ω_2 are smoothing weights, and $\text{corr}(f_i, f_j)$ denotes correlation of two features f_i and f_j.

Unfortunately, both separability ratio and feature correlation can be highly domain-dependant. As an example, we provide one possible design for measurements $\rho(\mathcal{F})$ and $\text{corr}(f_i, f_j)$ based on information of labeled background graphs. More specifically, let clustering $\mathcal{C} = \{C_1, \dots, C_\gamma\}$ denote the information of labeled background graphs, i.e., a cluster $C_i = \{G_1, \dots, G_n\}$ is a set of graphs that share the same class label $y_i \in \{-1, 0, 1, \dots \gamma\}$. Then given clustering $\mathcal{C}$, the separability ratio of a set of features $\mathcal{F}$, is defined as

$$\rho(\mathcal{F} \mid \mathcal{C}) = \max_{C \in \mathcal{C}} \left\{ \log \frac{\sum_{G_i \in C} ||\overrightarrow{\mathcal{F}(G_i)}||}{\sum_{G_j \notin C} ||\overrightarrow{\mathcal{F}(G_j)}||} \right\} \tag{3}$$

It is clear that the larger the value of $\rho(\mathcal{F} \mid \mathcal{C})$, the better "worst-case" class separable capability feature set $\mathcal{F}$ is.

Algorithm 1. Discriminative Feature Selection.

Input: a clustering $\mathcal{C}$ of labeled graphs, a feature set $\mathcal{F}$
Output: a subset of features $\mathcal{F}_d$
1: $\mathcal{T} = \emptyset$, flag=true;
2: **while** flag
3:    $f_k = \arg\max_{f \in (\mathcal{F} \backslash \mathcal{T})} \{R(\mathcal{T} \cup \{f\} \mid \mathcal{C}) - R(\mathcal{T} \mid \mathcal{C})\}$;
4:    **if** $R(\mathcal{T} \cup \{f_k\} \mid \mathcal{C}) - R(\mathcal{T} \mid \mathcal{C}) > 0$
5:       $\mathcal{T} = \mathcal{T} \cup \{f_k\}$;
6:    **else**
7:       flag=false;
8: **return** $\mathcal{T}$;

Similarly, correlation of two features f_i and f_j given clustering $\mathcal{C}$, is defined as

$$\text{corr}(f_i, f_j \mid \mathcal{C}) = \frac{|X - E(X)||Y - E(Y)|}{\sigma_X \sigma_Y} \qquad (4)$$

where the values of feature f_i and f_j over graphs in $\mathcal{C}$ are denoted as X and Y respectively.

With respect to Equ. 1, we propose a greedy algorithm to solve it, as shown in Algorithm 1, we start with an empty set $\mathcal{T} = \emptyset$, and in step k, iteratively add a feature f_k such that it maximizes the margin gain

$$f_k = \arg\max_{f \in (\mathcal{F} \backslash \mathcal{T})} \{R(\mathcal{T} \cup \{f\} \mid \mathcal{C}) - R(\mathcal{T} \mid \mathcal{C})\} \qquad (5)$$

Now given a graph G in a specific domain, we represent it as a feature vector using discriminative features. We also use the Euclidean distance to measure the dissimilarity of two graphs: $\text{dist}(G_i, G_j) = \sqrt{\|\overrightarrow{\mathcal{F}(G_i)} - \overrightarrow{\mathcal{F}(G_j)}\|^2}$.

3   Graph Classification

In the previous section, we have discussed the features used to describe and represent graphs. In this section, with graph representation using theoretical metrics, we explore its potential in the domain of graph classification. Once graphs are converted into feature vector space, one can use traditional classifier to classify graphs. As a start, we build two classifiers which are similar to distance weighted KNN classifier. Before we present the details of two classifiers, first let us introduce the notation of "KNN" and "RKNN".

For query graph G and the set of background graphs $\mathcal{G}$, the set of k-nearest neighbors of G is the smallest set KNN(G) such that KNN(G)$\subseteq \mathcal{G}$, $|\text{KNN}(G)| \geq k$, and $\forall G_i \subseteq \text{KNN}(G)$, $\forall G_j \subseteq (\mathcal{G} \backslash \text{KNN}(G))$: $\text{dist}(G, G_i) < \text{dist}(G, G_j)$. Similarly, we define the set of reverse k-nearest neighbors (RKNN [18]) of a graph G as $\text{RKNN}(G) = \{G_i \subseteq \mathcal{G} \mid G \subseteq \text{KNN}(G_i)\}$.

Note that k-nearest neighbors and reverse k-nearest neighbors are related but not exactly the same concept. We illustrate more using the following examples.

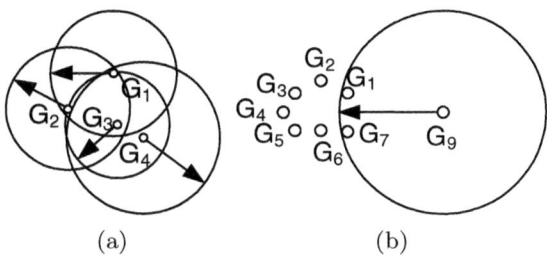

(a)   (b)

Fig. 1. An example of KNN and RKNN

Algorithm 2. Classifier

Input: a set of tuples $\mathcal{T}=< y, \{\overrightarrow{\mathcal{F}(G)}\} >$ and a vector $\overrightarrow{\mathcal{F}'}$ representing query graph

Output: a class label y' representing classification of $\overrightarrow{\mathcal{F}'}$

1: **if** KNN classifier

2:     $\mathcal{T}_{nn}$=KNN($\overrightarrow{\mathcal{F}'}$, $\mathcal{T}$);

3: **else** $\mathcal{T}_{nn}$=RKNN($\overrightarrow{\mathcal{F}'}$, $\mathcal{T}$);

4: initializing an array "vote" with length equal to number of classes γ;

5: **for** each tuple $t \in \mathcal{T}_{nn}$

6:     vote[t[1]]+=$\frac{1}{\text{dist}(\overrightarrow{\mathcal{F}'},t[2])+\epsilon}$

/* we use t[1] to denote the first entry of tuple t (i.e., the class label)
and t[2] to denote the second entry of tuple t (i.e., a feature vector) */

7: **return** y'=arg max$_{i=\,1\ \text{to}\ \gamma}$vote[i]

Example 1. Let us consider an example shown in Figure 1. Figure 1(a) shows 4 graphs G_1, G_2, ..., G_4, where each graph is assigned with a circle covering its two nearest neighbors. For example, 2NN(G_1)={G_2, G_3} and both G_2 and G_3 are enclosed in a circle with center G_1. Note that $G_i \in$KNN(G_j) does not necessarily imply $G_i \in$RKNN(G_j): $G_1 \in$2NN(G_4)={G_1, G_3}, but $G_1 \notin$R2NN(G_4)={G_3}. In addition, for any given query G with $k >= 1$, KNN(G)$\neq \emptyset$; while the set of reverse k-nearest neighbors RKNN(G) might be empty. As shown in Figure 1(b), 2NN(G_9)={G_1, G_7}, but R2NN(G_9)=$\emptyset$.   □

From an abstract point of view, given a query graph, in the KNN classifier, the algorithm finds the k most similar graphs in the training graph set. Then the weighted-based majority vote by the k most similar graphs, decides the classification. A RKNN classifier is built in a similar way except that the weighted-based majority vote is done by the reverse k nearest neighbors.

The details are shown in Algorithm 2. The feature vector representation of the query graph $\overrightarrow{\mathcal{F}'}$ and sets of training feature vectors from each class y are fed to the classifier. Next, for $\overrightarrow{\mathcal{F}'}$, we perform either KNN or RKNN search to find a set of closest feature vectors and their class labels in the training set, denoted as $\mathcal{T}_{nn}$ (lines 1–3). After that, a distance weighted score is calculated for classification (lines 4–7). In line 6, we use a very small number ϵ as a smoothing factor to avoid unbalance when the distance $\text{dist}(\overrightarrow{\mathcal{F}'}, t[2])$ is equal to 0. Note that for RKNN

Table 1. Statistics of graph data

Class	# of graphs	# of real graphs	average $|V|$	average $|E|$	average d_g
DUAL	139	22	11973	48481	0.138874
INTERVAL	84	10	19996	106001	0.009191
HOPI	77	8	4886	16575	0.155612

classifier, $\mathcal{T}_{nn}$ may be empty (line 3). In this sense, the classification performance is the same with a random guess.

4   Experimental Evaluations

4.1   Tasks and Data Sets

We choose an interesting task from the query optimization area to evaluate our theoretical metric-based method for graph classification. The task is "graph reachability index selection", which aims to decide a best index from a set of candidate reachability index. With the best index, graph reachability query performance could be significantly improved. The problem of "graph reachability index selection" can be formulated as a graph classification problem, where each class label is a type of graph reachability indexing. Specifically, one classifies two directed graphs into a same cluster if they share the same best reachability index. In our experiments, we choose INTERVAL approach [1], HOPI [4] and DUAL labeling [22] as the set of candidate reachability indexing.

We used both of real data and synthetic data in our experiments. All of the graphs are directed. For the real data, we used a set of real graphs obtained from a Web graph repository[2] and papers [21,22]. In addition, we generated a set of synthetic graphs with two different graph generators. The first generator[3] controls the graph properties with four input parameters: the maximum fanout of a spanning tree, the depth of a spanning tree, the percent of non-tree edges and number of cycles. The second generator[4] generates a scale-free power-law graph, whose degree distribution follows a power-law function. The total number of graphs used is 300. For each graph, we manually labeled it with the best reachability index as follows: we pre-built each index on the graph and compared the total reachability query time of one million random queries. The best index is the one such that the total query cost is minimal. The statistics of graphs we used are summarized in Table 1.

To measure the performance of our framework, we used the prediction accuracy A and micro precision P to measure the performance of our framework. The prediction accuracy A, is defined as $A=T_i/n$, where T_i is number of samples with right classification, and n is number of samples in total. The micro precision

---

[2] http://vlado.fmf.uni-lj.si/pub/networks/data/
[3] http://www.cais.ntu.edu.sg/~linhong/gen.rar
[4] http://www.cais.ntu.edu.sg/~linhong/GTgraph.zip

Table 2. Running time and memory usage of feature selection

Running time (seconds)		Memory usage (MB)	
Metric computation	Feature selection	Metric computation	feature selection
50.373	24.035	89.83	3.8

P is defined as $\sum_{i=1}^{\gamma} TP_i / \sum_{i=1}^{\gamma} (TP_i + FP_i)$, where γ is number of class labels, TP is the true positive, and FP is the false positive.

4.2 Results

Discriminative features. For the specific task of "reachability index selection", the set of discriminative features that returned by Algorithm 1 are: 1) Non-tree density; 2) Degree distribution; 3) Global efficiency; 4) Diameter; 5) Vertex compression ratio; 6) Coreness; 7) Global clustering coefficient; and 8) Graph density. Next we report the total running time and memory usage of two stages in discriminative feature selection: I) computing graph metrics (See Sec 2.1), and II) discriminative feature selection (See Algorithm 1). The results are shown in Table 2. It is observed that the memory usage is dominated by Stage I, i.e, the storage consumption to keep the whole graph in the memory. Interestingly, running time is also mainly taken up by the first stage. The reason is that we need to compute 14 metrics for each graph in the first stage and each metric computation needs to be done at least in linear time to the graph size.

Classification Performance. We randomly chose N_T number of test graphs both from real graphs and synthetic graphs, where N_T varies from 20 to 100. In addition, in average, the first i graphs are more similar to the training graphs than the fist j graphs ($i < j$). The total running time and memory usage of KNN and RKNN classifiers, are presented in Figure 2. The result shows that though KNN approach is faster than RKNN approach, the margin is quite small. Figure 2(b) reports the memory usage (excluding the memory consumption to keep graphs) comparison of two approaches. It implies that each approach is comparable to another in memory consumption.

Figure 3 gives the prediction accuracy and micro-precision when number of test graphs is varied. Despite KNN classifier gets higher accuracy on a small number of testing graphs that are highly similar to training graphs, in average, RKNN classifier obtains substantial improvement than KNN approach in classification accuracy. The results indicate that KNN classifier are more vulnerable to the quality of training graphs than RKNN classifier: when training graphs are noticeably similar to testing graphs, KNN classifier could get classification of good quality; when training graphs are relatively dissimilar to testing graphs, the performance of KNN drops off sharply. On another hand, we also observe that the performance of KNN is more stable than that of RKNN. The reason is that there is no guarantee for the existence of RKNN search results. Hence, the weighted major vote on an empty set would trigger a random guess (see Algorithm 2).

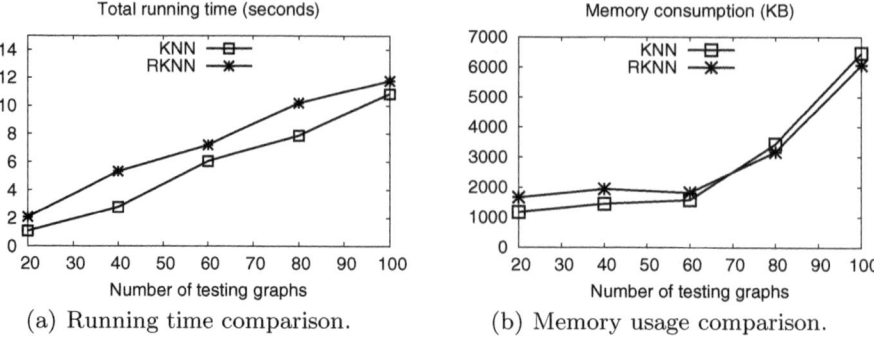

(a) Running time comparison.   (b) Memory usage comparison.

Fig. 2. Performance comparison

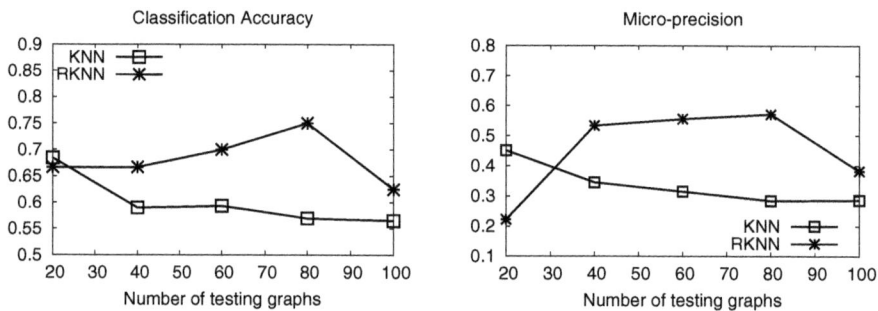

Fig. 3. Prediction accuracy and micro-precision comparison

We attain similar results of micro precision of these two classifiers. Based on Figure 3, we can draw a conclusion that graph classification using theoretical metric is feasible since the accuracy of both two classifiers is much higher than random guess (33.3%).

5   Related Work

Related works about graph classification are mainly in the field of subgraph pattern selection and graph pseudo-isomorphism testing. If we could have a set of subgraph patterns, graphs can be represented as feature vectors of patterns and graph classification problem is converted into traditional data classification problem. In subgraph pattern selection, the major difficulty of this approach is to find informative subgraph patterns efficiently. The pioneer work [6] starts to find frequent subgraph patterns to represent graphs. One drawback of frequent subgraph pattern representation is that one can find a tremendously large quantity of frequent patterns in a large graph, which leads to relatively poor scalability. Later, Leap [23], gPLS [16], CORK [19], graphSig [15] and GAIA [11] propose to search directly for discriminative subgraph patterns that can better assist graph classification. A recent work COM [10] makes use of co-occurrences of subgraph patterns to improve graph classification performance.

Another direction of graph classification relies on pseudo-isomorphism testing (graph similarity measure). Various degree of matching metrics such as graph-edit distance [3], maximum common subgraph [9], canonical sequence similarity [2] and so on have been proposed and used to assist graph classification. These approaches are theoretically sound and can guarantee optimal or near-optimal solutions in some sense. However, computing graph-edit distance or maximum common subgraph itself is really time-consuming due to their intractability. Recently, Zeng et al. [24] propose a solution to compute the lower bound and upper bound of graph-edit distance in polynomial time to facilitate graph similarity search as well as graph classification.

Unfortunately, the above works are mainly in the area of chemical and biological data. In addition, all of them are proposed for the isomorphism-based graph classification problem. Although the problem of property-based graph classification is important in practice for applications in social network analysis, database design and query optimization, we are not aware of any focused study of this problem.

6   Conclusions

In this paper, we formalized the problem of property-based graph classification. Upon this formalization, we studied the feasibility of graph classification using a number of graph theatrical metrics. We also proposed a greedy algorithm to select a set of discriminative metrics. Experimental results showed that our framework works good in terms of both accuracy and micro precision, which verified the possibility of graph classification using graph metrics.

References

1. Agrawal, R., Borgida, A., Jagadish, H.V.: Efficient management of transitive relationships in large data and knowledge bases. In: Proceedings of the 1989 ACM International Conference on Management of Data, pp. 253–262. ACM, New York (1989)
2. Babai, L., Luks, E.M.: Canonical labeling of graphs. In: Proceedings of the 15th Annual ACM Symposium on Theory of Computing, pp. 171–183. ACM, New York (1983)
3. Bunke, H.: On a relation between graph edit distance and maximum common subgraph. Pattern Recognition Letters 18(9), 689–694 (1997)
4. Cheng, J., Yu, J.X., Lin, X., Wang, H., Yu, P.S.: Fast computation of reachability labeling for large graphs. In: Ioannidis, Y., Scholl, M.H., Schmidt, J.W., Matthes, F., Hatzopoulos, M., Böhm, K., Kemper, A., Grust, T., Böhm, C. (eds.) EDBT 2006. LNCS, vol. 3896, pp. 961–979. Springer, Heidelberg (2006)
5. Coffman, T.R., Marcus, S.E.: Dynamic classification of groups through social network analysis and hmms. In: IEEE Aerospace Conference, pp. 3197–3205 (2004)
6. Deshpande, M., Kuramochi, M., Wale, N., Karypis, G.: Frequent substructure-based approaches for classifying chemical compounds. IEEE Transaction on Knowledge and Data Engineering 17(8), 1036–1050 (2005)

7. Diestel, R.: Graph Theory, 3rd edn., vol. 173. Springer, Heidelberg (2005)
8. Faloutsos, M., Yang, Q., Siganos, G., Lonardi, S.: Evolution versus intelligent design: comparing the topology of protein-protein interaction networks to the internet. In: Proceedings of the LSS Computational Systems Bioinformatics Conference, Stanford, CA, pp. 299–310 (2006)
9. Montes-y-Gómez, M., López-López, A., Gelbukh, A.: Information retrieval with conceptual graph matching. In: Ibrahim, M., Küng, J., Revell, N. (eds.) DEXA 2000. LNCS, vol. 1873, pp. 312–321. Springer, Heidelberg (2000)
10. Jin, N., Young, C., Wang, W.: Graph classification based on pattern co-occurrence. In: Proceeding of the 18th ACM Conference on Information and Knowledge Management, pp. 573–582. ACM, New York (2009)
11. Jin, N., Young, C., Wang, W.: Gaia: graph classification using evolutionary computation. In: Proceedings of the 2010 International Conference on Management of Data, pp. 879–890. ACM, New York (2010)
12. Kong, X., Yu, P.S.: Semi-supervised feature selection for graph classification. In: Proceedings of the 16th ACM International Conference on Knowledge Discovery and Data Mining, pp. 793–802. ACM, New York (2010)
13. Luce, R., Perry, A.: A method of matrix analysis of group structure. Psychometrika 14(2), 95–116 (1949)
14. Milgram, S.: The Small World Problem. Psychology Today 2, 60–67 (1967)
15. Ranu, S., Singh, A.K.: Graphsig: A scalable approach to mining significant subgraphs in large graph databases. In: Proceedings of the 2009 IEEE International Conference on Data Engineering, pp. 844–855. IEEE Computer Society, Washington, DC, USA (2009)
16. Saigo, H., Krämer, N., Tsuda, K.: Partial least squares regression for graph mining. In: Proceeding of the 14th ACM International Conference on Knowledge Discovery and Data Mining, pp. 578–586. ACM, New York (2008)
17. Seidman, S.B.: Network structure and minimum degre. Social Networks 5, 269–287 (1983)
18. Tao, Y., Papadias, D., Lian, X.: Reverse knn search in arbitrary dimensionality. In: Proceedings of the 30th International Conference on Very Large Data Bases, pp. 744–755. Very Large Data Bases Endowment (2004)
19. Thoma, M., Cheng, H., Gretton, A., Han, J., Peter Kriegel, H., Smola, A., Song, L., Yu, P.S., Yan, X., Borgwardt, K.: Near-optimal supervised feature selection among frequent subgraphs. In: SIAM Int'l Conf. on Data Mining (2009)
20. Thomason, B.E., Coffman, T.R., Marcus, S.E.: Sensitivity of social network analysis metrics to observation noise. In: IEEE Aerospace Conference, pp. 3206–3216 (2004)
21. University of Michigan: The origin of power-laws in internet topologies revisited. Web page, http://topology.eecs.umich.edu/data.html
22. Wang, H., He, H., Yang, J., Yu, P.S., Yu, J.X.: Dual labeling: Answering graph reachability queries in constant time. In: Proceedings of the 22nd International Conference on Data Engineering, p. 75. IEEE Computer Society, Washington, DC, USA (2006)
23. Yan, X., Cheng, H., Han, J., Yu, P.S.: Mining significant graph patterns by leap search. In: Proceedings of the 2008 ACM International Conference on Management of Data, pp. 433–444. ACM, New York (2008)
24. Zeng, Z., Tung, A.K.H., Wang, J., Feng, J., Zhou, L.: Comparing stars: on approximating graph edit distance. Proc. VLDB Endow. 2, 25–36 (2009)

A GML Documents Stream Compressor*

Yinan Yu[1], Yuzhen Li[2], and Shuigeng Zhou[3]

[1] Dept. of Computer Science & Technology, Tongji University, China
[2] Graduate School of Science and Technology, Chiba University, Japan
[3] Shanghai Key Lab of Intelligent Information Processing, Fudan University, China
{yuyinan1986,liyuzhen8}@gmail.com, sgzhou@fudan.edu.cn

Abstract. GML has become the standard format for geographical data transfer, exchange and storage. Usually, in GML documents there are many verbose tags and a large amount of coordinate data, which makes them be of extremely large volume. Thus, it is necessary to compress these documents to reduce storage and transmission cost. GML data is often stored and transferred in the form of multiple documents. Although some GML compressors have been developed recently, all of them can process only a single GML document at a time. In this paper, we propose a stream compressor for GML documents, called GDScomp, which can compress a stream of multiple GML documents effectively. It shares the structural information among multiple GML documents by a common dictionary to employ the dynamic compression method and uses the delta compression method for the coordinate data. Experimental results show that GDScomp can achieve satisfactory compression performance when compressing GML documents streams.

Keywords: GML; Documents stream; Common dictionary; Dynamic compression; Delta compression.

1 Introduction

With the popularity of Geography Information Systems (GISs) and their applications, many heterogeneous spatial data sources emerge rapidly. To promote the sharing and inter-operability of different spatial data sources and GIS applications, the Open Geospatial Consortium (OGC) proposed the Geography Makeup Language (GML) [1] to offer a general framework for formatting and representing geographical data. Nowadays, GML has become the *de fact* standard for geo-spatial data transfer, exchange and storage.

Although GML is flexible and extensible to use through the Internet, GML documents usually contain a lot of redundant information (e.g. verbose tags) and a large amount of coordinate data (representing the spatial objects), which makes the documents be of very large volume and thus causes much cost for

---

* This work was supported by National Natural Science Foundation of China under grants No. 60873040 and No. 60873070, and China 863 Program under grant No. 2009AA01Z135.

J. Xu et al. (Eds.): DASFAA Workshops 2011, LNCS 6637, pp. 65–76, 2011.

transmission and storage. Particularly, for the scenarios of mobile GIS applications (e.g. GPS and mobile tourism information systems) where the client-end devices (e.g. iPADs, iPhones and cell phones) have very limited storage space and communication bandwidth (compared to desk-top devices), reducing the cost of data transmission and storage is crucial to the success of these applications. Therefore, it is necessary to develop effective GML compressors to make GML documents be stored and transferred at acceptably low cost. However, existing research works on GML have mainly focused on data storage and query processing [2]. Not enough effort has been made to GML compression, though a few GML compressors do have been reported in the literature (e.g. GPress [3,4]).

Since GML is a special form of XML and GML documents are essentially text documents, so technically, GML documents can be compressed by three types of compression techniques. They are 1) general text compressors such as Gzip [5] and PPM [6], 2) XML-oriented compressors, representatives including XMill [7], XGrind [8] and so on, and 3) GML-specific compressors (e.g. GPress [3,4]). However, all these compressors above do not work with GML streams. There are also some XML stream compressors like XMLPPM [9], Millau [10] and XSC [11], but these compressors do not consider the uniqueness of GML documents and work only in a simple document by document style. Concretely, they can only treat GML documents as ordinary XML documents while neglecting the unique features of GML. And when compressing a stream of documents they will not exploit the common statistic information among these documents, thus they can not achieve optimal compression performance.

Typically, we argue that an effective GML stream compressor should have the following features:

- The data to be compressed comes as a stream that can not be kept entirely in the system during the compression process. At any time, only part (a window) of the stream is stored in the buffer for processing.
- The compressor will not scan the documents more than one time.
- The transmission speed of data stream may be very fast, so the compressor should work nearly in real time, and the decompression process can go simultaneously.
- The compressor should fully use the common information among the documents under compression.

With these features above in mind, in this paper we propose a new GML stream compressor for multiple documents, which is called GDScomp. It parses the GML documents into different streams that will be handled with corresponding methods, then shares the structure information as a common dictionary among different documents and uses the delta compression for the spatial data. Experimental results show that the new compressor can achieve better compression performance than some state of the art text, XML and GML compressors.

The rest of this paper is organized as follows. Section 2 surveys the related work. Section 3 introduces the architecture and technical details of GDScomp. Section 4 gives the experimental results and Section 5 draws the conclusion and highlights the future work.

2   Related Work

In what follows, we will present a brief survey on major related works of XML (especially XML stream) compression and GML compression.

XMLPPM [9] was proposed by Cheney many years ago, but the principle is still worth of being paid attention to when compressing XML streams. XMLPPM uses a multiplexed hierarchical PPM Model called MHM, which is based on SAX [12] events and does prediction by partial match (PPM) [6]. It predefines four models based on different contexts (elements, characters, attributes and miscellaneous symbols) in the documents. So during the paring phase, it can group the data according to the defined models. These data from the four models will eventually be combined and compressed by PPM to form the compressed document. Although it can get a good compression ratio, its compression time is a little too long due to the time-consuming PPM method.

Millau [10] is another XML stream compressor for efficient encoding and streaming of XML documents. It uses complicated encoding techniques for XML compression. It employs Differential DTD Tree (DDT) compression that makes full use of DTD associated with the document to encode data values and structural information into two data streams, thus gets a good compression ratio. Millau also extends its encoding format to adapt it for business applications.

XMill [7] is the first XML conscious compressor and some innovative principles introduced by it were followed by other compressors. It separates the structural information from the text data and compresses them separately. The text data is grouped by semantics. Finally, all data is compressed by a general compressor Gzip.

Gpress [3,4] may be the first GML-specific compressor that has good compression performance. It follows the principles of XMill to separate the structural information from data, and group the data into different containers, and finally compress them by Gzip. The innovative contribution of GPress is that it transforms the spatial coordinate strings back to numeric values, which is then compressed by delta compression. It has better compression ratio than XMill.

All these compressors mentioned above are designed only for working in a document by document style, no matter whether the documents come in the form of stream or separately. So when compressing a documents stream, they can only utilize information of individual documents, which results in non-optimal compression results. Although dictionary-based compression [13] is a classical method, in this paper we use dictionary for document streams compression.

3   The GDScomp Method

In this section, we will first introduce the architecture of GDScomp, and then present the core techniques for implementing GDScomp, including SAX event handling, dynamic structure compression and delta compression.

3.1 GDScomp Architecture

Figure 1 shows the architecture of the GDScomp compressor. When a GML documents stream comes, GDScomp first parses each document in the stream sequentially by SAX, then outputs the parsed events to the events handler for further processing. The events handler will filter the coming information according to data semantics into different sub-streams, which will be compressed by different methods.

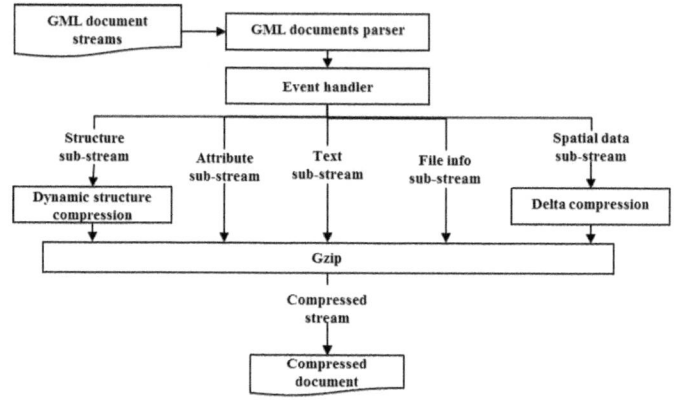

Fig. 1. The architecture of GDScomp

The structure sub-stream contains the tags in GML documents. Since similar tags may appear in the same or different GML documents repeatedly, the dynamic compression method is used to compress the structural information by creating a common dictionary to store all tags have appeared. As the stream goes by, re-appearing tags will be replaced with the corresponding indices in the dictionary to reduce storage space.

The attribute sub-stream, spatial data sub-stream and text sub-stream (here "text" indicates the text information except for tag names, attribute values and spatial coordinates) all contain text content. The attribute sub-stream contains attribute values. The text sub-stream has text content under different tags except for the spatial coordinate tags. The spatial data sub-stream holds the coordinate data. It is worthy of being mentioned that the spatial coordinates are used to represent spatial objects, and for nearby spatial objects, their coordinates are not much different. So the delta compression technique is adopted to compress the coordinate data. That is to store the differences, instead of the original values. Each file's (here "file" and "document" are used interchangeably) information in the same stream is stored in the file information sub-stream. Thus, when decompressing the compressed stream, the original file information can be recovered.

```
<gml:surfaceMember>                        <gml:surfaceMember>
 <gml:Polygon gml:id="UUID_2ab61">         <gml:Polygon gml:id="UUID_d755f">
 <gml:exterior>                            <gml:exterior>
  <gml:LinearRing gml:id="UUID_3_0">        <gml:LinearRing gml:id="UUID_d_0">
  <gml:posList srsDimension="3">            <gml:posList srsDimension="3">
   23356.753 21114.414 1.0                   23186.088 21091.724 0.0
   23354.921 21131.342 1.0                   23186.087 21091.723 16.0
   23358.44  21131.722 1.0                   23186.088 21091.724 16.0
   23360.271 21114.794 1.0                   23186.088 21091.724 0.0
   23356.753 21114.414 1.0                  </gml:posList></gml:LinearRing>
  </gml:posList></gml:LinearRing>          </gml:exterior>
 </gml:exterior> </gml:Polygon>            </gml:Polygon>
</gml:surfaceMember>                       </gml:surfaceMember>
```

Fig. 2. GML document samples

The outputs of all sub-streams above are input to a Gzip compressor. Based on the architecture in Figure 1, GDScomp needs only to store the dictionary, the reference coordinates of spatial data and file names in the memory.

3.2   Event Handler

Figure 2 shows two samples of GML document. It can be seen that GML documents are a special type of XML documents. Many similar tags in different documents, especially when the documents are formatted under the same schema. Moreover, there are lots of coordinates representing spatial objects where the difference among the coordinates of adjacent spatial objects is very small.

GDScomp uses SAX to parse every document in the coming stream, and the result corresponds to a sequence of events. When analyzing each document, if the current event is a start tag or end tag event, GDScomp forwards it to the structure sub-stream, which will be compressed by the structure compression module. If the start tag is a spatial data tag, then GDScomp forwards it to the spatial data sub-stream that will be compressed by the delta compression module. If the event is an attribute or a PCDATA, GDScomp will send the data to the attribute sub-stream or the text sub-stream. If the current event refers to the end of the current document, GDScomp will push the filename into the file information sub-stream.

The implementation of the event handler is outlined in Algorithm 1.

3.3   Dynamic Structure Compression

As there are many re-appearing tags in the structure sub-stream, but the numbers and probabilities are unknown in advance, so we use the dynamic compression method to compress the structure sub-stream at word level. A *dynamic dictionary* is established to store the appeared tags from different documents in the stream under compression. When "old" tags reappear, just their indices in the dictionary are used to replace them so as to reduce storage space. Note that in the dictionary, words instead of characters are indexed.

Algorithm 1. Event handler

1: **Input: SAX event stream**
2: **Output: different sub-streams**
3: **Begin**
4: **while** SAX event stream is not null **do**
5:    **if** SAX-event is StartElement **then**
6:       *name*:= GetElementName
7:       send *name* to the structure sub-stream
8:       **if** *name* is a coordinate element **then**
9:          *value* := GetTextValue
10:         send the *value* to the spatial data sub-stream
11:      **end if**
12:      **if** the element has an attribute **then**
13:         send the attribute's *value* to the attribute sub-stream
14:      **end if**
15:   **else if** SAX-event is EndElement **then**
16:      send its *name* to the structure sub-stream
17:   **else if** SAX-event is PCDATA **then**
18:      send its *value* to the text sub-stream
19:   **else if** SAX-event is end-file event **then**
20:      send *filename* to the file information sub-stream
21:   **end if**
22: **end while**

Algorithm 2 describes the detailed structure compression process. At first, the dictionary is initialized to null. The dictionary is organized as a table with two columns: the first column is the key, and the second column is the value, which correspond to a tag's index and name respectively. We do not use schemas or DTDs because in real applications these schemas or DTDs perhaps are not available.

Once a new tag (not in the dictionary) emerges while parsing the GML documents, the new tag is inserted into the dictionary, and its name is output to the structure sub-stream to help decompression later. Otherwise, if an "old" tag comes, the tag is put to the work buffer while waiting for its succeeding tag(s). If the combination of consecutively appearing tags in the buffer does not exist in the dictionary, their indices of these tags in the buffer are output, and the new combination of tags is inserted to the dictionary as a new entry, which can be utilized in the future.

The advantage of this method lies in that tags instead of characters are used as dictionary entries, so the lengths of words that the entries in the dictionary represent can be maximized. Creating a dynamic common dictionary in the compression phase, and outputting tags' names and their corresponding indices in the dictionary alternately can make the decompression process go simultaneously.

The buffer used to store "old" tags provides the possibility of combining existing tags to generate new and longer tags. For example, in the left GML document sample illustrated in Figure 2, there are entries like "< *gml* : *surfaceMember* >"

and "$< gml : Polygon >$", which will be inserted into the dictionary. While processing the right sample, "$< gml : surfaceMember >< gml : Polygon >$" will be treated as a new entry added to the dictionary so that if similar combination reappears later, we can use this combination's index to represent it.

Algorithm 2. Dynamic structure compression

```
 1: Input: in-stream — GML structure sub-stream
 2: Output: out-stream — compressed GML structure sub-stream
 3: Begin
 4: Initialize dic, buffer, tag := null
 5: while in-stream is not null do
 6:     while tag:= Readtag() is not null do
 7:        if dic does not contain tag then
 8:           if buffer is not empty then
 9:              add the whole word in the buffer to dic
10:           end if
11:           output tag.value to out-stream
12:           put tag into dic
13:        else
14:           if dic contains (buffer.concat(tag)) then
15:              buffer.append(tag)
16:              continue
17:           else
18:              output the index of the whole word in buffer to out-stream
19:              buffer.clear( )
20:              buffer.append(tag)
21:              continue
22:           end if
23:        end if
24:        output the index of each indexed entry in buffer to out-stream
25:        buffer.clear()
26:     end while
27: end while
```

We use the dictionary to store common structural information because more or less, there are similar tags in different GML documents, especially when the documents use or conform to similar schemas.

3.4   Delta Compression

GMl documents describe spatial objects such as rivers, bridges, and roads in the real world. These objects are represented by a series of coordinates under some specific elements like $< gml : X >$ and $< gml : Y >$ in GML version 2.0, or $< gml : pos >$ and $< gml : posList >$ in GML version 3.0.

However, the data arrangements under different tags are quite different. For example, the data with the tags like $< gml : X >$ and $< gml : Y >$ refers to the X and Y coordinates of a certain spatial object. The data with the tags like $< gml : posList >$ or $< gml : coordinates >$ refers to a collection of coordinates standing for a certain spatial object extending from one point to another point. The dimension of the coordinates can be seen from the attributes of the current tag. As shown in Figure 2, each row under the tag $< gml : postList >$ stands for the coordinates of a 3-dimensional point.

As there are many coordinates under a tag representing the same spatial object, and the difference between neighboring coordinates is very small. These coordinates are usually monotonous or piece-wisely monotonous in value, so we can use the difference (delta) between adjacent coordinates instead of the original coordinates (except for the starting coordinate that will be used as the reference coordinates) to represent a series of coordinates. In real applications, even in the same GML document, the difference between coordinates of different spatial objects may be still large, so several reference coordinates should be used. In such a way, the coordinates in GML documents can be transformed to several reference coordinates and the differences between the other coordinates with these reference coordinates.

Furthermore, we notice that the significant difference is originated from the integer part of a coordinate value. As for the decimal part, its length is diverse and the numbers at different positions are quite different. Based on this observation, GMScomp stores the delta of the integer parts of two adjacent coordinates and the original values of the decimal parts. As all data in GML documents appears as plain text, and the spatial data is usually with so high precision that usually a string of 8 to 10 digits is needed to represent (each digit needs 1 byte to store). So the text strings are transformed back to numeric numbers that are stored by binary format.

The process of delta compression in GDScomp is as follows. It first establishes an array of 3 elements to store the reference coordinates, each element represents the value of one dimension. When processing the coordinates of a document in

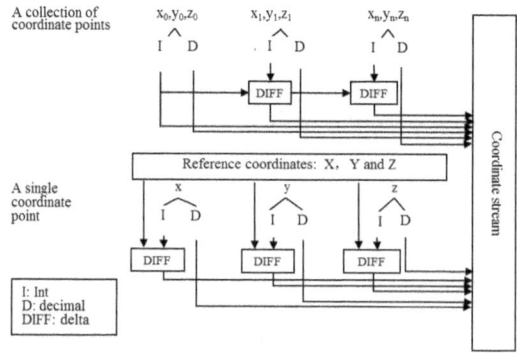

Fig. 3. Delta compression in GDScomp

the stream under compression, it stores the coordinates of the first point in the array, and forwards the original values to the spatial data sub-stream. For each following point, it will output the differences of its integer part values with that of the preceding point and the values of its decimal part in binary format. As the process goes ahead, the difference values may be so large that overpass the predefined threshold (relying on how many bytes are used to store the difference values), the reference coordinates will be updated. That is to update the values in the array. The process above continues till all documents in the stream are finished. Fig. 3 illustrates the whole process. Here, two cases are demonstrated: one is that coordinate points with tags like $< gml : X|Y|Z >$ are processed one by one, another is that a collection of coordinate points with tags like $< gml : pos >$ and $< gml : posList >$ are processed together.

4   Experimental Evaluation

In this section, we evaluate the compression performances of GDScomp over three groups of GML documents. All testings are conducted on a PC with a 2.6GHz CPU and 2GB memory. We compare the compression performance of GDScomp with that of three existing compressors in aspects: compression ratio, compression time and decompression time.

- CR (compression ratio). we use the following formula to evaluate compression ratio:

$$CR = \frac{8 * (compressed\ file\ size)}{(original\ file\ size)} bpc \qquad (1)$$

  Because all the other three compared compressors can only compress a group of GML documents one by one, we use the following formula to calculate the average compression ratio:

$$averageCR = \frac{8 * (total\ compressed\ file\ size)}{(total\ original\ file\ size)} bpc \qquad (2)$$

  From the formulas above, we can see that the smaller the CR is, the better the compressor is.
- Compression time and decompression time: we record the time that a compressor takes to compress all GML documents in a group and the time to decompress the compressed documents. The larger the time is, the less efficient the compressor is.

According to the experimental results of Sheriff [14], we choose two compressors that have lower CR, XMill and XMLPPM, and one compressor with the smallest compression time, Gzip, for comparison. Three groups of GML documents are used for testing: CityGML [15], ALKISATKIS [16] and the third group that is transformed from Oracle Spatial datasets. Table 1 shows the statistics of the testing documents, including document size, the number of tags in every document, and the percentages of tags and spatial data contained in every document.

Table 1. Statistics of testing GML documents

Datasets	File name	Size (Kb)	#Tags	Tags (%)	Spatial data (%)
CityGML	080305SIG3D_Breakline_Levkreuz.xml	4150	72398	0.646	0.339
	080305SIG3D_LSENoise_Levkreuz.xml	316	3675	0.507	0.407
	080305SIG3D_RailwayATKIS_Levkreuz.xml	224	2334	0.447	0.505
	080305SIG3D_RoadNoise_Levkreuz.xml	2327	41317	0.762	0.105
	Berlin_Pariser_Platz_v0.4.0.xml	4178	30796	0.281	0.337
	Berlin_Pariser_Platz_v1.0.0.xml	4181	30796	0.281	0.337
	CityGML_British_Ordnance_Survey_v1.0.0.xml	4219	36163	0.326	0.250
ALKISATKIS	20070227_B0010_doeteberg51.xml	11746	223780	0.585	0.044
	E.Ben.2005_1001.0001.xml	6803	122291	0.538	0.085
Oracle spatial	admin3.xml	965	24308	0.455	0.258
	admin4.xml	542	13685	0.449	0.263
	admin7.xml	2856	71564	0.467	0.246
	arc.xml	4514	111923	0.507	0.203
	ROAD1.xml	2643	65455	0.451	0.248

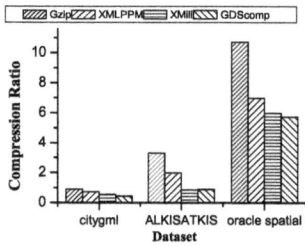

Fig. 4. Compression ratio   **Fig. 5.** Compression time

4.1 Compression Ratio

Fig. 4 shows the comparison of compression ratio among GDScomp and the other three compressors. We can see that GDScomp has the smallest CR. The reason lies in two aspects: 1) GDScomp utilizes the common structural information of different GML documents, and 2) GDScomp uses the delta compression method to compress the spatial data. Though XMill and XMLPPM have good performance in compressing general XML files, they perform worse than GDScomp on GML documents. On one hand, they compress each document separately, so do not utilize common structural information of different documents in the same group (stream). On the other hand, the two compressors were developed for general XML documents, they do not consider the uniqueness of GML documents.

4.2 Compression Time and Decompression Time

Fig. 5 and 6 illustrate the comparison of compression time and decompression time among the four compressors, respectively. It is obvious that XMLPPM has the longest compression and decompression time because it uses PPM as its backend compressor that predicts the next character by the priori probability, whose computation process is very complicated by using Markov chains. Gzip

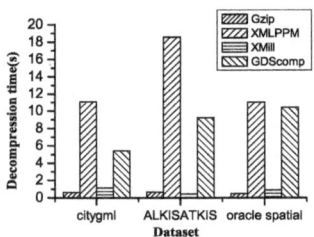

Fig. 6. Decompression time

has the smallest compression and decompression time because it uses a 32KB slide window to complete the compression, the switching speed is very fast and it takes little time in finding the characters ahead.

The reason why XMill is faster than GDScomp is that GDScomp needs to establish a dictionary, collect structural information in the documents, and find the indices in the dictionary. Furthermore, the delta compression needs to split spatial coordinates from the documents and compute the difference of neighboring coordinates.

5   Conclusion and Future Work

GML has become the *de fact* standard of geographic data representation and exchange, and GML data is usually transferred in the form of streams via network. However, there are many redundant tags and hyperlinks, spacious data coordinates in GML documents, which makes them be of extremely large volume and causes lots of cost for transferring and storage.

In this paper, we propose a new stream compressor GDScomp for GML documents. It explores the common information and unique features of GML documents to improve compression performance. For this purpose, it uses a dictionary to store the common structural information of different documents and employs the delta compression method to deal with spatial data. Experimental results show that GDScomp can achieve better compression performance than some state of the art compressors.

Considering that GDScomp groups the information of GML documents into different sub-streams and compresses these sub-streams separately, thus it can not directly support queries over the compressed documents. Therefore, in the future we will try to develop a queriable version of GDScomp.

References

1. Open GIS Consortium. Open GIS Geography Markup Language (GML) Implementation Specification Version 3.1.0, http://www.opengis.net/gml
2. Corcoles, J., Gonzalcz, P.: A specification of a spatial query language over GML. In: Proceedings of ACM-GIS 2001, pp. 112–117 (2001)

3. Guan, J., Zhou, S.: GPress: Towards Effective GML Documents Compression. In: Proceedings of the 23rd International Conference on Data Engineering (ICDE 2007), pp. 1473–1474 (2007)
4. Guan, J., Zhou, S., Chen, Y.: An Effective GML Documents Compressor. IEICE Transactions on Information and Systems E91-D(7), 1982–1990 (2008)
5. Gzip (2000), http://www.gzip.org/
6. Moffat, A.: Implementing the PPM data compression scheme. IEEE Transactions on Communications 38(11), 1917–1921 (1990)
7. Liefke, H., Suciu, D.: XMill: An efficient compressor for XML data. In: Proceedings of the 2000 ACM SIGMOD International Conference on Management of Data (SIGMOD 2000), pp. 153–164 (2000)
8. Tolani, P., Haritsa, J.: XGRIND: A query-friendly XML compressor. In: IEEE Proceedings of the 18th International Conference on Data Engineering (ICDE 2002), pp. 225–234 (2002)
9. Cheney, J.: Compressing XML with multiplexed hierarchical PPM models. In: Proceedings of the IEEE Data Compression Conference (DCC 2001), pp. 163–172 (2001)
10. Girardot, M., Sundaresan, N.: Millau: An encoding format for efficient representation and exchange of XML over the Web. In: Proceedings of the 9th International WWW Conference, pp. 747–765 (2000)
11. Gao, J., Yang, D., Tang, S., et al.: XPath Evaluation Oriented XML Data Stream Compression. Journal of Software 16(2), 123–232 (2005)
12. SAX, http://www.saxproject.org/
13. Ziv, J., Lempel, A.: A universal algorithm for sequential data compression. IEEE Transactions on Information Theory 23(3), 337–343 (1977)
14. Sakr, S.: An Empirical Evaluation of XML Compression Tools. In: Proceedings of DASFAA Workshops 2009, pp. 49–63 (2009)
15. CityGML (2002), http://www.citygml.org/
16. German Working Committee of the Surveying Authorities. Authoritative Real Estate Cadastre Information System (ALKIS), http://www.lv-bw.de/alkis.info/nas-bsp.html

A Query-Friendly Compression for GML Documents⋆

Qingting Wei[1,2]

[1] Dept. of Computer Science and Technology, Tongji University, Shanghai, China
[2] School of Software, Nanchang University, Nanchang 330047, China
qtwei@ncu.edu.cn

Abstract. Geography Markup Language (GML) has become a standard encoding format for exchanging geographic data among heterogeneous Geographic Information System (GIS) applications. Whereas, the iteration of document structure and the textual expression of geographic data often cause the huge size of GML documents. In this paper, a query-friendly GML compression method is proposed, where the GML documents in SAX document parsing are transformed to a compact representation encompassing an event dictionary, the events hierarchy in balanced parentheses, a binary event wavelet tree and the document content blocks before compressed using a general compression utility. The proposed compression method supports direct path queries and spatial queries over the compressed files without the requirement of a full decompression. The compression model, the query resolution process and the compression algorithm are detailed in this paper, though the presentation is a preliminary investigation and it remains to carry out experiments to validate the proposed compression method on real GML documents.

Keywords: GML; compression; query; model; algorithm.

1 Introduction

Geography Markup Language (GML) has become a standard encoding format for exchanging geographic data among heterogeneous Geographic Information System (GIS) applications [1]. Like general XML documents, GML documents are comprehensible and flexible because they are actually textual data encoding files consisting of structures (elements and attribute names) and contents (attribute values and text segments). Nevertheless, GML documents are more redundant than general XML documents. On one hand, GML document structures are extreme regular, iterating their occurrences in a high repeating frequency. On the other hand, GML document contents contain rich geographic data and

---

⋆ This work was supported by the National Natural Science Foundation of China (NSFC) under grant No. 60873040 and China 863 Program under grant No. 2009AA01Z135.

J. Xu et al. (Eds.): DASFAA Workshops 2011, LNCS 6637, pp. 77–88, 2011.

there often exist much high-precision geometric coordinates in text format occupying more storage units than in binary format. As a result, GML documents are usually of huge sizes and costly in storage and transport.

The problem necessitates the compression of GML documents. Since GML documents are actually a special kind of XML-grammar text files, the existing general text compression and XML compression methods are all applicable to GML compression. However, the popular general text compression methods [2,3,4,5,6] do not distinguish document structure and document content, processing all symbols in GML documents equally without exploiting the structure and data characteristics, so usually perform poor in the compression of GML documents.

The well-known XML compression methods are conscious of document structure in compression. According to the ability of supporting query, the XML compression methods are often classified into two groups: non-queryable and queryable. The first group archives one or more XML documents into an encoding file as small as possible [7,8,9,10,11]. They can achieve a satisfying document compression ratio, while do not support direct queries over the compressed files (i.e., a full decompression is required before executing queries). On the contrary, the second group aims to support direct queries over the compressed files [12,13,14,15]. For instance, XGrind [12] employs homomorphic transformations from original documents to the compressed files, so direct path queries can be executed by seeking the compressed files for the matched tag sequences. However the queries based on homomorphic transformations require scanning the compressed files from the beginning to the end, which costs a long time. Take another example, XQueC [14] preserves a tree-structured index that summarizes the original document structure in the compressed files to provide a query support based on tree navigation. However, the tree-like index consumes much storage space on saving the pointers linking the tree nodes, which worsens compression ratio. Moreover, the existing queryable XML compression methods only support path queries (searching for document elements according to their root-to-node paths in the document tree), without considering spatial queries (searching for spatial objects according to their geometric shapes, spatial locations or spatial relationships). Path queries and spatial queries are both important in GML-based GIS applications. It is very common to execute a query over a GML document to find a spatial object with a given name or in a selected region. Therefore, the existing XML compression methods do not suit queryable GML compression well.

To the best of our knowledge, there is little work on GML compression in the literature except for GPress [16,17], GMill [18] and GQComp [19]. GPress [16,17] is the first proposed GML specific compression method, which exploits the structure and data characteristics of GML documents to help compression and applies delta compression to coordinate data the first time. GMill [18] is a GML compression approach based on on-line semantic clustering, where document content with similar semantics is assembled into the same cluster and compressed together. GPress and GMill outperform the other compression methods in compression ratio when compressing GML documents, while neither of them provides a support of direct queries over the compressed files. GQComp [19] is a query-supported GML compressor, which compresses a GML document via encoding the document structure as a *Feature Structure Tree* (FST) and the result of matching the document structure against the FST, encoding attribute data and spatial data separately, and constructing spatial index for spatial data. Although GQComp claims to support attribute query and spatial query over the compressed files, its compression efficiency is worrying because the procedure of matching the document structure against the FST is time-consuming when the document structure is nested deeply.

Thereupon, in this paper, an query-friendly GML compression method GQueC, which supports both direct path queries and spatial queries over the compressed files and aims to achieve a tradeoff between compression ratio and compression efficiency is proposed. The paper details the compression model, the query resolution process and the compression algorithm of the proposed method.

The rest of this paper is structured as follows. Section 2 provides a background into the problem. Section 3 describes the compression model, the query resolution process and the compression algorithms, while section 4 concludes the paper.

2   Background

Geography Markup Language (GML), introduced by Open Geospatial Consortium (OGC), offers a solution to exchange geographic data among heterogeneous Geographic Information System (GIS) applications. As an XML-based geographic modeling language, GML describes geo-spatial objects by collections of hierarchical features (geographic entities) with a list of properties (e.g., name, type and value) and geometries composed of basic geometry building blocks (e.g., points, lines, curves, surfaces and polygons). Practically, querying GML documents are equivalent to retrieving features according to the given properties or geometries. The following subsections show the examples of GML documents and GML queries.

2.1   An Example of GML Documents

Example 1. Consider the following GML document "topography.xml" which provides a topographic description of two areas:

```
<osgb:topographicMember>
  <osgb:TopographicArea fid="osgb1000000334340437">
    <gml:Polygon>
      <gml:outerBoundaryIs>
        <gml:LinearRing>
          <gml:coordinates>
            277792.850,186180.850 277864.200,186139.900
            277994.160,185959.510 277822.850,186163.950
            277818.700,186167.950 277792.850,186180.850
```

```
          </gml:coordinates>
        </gml:LinearRing>
      </gml:outerBoundaryIs>
    </gml:Polygon>
  </osgb:TopographicArea>
  <osgb:TopographicArea fid="osgb1000000334340499">
    <gml:Polygon>
      <gml:outerBoundaryIs>
        <gml:LinearRing>
          <gml:coordinates>
            277804.500,185706.100 278302.600,185736.600
            278244.370,186000.000 277889.150,186167.250
            278005.150,185534.780 277804.500,185706.100
          </gml:coordinates>
        </gml:LinearRing>
      </gml:outerBoundaryIs>
    </gml:Polygon>
  </osgb:TopographicArea>
</osgb:topographicMember>
```

There are two `osgb:TopographicArea` elements representing two topographic area objects. Each `osgb:TopographicArea` element holds a `fid` attribute, the attribute value indicating the area's identifier, and contains a `gml:Polygon` subelement, which describes the area's shape. Every `gml:Polygon` element has a `gml:outerBoundaryIs` subelement standing for the outer boundary, portrayed by the subelement `gml:LinerRing` and the sub-subelement `gml:coordinates`. Each `gml:coordinates` element encloses the text segments of coordinates.

2.2 An Example of GML Queries

There exist two types of queries on GML documents: path queries and spatial queries. The former involves the properties of geographic objects, while the latter is related to the geometries of geographic objects. The common operations (including their names and descriptions) of path queries and spatial queries are listed as follows:

- Common path query operations
  - **label**(i): return label of node i;
  - **child**(i): return children of node i;
  - **descendant**(i): return descendants of node i;
  - **parent**(i): return parent of node i;
  - **ancestor**(i): return ancestors of node i;
  - **nextsibling**(i): the first child of node i's parent that occur after node i;
  - **previoussibling**(i): the last child of node i's parent that occur before node i.

- Common spatial query operations
  - **isequal**(i,j): whether geometry i and geometry j is equal;
  - **isdisjoint**(i,j): whether geometry i and geometry j is disjoint;
  - **intersects**(i,j): whether geometry i intersects geometry j;
  - **iswithin**(i,j): whether geometry i is within geometry j;
  - **touches**(i,j): whether geometry i touches geometry j;
  - **crosses**(i,j): whether geometry i crosses geometry j;
  - **contains**(i,j): whether geometry i contains geometry j;
  - **overlap**(i,j): whether geometry i overlaps geometry j;
  - **distance**(i,j): the shortest distance between geometry i and geometry j.

The operations above are usually stated in a specific GML query languages [20,21,22]. GQL [22] is a query language specification to support spatial query over GML documents by extending XQuery [23]. An example of GML query statements in GQL is as follows:

Example 2. A query on the GML document in Example 1 to retrieve the topographic areas covering a point (278000,186000) in geometry is written in GQL as follows:

```
<AreaCollection>{
  for $a$ in document("topography.xml")//osgb:TopographicArea
  where iswithin(Point(278000,186000),$a$/gml:Polygon)=1
  return $a$}
</AreaCollection>
```

According to this query statement, the second `osgb:TopographicArea` element in Example 1 is returned:

```
<AreaCollection>
  <osgb:TopographicArea fid="osgb1000000334340499">
    <gml:Polygon>
      <gml:outerBoundaryIs>
        <gml:LinearRing>
          <gml:coordinates>
            277804.500,185706.100 278302.600,185736.600
            278244.370,186000.000 277889.150,186167.250
            278005.150,185534.780 277804.500,185706.100
          </gml:coordinates>
        </gml:LinearRing>
      </gml:outerBoundaryIs>
    </gml:Polygon>
  </osgb:TopographicArea>
</AreaCollection>
```

In this query, the operations involved include **descendant**, **label**, **child** and **iswithin**. Firstly, the **descendant** operation is carried out to obtain the descendants of the root. Then the **label** operation is used to get the descendants' labels, which are compared with the restriction condition "osgb:TopographicArea". When a node labeled `osgb:TopographicArea` is matched, the **child** and **label** operations are employed to search for one of that node's children labeled `gml:Polygon`, then the **iswithin** operation is adopted to test whether the `gml:Polygon` node covers a point (278000,186000) in geometry. The matched `osgb:TopographicArea` node is returned finally.

3   Query-Friendly GML Compression

To provide a query-friendly compression, GQueC transforms a GML document under compression into a compact representation before applies the general compression utility. *Simple API for XML* (SAX) is chosen to be the document parser because it does not require loading the whole GML document under compression into the main memory. In a SAX document parsing, GQueC separates the document structure from the document content. The document structure is replaced by a SAX event dictionary, the events hierarchy in balanced parentheses and a binary event wavelet tree. The document content is grouped by events into different memory blocks where each coordinate block has a head including the corners of the geometric bounding box specially. Finally, GQueC forwards this compact representation of the GML document to the popular general compressor gzip [24] (other general compressors are also applicable here). The following subsections describe the detail of the compression model, the query resolution process and the compression algorithm.

3.1   Compression Model

The compressed file produced by GQueC consists of four parts: 1) a SAX event dictionary, 2) the SAX events hierarchy, 3) a SAX event wavelet tree, and 4) the document content blocks. These four parts are compressed further using the general compression utility gzip before written to the output file. Figure 1 shows the constitution of the compressed file in compressing the GML document in Example 1.

SAX Event Dictionary. The first part of the compressed file is a SAX event dictionary. SAX events refer to the events of beginning or ending a document component in parsing a GML document by the SAX parser. In this paper, the beginning of an element is called the *element start event*, represented by the element names; the beginning of an attribute is called the *attribute start event*, denoted by '@' prepending the attribute names; the beginning of an attribute value or a text segment is called the *content start event*, marked as '#' prepending the name of the previous event; and the ending of any document components is called the *end event*, symbolized as '/'. For instance, for the GML document in Example 1, the content start event of beginning the value of the attribute "`fid`", is represented by "`#@fid`".

GQueC employs a dictionary with a default end event '/' to collect distinct SAX events. In the document parsing, if an event does not occurr before, then

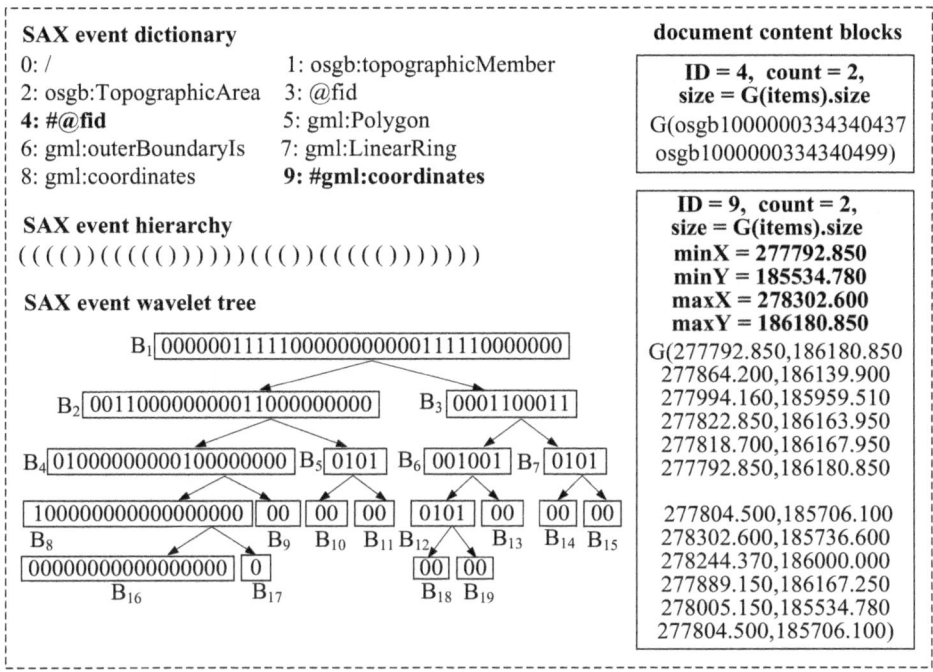

Fig. 1. The compression model for the GML document in Example 1

that event will be inserted to the dictionary. All dictionary items are ordered by the turn inserted to the dictionary. As shown in Fig. 1, ten distinct events are inserted into the dictionary in the SAX parsing of the GML document in Example 1. The fourth and the ninth event are the content start events, denoted by "#@fid" and "#gml:coordinates" respectively.

SAX Events Hierarchy. The second part of the compressed file is the SAX events hierarchy. In this paper, the hierarchy of two start events A and B refers to the ancestor-descendant relationship between A and B. If B occurs after A and before the counterpart end event of A, then A is the ancestor of B and B is the descendant of A.

Balanced parentheses are adopted to indicate whether an event is a start event or an end event. The open parenthesis '(' symbolizes the start events and the close parenthesis ')' denotes the end events. Then the hierarchy of an events sequence can be represented by a sequence of parentheses. For two start events A and B, their hierarchy depends on the *excess* of parentheses in scanning the parentheses sequence from A's position to B's position (B is excluded). When an open parenthesis is met, 1 is added to the excess; when a close parenthesis is met, 1 is reduced from the excess. If the excess is a positive integer, then A is B's ancestor and B is A's descendant. Especially, if the excess is equal to 0 then A and B are siblings, and if the excess is equal to 1 then A is B's parent and B is A's child.

Take an example of the document in Example 1, the SAX events in parsing the GML segments "fid=`osgb1000000334340437`'" include the attribute start event "@fid", the content start event "#@fid" and the two end events '/'. Their hierarchy can be represented as a sequence "(())". Since the excess from the first parenthesis to the second parenthesis is 1, the corresponding attribute start event "@fid" is the parent of the content start event "#@fid".

SAX Event Wavelet Tree. Wavelet tree is a kind of text index for executing efficient rank and select operations in a sequence S of symbols, which come from an arbitrary alphabet $\sum$ of the size n ($|\sum| = n$) [25]. Actually a wavelet tree is a binary tree with $\lceil \log_2 n \rceil + 1$ levels. Each wavelet tree node has a reference alphabet $\sum'$ that is a subset of $\sum$ ($\sum' = \sum$ only in the root node). If a node with a reference alphabet $\sum'$ is a interior node, then its left child succeeds to the first half of $\sum'$ and its right child succeeds to the second half of $\sum'$. Moreover each tree node stores a bitmap for the symbols that comes from the current reference alphabet and occurs in S, where the symbols from the first half of $\sum'$ is mapped to the bit '0' while the symbols from the second half of $\sum'$ is mapped to the bit '1'.

In GQueC, the third part of the compressed file is a SAX event wavelet tree. Here, the SAX events are regarded as the symbols to be indexed and the event dictionary is regarded as the alphabet $\sum$. As shown in Fig. 1, the alphabet $\sum$ of the wavelet tree contains ten symbols, i.e., ten distinct SAX events, and has five levels of nodes. The root node in the 0-th level stores a bitmap B_1 for the sequence of 34 SAX events, where the events from the first half of the event dictionary is mapped to '0' while the events from the second half of the event dictionary is mapped to '1'. The ten leaf nodes store the bitmaps B_{16}, B_{17}, B_9, B_{10}, B_{11}, B_{18}, B_{19}, B_{13}, B_{14}, B_{15} corresponding to the subsequences of ten distinctive SAX events respectively.

Document Content Blocks. The last part of the compressed file is the document content blocks. Since the document content parsed in the same kind of SAX events is usually relevant and composed of similar texts, the compression of them as a whole requires less storage space to construct the compression model than the compression of mixing them with other irrelevant texts. Therefore, in GQueC, different memory blocks of the same size are assigned to the document content parsed in different kinds of content events. Moreover, the event IDs, the item count and the total size of the compressed items are recorded in the head of each block. In addition, for speeding up spatial query, the corners (the values of minimum X, minimum Y, maximum X and maximum Y) of the box bounding the block in geometry are saved in the block head when the content block is a coordinate block associated with the events "#gml:coordinates", "#gml:pos", "#gml:poslist", "#gml:X" or "#gml:Y".

As shown in the Fig. 1, the attribute values read in the event "#@fid" and the text segments read in the event "#gml:coordinates" are placed into two different memory blocks, the second of which is a coordinate content block.

3.2 Query Resolution Process

Based on the compression model described in the previous subsections, GQueC
provides supports of both direct path queries and spatial queries over the compressed GML files via searching the SAX event dictionary, the SAX events hierarchy in balanced parentheses, the binary wavelet tree and the document content
blocks.

Take an example, the following steps are executed to resolve the query string
"//osgb:TopographicArea/gml:Polygon" in Example 2:

1. Get the identifier of the event "osgb:TopographicArea" in the SAX event
   dictionary (here $ID=2$).
2. Locate the No. $n=ID+1$ left wavelet tree leaf node B_x (here $n=3$ and $x=9$).
3. Get the item count c (here $c=2$) of B_x.
4. For a integer i from 1 to c do:
   (a) Set $y=x$, $j=i$.
   (b) Locate the parent B_z of B_y, here $z=\lfloor y/2 \rfloor$.
   (c) Get the position $p \geq 1$ of the j-th bit with a value equal to the residue
       of $y/2$ in the bitmap stored in B_z.
   (d) Set $y=z$, $j=p$ and recurse to step (b) unless B_z is the root.
   (e) Locate the parenthesis at the position p in the SAX events hierarchy.
   (f) Get the next position k $(k > p)$ where the parenthesis is an open paren-
       thesis and the excess from the position p to k is equal to 1.
   (g) Set $p'=k$, $m=1$
   (h) Locate the wavelet tree node B_m
   (i) Get the bit b at the position p' in the bitmap stored in B_m.
   (j) Get the rank r of the bit b compared with all bits of the same value in
       B_m.
   (k) If $b=0$ then $m'=2 \times m$, else $m'=2 \times m+1$
   (l) Set $p'=r$, $m=m'$ and recurse to step (h) unless m' is greater than the
       total number of the wavelet tree nodes.
   (m) Get the order ID' of B_m in all wavelet tree leaf nodes from left to right.
   (n) If $ID'=5$ (representing the event "gml:Polygon"), then return the as-
       sociated element name "<gml:Polygon>" and all descendant events's
       associated document component names.
   (o) Recurse to step (f).

3.3 Compression Algorithm

The compression procedure is outlined in Algorithm 1, which accepts as inputs
a GML document (an example is shown in Example 1), and outputs the compressed file (as shown in Fig. 1).

In the parsing of the document, different actions are adopted when different
events occurr. If the current event is a start event that does not occurr before,
the event name will be inserted to the SAX event dictionary according to the
event type. Moreover, the current event ID is recorded into a sequence of integer,
and the events hierarchy is encoded as a sequence of bits '0' (representing open

---

Algorithm 1. GQueC(G,G') to execute a query-friendly compression of a GML document

---

input : GML document G
output: compressed file G'
Let D be a SAX event dictionary $\{$ "/" $\}$, i be the size of D, $harray$ be a bit array to save sequences of the SAX events hierarchy, e be the current SAX event occurring in the parsing of D, $last$ be the name of the last event, c be the ID of the last event, $carray$ be an integer array to save the event ID sequence, j be the length of $carray$, $block$ be the content block with a head (ID, $count$, $size$) and a body $items$, $BLOCKCAPACITY$ be the const memory size allocated to a block, and $root$ be a wavelet tree where nodes are labeled by a bit string.
$i \leftarrow 1$
$j \leftarrow 0$
while $e! = EOF$ **do**

    if e is an element start event **or** an attribute start event **then**
        if e is an element start event **then** $last \leftarrow$ nameof(e)
        else $last \leftarrow$ strcat("@",nameof(e))
        $harray[j] \leftarrow 0$
        if $last \in D$ **then** $c \leftarrow$ getEventID($last,D$)
        else
            $D \leftarrow D \cup \{last\}$
            $c \leftarrow i$
            $i \leftarrow i + 1$
    else if e is an event of beginning content **then**
        $last \leftarrow$ strcat("#",$last$)
        $harray[j] \leftarrow 0$
        if $last \in D$ **then**
            $c \leftarrow$ getEventID($last,D$)
            $block \leftarrow$ getBlockbyID(c)
        else
            $D \leftarrow D \cup \{last\}$
            $c \leftarrow i$
            $block \leftarrow$ creatBlockofID(c)
            $i \leftarrow i + 1$
        if $block.size +$sizeof(e)$> BLOCKCAPACITY$ **then**
            if $last$ contains "coordinates", "poslist", "pos", "X" or "Y" **then**
                fwrite(G',$block.ID$,$block.count$,sizeof(gzip($block.items$))),
                minX($block.items$),minY($block.items$),maxX($block.items$),
                maxY($block.items$),gzip($block.items$))
            else
                 fwrite(G',$block.ID$,$block.count$,sizeof(gzip($block.items$))),
                gzip($block.items$))
            $block.count \leftarrow 1$
            $block.items \leftarrow \{e\}$
        else
            $block.count \leftarrow block.count + 1$
            $block.items \leftarrow block.items \cup \{e\}$
    else
        $harray[j] \leftarrow 1$
        $c \leftarrow 0$
    $carray[j] \leftarrow c$
    $j \leftarrow j + 1$
    $e \leftarrow e$.next
TreeLabel($root,carray$)
fwrite(G',gzip(D))
fwrite(G',gzip($harray$))
fwrite(G',gzip($root$))

---

parentheses) or bits '1' (representing close parentheses). In addition, if the current event is a content start event, GQueC will search for its associated memory block or allocate a memory block to it, then stores the content into the block. When a block is full, the compressor gzip is called to compress the block's items, and the output of gzip and the block head (including the event ID, the item count and the size of the compressed items) are written to the compressed file, then the block is dumped. Especially, if the block caches the content of coordinates, the corners (the values of minimum X, minimum Y, maximum X and maximum Y) of the box bounding the block in geometry are compressed as a part of the block head too.

When the document parsing comes to the end, GQueC generates a labeled wavelet tree based on the event ID sequence using the *TreeLabel* algorithm proposed by P. Ferragina [26], then compresses the event dictionary, the wavelet tree nodes and the bit-encoded events hierarchy sequence using the compressor gzip and inserts the output of gzip to the compressed file finally.

4   Conclusion

This paper proposes a query-friendly GML compression method, where the compression of a GML document is reduced to the general text compressions of the SAX event dictionary, the SAX events hierarchy in balanced parentheses, the SAX event wavelet tree and the document content blocks. The proposed method supports both direct path queries and spatial queries over the compressed files. And the paper details the compression model, the query resolution process and the compression algorithms for explaining the proposed method.

In order to demonstrate the effectiveness and efficiency of the proposed method, extensive experiments are needed to evaluate the proposed method and to compare it with the existing compression methods. This is the future work.

References

1. Geospatial information – Geography Markup Language (GML). ISO 19136:2007 (2007)
2. Huffman, D.A.: A Method for the Construction of Minimum-Redundancy Codes. Proceedings of the IRE 40(9), 1098–1101 (1952)
3. Ziv, J., Lempel, A.: A universal algorithm for sequential data compression. IEEE Transactions on Information Theory IT-23, 337–343 (1977)
4. Witten, I.H., Neal, R.M., Cleary, J.G.: Arithmetic coding for data compression. Communications of the ACM 30(6), 520–540 (1987)
5. Cleary, J.G., Witten, I.H.: Data compression using adaptive coding and partial string matching. IEEE Transactions on Communications 32(4), 396–402 (1984)
6. Burrows, M., Wheeler, D.J.: A block-sorting lossless data compression algorithm. Technical report SRC-RR-124, Hewlett-Packard Company (1994)
7. Hartmut, L., Suciu, D.: XMill: an efficient compressor for XML data. In: ACM SIGMOD 2000, pp. 153–164. ACM Press, New York (2000)

8. Girardot, M., Sundaresan, N.: Millau: an encoding format for efficient representation and exchange of XML over the Web. Computer Networks 33(1-6), 747–765 (2000)
9. Cheney, J.: Compressing XML with multiplexed hierarchical PPM models. In: DCC 2001, pp. 163–172. IEEE Press, New York (2001)
10. League, C., Eng, K.: Type-based compression of XML data. In: DCC 2007, pp. 272–282. IEEE Press, New York (2007)
11. Skibiński, P., Grabowski, S., Swacha, J.: Effective asymmetric XML compression. Software: Practice and Experience 38(10), 1024–1047 (2008)
12. Tolani, P.M., Haritsa, J.R.: XGrind: a query-friendly XML compressor. In: ICDE 2002, pp. 225–234. IEEE Press, New York (2002)
13. Min, J., Park, M., Chung, C.: XPress: a queriable compression for XML data. In: ACM SIGMOD 2003, pp. 122–133. IEEE Press, New York (2003)
14. Arion, A., Bonifati, A., Costa, G., D'Aguanno, S., Manolescu, I., Pugliese, A.: XQueC: Pushing queries to compressed XML data. In: VLDB 2003, pp. 1065–1068 (2003)
15. Lam, W.Y., Ng, W., Wood, P.T., Levene, M.: XCQ: A queriable XML compression system. Knowledge and Information Systems 10(4), 421–452 (2006)
16. Guan, J., Zhou, S.: GPress: Towards effective GML documents compresssion. In: ICDE 2007, pp. 1473–1474. IEEE Press, New York (2007)
17. Guan, J., Zhou, S., Chen, Y.: An effective GML documents compressor. IEICE Transactions on Information and Systems E91-D(7), 1982–1990 (2008)
18. Wei, Q., Guan, J.: A GML Compression Approach Based on On-line Semantic Clustering. In: Geoinformatics 2010, pp. 1–7. IEEE Press, New York (2010)
19. Dai, Q., Zhang, S., Wang, Z.: GQComp: A Query-Supported Compression Technique for GML. In: 9th IEEE International Conference on Computer and Information Technology, pp. 311–317. IEEE Press, New York (2009)
20. Vatsavai, R.R.: GML-QL: A spatial query language specification for GML. Department of Computer Science and Engineering, University of Minnesota, http://www.cobblestoneconcepts.com/ucgis2summer2002/vatsavai/vatsavai.htm
21. Boucelma, O., Colonna, F.M.: GQuery: a query language for GML. In: 24th Urban Data Management Symposium (2004)
22. Jihong, G.: GQL: Extending XQuery to query GML documents. Geo-spatial Information Science 9(2), 118–126 (2006)
23. XQuery 1.0: An XML query language, http://www.w3.org/XML/Query/
24. GZip 1.2.4, http://www.gzip.org
25. Grossi, R., Gupta, A., Vitter, J.S.: High-order entropy-compressed text indexes. In: SODA 2003 (2003)
26. Ferragina, P., Giancarlo, R., Manzini, G.: The myriad virtues of wavelet trees. Information and Computation 207(8), 849–866 (2009)

Storing GML Documents: A Model-Mapping Based Approach[*]

Fubao Zhu[1,2], Qianqian Guo[2], and Jinmei Yang[2]

[1] School of Computer Science, Fudan University, Shanghai 200433, China
fbzhu@fudan.edu.cn
[2] School of Computer and Communication Engineering,
Zhengzhou University of Light Industry, Zhengzhou 450002, China

Abstract. Geography Markup Language is a de facto standard developed by OGC to standardize the representation of geographical data in XML, which makes the exchanging and sharing of geographical information easier. With the popularity of GML technology, more and more geographical data is presented in GML format. This causes the problem of how to efficiently store GML data to facilitate its management and retrieval. An approach to store and query GML document based on model-mapping is proposed in this paper. The proposed approach mainly focus on non-schema GML documents, but it is also applicable to GML documents with corresponding schemas. A GML document is first parsed, and a document tree is generated. Then the tree nodes are analyzed and processed, and the schema mapping is established for storing GML documents into object-relational database with structural information preserved. Spatial data analysis and non-spatial data query are supported on database for document. Experiments show that the proposed approach is feasible and efficient.

Keywords: Geography Markup Language; Schema; Model-mapping; Storage; Object-relational Database.

1 Introduction

The Geography Markup Language (GML) is the XML grammar defined by the Open Geospatial Consortium (OGC) for expressing geographical features. It is an international standard for data encoding, transmission, storage and release. It is applied to geographical data sharing, exchanging and integration on Internet. With the development of GML, the growing ability of GML has solved inconsistent formats of spatial data providing the data expression including data structure and semantics. It conforms to the requirements of Web and makes it much easier for data exchange, integration, and sharing between different systems.

---

[*] This work was supported by the National Natural Science Foundation of China (No. 60873040), China 863 Program (No. 2009AA01Z135), Key Science and Technology Project of Education Department of Henan Province (No. 2010B520033), and Doctoral Research Fund of Zhengzhou University of Light Industry (No. 2008BSJJ012).

J. Xu et al. (Eds.): DASFAA Workshops 2011, LNCS 6637, pp. 89–100, 2011.

With the extensive application of GML, a growing number of geographic information is described by GML format. Like the XML encoding, GML uses text format to express geographic information. Because text file of GML is quite large, it is impossible to manage geographical data well with good spatial information query, spatial data analysis, access and concurrency control, etc. The effective management of GML data becomes an urgent problem which is aimed to facilitate analysis of spatial operations and promote the sharing of spatial data and GIS interoperability. GML document will be stored in the object-relational database where GML non-spatial attributes act as common fields, spatial attributes act as objects. It is an available solution to manage GML data by object-relational databases.

GML application schema provides the format specification and semantic constraints for the preparation of GML document. The document which has schema can implement storage through mapping mechanism between GML schema and database schema. This method is called structure-mapping [1]. However, in many applications, the schema of GML document is not clear, or even has no schema. The major problem here is how to store GML document in the database and meet different query requirements. Such method is called model-mapping [3,4,5,6]. It does not use any schema of stored documents. Instead, it parses the original document into a document tree in memory, creating a generic (object) relational schema to store the document's structure and data. Therefore we propose a GML document storage method based on model mapping. GML document is regarded as a tree composed of the element nodes, attribute nodes and geometry nodes. The process of tree nodes and edges help to store GML document into the database and to support spatial information analysis and non-spatial information inquiries.

2   Related Work

At present, a lot of work focused on the XML documents based on model-mapping and the storage management of database [3,4,5,6]. The work parsed the XML document into a tree, and processed differently nodes and edges of document tree.

The method of XParent [6] summed up the three methods of Edge [3], Monet [4] and XRel [5]. It divided those into the Edge-Oriented [3,4] method and the Node-Oriented [5] method. And it also used four tables to represent the data and structure information in XML document, which included path table, data path table, element table and data table. It made full use of advantages of Edge, Monet and XRel, and provided a good query capability.

Córcoles [7] analyzed the performance of storing GML documents of LegoDB [2], Monet and XParent. And it proved that LegoDB had the optimal performance in storage and query, as well as LegoDB was better to support the spatial information processing. But LegoDB was based on the structure mapping and it was not suitable for non-schema GML document storage.

All the above methods [2,3,4,5,6] are designed for XML document storage, and without considering the spatial data information of GML document, thus they were not suitable for GML document storage. At present, there is no literature related to storage management technologies and methods about the non-schema GML document. Since GML is based on XML, methods of XML storage management can help a lot on the study of non-schema GML document storage management. This paper makes full use of these advantages of non-spatial information storage, and expands its storage capability of spatial information. It has proposed the model-mapping method which is based on nodes and edges to store non-schema GML documents.

3   Model-Mapping Storage Method Based on Nodes and Edges

This section details the architecture of GML document storage and GML document data model. According to an illustration of GML document, we describe the schema mapping method from nodes and edge of GML document tree to database. The storage time-consuming of different document size and different data type is presented to validate the proposed storing method.

3.1   GML Document Storing Architecture

The types of data in the GML document are point, line, polygon and so on. Separating different data effectively and storing data into the database by the processor is the main issue of the paper. The architecture is shown in Fig. 1.

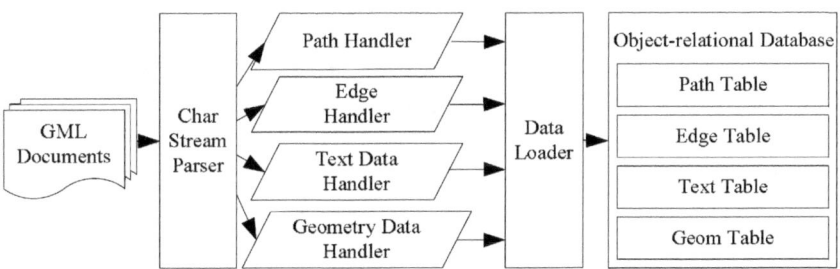

Fig. 1. Architecture of GML Storage

The path data, edge data, text data and geometry data of the document are parsed by the character stream parser from the GML document. The data is processed by the corresponding processor, and then stored in the corresponding table in object-relational database by the data loader, to realize analysis and computation of spatial data and query and manipulation of non-spatial data.

3.2 GML Document Data Schema

A complete GML document can be viewed as a tree composed of a number of nodes. Figure 2 shows the structure of document tree of document instance described in Prev verse, including root node, element nodes, attribute nodes, text nodes and geometry nodes. Root node is a virtual node which points to root element node of GML document. Element node represents element in the GML document which is named after the label of element and contains more than one text node and child element node. Text node only contains string information which represents non-spatial data in GML document and does not contain any child nodes. Elements in GML document can contain a number of attributes which are shown as attribute nodes in the document tree. Attribute nodes are composed by attribute name and attribute values which have no child nodes. Geometry is a special element node which is used to represent the geometric information of spatial data in GML document. In Fig. 2, symbols $\triangle$, $\bigcirc$, $\square$, $\Diamond$ separately represent the element node, attribute node, text node and geometry node. Note that in GML document tree, we do not draw a specific spatial information which is represented in geometry node. When the parser encounters a geometry node, it will deal with it as an object.

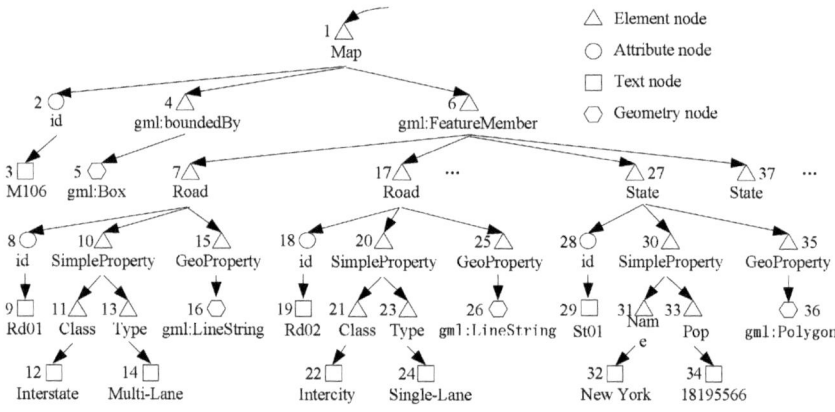

Fig. 2. GML document tree

3.3 Constructing of GML Document Tree

GML document is stored as a text file. We use the follow steps to store the data and structural information of the document into the database. The first step, it should parse the GML document. The second step, use the result of parsing to build the document tree. Next, use IDs to number nodes of different types of the tree. Last, determine the order of nodes which have the same parent and the same label in siblings. The algorithm of document tree construction is shown in Algorithm 1.

Algorithm 1. Constructing GML document

Result: CreateGMLDocTree(*gb*)

input   : GMLDocument *gd*

output: GMLDocumentTree *gdt*

parse GML document *gd* as DOM tree *dt*;

get the root node *root* of DOM tree *dt*; let *nID* and *ord* be id and ordinal of
nodes in *dt*, respectively

GenerateDocTree (*root*); // Generate document tree

Procedure GenerateDocTree(*node*)

if *the node type of node is not Geometry* **then**
> assign node id *nID* to *node*;
> //Given node to node numbering
> *nID=nID+*1;
> //Calculated number for the next node
> let *pn* be parent node of *node*;
> //Get the parent node
> let *ord* be ordinal of sibling node which has same name;
> **if** *pn is not null* **then**
> > //If the parent node is not null
> > calculate the ordinal *ord* of *node* in its same-tag-sibling
> > assign *ord* to *node*; //computer node's ordianal value
>
> **else**
> > let *cns* be the child node list of *node*;//Get the child node
> > for each cnsi in cns do GenerateDocTree(cnsi);
> > //Recursively process each child node
>
> **end**

else
> construct Geometry g of node;
> //Construct geometry object
> calculate nID and ord for g;
> //Calculate the number and order for geometry

end

3.4   GML Document Database Model

When GML document has been stored in the database, it should support querying and processing of non-spatial information, and analyzing and calculating of
spatial data. To realize the object, an approach of node and edge based Model-
mapping is proposed in the paper (Node and Edge based Model-mapping, NEM).
NEM defines the label path table, edge table, text table and geometric table in
processing nodes and edges of the document tree. The structure of the four tables
is shown as follows:

```
LabelPathpathID, pathExpr
EdgedocID, pID, cID
TxtDatadocID, pathID, ordinal, nID, type, value
GeoDatadocID, pathID, ordinal, nID, type, shape
```

Algorithm 2. Storing GML document

Result: StoringGMLDocuments(*gdt*)
input   : GMLDocumentTree *gdt*
output: four tables
let *docID* be the maximum document id of current database;
//Get root node
if *docID is not null* **then**
  | //Calculated the number of docID
  | *docID* = *docID* + 1;
else
  | *docID* =1;
end
let *root* be root node of *gdt*;
//Get the root
let *pathID* be the maximum path id of current database;
//Get the pathID
if *pathID is not null* **then**
  | //Calculate the number of pathID
  | *pathID* = *pathID* + 1;
else
  | *pathID* =1;
end
TraversalNodes(*root*);
Process TraversalNodes(Node *n*)
if *type of node n is Element* **then**
  | Let *pathExpr* be label path from root to current node;
  | **if** *pathExpr does not exist in table LabelPath* **then**
  | | *pathID* = *pathID* + 1; insert *pathID* and *pathExpr* into table
  | | LabelPath;
  | **else**
  | | endif
  | **end**
else
  | endif
end
let *pn* be parent node of *n*;
//Get the parent node
if *pn is not null* **then**
  | ///if the parent node is not null
  | let *pID* and *cID* be node id of *pn* and *n*, respectively;
  | insert *docIDC*, *pID* and *cID* into table Edge;
else
  | endif
end
if *type of n is Text or Attribute* **then**
  | calculate pathID and get ordinal, type and value of node *n*;
  | insert the above values into table TxtData;
else

end
if *type of n is Geometry* **then**
  | calculate pathID and get ordinal, type and value of node *n*; insert the above
  | values into table into table GeoData;
else
  | endif
end
let *cns* be the child node list of *n*;
for each *ni* in *cns* do TraversalNodes (*ni*);
return LabelPath, Edge, TxtData and GeoData;

Table 1. LabelPath table

pathID	pathExpr
1	/Map
2	/Map/id
3	/Map/boundedBy
4	/Map/boundedBy/gml:Box
5	/Map/FeatureMember
6	/Map/FeatureMember/Road
7	/Map/FeatureMember/Road/id
8	/Map/FeatureMember/Road/SimpleProperty
9	/Map/FeatureMember/Road/SimpleProperty/Class
10	/Map/FeatureMember/Road/SimpleProperty/Type
11	/Map/FeatureMember/Road/GeoProperty
12	/Map/FeatureMember/Road/GeoProperty/gml:LineString
13	/Map/FeatureMember/State
14	/Map/FeatureMember/State/id
15	/Map/FeatureMember/State/SimpleProperty
16	/Map/FeatureMember/State/SimpleProperty/Name
17	/Map/FeatureMember/State/SimpleProperty/Pop
18	/Map/FeatureMember/State/GeoProperty
19	/Map/FeatureMember/State/GeoProperty/gml:Polygon

Table 2. Edge table

docID	pID	cID	docID	pID	cID
1	1	2	1	17	18
1	1	4	1	17	20
1	1	6	1	17	25
1	2	3	1	18	19
1	4	5	1	20	21
1	6	7	1	20	23
1	6	17	1	23	24
1	6	27	1	25	26
1	7	8	1	27	28
1	7	10	1	27	30
1	7	15	1	27	35
1	8	9	1	28	29
1	10	11	1	30	31
1	10	13	1	30	33
1	11	12	1	31	32
1	13	14	1	33	34
1	15	16	1	35	36

LabelPath table records the label path information of the document tree node, the pathID is the LablePath table's ID and the pathExpr is the corresponding path's expression. Edge table records the document tree edge, docID, pID and

Table 3. TxtData table

docID	pathID	ordinal	nID	type	value
1	2	1	3	T	M16
1	7	1	9	A	Rd01
1	9	1	12	T	Interstate
1	10	1	14	T	Multi-Lane
1	7	2	19	A	Rd02
1	9	2	22	T	Intercity
1	10	2	24	T	Single-Lane
1	14	1	29	A	St01
1	16	1	32	T	Clark Fork
1	17	1	34	T	Columbia

Table 4. GeoData Table

docID	pathID	ordinal	nID	type	shape
1	4	1	5	Polygon	GEOM (Box)
1	12	1	16	Line	GEOM (LineString)
1	12	2	26	Line	GEOM (LineString)
1	19	1	36	Polygon	GEOM (Polygon)

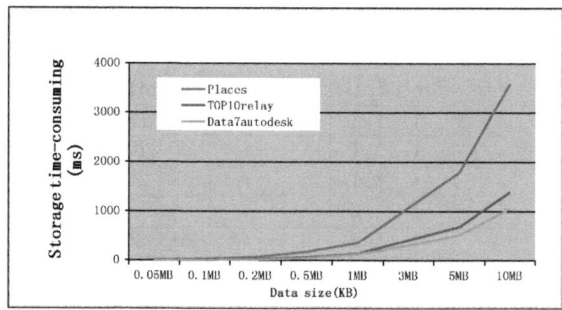

Fig. 3. Model-independent GML data storage time-consuming

cID which separately represent IDs of document identification, parent node and child node in the document. TxtData table records the information of text node and attribute node, ordinal represents the order of child nodes which have the same label path and parent node, nID represents the node's number, type represents the node's type (text node or attribute node), value represents the node's value, GeoData table records the information of the geometry node, type represents the type of the geometry node, shape represents the information of the geometry node. The algorithm of the data in the document tree stored in the database is shown in Algorithm 2. When the document tree in Fig. 2 stored in the database, you can use the node, edge and path information in Table 1 to Table 4 to describe.

3.5   The Experimental Analyzing of GML Document Data Storing Time

We have done the experiments on 24 different GML documents with its size range from 50KB to 10MB, and figured out the time-consuming of storage. The GML documents contains 8 places documents of Point Layer, 8 TOP10relay documents of Line Layer and 8 Data7autodesk documents of Plane Layer. It can be seen from experiments' results, shown in Fig. 3, that the proposed storage method has an obvious advantage on the time-consuming of storage.

4   GML Query Processing

NEM supports non-spatial data query and spatial data analysis and computation. We use GML-QL [8] to write query language. But GML-QL is just a query language used on GML document, and cannot be directly executed on the database, so it must be converted to the corresponding SQL query language. In the paper, Oracle Spatial is used to store GML documents. In order to determine the relationship between grandparent and grandchild nodes quickly, we use a self-defining function UDF_PARENT. Querying-transforming is a complex process [9], and it is closely related to the data storage schema in database [10,11]. The paper defines four tables, separately storing node paths, parent-child relationships between nodes, the paths and the values of text nodes and geometry nodes. The four tables are suitable for the conversion of the XQuery based on path expression and SQL query extended to the database. Through the multi-geometry spatial operation and analysis, GML query queries the non-spatial information which satisfy spatial relations. GML query is shown in the following part.

```
FOR $r IN document("Map.xml")//Road,
    $s IN document("Map.xml")//State
WHERE Cross($r/GeoProperty/gml:LineString,
        $s/GeoProperty/gml:Polygon) == 1
RETURN
<RiverStates>
 <Rid>$r/@gml:id</Rid>
 <sname>$s/SimpleProperty/Name</sname>
</RiverStates>
```

The corresponding conversed SQL query is shown as following:

```
SELECT t1.value, t2.value
FROM TxtData t1, TxtData t2, GeoData t3, GeoData t4,
      Path p1, Path p2, Path p3, p4, Edge e1, Edge e2
WHERE p1.pathExpr LIKE '%/Road/gml:id'
AND p2.pathExpr LIKE '%/State/SimpleProperty/Name'
AND p3.pathExpr LIKE '%/Road/GeoProperty/gml:LineString'
```

```
AND p4.pathExpr LIKE '%/State/GeoProperty/gml:Polygon'
AND e1.cID = t1.nID AND e2.cID = t2.nID
AND t1.type = 'A' AND t1.pathID = p1.pathID
AND t2.pathID = p2.pathID
AND t3.type = 'Line' AND t3.pathID = p3.pathID
AND t4.type = 'Polygon' AND t4.pathID = p4.pathID
AND SDO_RELATE(t3.shape, t4.shape,
'mask=ANYINTERACT querytype=WINDOW') = 'TRUE';
```

5   Experimental Analysis

GML document contains both non-spatial data and a large number of spatial data. Parsing the spatial data and storing the data as spatial data type in the database are key issues of storing GML document [12]. So we use Oracle xmlparserv2 and sdoutl in Oracle Spatial to realize the storing method proposed in the paper. The architecture of GML storage is shown in Fig. 4.

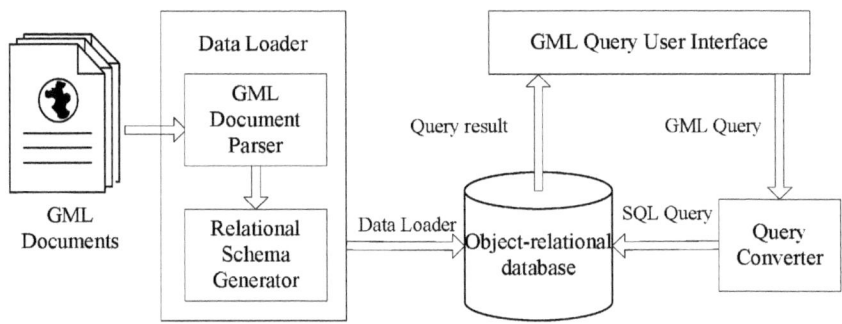

Fig. 4. The architecture of model-driven GML document storage

Data Loader consists of GML document parser and relational schema generator. Parser parses the input GML documents, and constructs document tree. At the same time, DOMParser and JGeometry components are used, DOMParser parses the document as a DOM tree, and JGeometry constructs the geometry nodes in DOM tree as a spatial data object which can be stored directly in database. Relational schema generator traverses the document tree, and separately writes the tree nodes, node direct edges, node path information into the corresponding data table. Query converter converts the input GML query into the SQL query which can run directly in database.

In the experiment, we compare the information in database and documents which contains different sizes and different spatial types. Through the analysis of the data from the instance in ArcView 3.2 and MapInfo 6.0, the analysis of querying time is shown in Fig. 5 and Fig. 6.

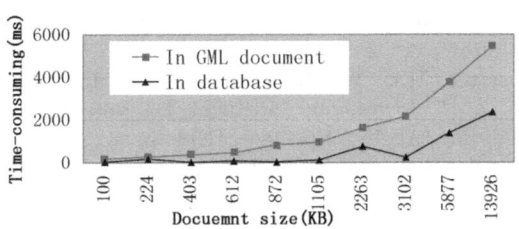

Fig. 5. The same query in GML document and database

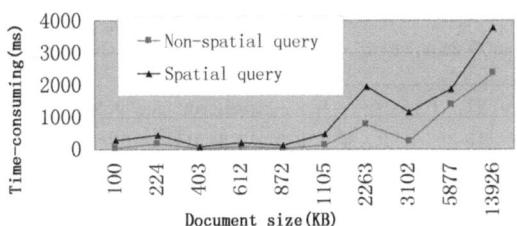

Fig. 6. Spatial query and non-spatial query in database

6   Conclusion

More and more GML documents are created with the wide application of GML technology, and it is important to manage these GML documents. An approach based on node and edge is proposed in the paper in order to store non-schema GML document into object-relational database and model-mapping storage. Firstly document parser converts the GML document into the corresponding document tree, and then constructs schema mapping from GML document to the object-relational database through analyzing and processing the node path, node type, and the parent-child relationship between nodes. The structural information of the document is stored as well. Spatial data analysis and non-spatial data query are the supported through join operations between tables.

References

1. Cox, S., Daisey, P., Lake, R., Portele, C., Whiteside, A.: Geography Markup Language (GML) Implementation Specification, OpenGIS Consortium, version 3.0 (2003)
2. Bohannon, P., Freire, J., Roy, P., Siméon, J.: From XML schema to relations: A cost-based approach to XML storage. In: Proceedings of ICDE, San Jose, California, USA, pp. 64–75 (2002)

3. Florescu, D., Kossmann, D.: A performance evaluation of alternative mapping schemes for storing xml data in a relational database. Technical Report, No. 3680, INRIA, France (1999)
4. Schmidt, A., Kersten, M.L., Windhouwer, M., Waas, F.: Efficient relational storage and retrieval of XML documents. In: Suciu, D., Vossen, G. (eds.) WebDB 2000. LNCS, vol. 1997, pp. 137–150. Springer, Heidelberg (2001)
5. Yoshikawa, M., Amagasa, T.: XRel: A path-based approach to storage and retrieval of XML documents using relational databases. ACM Transactions on Internet Technology 1(1), 110–141 (2001)
6. Jiang, H., Lu, H., Wang, W., Yu, J.X.: Path materialization revisited: An efficient storage model for XML data. In: Proc. of ADC, Melbourne, Victoria, Australia, pp. 85–94 (2002)
7. Córcoles, J.E., González, P.: Analysis of different approaches for storing GML documents. In: Proc. of ACM GIS, McLean, Virginia, USA, pp. 11–16 (2002)
8. Vatsavai, R.R.: GML-QL: A spatial query language specification for GML (2002), http://www.cobblestoneconcepts.com/ucgis2summer2002/vatsavai/vatsavai.htm
9. Krishnamurthy, R., Kaushik, R., Naughton, J.: XML-to-SQL query translation literature: The state of the art and open problems. In: Proc. of XML Database Symposium, Berlin, Germany, pp. 1–18 (2003)
10. Sun, H., Zhang, S., Zhou, J.: XQuery-to-SQL translating algorithm with little dependence on schema mapping between XML and RDB. In: Proc. of CSCWD 2004, Xiamen, China, vol. 1, pp. 526–531 (2004)
11. Grinev, M., Pleshachkov, M.: Rewriting-based optimization for XQuery transformational queries. In: Proc. of IDEAS 2005, pp. 163–174 (2005)
12. Long, W.X., Hu, C.: An Effective Storage Mode for GML Spatial. In: Proc. of Urban Geotechnical Investigation Surveying, pp. 31–35 (2009)

GML Data Management: Framework and Prototype

Weili Wang[1,2], Fabiao Wang[1], Zhiping Qian[1], and Long Zhang[1]

[1] Dept. of Computer Science and Technology, Tongji University, Shanghai, China
[2] School of Info. Eng., Nanchang University, Nanchang, China
{ken.wlwang,wang.fabiao2010}@gmail.com

Abstract. Geography Markup Language (GML) with the feature of XML-encoding and full-formed specification has been widely used in various application systems as internal representation of geo-spatial data for its efficient interoperability in heterogeneous systems. This arises the problem of how to effectively manipulate GML data which is different from pure XML data with both non-spatial information as properties of geographical features and spatial information as geometry. The GML researches focus on separate processing technology which cannot provide a whole geo-spatial data management service for users or developers. In this paper, we propose a GML data management framework and analyze how the related technology cooperate to fill the need of geo-spatial data expressing, processing and exchanging. Storage, query, index and user interface modules of the framework are presented based on GML/XML technology. These modules were implemented in the prototype of GML Data Management System (GDMS).

Keywords: Geo-spatial data, GML, GML data management.

1 Introduction

Geo-spatial data which exists widely in various application systems is textual expressing of objects in space, especially geographical space, including basic geometry like point, line, polygon and geographical object consisted of basic geometry. With the amount of geo-spatial data continuing growing, geo-spatial data organization, query and processing have become the challenge in data management field and have triggered wide and deep research.

Geography Markup Language (GML) is XML-encoding tag language defined by Open Geospatial Consortium (OGC)[1]. As the OGC GML specification was accepted by a mass of industrial companies and research institutions, GML gradually acts as de fact standard in spatial data processing and exchanging, which makes GML data management become mainstream in geo-spatial data management.

Though GML has similar properties as XML in data structure and model, the approaches of XML storage, transformation, query and index cannot be

---

[1] http://www.opengeospatial.org

J. Xu et al. (Eds.): DASFAA Workshops 2011, LNCS 6637, pp. 101–111, 2011.

directly applied to GML for the spatial information existing in GML. GML includes geometry and properties of geographic features, that combine the spatial and non-spatial parts of GML respectively. Non-spatial part is same as XML and can be parsed by the XML technology, but spatial part need be processed by spatial topology technology despite it is expressed as same tag structure as XML. The key of GML processing is how to integrate XML technology with spatial technology. The study of GML concentrates on 3 aspects: storage, parsing and query [1]. Since GML file is textual document, simple files system and text parsing approaches are applicable in GML processing. GML can be parsed by the same methods as XML, like DOM[2] and SAX[3] with the only difference that spatial data nodes in GML are parsed in whole rather than decomposed one by one into nodes or elements. Compared with files system, database is a better alternative mechanism in data storage and has higher performance for its facilities in indexing and querying mass data. GML has different structure from relational model of database, hereby model mapping [2] and schema mapping [3] are two approaches to convert data model in storing GML data into database. Different database structures can be used to store GML data, like native XML database, relational database [4], object-relational database [2,3] and spatial databases [1,5]. In XML query, many query languages were proposed, like Lore [6], XQL [7], XML-GL [8], Quit [9], XQuery[4] where XQuery defined by W3C has become the standard in XML query. MonetDB/XQuery [10], Saxon [11] and BaseX [12] are well known XQuery implementation. With spatial operator extension to XQuery, GML query languages were proposed, and in these languages, GQuery [13] added external spatial operator into XQuery, GQL [14] presented complete data model and definition of GML query and XML/GML prefilter [15] was proposed in building native GML processor. These incomplete related approaches in storage, parse and query cannot be used to provide a whole gml-cored geo-spatial data management service unless they are integrated in a framework and cooperate in answering users requests.

In this paper, we propose a GML organization and search framework for geo-spatial data management. The Framework has three layers, GML storage, processing and interface for users or applications that need geo-spatial data manipulation service. In GML storage, data is stored by alternative way of DB-based and file-based approaches. The model-mapping and schema-mapping in DB-based storage are provided to support user's flexible selection for their application. Besides mapping approaches, two query mechanism are design in processing layer for different storage strategies, database and files system. The framework provides two forms of user interface, GUI and API. The prototype system was implemented based on the proposed framework and can effectively manipulate GML geo-spatial data.

---

[2] http://www.w3.org/dom
[3] http://www.saxproject.org
[4] http://www.w3.org/xquery

The rest of this paper is organized as follows. Section 2 analyzes GML data structure and corresponding GQL query. Section 3 introduces the framework and its modules where the core module Processing Center is described in detail in Section 4. Prototype implementation is given in section 5. Finally, Section 6 concludes the paper and highlights future works.

2 GML Structure and Query Language

2.1 GML Structure and Model

GML is XML-encoding tag language defined by OGC to express geographical information. The latest OGC GML specification[5] has 28 kernel schemas, including basic GML, geometry, topology, coordinate reference system, etc. GML data in specific geographical space is described using pre-defined application schema based on kernel schemas. Fig. 1 shows a GML fragment example about a river named "Cam" in GML file named "cambridge.xml" and its tree model where the subtree "LineString" with labeled star is spatial data node. In this example, the node "River" has two non-spatial text child nodes: "description", "name" and one spatial child node "LineString" where "gml" denotes the prefix of namespace declared with string "http://www.opengis.net/gml".

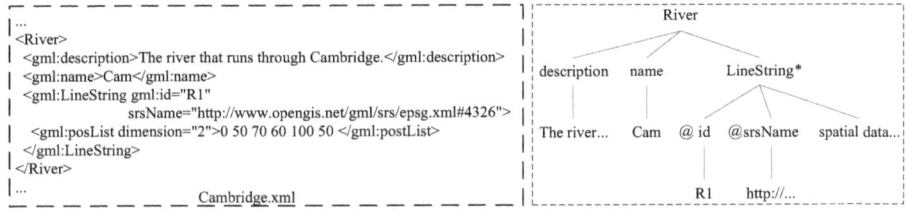

Fig. 1. A GML example and its tree model

2.2 GML Query Language

XQuery is W3C standard to query XML data. GML Query Language (GQL) [14] extends XQuery by adding spatial data model and operators. GQL effectively support to query non-spatial and spatial information in GML document, that makes GQL widely accepted in writing standard GML query expression.

GQL supports spatial operators by adding spatial processing functions into XQuery. The GQL query expression is similar to XQuery expression and fully supports the XQuery syntax, such as FLWR, namespaces, variate declaration, etc. We give an example of a simple GQL query expression on GML file "cambridge.xml" in Fig. 1 as below.

---

[5] http://www.opengeospatial.org/standards/gml

```
for $x in doc("cambridge.xml")//Road
let $y := doc("cambridge.xml")//River
where  $x/number = 11 and
       gmlfn:crosses($x//gml:LineString, $y//gml:LineString) = 1
return
  <RoadInfo>
   <Name>{$y/gml:name/text()}</Name>
   <Length>{gmlfn:length($y//gml:LineString)}</Length>
   <Boundary>{gmlfn:boundary($y//gml:LineString)}</Boundary>
  </RoadInfo>
```

The Query is to search the rivers crossing the road No.11 and return their names, length and boundaries. In the query expression, the functions "crosses", "length", "boundary" with prefix "gmlfn" are predefined spatial process functions described in detail in GQL specification [14].

3   The Framework of GML Data Management

It is essential to integrate various GML technology into a GML-cored geo-spatial data management system for providing geo-spatial data service. We designed a framework for GML data management covering modeling, storage, query and index. Fig. 2 shows the framework we designed which includes four modules: Data Storage, Processing Center, User Interface, External Tools.

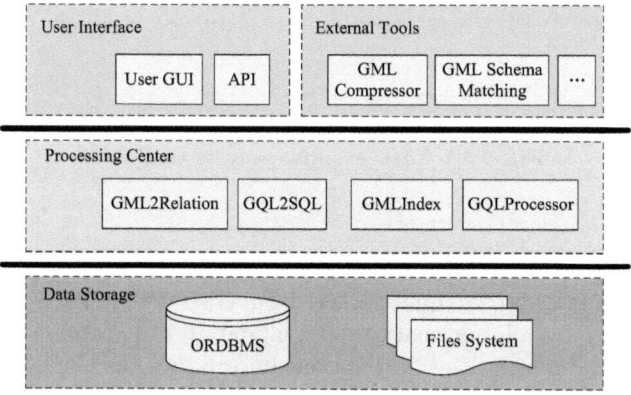

Fig. 2. GML data management framework

Data Storage: we designed two storage strategies for flexibility in the bottom layer of the framework. The framework uses object-relation database system (ORDBMS) and files system to store GML data in bottom layer. The features of ORDBMS can manage GML data efficiently where non-spatial data is stored as relational model and spatial data as objects. In this approach, GML data has to be converted from tree model to object-relational model by invoking GML2Relation in module Processing Center. The second selection that GML data is stored as text document is easy to implement utilizing files system of Operating System (OS) to manage original GML files.

Processing Center: the Processing Center is core module in the middle layer of the framework. It accepts and analyzes the user request from upper layer, then invokes and executes corresponding submodules to generate and return the result back to upper layer. The processing center contains four submodules: GML2Relation, GQL2SQL, GQLProcessor, GMLIndex, that are introduced as below and described in detail in next section.

- GML2Relation: convert GML data into object-relational model by selecting one of two approaches: model mapping and schema mapping;
- GQL2SQL: convert GML query expression written by GQL to SQL and execute the query in database;
- GQLProcessor: execute the GQL query to GML data expressed as text documents;
- GMLIndex: summary GML data and generate index structure for accelerating query execution.

User Interface: at the top layer, the framework provides GUI for terminal user and API for other GML-cored application system. User Interface accepts user request, then delivers the requests downward into Processing Center.

External Tools: the framework also provides tools in the top layer to user to conduct their GML data such as: GML Compressor, GML Schema Matching, etc. The external tools also provides GUI interface and API. The External Tools module is flexible and open, user can define their own processing tools utilizing the provided API.

4   The Processing Center

4.1   Storing GML Data in Object-Relational Database

The first step to parse XML-encoding textual GML data is to read the GML file from OS. IO processing is less efficiency compared with data processing, and the situation may become worse facing mass data. DBMS can guarantee the high efficiency in data management, while GML data conversion is needed. We presented two approaches that convert GML data from nested tags form to relational model. The two approaches-model mapping and schema mapping-are used in the framework to convert GML data.

Model mapping is the method that converts GML to a predefined and fix relational model which is suit for arbitrary GML application schema. XML is nested tag structure, so XML can be considered as combination of tree structure and its value, where tree structure is composed of parent tags, child tags and the edges connecting them, and the value is text or attributes in these tags. GML has the similar form to XML with the difference that GML has spatial value as geometry node which is described using schema and includes some XML nodes. We predefined the relational DB schema to store GML data based on above analysis. The DB schema includes four tables: LabelPath, Edge, TextData and GeoData, and they are described in detail as below.

- LabelPath (pathID, pathExpr): store all paths that is the element label sequence from the root to current node with pathId as the key to identify different path.
- Edge(pID, cID): store the edges connecting parent node and child node where pID and cID denote the parent node ID and child node ID respectively.
- TxtData(pathID, ordinal, nID, type, value): store non-spatial textual data in text node and attribute where pathID denotes its label path, ordinal denotes the order number if there are same label paths, nID is unique global node ID as the key, type is one of two types: text or attribute denoting the different places of non-spatial data occurrence, value is the string value.
- GeoData(pathID, ordinal, nID, type, shape): three models above are designed for conventional XML data storage. The spatial value storage of GML is critical issue of GML management. We designed a similar structure as non-spatial model for consistency with the difference in type and shape fields. The type in GeoData is store the type of geometry, such as point, linestring, polygon. The shape is used to store spatial object converted from textual GML spatial fragment. The mainstream databases like Oracle, DB2, and open-source PostgreSQL provide spatial feature or plugin to store spatial value. So it is not difficult to store spatial information defined as spatial object in ORDBMS.

The part above the dashed line in Fig. 3 shows the mapping process explained above.

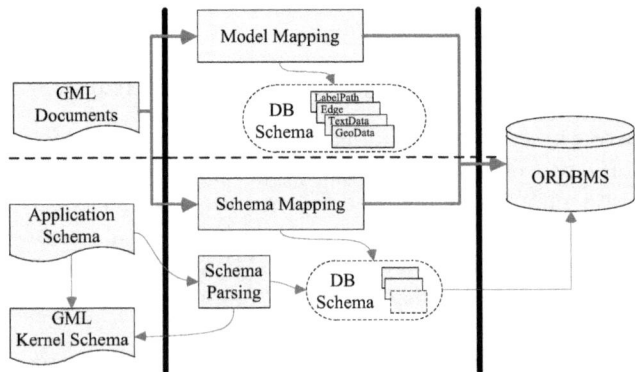

Fig. 3. GML storage: model mapping (above the dashed line) and schema mapping (below the dashed line)

Schema mapping is another method to convert GML to relational model considering its application schema. The conversion process is composed of application schema parsing, relational model structure generation and GML data conversion from text to object-relational data. The part below the dashed line in Fig. 3 shows its process. GML application schema is based on GML kernel schema for the specific field. The schema parser reads the application schema and corresponding GML kernel schema to generate relational model as DB schema to create in ORDBMS, then the GML data is converted to specified data structure to store in ORDBMS according to the created DB schema.

We designed two mapping methods in framework to fill different needs. The schema mapping can retain the semantic information of geo-spatial data stored in GML application schema which is useful in spatial data understanding and mining, but it is less performance in parsing and process GML data for its complicated and varied relational model compared with model mapping by omitting semantic information to define a simple and fix relational model.

4.2   GQL Query in Object-Relation Database

GQL is query language executed on GML data text. Though GQL clauses have the similar semantic meaning as SQL's, they cannot be directly executed in DBMS. The other GML query languages that are mainly implemented by adding spatial operator to XQuery cannot query spatial data stored in DBMS. SQL with spatial extension can query GML data stored in spatial database, like PostgreSQL with additional component PostGIS providing spatial features. So it is essential to convert GQL expression to SQL expression to support mapping from GML query to DBMS query.

We designed the module that convert the GQL given from user interface to SQL used in database in bottom layer, and execute the SQL query in ORDBMS to generate and return result back to user interface. The conversion process in module GQL2SQL has four steps: 1) parse and partition query expression into XPath[6] expressions, set expressions, condition expression, return expression corresponding For, Let, Where, Return clauses in original GQL expression; 2) convert subexpressions into SQL by conversion rules based on specific database schema; 3) execute each SQL expression; 4) assembly the result from each SQL expression, and reconstruct the data into xml-encoding GML text. The detail design and performance analysis will be present on our future paper.

The GQL2SQL module is a preliminary implementation of DB-based GML data query, and the algorithm we designed just cover the GML data in DBMS by model mapping for the relational schema gained by schema mapping is varied and complicated.

4.3   GQL Processor

GQL is based on XQuery, so we design a GQL processor by extending XQuery processor. MonetDB/XQuery [10], Saxon [11], BaseX [12] are well known

---

[6] http://www.w3.org/TR/xpath

implementations of XQuery processors respecting W3C XQuery specification. Since BaseX is open-source and released under the GPL, we chose BaseX from three to build GQL processor by adding spatial features of GQL. We extend the BaseX at two aspects: adding geometry types in type checking and integrating spatial functions in query item processing. We defined a class inherited from "org.basex.query.item.Item" in BaseX classes package to represent geometry type. The class Item in BaseX is used to express current processing item in query. In GQL specification, 12 additional types need be defined, including: Geometry, Coord, Coordinates, Point, LineString, LinearRing, Polygon, Box, GeometryCollection, MultiPoint, MultiLineString, MultiPolygon. Tag "Coord" and "Coordinates" aren't recommended in the latest OGC GML specification of version 3.2.1 and type "GeometryCollection" is the base type of 3 multiply geometry types, so "Coord", "Coordinates", "GeometryCollection" are omitted for simplicity. We defined 8 fields in class "Geometry" to distinguish different virtual types of abstract geometry type.

Spatial operator is implemented using spatial function extended from "org.basex.query.func.Fun" in BaseX classes Package. The class "Fun" is an abstract process of function call. We define 3 classes named "FnGEORelation", "FnGEOSimple", "FnGEOAnalysis" representing spatial relation between two geometric objects, base processing on simple geometric objects (Point, LineString, LinearRing and Polygon) and spatial analysis respectively. Spatial functions use new token "gmlfn" as prefix for differentiating from the prefixes existing in XQuery and BaseX. Functions in 3 classes are showed in Fig. 4, and detail description about every functions in paper [14].

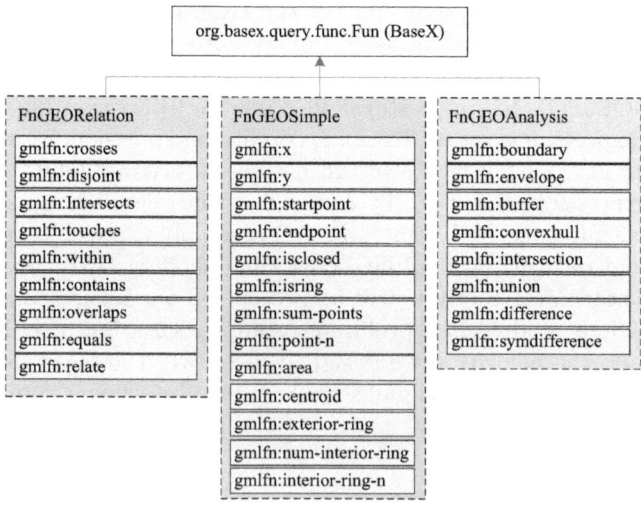

Fig. 4. Spatial processing functions in GQL processor

4.4   GML Indexing

The problem of parsing GML documents stored as files is low performance facing mass GML data. Although processing GML data stored in DBMS has high efficiency utilizing the DMBS features, the conversion is expensive. GML documents query is included in our framework for another selection to avoiding expensive conversion processing. The mechanism improving performance of GML documents query is needed to the problem facing mass GML data.

GML index is the method to accelerate query by directly locating the target data not need traverse all data. We designed a structure indexing GML document by extends XML index with additional spatial index. The GML index structure in our framework is the combination of text index, attribute index, path index and spatial index. The first 2 indexes are same as XML index where text index is used to locate text value and attribute index to attribute name and its value. The path index is similar as XML index with the only difference that the spatial geometry in GML documents is regard as a whole node rather than decomposing into their internal coordinate nodes. The text, attribute and path index is easy implemented with classical B+tree. The last one is specific structure for indexing geographic information of GML. We constructed the spatial index with R+tree.

The process of construct GML spatial index is to sequentially read the whole document and extract all geometry elements, and then wrap them as geometry objects and put into R+tree based on their spatial range. In the process of GQL query execution, some spatial operators can be accelerated with GML index. We firstly query in B+tree with text, attribute or specific path and in R+tree with geometry object or spatial range given by GQL query expression and return the target sets, then we reconstruct the GML documents with the target sets which have omitted the useless parts of the original GML document, finally we execute the GQL query on the reconstructed GML data. The node of the reconstructed GML is no more than original document, so it can decrease the number of items need be processed in the process of query evaluation, especially facing mass GML data.

5   The Prototype

We have implemented prototype system of the framework described above. PostgreSQL[7] is chosen as the ORDBMS to store GML data for the PostgreSQL is open-source and have spatial plugin PostGIS for geometry object processing. In GQL Processor, we extended the open-source XQuery implementation BaseX [12] with JTS Topology Suite[8] to provide spatial predicates and functions. The System is coded using Java and built by JavaSE 6.

The prototype is consisted of several views. In storage view, user can choose the mapping mode, then load the GML documents or its corresponding application schema if need when schema mapping is selected. In search view, user

---

[7] http://www.postgresql.org

[8] http://www.vividsolutions.com/jts

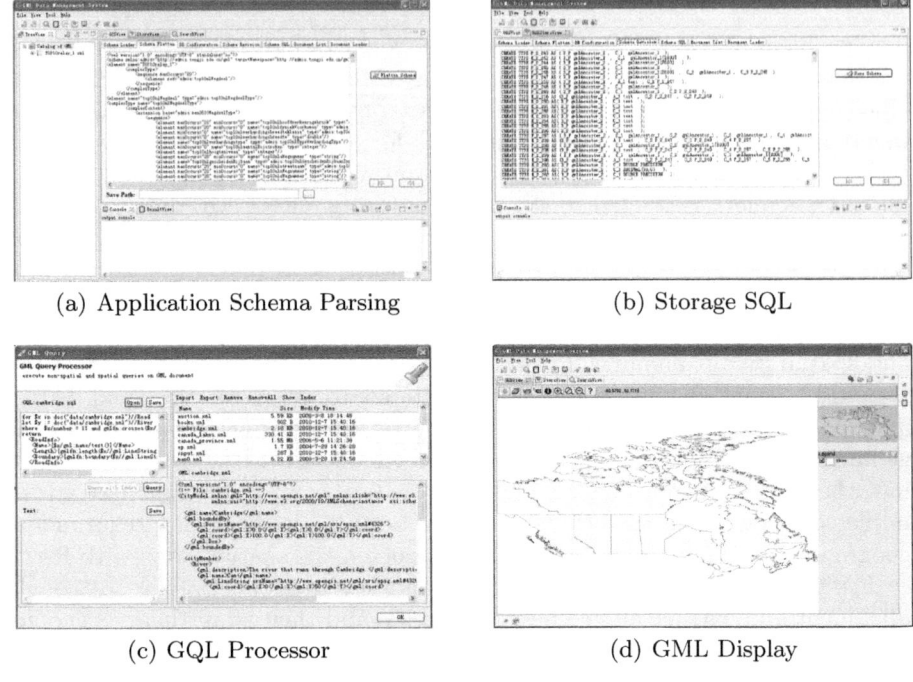

(a) Application Schema Parsing   (b) Storage SQL

(c) GQL Processor   (d) GML Display

Fig. 5. The prototype: (a) application schema parsing in storage view; (b) SQL in DB storage view; (c) GQL processor view; (d) GML display view

can query any GML information stored in DB or in documents with GQL expression. In Display view the geographical information can be returned by query processing on data storage and displayed in GUI as svg-type geographical graphs transformed by user interface. The other views provide different functions, such as database setting, external tools, etc. Fig. 5 shows main views in the implementation of prototype where Fig. 5(a) is application schema parsing view in storage module, Fig. 5(b) is SQL generator view in DB storage, Fig 5(c) is GQL processor view and Fig 5(d) is GML display view.

6   Conclusion and Future Work

GML manipulation become increasingly important for mass using of GML in geospatial data expressing and exchanging. A holistic structure integrating various GML technologies is need to fill the requirements of effective GML geo-spatial data management. In this paper, we proposed a GML data management framework include storage, query, index and user interface. The model-mapping and scheme-mapping in GML storage, conversion from GQL to SQL, GQL processor on GML documents in GML query and GML index structure based on B+tree and R+tree were implemented in prototype of proposed framework. How to efficiently support GML update and construct GML database is the direction of our future work.

Acknowledgment

This work was supported by China 863 Program under grant No. 2009AA01Z135 and the National Natural Science Foundation of China (NSFC) under grant No. 60873040.

References

1. Sripada, L.N., Lu, C., Wu, W.: Evaluating GML support for Spatial Databases. In: COMPASAC, Hong Kong, pp. 74–77 (2004)
2. Zhu, F., Guan, J., Zhou, S.: Storing and Querying GML Documents based on Model-mapping. Journal of Computer Research and Development 43(suppl.) (2006)
3. Zhu, F., Guan, J., Zhou, J., Zhou, S.: Storing and Querying GML in Object-relational Databases. In: ACM-GIS, Arlington, pp. 107–114 (2006)
4. Corcoles, J.E., Gonzalez, P.: Analysis of Different Approaches for Storing GML Documents. In: ACM-GIS, McLean, pp. 11–16 (2002)
5. Li, Y., Li, J., Zhou, S.: GML Storage: A Spatial Database Approach. In: Wang, S., Tanaka, K., Zhou, S., Ling, T.-W., Guan, J., Yang, D.-q., Grandi, F., Mangina, E.E., Song, I.-Y., Mayr, H.C. (eds.) ER Workshops 2004. LNCS, vol. 3289, pp. 55–66. Springer, Heidelberg (2004)
6. Abiteboul, S., Quass, D., Mchugh, J., et al.: The Lorel Query Language for Semistructured Data. International Journal on Digital Libraries 1(1), 68–88 (1997)
7. Robie, L.J., Schach, D.: XML Query Language (XQL). In: The Query Languages Workshop, Boston (1998)
8. Ceri, S., Comai, S., Damiani, E., et al.: XML-GL: A Graphical Language for Querying and Restructuring WWW Data. In: WWW, Toronto, pp. 1171–1188 (1999)
9. Chamberlin, D., Robie, J., Florescu, D.: Quilt: An XML Query Language for Heterogeneous Data Sources. In: WebDB, Dallas, pp. 1–25 (2000)
10. Boncz, P., Grust, T., van Keulen, M., Manegold, S., Rittinger, J., Teubner, J.: MonetDB/XQuery: a Fast XQuery Processor Powered by a Relational Engine. In: SIGMOD, Chicago, pp. 479–490 (2006)
11. SAXONICA XSLT and XQuery Processor, http://saxonica.com
12. BaseX XPath/XQuery processor, http://www.inf.uni-konstanz.de/dbis/basex
13. Boucelma, O., Colonna, F.M.: Querying GML Data with GQuery. Technical report (2003)
14. Guan, J., Zhu, F., Zhou, J., Niu, L.: GQL: Extending XQuery to Query GML Documents. Geo-spatial Information 9(2), 118–126 (2006)
15. Huang, C., Chuang, T., Deng, D., Lee, H.: Building GML-native Web-based Geographic Information Systems. Computers & GeoScineces 35(9), 1802–1816 (2009)

An Efficient Multi-layer Grid Method for Skyline Queries in Distributed Environments

He Li, Sumin Jang, and Jaesoo Yoo

Department of Information and Communication Engineering,
Chungbuk National University, Cheongju, 361-763, Korea
{lihe,jsm,yjs}@cbnu.ac.kr

Abstract. The skyline query has been received much attention as an important operator in database systems for multi-preference analysis and decision making. Most of the previous works have focused on processing skyline queries on centralized data sets. However, the related data of real applications are practically scattered at several different servers. The skyline query computation in distributed environment is needed to gather a large number of data from the connected servers. The existing methods for a skyline query in distributed environment have two problems: (i) They have slow processing time for a skyline query. (ii) Most of the transferred data among servers in the network are unnecessary. In this paper, we propose a multi-layer grid method for efficiently processing skyline queries in distributed environments (MGDS). The proposed method minimizes the unnecessary transferred data using the grid mechanism. Experiments based on various data sets show that our proposed method outperforms the existing methods.

Keywords: Skyline query, distributed skyline query, grid method, distributed data.

1 Introduction

In the recent years, the interests on skyline query processing have been significantly increased since the skyline results can be used in many applications with multi-dimensional data set. Given a data set D containing objects $D=\{p_1, p_2 \ldots p_n\}$, the skyline operator returns all objects p_i such that p_i is not dominated by another object p_j.

Most of the previous skyline literatures [1], [2], [3] have primarily focused on providing efficient skyline algorithms on centralized data set. In practice, however, the vast numbers of independent data are often collected from multiple sources that stored in distributed servers. Figure 1 shows an example of skyline query in distributed environment. Assume a set of distributed servers $\{S_1, S_2, S_3,$ and $QS\}$. Each server stores a set of tuples that is a fraction of the entire data set. The skyline query is initiated by users through query server (QS). The result of the skyline query is evaluated by gathering all of local data from the connected servers. A real-life example of the skyline queries in distributed environments is the online comparative shopping, in which a user needs to get good bargains from many different shopping sites according to

J. Xu et al. (Eds.): DASFAA Workshops 2011, LNCS 6637, pp. 112–119, 2011.
© Springer-Verlag Berlin Heidelberg 2011

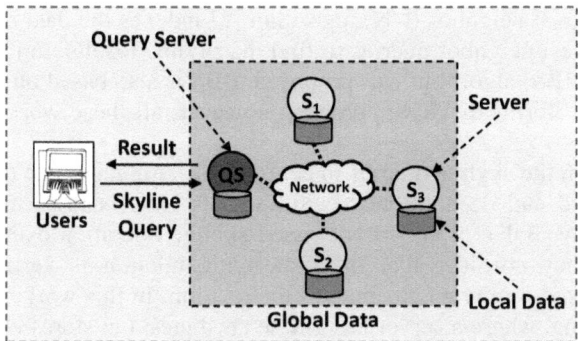

Fig. 1. The skyline queries in distribute environments

multiple criteria like price, quality, guarantee, etc. Such multiple criteria can be captured by a skyline query. In such cases, directly applying existing techniques would incur large overhead. Then, a distributed architecture is needed.

A naïve approach to process the distributed skyline queries is to send the skyline queries to all of the connected servers which in turn process the skyline queries locally and report the results to QS. The QS merges the received results and evaluates the global skyline result. This approach needs to transmit and process an excessive unnecessary data which is local skyline data but not global skyline data. Joao B. et al. in [4] proposed a grid-based strategy for distributed skyline query processing (AGiDS), which use a grid-based data structure to capture the data of each server. Instead of sending the local skyline data, the local cell information which contains local skyline data are firstly sent to the QS. The dominated local cells at QS are eliminated. Then only the local skyline data within the non-dominated cells are transferred to the QS. However, if the cells of the local servers transferred to QS are overlapped, lots of unnecessary data need to be processed.

In this paper, we propose a multiple layer grid method for skyline queries (MGDS) in the widely distributed environments. The proposed method assumes that each server shares a common grid structure. The QS first gathers the cell information which contains local skyline data. If the cell information is overlapped, we propose to generate a multiple layer grid based on the overlapping cells. The dominated cells of the multiple layers grid are eliminated. According to the multiple layers grid mechanism, more unnecessary local data are filtered out before transferred to the query server for processing.

The remainder of the paper is organized as follows. Section 2 surveys the previous related works. Section 3 presents the details of the proposed method. Section 4 contains an experimental evaluation that demonstrates the superiority of our proposed MGDS method. Finally, Section 5 concludes this paper.

2   Related Work

In [1], Borzsonyi et al. first introduced the skyline operation in database systems and proposed two solutions based on Block Nested Loop (BNL) and Divide and Conquer

(D&C). The nearest neighbor (NN) algorithm [2] indexes the data set with an R-tree. NN utilizes nearest neighbor queries to find the skyline results. In [3], the branch and bound skyline (BBS) algorithm was proposed. BBS is also based on nearest neighbors search and outperforms the NN approach. However, all these works assume centralized data storage.

Different from the skyline queries in centralized setting, skyline queries processing in the distributed and decentralized environments have been received considerable attention recently. Balke et al. [5] addressed skyline operation over multiple distributed sources, they consider that the underlying relation is vertically partitioned, i.e. each server keeps only an attribute of the relation. In this work, we focus on horizontal partitioning, where a server has all the attributes, but stores only a subset of all the tuples. Wang et al. [6] developed a skyline space partitioning (SSP) approach to compute skyline on a tree-structured p2p platform BATON. For this method, a server cannot freely decide the tuples in its own storage. Our techniques allow arbitrary horizontal partitioning. Cui et al. [7] study skyline queries in a distributed environment. They propose the use of MBRs (Minimum Bounding Regions) to summarize the data stored at each server. According to the MBRs of all servers, incomparable groups are assigned. The skyline is computed within each group using specific plans. In [8], a feedback-based distributed skyline (FDS) algorithm is proposed, which computes skyline in the no particular overlay network with economical bandwidth cost. The FDS algorithm is bandwidth efficient as the querying computer transmits to each server the precious information that prevents the delivery of a large number of non-skyline points. However, it requires several round-trips to compute the skyline, which incurs high response time. Joao B. et al. in [4] proposed a grid-based strategy for distributed skyline query processing (AGiDS). The response time of AGiDS is fast as it adopts the parallel computing over the distributed servers. However, if the cells of the local servers transferred to QS are overlapped, this method cannot efficiently reduce the unnecessary local data that are transferred from local servers for processing the global skyline.

3   The Proposed Method

3.1   Motivation

As mentioned before, when the cells of local servers transferred to the QS are overlapped, the AGiDS method leads to the transmission of unnecessary data. Therefore, we propose a new multi-layer grid method which processes skyline queries in distributed environments (MGDS). We assume that each server shares a common grid and the grid can include the entire data set. Given a set of distributed servers S= {S_1, S_2, S_3..., S_i}. Each server S_i stores a set of tuples that is a part of the entire data set and has the capability of computing the local skyline set based on the stored data points. The server who produces a skyline query is called query server (QS). Without loss of generality, we assume that smaller values are preferred in the skyline operator. In order to evaluate skyline queries efficiently, the proposed MGDS method uses three kinds of dominance relationships among the cells of grid. We consider that each cell of a 2-dimensional grid has a lower left corner coordinate value and a top right corner coordinate value.

Three kinds of dominance relationships among cells of grid

- $cell_i$ is dominated by $cell_j$, if the lower left corner coordinate of $cell_i$ is dominated by the top right corner coordinate of $cell_j$.
- $cell_i$ is overlapped with $cell_j$, if the lower left corner coordinate of $cell_i$ equals the lower left corner coordinate of $cell_j$.
- $cell_i$ dominates $cell_j$, if the top right corner coordinate of $cell_i$ dominates the lower left corner coordinate of $cell_j$.

If $cell_i$ is dominated by $cell_j$, which means that all the data points of $cell_i$ is dominated by any data point of $cell_j$. We define the cells which contain skyline data as region-skyline, as shown in Figure 2, the shaded area. If the region-skyline overlaps at different servers, we define these regions as overlap region-skyline, e.g. the cell A and B of Figure 2. If the overlap region-skyline contains more data points (*rCount*) than the predefined threshold value k (the value of k is defined according to the practical application), it is defined as hot overlap region-skyline, e.g. the cell B of Figure 2.

3.2 The Processing of MGDS Algorithm

The proposed MGDS method is comprised of three basic stages: planning, analyzing and execution. At the beginning of the planning stage, a skyline query can be initiated by a query server (*QS*). When receive the skyline query, each server S_i computes its local region-skyline by using an existing grid algorithm. The *QS* contacts all the connected servers and obtains the region-skyline information. In the analyzing phase, the received cells are analyzed and the global region-skyline is evaluated at *QS*. If the hot overlap region-skyline is occurred at the connected servers, it can be handled by creating an upper layer grid. As shown in Figure 2, if cell B is hot overlap region-skyline both at server S_1 and S_2, it is converted to an upper layer grid. The data points of cell B are managed by the upper layer grid and the local region-skyline of the upper layer grid is computed. Then, *QS* requests the local region-skyline of the upper layer grid and computes the global region-skyline. If the hot overlap region-skyline still exists on the upper layer grid, more layers grid can be generated. Notice that, though cell A is a overlap region-skyline both at S_1 and S_2, it is not converted to a upper layer grid, this is because there are only a small amount of data in cell A both at S_1 and S_2, and processing these data is efficient than processing the upper layer grid. In the execution stage, only the data points existed in the global region-skyline of each server are requested for the final global skyline computation. Since most of unnecessary local skyline data points are filtered out, the MGDS method reduces both the communication and processing cost.

Next, we illustrate the MGDS method by means of an example depicted in Figure 2. Assume that each server has a 2-dimensional data set and the grid consists of 3*3 cells. In the planning stage, the cells A, B, D of S_1's grid and A, B, C of S_2's grid are evaluated as local region-skyline and sent to the *QS*. In the analyzing stage, the received local region-skylines which dominated at *QS* are eliminated. Because cell B is hot overlap region-skyline, it needs to be converted to an upper layer grid, the data within cell B is managed by the upper layer grid and the local region-skyline B-1, B-2, B-3, B-4 and B-5 of the upper layer grid is evaluated at S_1 and S_2 respectively.

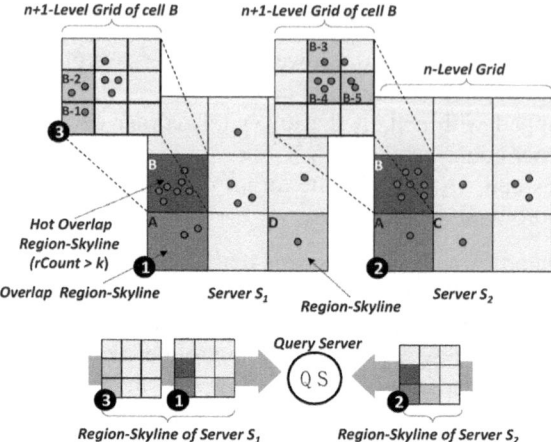

Fig. 2. The processing of the proposed MGDS method

The QS gathers the local region-skyline of the upper layer grid and evaluated the global region-skyline. As the cells B-3, B-4 and B-5 of S_2's grid are dominated by the cells B-1 and B-2 of S_1's grid which is not global region-skyline. Therefore, in this example, the global region-skyline is cells A, D, B-1, B-2 of S_1's grid, and cells A, C of S_2's grid. In the execution stage, only the local skyline data within cells A, D, B-1, B-2 of S_1's grid, and cells A, C of S_2's grid are transferred to the QS for final global skyline computation.

4 Experiment Evaluation

4.1 Experimental Environment

In this section, we study the performance of our proposed method. We present the experimental results comparing AGiDS method [4] and the naive method with our proposed MGDS method. We conducted our experiments on a desktop PC running on Windows XP professional. The PC has an Intel Core2 Duo 2.66GHz CPU and 1GB memory. All of the experiments were coded in Java. Table 1 shows the parameters for experiments evaluation.

In this experiment, there are two critical types of data distributions that stress the effectiveness of skyline methods have been used, the anti-correlated and uniform data set. The anti-correlated data represent that the data points that have a high value in one dimension and have a low value in one or all of the other dimensions. And the uniform data set is the data distributed in the arbitrary work space. All of the experiments are evaluated based on these two synthetic data sets.

4.2 Experimental Results

We consider that all of the distributed servers are connected and each server posses an equal number of data points, the data points distributed in 30% range of the grid at each server. In the first experiment, the performance is measured by the amount of data transmitted over the network. We examine the performance of the methods by varying the number of data tuples from 10K to 200K at each server and the dimension of the data is set to 2. Figure 3 shows the results for the amount of transmitted data with respect to the number of data tuples. The skyline queries in anti-correlated data set transmit more data than in uniform data set. The reason is that the skyline data of anti-correlated data set is larger than the uniform data set. The MGDS method outperforms the AGiDS method and the naive method since more non-promising data points are filtered out by the multiple layer grids mechanism.

Figure 4 shows the response times of the three schemes according to the data size. In this experiment, the dimension of the data is set to 2 and the data size is varied from 10,000 to 200,000 at each connected server. From the results we can see that, the response time increases sharply when the data tuples are increased. This is because the increasing data size needs high processing cost which prolong the processing time.

Table 1. The parameters for experiments evaluation

Parameter	Values
The number of dimensions	2~5
The number of servers	10
The amount of data at each server	10K~ 200K
The size of the grid	10*10

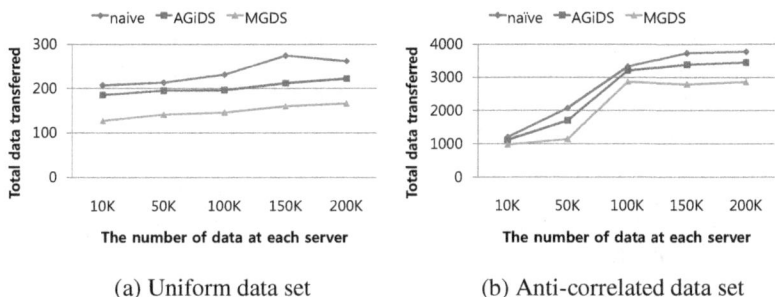

(a) Uniform data set   (b) Anti-correlated data set

Fig. 3. Evaluation of the total transferred data for various data size

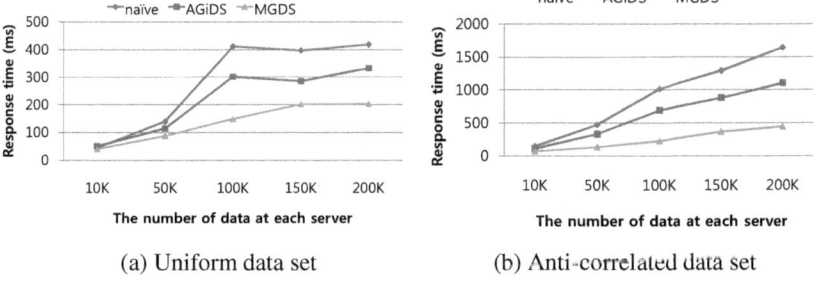

(a) Uniform data set   (b) Anti-correlated data set

Fig. 4. Evaluation of the response time for various data size

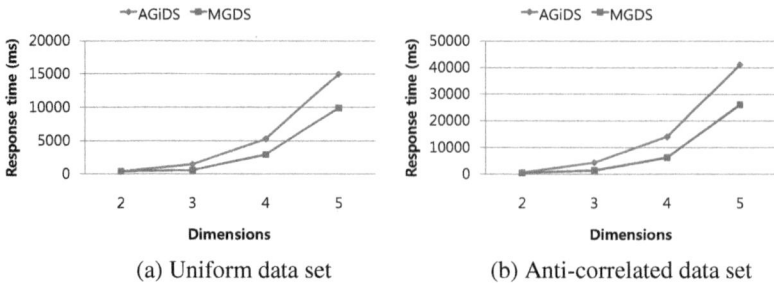

(a) Uniform data set   (b) Anti-correlated data set

Fig. 5. Evaluation of the response time for various data dimensions

The response time of MGDS is better than the response time of naïve and AGIDS for anti-correlated data and uniform data. This is because less data is processed.

As shown in Figure 5, the response time of the AGiDS method and MGDS method are compared according to the number of dimensions. In this experiment, the data size is set to 10,000. We study the results of the different methods with increasing the dimensions of the data set at each server. The response time of AGiDS increases more rapidly than the MGDS method when the dimensions increase. The reason is that with the increase of the dimensions more data points need to be processed.

5 Conclusions

This paper studies skyline queries over the horizontally partitioned data set. As the relevant data are scattered at several servers, the skyline query in distributed environment requires gathering a large number of data from the distributed servers that connected by the network. The proposed MGDS method employs a multi-layer grid mechanism to process the distributed data. It can filter out much of the unnecessary data which need to be transmitted to the querying server for final skyline result computation. The experimental results have shown that the proposed method is efficient than the existing methods.

Acknowledgments. This work was supported by the Ministry of Education, Science and Technology Grant funded by the Korea Government" (The Regional Research Universities Program/Chungbuk BIT Research-Oriented University Consortium).

References

1. Borzonyi, S., Kossmann, D., Stocker, K.: The skyline operator. In: 17th International Conference on Data Engineering, pp. 421–430. ICDE Press, Heidelberg (2001)
2. Kossmann, D., Ramsak, F., Rost, S.: Shooting stars in the sky: An online algorithm for skyline queries. In: 28th International Conference on Very Large Data Bases, pp. 181–184 (2002)
3. Papadias, D., Tao, Y., Fu, G., Seeger, B.: An optimal and progressive algorithm for skyline queries. In: ACM-SIGMOD International Conference on Management of Data, pp. 467–478. ACM Press, San Diego (2003)

4. Rocha-Junior, J.B., Vlachou, A., Doulkeridis, C., Norvag, K.: AGiDS: A Grid-based Strategy for Distributed Skyline Query Processing. In: Data Management in Grid and Peer to Peer Systems, Second International Conference, Globe 2009, Linz, pp. 12–23 (2009)
5. Balke, W.T., Güntzer, U., Zheng, J.X.: Efficient distributed skylining for web information systems. In: Hwang, J., Christodoulakis, S., Plexousakis, D., Christophides, V., Koubarakis, M., Böhm, K. (eds.) EDBT 2004. LNCS, vol. 2992, pp. 256–273. Springer, Heidelberg (2004)
6. Wang, S., Vu, Q.H., Ooi, B.C., Tung, A.K.H., Xu, L.: Skyframe: A framework for skyline query processing in peer-to-peer systems. VLDB Journal 18, 345–362 (2009)
7. Cui, B., Lu, H., Xu, Q., Chen, L., Dai, Y., Zhou, Y.: Parallel distributed processing of constrained skyline queries by filtering. In: 24th International Conference on Data Engineering, ICDE 2008, Cancun, pp. 546–555 (2008)
8. Zhu, L., Tao, Y., Zhou, S.: Distributed skyline retrieval with low bandwidth consumption. In: TKDE, pp. 384–400 (2009)

3D Indoor Route Planning for Arbitrary-Shape Objects

Wenjie Yuan and Markus Schneider*

Department of Computer & Information Science & Engineering
University of Florida
Gainesville, FL 32611, USA
{wyuan,mschneid}@cise.ufl.edu

Abstract. Route planning, which is used to calculate feasible routes in a given environment, is one of the key issues in navigation systems. According to different constraints in different given space, various route planning strategies have been developed in recent years. Current route planning models for indoor space focus on providing routes for pedestrians or fix-sized users, like robots and persons in wheelchairs. None of the existing model can provide feasible routes for arbitrary-shape users, which appears to be more and more useful in many situations, like users driving small indoor autos or moving carts with products. This paper proposes a two-phase route planning model which can support route planning for users with arbitrary shapes. In the first phase, the LEGO model represents the entire space by using different types of cubes. These cubes are further merged in the second phase to form the maximum accessible blocks. By computing the maximum accessible widths and lengths between blocks, a LEGO graph is built to perform route searching algorithms.

1 Introduction

A navigation system consists of two main parts: localization and route planning. Localizations refer to the determination of the locations with the aid of some Equipment. Route planning strategies are used to compute feasible routes between two specific locations. There are a lot of route planning strategies used in different route planning models. The most important feature they have to have is to provide the user a feasible route so that the user can go to the desired place without colliding with any obstacles in the given space.

The design of route planning strategies depends on multiple constraints in the given environment. One of the most common constrains is the structure of the environment. For example, outdoor space has the network structure while indoor space is based on cells. Therefore, route planning for outdoor space is different from the one applied to indoor space. The type of users is another vital

---

* This work was partially supported by the National Science Foundation under grant number NSF-IIS-0915914.

J. Xu et al. (Eds.): DASFAA Workshops 2011, LNCS 6637, pp. 120–131, 2011.

constraint. From one location to another, the routes provided to vehicles and to pedestrians are may be different. In indoor space, most route planning models are designed for pedestrians only. They approximate pedestrians into points without considering their volumes. However, besides pedestrians, there are some other kinds of users, like persons in wheelchairs, small indoor autos, robots and carts carrying products. These types of users cannot simply be approximated into points because their volumes may affect the accessibility in indoor space. However, to the best of our knowledge, there is no route planning model that can support users with arbitrary-shapes.

The purpose of this paper is to propose a model that can provide feasible routes for users with arbitrary shapes in indoor space. Our solution consists in a two-phase method that includes a representation phase followed by an accessibility checking phase. In the first phase, a LEGO-based representation model is proposed to efficiently represent the 3D structure as well as all the obstacles in the indoor space. The entire space is approximated by several LEGO cubes. All the cubes have the same-sized basal area, and each of them has its own height and type according to the object it represents. In the second phase, LEGO cubes are merged into blocks that can be used to evaluate the maximum accessible space constrained by obstacles. At last, a LEGO graph is built to support shortest path search algorithms.

The rest of the paper is organized as follows. Section 2 discusses the existing approaches of indoor navigation models. Section 3 introduces the LEGO representation model we used to represent the indoor space. The accessibility checks for arbitrary-shape objects are discussed in Section 4. In Section 5, a LEGO graph is introduced to support efficient path searching algorithms. Finally, Section 6 draws some conclusions and depicts future work.

2   Related Work

The existing route planning models can be classified into two main categories: path-based models and grid-based models.

Most of the models designed for pedestrians are path-based models. In these models, users are approximated by moving points. The earlier models [1,2] use center points to represent cells and build routes based on the reachability of the cells. Since they do not consider architectural constrains, the generated routes are very coarse, and may be circuitous. *CoINS* model in [3,4] simplify the route by eliminating some unnecessary nodes and recalculating the segments between different nodes. Later, models take more architectural constrains into account (see [5,6,7,8,9]). The model proposed in [5] captures the relationships between the cells and the exits. After that, it organizes the relationships between cells and exits in a hierarchical structure according to their reachability. The model introduced in [7] employs some representative points to represent rooms, corridors and exits. The calculation of the path is processed based on the connections between these representative nodes. Later, in [8], this model is extended by decomposing cells into several convex regions to provide users better route instructions.

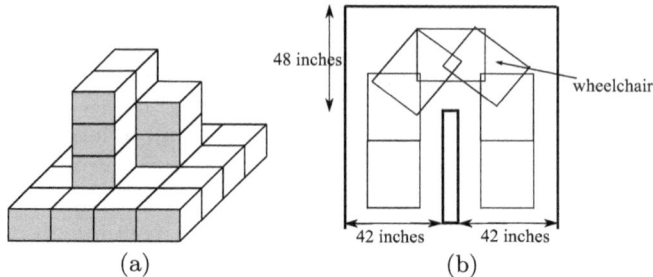

Fig. 1. Representing space by using cubes (a)

However, since it uses points to represent cells, the generated routes are not guaranteed to be the shortest paths. The iNav model proposed in [25] considers the shapes and all the exits of cells and develops a novel strategy to find the shortest path between two exits. The above mentioned models are all 2D models. Lee proposes a 3D model in [10,11]. In this model, Poincaré duality combined with a hierarchical network structure are used to explore the relations between objects. One drawback of this model is that the 3D information is only used to distinguish different floors. In each floor, the route planning still focuses on 2D information. The same problem happens in the models in [12,13]. The model in [14] is used to generate the evacuation routes. This model takes into account different features of the interiors, such as the types of the passing (e.g. *uni-* or *bi-directional*) and the types of the boundaries (e.g. persistent boundaries like walls and virtual boundaries like openings). By using these features, this model is able to distinguish the accessible parts and non-accessible parts in the indoor space. The drawback of this model is that it focuses on the surroundings, but ignores the structure of the floor plane.

The grid-based models are more suitable for fix-size users. They usually decompose the space into cells and compute routes by exploring the connectivity of these cells. Most of the grid-based models [15,16,17] represent the available space by unified shapes (e.g., rectangles). The union of the generated cells may not be exactly the available space, especially for the space on the boundaries. However, since they use simple and unified representative units, they are usually more efficient for the route planning. The model proposed in [15] is one of the most popular grid-based models. It represents indoor space by equal-size cells marked as *obstacles* and *non-obstacles*. Available routes are computed by checking the availability of the movements from cells to their 8 neighbors. His model also support navigation in 3D space by filling out the indoor space with the *obstacle* and *non-obstacle* cubes (as shown in Figure 1a). The obstacle cubes are further classified into insurmountable and surmountable ones to facilitate the 3D navigation. In [17], topological maps are formed by merging cells into a hierarchical structure. This model is useful when the number of cells is large. The model proposed in [18] discussed the accessibility of wheelchairs in indoor space. It computes the minimum requirements for a wheelchair to make turns and provide possible routes for them. However, this model is only suitable

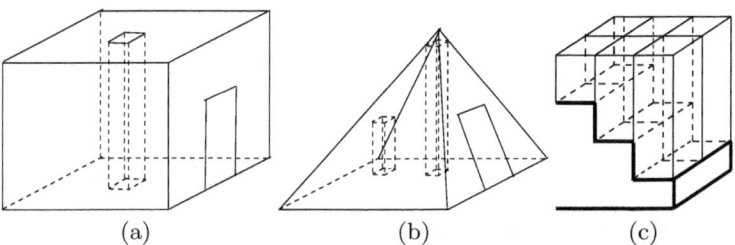

Fig. 2. A cube in a cell with regular shape (a), cubes in a pyramid shaped cell (b), and cubes representing stairs (c)

for common sized wheelchairs. The minimum requirements are predetermined, not dynamically computed according to the size of the objects (as shown in Figure 1b.

3   The LEGO Model

The indoor environment becomes more and more complicated in recent years. Therefore, the representation models designed for route planning in indoor space are required to not only efficiently handle a large number of data, but also benefit route planning approaches. In [19], we proposed a data model, called LEGO model, to represent the 3D structure in indoor space and support route planning.

In the LEGO model, the entire indoor space is represented by several cubes. These cubes have the same size of the basal area, but their heights and types may different depending on the objects they represent. Figure 2a and b show two examples of the cubes in indoor space, and the heights of the cubes in Figure 2b are different depending on the distances between the floor and the ceiling. In our model, according to the objects they represent, cubes are classified into three categories: *plane_cubes*, *stair_cubes* and *obstacle_cubes*.

The cubes used to represent planes in indoor space are called *plane_cubes*. When a floor and a ceiling are flat, and there is no obstacle between them, the accessible space between the floor and the ceiling will be represented by a cube whose height is the distance between the floor and the ceiling (as shown in Figure 2a). When a floor or a ceiling is not flat, the available space can be approximated into multiple LEGO cubes with different heights (as shown in Figure 2b).

The cubes used to represent stairs are called *stair_cubes*. Similar to the approximation of the sloping planes, a stair is represented by a set of accenting or descending LEGO cubes (as shown in Figure 2c).

The cubes used to represent obstacles are called *obstacle_cubes*. Obstacles refer to the objects whose occupied areas are not available for users. They can be walls, tables, chairs and other objects. If the obstacle is too high to be passed over, the cubes representing it will reach the ceiling, which means the space from

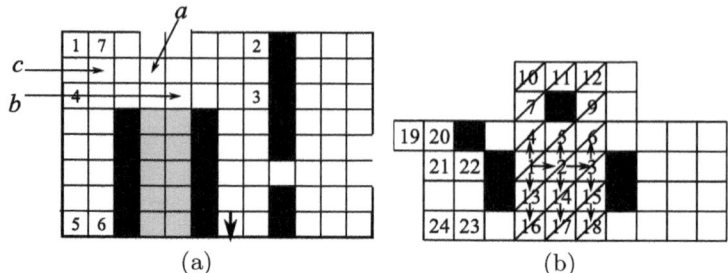

Fig. 3. Examples of merging cubes to generate larger blocks

the floor to the ceiling in this location is unavailable. If the obstacles are so low that pedestrians can pass over them, they are represented by stair_cubes instead. Some obstacles are lying in the air, and the spaces below them are available for users. This kind of obstacles will limit the height of the available space below them. The space from the floor to the ceiling in this area will be represented by two different LEGO cubes. The bottom one represents the available space and the upper one represents the obstacle. The top area of the bottom cube will be the basal area of the upper cube.

4   Checking the Accessibility for Arbitrary-Shape Objects

The goal in our paper is to provide feasible routes for arbitrary-shape users. The accessibility is affected by multiple facts like the walls, the exits and the obstacles inside rooms. We will introduce how we check the accessibility of users' widths, heights and lengths.

4.1   The Maximum Widths

In order to check the accessibility of the widths, we have to find the maximum available widths in any places. The maximum widths in different places are restricted by obstacles. For example, Figure 3a is the 2D projection of a cell represented by LEGO_cubes. The white, black and grey cubes represent the available space, obstacles and stairs respectively. From the figure, we can learn that the maximum accessible horizontal width in the location of the cube ab and c is the same. This maximum width can be obtained by merging the plane cubes in horizontal direction until we meet an obstacle cube. The merging approach is introduced in [19]. Due to the space limit. The details are not discussed in this paper. Interested readers are referred to [19].

The result of the merging process is a set of blocks. For each block, there will be at least one obstacle cube beside each side of its boundary. As shown in Figure 3b, the rectangle (4,6,18,16) is one generated block. It is the maximum block in the corresponding location because it is impossible to further extend its boundary to form a lager rectangle.

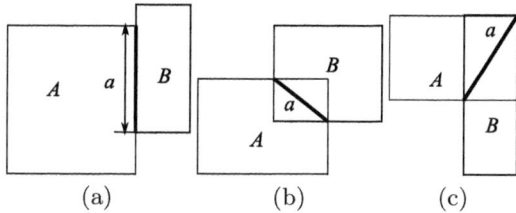

Fig. 4. Construct connectors according to the relationships between blocks

There are three possible relationships between two blocks: disjoint, adjacent and overlap. For adjacent and overlap blocks, the maximum accessible widths between them are determined by their connecting area (called *connectors*). The accessible width of a connecting area is different from the accessible width of a block. The maximum accessible width of a block is determined by two connecting part (the one entering the block and the one exit the block). For example, in Figure 4c, although the connector in bold is wider than the minimum width of the block B (the bottom side of B), if we want to go through block B from its bottom side, the maximum width will be restricted by the connector on the bottom side.

4.2   The Maximum Heights

The accessible height is the second condition we have to check. Although the heights of LEGO cubes can reflect the heights of the available space inside cells, the accessible height of a cell is actually controlled by the heights of the exits. Assuming that the default height of one cell is the maximum height among all its exits, the area higher than the default height can be accessed without restriction. Therefore, in our merging process, the LEGO cubes higher than the default height will be merged together. For the LEGO cubes lower than the default height, only the cubes of the same height can be merged together. This makes sure that all the cubes in one block either have the same height, or higher than the default height. For the former case, the height of the cubes is the height of the block, and for the latter one, the default height becomes the height of the block. For each pair of adjacent or overlap blocks, the maximum accessible height is the minimum height of the two blocks.

4.3   The Maximum Lengths

The process of checking accessible lengths is very complicated. Turns, obstacles and user's volumes all have impact on the maximum accessible length (examples are shown in Figure 5a, b and Figure 6a). In our model, we propose an approach to provide users a feasible way to their destinations. The generated route may be not optimal, but it is guaranteed to be feasible.

There are several reasons why we cannot provide the optimal ways. First, the shapes of users may be different. It is hard to check the availability for every

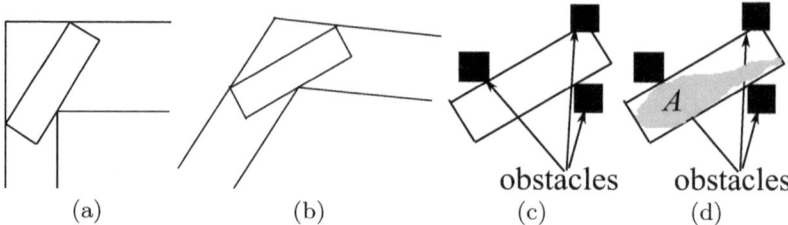

(a)   (b)   (c)   (d)

Fig. 5. Examples showing difficulties to find the optimal routes

part of the object in all places. For example, in Figure 5c, the rectangle cannot pass though these obstacles. However, in Figure 5d, although object A shares the same rectangle, it is able to go through the path. Even if we can find a way to check the accessibility for every part, it is inefficient and unpractical to do that. Second, in real world scenario, users may prefer *comfortable* ways rather than the optimal one. For example, it will not be a realistic route if users have to make several tries to find the right angle to make a turn. Thus, it is better to provide a route with enough space.

As we know, users' movements can be varying. However, if we decompose the movements into small steps, they can be classified into two main categories: going straight and making turns. To check the accessibility of the length for a straight path, we only need to compare the length of the object with the length of the straight path. However, checking the accessibility for turns is difficult. In our paper, we will first introduce an approach that can check the accessible lengths for all general cases in indoor space. Then refine the approach for special cases.

Let's start from a simple case. Figure 6a shows a typical corridor turn. One important observation is that if the minimum bounding circle of the object can be contained in the turning corner (the grey area), this object is able to make the turn. This *corner* concept can be applied to our LEGO representation model. In Section 4.1, we have discussed how to find the maximum accessible widths by merging cubes into larger blocks. Any two blocks may be disjoint, adjacent or overlap (as shown in Figure 4). Users may need to make turns only when they are going from one block to another through the corresponding connector.

The scenario for overlap blocks (as shown in Figure 4c) is similar to the typical corridor turns. If the minimum bounding circle of the object can be contained in the overlap area, this object is able to make the turn in this connector. However, for the corner in the typical corridor turn, the shape and the size is restricted by the walls, while for the overlapped area of two blocks, its boundary may not be restricted by obstacles. One possible solution is to extend its boundary to find the maximum corner areas. As shown in Figure 6c, block A and B are generated according to the layout of the obstacles. In this scenario, we can extend the boundary of the overlap area to form a larger accessible space indicated by the dashed lines in Figure 6c. In fact, this extension process is unnecessary; because the merging process of the LEGO model guarantees that any block with the maximum accessible space will be generated. Therefore, any possible overlap

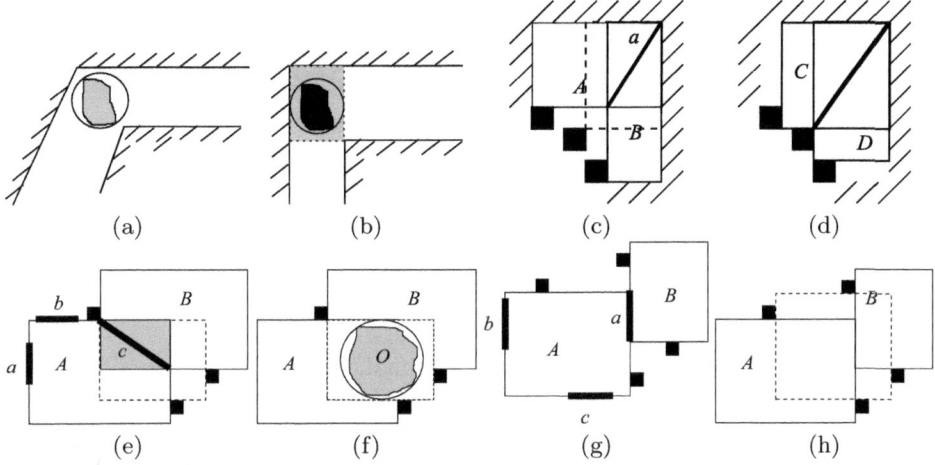

Fig. 6. Demonstrations of checking accessibility in different scenarios

areas can be captured by our model. For example, the area indicated by the dashed boundary in Figure 6c is the overlap area of the block C and D. We can find the maximum alternative corner area of A and B by looking for the largest corner area that contains the current one.

Figure 4b shows another kind of overlap relationship between two blocks. For this kind of scenario, according to the locations of other connectors, users may or may not have to make turns. Taking Figure 6e as an example, A and B are two adjacent blocks, and c is the connector between A and B. Assuming a and b are two connectors connecting A and other blocks, if one user goes from a to c, she can go straight to B. However, if she goes from b to c, then probably she will need to take a turn. Our approach to handle this scenario is to find the maximum overlap area for the two blocks. If the minimum bounding circle of the object can be contained in the maximum overlap area, the object can go through the two blocks without any problem. For example, in Figure 6e, the grey part is the overlap area between A and B. According to the locations of the obstacles, this area can be extended to the area indicated by the dashed lines. In figure 6f, since the minimum bounding circle of the object O can be contained in this extended overlap area, O can successfully go from A to B.

The situation of two adjacent blocks shown in Figure 4a is similar to the overlap blocks in Figure 4b. As shown in Figure 6g, A and B are two adjacent blocks, and c is the connector between A and B. Assuming a and b are two connectors connecting A with other blocks, if one user goes from a to c, she can go straight to B. However, if she goes from b to c, then probably she will need to take a turn. To simplify the two cases, we have observed that if we extend the connector c to form a larger block (as shown in Figure 6h), the scenario becomes the same as the overlap blocks we discussed before. Therefore, we apply the same strategy to check the accessibility of the adjacent blocks.

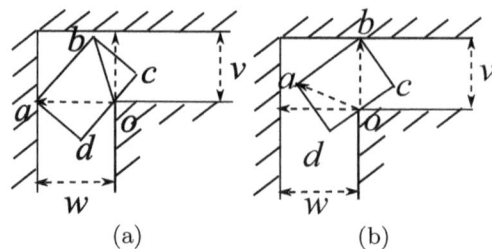

Fig. 7. An example of the refinement for 90° corners

This minimum bounding circle approach makes sure that the provided routes are feasible for users. However, this approach is not precise enough. For some particular scenarios, we are able to refine the accessibility check.

In indoor space, corridor corners are one of the most common areas where users may have problems to successfully go through. For the traditional 90° corners, we have developed an approach to refine the accessibility checking. As shown in Figure 7a, the rectangle (a, b, c, d) is the minimum bounding box of an object (user). The segment (o, a) is parallel to one side of the corner, and (o, a) has the same width of w. We have noticed that if the length of (o, b) equals or is less than v, then this rectangle can make the turn. Point b has the same situation. In Figure 7b, the length of (o, b) equals to v. If the length of (o, a) equals or is less than w, then this rectangle can make the turn.

Therefore, our approach contains two steps to check the accessibility of a 90° corner. First, the minimum bounding rectangle (MBR) of the user is constructed (e.g. (a, b, c, d) in Figure 7a). Second, assuming the boundary (a, b) and (c, d) are longer than (a, d) and (b, c), find a point o on the boundary (c, d) so that the length of (a, o) equals the width of one side of the corner (e.g., w in Figure 7a). If the length of (b, o) equals or less than the width of the other side of the corner (e.g., v in Figure 7a), then the user can successfully make the turn.

We can perform the second step in another way that we try to find a point Q that the length of (b, Q) equals to the width of one side of the corner. If the length of (a, Q) equals or less than the other side of the corner, the user can make the turn. Otherwise, this corner is not feasible for her.

5   The LEGO Graph

Most of the existing path searching algorithms (e.g., the shortest path search and the A* algorithm) are graph-based algorithms. In this section, we will discuss how to build a graph to support route searching algorithms.

As discussed in previous sections, the indoor space is approximated by LEGO cubes, which are further merged to form larger blocks. Users can walk blocks by blocks to reach their targets. The accessible widths, heights and lengths are restricted by these blocks and the connectors between them. In order to support the accessibility checks, these information must be stored in the graph.

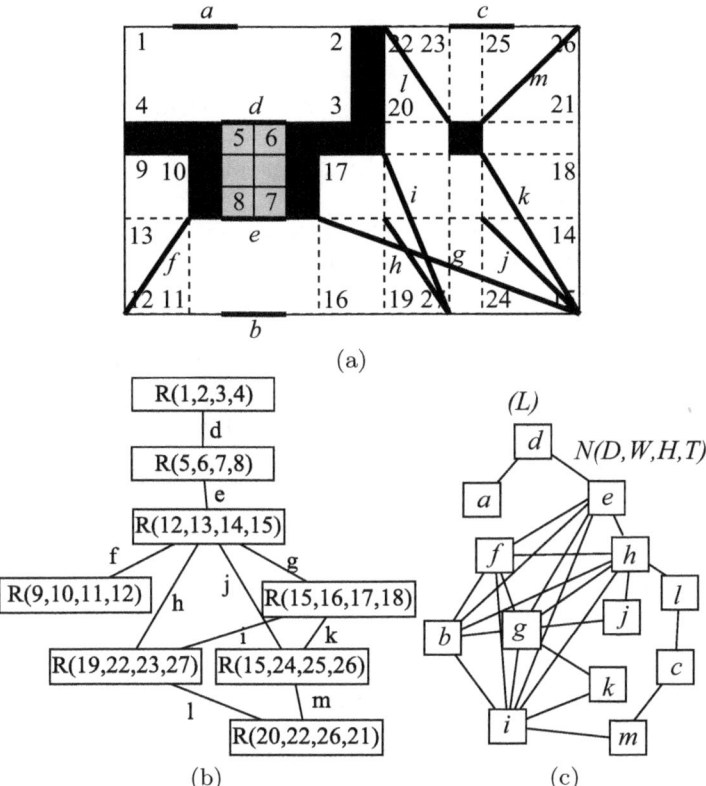

(a)

(b)   (c)

Fig. 8. An example of a floor plane with obstacles and stairs (a), the graph reflecting the connectivity of the blocks (b), and the corresponding LEGO graph (c)

One solution is to build a graph in which nodes denote blocks and edges represent connectors. Figure 8b is such a graph consisting all the blocks and connectors for the scenario shown in Figure 8a. One big problem of this graph is that the distance of each path is stored in nodes instead of edges. Therefore, it is difficult to apply the shortest path algorithms.

Actually, the process of walking blocks by blocks is the same as the process of walking connectors by connectors. A better solution is to build a graph in which nodes denote all the connectors and edges represent their distances. In our model, this kind of graph is called LEGO graph. Definition 1 is the formal definition of the LEGO graph.

Definition 1. *A LEGO graph* $LG = (V, E)$ *is a graph which reflects all possible paths with different accessible widths, heights and lengths in a given indoor space scenario.* V *is a set of connectors with the information of the supportable lengths* $< L >$. *E is a set of implicit paths in the format of* $< N(D, W, H, T) >$, *where* N *is the name of the edge,* D *is the distance between two connected nodes,* W, *and H are the maximum accessible width and heights. T is the type of the edges, which can be plane, obstacle or stair.*

Now, let's discuss how to determine the values attached to each edge:

- **D** : The length of an edge in a LEGO graph is the distance between the center points of the two end nodes.
- **W** : The accessible width of an edge depends on the maximum widths of the two end nodes. It will be set to be the minimum width of the two nodes.
- **H** : As discussed in previous sections, our generated blocks are always rectangles, and the connectors are either on the boundary or inside the block. Thus, the path between two connectors is always inside the corresponding block. The accessible height of an edge is the height of the block.
- **T** : Since each edge is inside one block, there is only one type for each edge. For example, if the cubes in one block are all plane_cubes, the path is *plane*.
- **L** : The accessible length is maintained in nodes, which is the diameter of the maximum circle introduced in Section 4.3. The reason why we don't check length in edges is because if the minimum bounding circle of the user can be contained in the extended connecting area, there must be enough space to fit for the user's length.

6   Conclusions and Future Work

Providing feasible routes for different kinds of users is an essential requirement for route planning models. Current existing models either only consider pedestrians' movements, or assume that users have fixed shapes and sizes. In this paper, we propose a model to provide feasible routes for arbitrary-shape users. Our model consists of two phases. In the first phase, we have introduced how to use LEGO cubes to represent indoor space. These cubes are then merged together to form larger blocks in the second phase. By checking the adjacent or overlap areas between different blocks, we have developed a novel approach to evaluate the accessibility of users in different places. At last, we have shown how to build the LEGO graph to record all the necessary information so that the feasible routes can be calculated by the traditional path search algorithms.

As we mentioned in the paper, the route this model can provide may not be the optimal one. In addition, in order to handle all the scenarios, our approach is not precise enough. Similar to the refinement we have introduced in this paper, a lot of refinements can be explored on this model.

References

1. Gilliéron, P.Y., Merminod, B.: Personal Navigation System for Indoor Applications. In: 11th IAIN World Congress, pp. 21–24 (2003)
2. Urs-Jakob, R.: Wayfinding in Scene Space: Transfers in Public Transport. PhD thesis, University of Zürich (2007)

 3. Lyardet, F., Grimmer, J., Muhlhauser, M.: CoINS: Context Sensitive Indoor Navigation System. In: ISM 2006: Eighth IEEE International Symposium on Multimedia, pp. 209–218 (2006)
 4. Lyardet, F., Szeto, D.W., Aitenbichler, E.: Context-Aware Indoor Navigation. In: Aarts, E., Crowley, J.L., de Ruyter, B., Gerhäuser, H., Pflaum, A., Schmidt, J., Wichert, R. (eds.) AmI 2008. LNCS, vol. 5355, pp. 290–307. Springer, Heidelberg (2008)
 5. Hu, H., Lee, D.L.: Semantic Location Modeling for Location Navigation in Mobile Environment. In: Proc. Of the IEEE International Conference on Mobile Data Management (MDM), pp. 52–61 (2004)
 6. Werner, S., Krieg-Brückner, B., Herrmann, T.: Modelling Navigational Knowledge by Route Graphs. In: Habel, C., Brauer, W., Freksa, C., Wender, K.F. (eds.) Spatial Cognition 2000. LNCS (LNAI), vol. 1849, pp. 295–316. Springer, Heidelberg (2000)
 7. Lorenz, B., Ohlbach, H., Stoffel, E.P.: A Hybrid Spatial Model for Representing Indoor Environments. In: Carswell, J.D., Tezuka, T. (eds.) W2GIS 2006. LNCS, vol. 4295, pp. 102–112. Springer, Heidelberg (2006)
 8. Stoffel, E.P., Lorenz, B., Ohlbach, H.: Towards a Semantic Spatial Model for Pedestrian Indoor Navigation. In: Advances in Conceptual Modeling C Foundations and Applications, pp. 328–337 (2007)
 9. Yuan, W., Schneider, M.: inav: An indoor navigation model supporting length-dependent optimal routing. In: 13th AGILE Int. Conf. on Geographic Information Science. Springer, Heidelberg (2010)
10. Lee, J.: A Spatial Access-Oriented Implementation of a 3-D GIS Topological Data Model for Urban Entities. GeoInformatica 8(3), 237–264 (2004)
11. Lee, J.: A Combinatorial Data Model for Representing Topological Relations among 3D Geographical Features in Micro-Spatial Environments. International Journal of Geographic Information Science 19(10), 1039–1056 (2005)
12. Thomas Becker, C.N., Kolbe, T.H.: A Multilayered Space-Event Model for Navigation in Indoor Spaces. In: Advances in 3D Geoinformation Systems, pp. 61–77. Springer, Heidelberg (2009)
13. Li, Y., He, Z.: 3D Indoor Navigation: a Framework of Combining BIM with 3D GIS. In: 44th ISOCARP Congress (2008)
14. Meijers, M., Zlatanova, S., Pfeifer, N.: 3D Geo-Information Indoors: Structuring for Evacuation. In: Proceedings of Next generation 3D City Models, pp. 21–22 (2005)
15. Bandi, S., Thalmann, D.: Space Discretization for Efficient Human Navigation. In: Proc. Computer Graphics Forum, vol. 17, pp. 195–206 (1998)
16. Thrun, S., Bücken, A.: Integrating Grid-Based and Topological Maps for Mobile Robot Navigation. In: Proceedings of the AAAI Thirteenth National Conference on Artificial Intelligence, Portland, Oregon, pp. 944–950 (1996)
17. Bandera, A., Urdiales, C., Sandoval, F.: A Hierarchical Approach to Grid-based and Topological Maps Integration for Autonomous Indoor Navgation. In: Proceedings of the IEEE/RSJ International Conference on Intelligent Robots and Systems, pp. 883–888 (2001)
18. Han, C.S., Law, K.H., Jean-claude Latombe, J.C., Kunz, J.C.: A Performance-Based Approach to Wheelchair Accessible Route Analysis. Advanced Engineering Informatics 16, 53–71 (2002)
19. Yuan, W., Schneider, M.: Supporting 3d route planning in indoor space based on the lego representation. In: 2nd ACM SIGSPATIAL Int. Workshop on Indoor Spatial Awareness (ISA), pp. 16–23. Springer, Heidelberg (2010)

A Web-Based Visualisation Tool for Analysing Mouse Movements to Support Map Personalisation

Ali Tahir[1], Gavin McArdle[2], and Michela Bertolotto[3]

[1,3] School of Computer Science and Informatics,
University College Dublin, Belfield, Dublin 4, Ireland
[2] National Centre for Geocomputation,
National University of Ireland Maynooth, Maynooth, Co. Kildare, Ireland
{ali.tahir,michela.bertolotto}@ucd.ie, gavin.mcardle@nuim.ie

Abstract. Information overload is a well known issue across many domains. Due to an increase in the quantity of information associated with geographic data, information overload is now also prevalent in the spatial domain. This makes interacting with maps tedious and difficult, as extracting relevant information becomes laborious. Map personalisation offers a solution to this problem. By implicitly monitoring user behaviour and interaction with maps, common patterns, preferences and interests can be identified. Using this approach, personalised maps can be generated which match user preferences and contribute to resolving information overload in the spatial domain. Traditionally data mining techniques are used to identify preferences however, visual analytics has proven useful in detecting interests and patterns not apparent via data mining. This paper presents a visual analysis tool called VizAnalysisTools, which can be used by developers and analysts to detect patterns in Web map usage among groups of users. The knowledge gained through this visual analysis can be used to strengthen map personalisation techniques.

Keywords: Web GIS, Geo-Visualisation, Map Personalisation, Geo-Visual Analysis, Human-Computer Interaction.

1 Introduction

The increasing mobility of individuals, the advances in ubiquitous technologies and the growing volume of location specific information obtained from sources such as Global Positioning Systems (GPS) have contributed to the demand for more sophisticated personalised Web applications. Map personalisation is a recent development which adapts a map to reflect user preferences and interests, thus alleviating information overload by filtering out irrelevant content. Previous studies have shown that mouse movements act as an implicit indicator of user interests with map data [17]. Visualising user interactions provides system designers with more insight into user behaviour which can be used to fine tune the personalisation algorithm described in [5]. To achieve this, we have created

J. Xu et al. (Eds.): DASFAA Workshops 2011, LNCS 6637, pp. 132–143, 2011.
© Springer-Verlag Berlin Heidelberg 2011

a Web application using Open Source technologies for visual analysis of mouse movement data. This framework can easily be extended to analyse other movement datasets which forms an element of the future work.

Our previous work implemented a Web architecture for map personalisation using Open Source technologies [18]. The benefit of using such an architecture is that users do not have to install or download additional software. The architecture has now been augmented with a component to analyse mouse movements through their visualisation on the underlying map data and the production of summary statistics. This is achieved by implementing different visual and reporting techniques such as heat maps, Keyhole Markup Language (KML) overlays and statistical charts. From a technical point of view, the approach demonstrates the typical exchange of data using interoperable Web services. The technologies employed in VizAnalysisTools use Web 2.0, which has emerged as a powerful extension of the Web and presents interactive information sharing, ensures interoperability, focuses on user-centered design, and provides a platform for collaboration over the Web [20]. This improvement in Web technologies, has permitted many applications to move from stand alone to Web based applications [13]. The Web tool will assist with analysing mouse movement datasets which reveal how users perform specific spatial choices and also helps to better understand map personalisation and user profiling techniques [22].

The remainder of this paper is organised as follows, Section 2 outlines the related work in the field. Section 3 describes the system and its functionality by providing some examples of the tool in use. The system architecture and the associated technologies are discussed in Section 4. Section 5 presents on going developments and describes the approach for evaluating the techniques, while some concluding comments are presented in Section 6.

2   Related Work

Mouse analysis is a standard way to interpret mouse movements within the Web interface. This can reveal user interests in terms of their high or low activity with different elements of a Web interface. Web applications such as [4] and [19] use mouse location and pointers on the Web pages to visually determine interest indicators. These implicit indicators have been widely studied in the Human-Computer Interaction (HCI) domain. Google Analytics [1] is another example that supports non-spatial applications and contributes rich insights into the website traffic and marketing effectiveness by providing a suite of reporting interfaces for analysis.

Studying user interaction with interactive maps is significant in the geospatial domain in order to understand user behaviour. Mouse activity acts as an implicit interest indicator in this area. In this regard, mouse movements on spatial interfaces were employed and explored by [15]. Preliminary work for analysing mouse movement data in the spatial domain has been performed by [16], who developed a visualisation tool called Geospatial Interactions Visualizer (GIViz),

---

[1] google.com/analytics

that analyses user behavior with geospatial datasets. The GIViz approach to analyse user interactions with a map was initially based on non-spatial implicit interest indicators such as bookmarking a Web page, clicking on a hyper link or saving a file, as discussed by [7].

CommonGIS, is another visualisation tool to analyse different movement datasets by providing extreme functionality [3]. This system is stable and mature in terms of providing different visualisation techniques, data transformation as well as spatial decision support capabilities. Although VizAnalysisTools and CommonGIS both provide analysis of spatial and spatio-temporal data, there are key differences in how this is achieved and the level of granularity offered. For example, VizAnalysisTools is multiplatform, unlike CommonGIS it is not limited to stand-alone applications where the user has to install the packages before running the tool or restrained to a Web Applet. Furthermore, while CommonGIS provides a set of controls for analysing time intervals, it does not consider movement datasets with small time intervals such as mouse movements which are recorded in milliseconds. However, VizAnalysisTools does not offer the full functionality of CommonGIS because this is not required for our purpose.

As the volume of data is increasing, visualisation techniques are becoming more powerful and meaningful for analysis and interpretation. [14] presents a classification of information visualization and visual data mining techniques for improved analysis. Geovisualisation, software visualization, and visual analytics have now become specialised fields to deal with information visualisation in their own domains [6]. These fields introduce new techniques for gaining an insight and deep understanding of the datasets being studied. For example, heatmaps have become popular on the Web where the colour intensity shows the amount of interaction that took place on a particular section of the Web page. One such system is [23], which describes a tool that increases user awareness by visualising their own navigation movements within Web pages. HotMap [10] is another system that generates a heat map which reflects users' attention to the map, for example which map tiles they download frequently.

The research presented in this paper provides a Web interface by using open Web services and standards in order to build a shared and accessible platform. The approach improves upon existing systems for analysing mouse movements by incorporating a Web-based dimension and also by providing an increased range of visual analysis options. The robust and open nature of the tool will permit it to be extended to incorporate other movement datasets in future experiments to resolve information overload issues.

3   System Description

VizAnalysisTools is a suite of tools to analyse mouse movements in order to identify specific usage patterns and behaviours which indicate important user intentions while highlighting their interests. The Web application performs two main tasks. Firstly, the interactive Web interface provides analysts with functionality to visualise mouse movements which are shown as an overlay on the

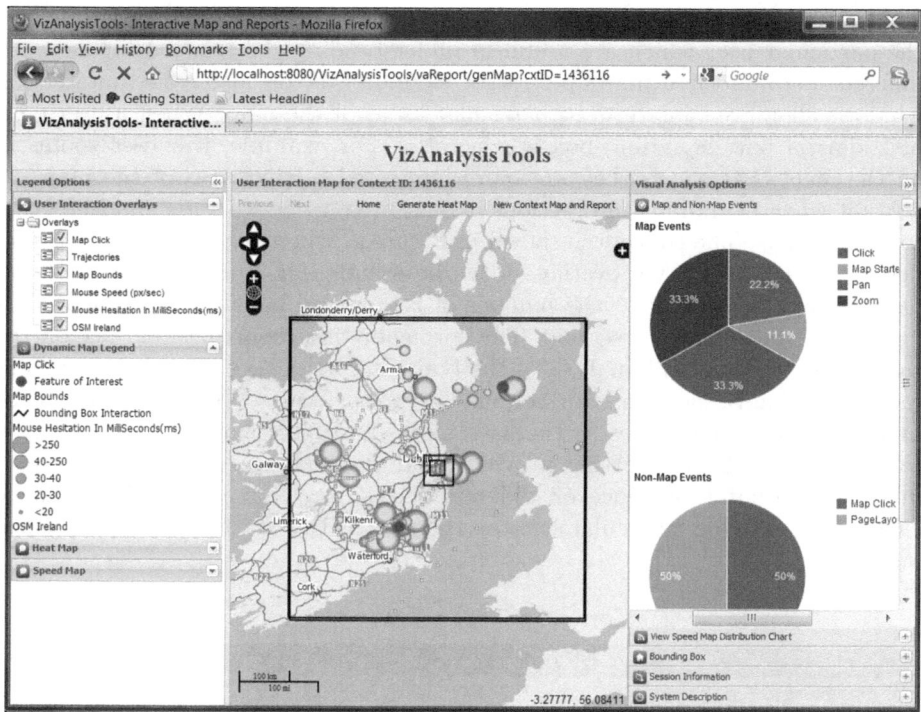

Fig. 1. User interaction showing mouse hesitations, map clicks as overlays and pie charts of map and non map events

base map. Secondly, interactive reporting features are embedded in the Web interface to examine the trends in the user interactions for detailed analysis. The key features of VizAnalysisTools are described in detail as follows.

The prototype has been deployed as a Web application which shows details of a user session. Each user session includes the interaction history of the user. As seen in Figure 1, the interface is delivered via a Web page which has two principal components. One component renders overlays of different interactions with a Web map, while a second component provides a statistical report of user interaction. Additionally, the interface contains the necessary tools to support the functionality which it provides. A typical user session is shown in Figure 1 where the Web interface is highlighting mouse hesitations as an overlay, rendered using the KML format. A symbol legend is shown in the panel to the left of the main map. This legend indicates that the circles shown on the map are indicative of mouse hesitations. The longer the mouse hesitation, the larger the circle on the map appears. Map clicks are also shown as a spatial overlay with corresponding symbols to the left of the dynamic map legend. These user interactions (mouse hesitations and map clicks) clearly show some spatial interest of the user in question. The reporting interface shows pie charts of both map and non map events which are generated using the Google Visualisation API.

The statistics show the amount of clicks, panning and zooming performed by the particular user which are useful in understanding the relationship between different map and non map operations. Bounding boxes also called interaction windows show user interests in particular areas of the map. When a user pans or zooms, a new bounding box is generated. For example, if a user zooms in on the map, that shows they are more interested in that area of the map. As a result when examining interactions in that area, more weight should be given to the findings. The concept is shown in Figure 2, where a bounding box is rendered in black as a KML overlay on the base map. The bounding boxes shown on the map correspond to the number of zooms and pans the user made during their interaction in a session. The corresponding reporting interface on the right shows a statistical analysis of the bounding box usage. This interface consists of bar chart showing map bounds on various scales. The x-axis shows the number of bounding boxes as the user zooms and pans while the y-axis shows the map scale. This reporting provides analysts with an overview of the user interactions that took place on different scales. Any bar in the chart can be clicked to move to a particular scale on the map as the chart and map are linked interactively.

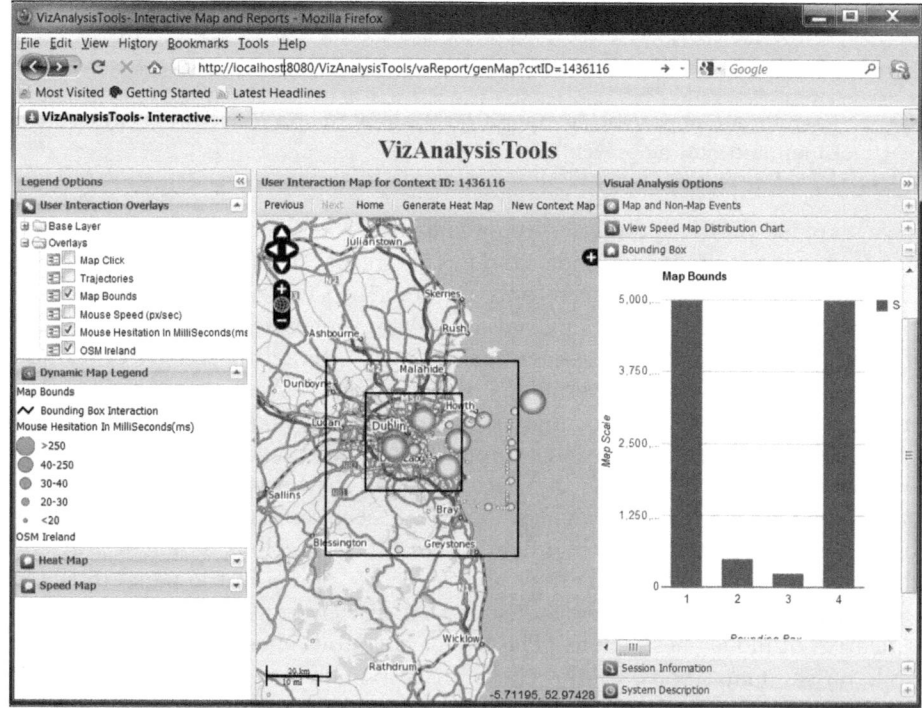

Fig. 2. User interaction within bounding box and corresponding bar chart

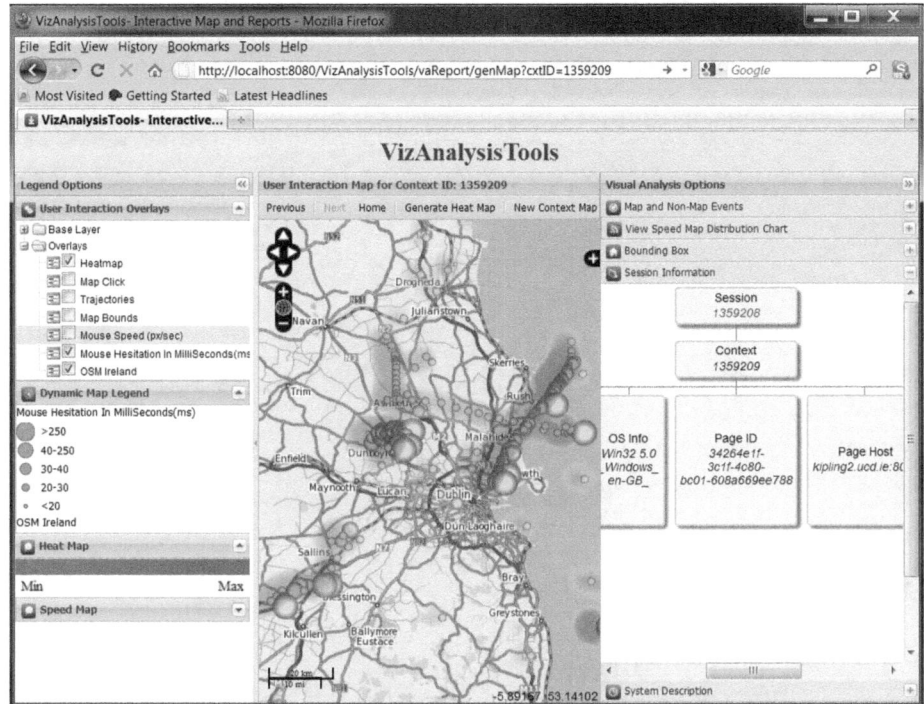

Fig. 3. User interaction heat map with map overlay

A heat map is another visual technique which is generated based on user mouse interactions and rendered as an overlay on the base map. This visualisation allows analysts to identify areas of interest and also where most interaction took place on the map. Figure 3, shows a mouse interaction heat map for a particular user. The mouse hesitation layer is also shown as an overlay in order to justify the heat map technique. The heat map has a colour range, as seen in the legend in Figure 3. In this case, an intense red colour indicates high activity whereas the blue colour represents low map activity. It can be clearly seen that the amount of interaction corresponds to the amount of heat generated as a result. Importantly, an analyst can zoom in and out to see how the relative heat changes at different map scales for a more detailed insight into certain areas of the map. Mouse speed is also visualised. When considered together with the trajectory of the mouse, the acceleration and deceleration of the cursor reveals a change in user behaviour which may indicate a shift in their intentions over time [16]. This functionality has been incorporated into the Web tool by categorising the mouse speed measured in pixels per second (px/sec). Each speed category is assigned a different colour ranging from a minimum to maximum speed value as shown in Figure 4. A red colour shows the maximum speed while a blue colour shows the minimum speed.

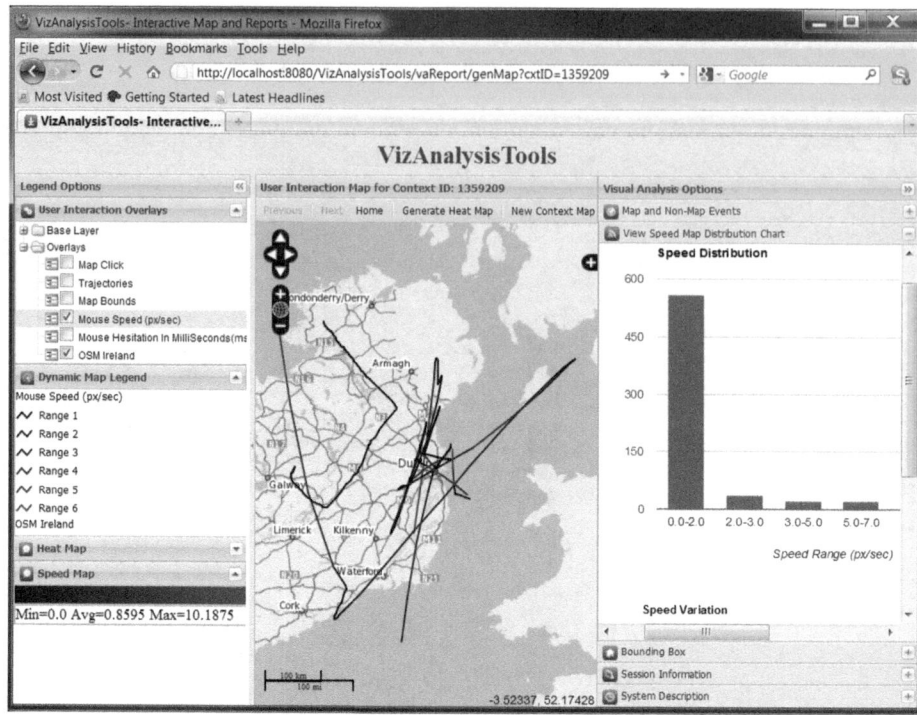

Fig. 4. Mouse speed and corresponding speed distribution and variation charts

The corresponding reporting interface in Figure 4 shows two speed charts which gives some information about the generated speed map. Firstly, the speed distribution shows a column chart which plots the speed range categories against the number of mouse movements. The trend shows that most of the time mouse movements were observed in the first speed range (0.0 to 2.0 px/sec). The speed categories are dynamically generated based on the minimum, average and maximum speed of a particular user interaction. By visualising the trend of the same user over multiple sessions, the average mouse speed can highlight the habits of the users and their level of interactions for example beginner or advanced users. The second chart shows the speed variation as the user session proceeds. The x-axis shows the number of mouse movements while the y-axis shows the mouse speed. This type of chart is useful to see any outliers in the user session and observing any corresponding changes on the interactive map.

When the features described above are used in conjunction with each other, this allows the analyst to reveal trends and patterns in the behaviours of users. Such patterns would be difficult to detect using data mining approaches alone. It is the domain knowledge of the analysts which adds a useful insight into interpreting the interactions which the tool produces. In section 5, further features to assist the analyst to interpret the visualisation are described.

4   System Architecture and Technologies

The system architecture of VizAnalysisTools is an extension of a Web architecture described in [18], and based on Open Source interoperable technologies. VizAnalysisTools provides components which utilise the dynamics of interactive JavaScript API's and the core business logic to perform visual analysis and geo-computation using a visualisation engine on the server side. The architecture is based on Grails [2], which is an Open Source framework to develop rapid, dynamic and robust Web applications. The system architecture is shown in Figure 5 and the core components of the architecture are described below. Web

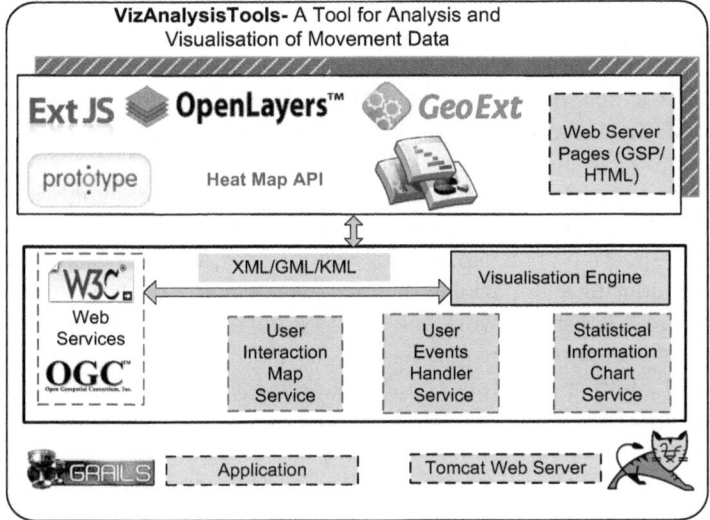

Fig. 5. System architecture

services have emerged as a framework to exchange information over the Web between applications using Extensible Markup Language (XML), which is a simple but an interoperable format. The Web service framework specifications include communication protocols such as Simple Object Access Protocol (SOAP), service descriptions as Web Services Description Language (WSDL), and service discovery as Universal Description, Discovery and Integration (UDDI) [8]. VizAnalysisTools requests data from the system described in [18], which is an application to log and record user interactions. The exchange of information takes place using SOAP, which returns an XML of events for a particular user session. A user session that includes the interaction history is returned as XML events. This information is passed to the visualisation engine which generates the map overlays and produces the statistical reports. Once the data is received in response to the Web service request by the visualisation engine, it starts processing this information using an internal KML service. This service parses the

---

[2] grails.org

data and generates various KML files describing user interactions. KML [3] is an OGC standard whose data is mainly composed of geographic visualisations including annotation of maps and images.

Various technologies are used to process the data in order to generate the map and statistical visualisations. For example, OpenLayers [4] is an Open Source JavaScript library for displaying and manipulating spatial data in Web browsers, without server-side dependencies and used for building rich Web-based geographic applications. OpenLayers widely supports OGC standards such as Web Map Service (WMS) and KML which are used in this Web application to render OpenStreetMap layer and dynamic overlays respectively. The heat map described in section 3 is generated using Heat Map, a client side JavaScript library which works in conjunction with OpenLayers. This API inputs geographical locations and outputs intensity in different colours. The actual heat map is then generated as an overlay on the base map to show areas of interests.

GeoExt [5] which also operates with OpenLayers provides a powerful way of developing highly interactive and dynamic Web GIS applications with the use of JavaScript. With GeoExt, Ext JS is used to build and design the layout which consists of a map, dynamic legends, and interactive visual analysis options. The advantage of using ExtJS dynamic layouts is that more functionality can be embedded irrespective of the space. For this project, GeoExt has been used extensively for rendering the spatial and non-spatial features of the interactive interface.

Prototype [6] is a framework that is used with the JavaScript development in the Web environment. VizAnalysisTools uses prototype to implement the system functionality and to generate the interaction maps and reporting interface. The dynamic reporting interface, provided within the Web interface produces charts summarising the main interactions. This is achieved using the Google Visualisation API [7]. The API is a JavaScript library that provides interactive reporting features within a Web application that includes a range of visualisation charts. This API is used for drawing several interactive charts for different analysis purposes. Embedding such a reporting facility is a powerful technique within a Web environment to assist developers to perform analysis while visualising the interaction map simultaneously.

By adhering to standards and using predominantly Open Source technologies, the functionality of the system can be easily extended through the addition of new modules. In the next section, an extension of the framework to include techniques for summarising the interactions are described.

5   Discussion and On-Going Developments

The functionality to monitor mouse interactions, offered by the VizAnalysisTools, is extremely beneficial for analysts to determine how a single user is

---

[3] opengeospatial.org/standards/kml
[4] openlayers.org
[5] geoext.org
[6] prototypejs.org
[7] code.google.com/apis/charttools

interacting with spatial data. The examples presented in the paper, assume that the interactions of an individual user over a single session are visualised. However, the approach can be made more powerful by visualising interactions over multiple sessions and furthermore, visualising the interactions of several users. Such visualisations can identify common trends and salient patterns which would otherwise be missing. This type of geo-visual analysis can assist developers to improve interaction paradigms while also augmenting the information required for personalising the map data and accompanying tools.

While it is feasible to add the additional interactions, from multiple users and multiple sessions, to the existing visualisation, this would cause cluttering and occlusion to occur, making any interpretation of the data difficult. To combat this, it is necessary to assist the analyst to identify patterns and trends within the data being studied by extending the functionality to include aggregation and clustering techniques [2], [9]. These methods can be applied to the mouse trajectory data in order to combine similar trajectories using clustering. This can be used to create a generalised view which resolves the problem of trajectories obstructing the analyst's view of nearby trajectories. There are several approaches for aggregating movement data and measuring the similarity of trajectories. The approaches vary, depending on the analyst's focus. For example, trajectories can be aggregated based on their similarity in geographic, temporal or attribute space. Other approaches focus solely on the origin and destination of a trajectory [9]. The challenge is to determine appropriate similarity metrics which accurately group similar mouse trajectories.

The system is currently being expanded to offer support for visualising the temporal aspects of mouse movement and interaction data. In particular the use of a space-time cube [12] will enable analysts to clearly see the temporal ordering and sequence in which mouse events occurred. Space-time cubes are generally used to show the movements of entities in geographical space, in which the x and y plane of the cube, represent the spatial components while time is plotted on the z-axis. This approach can easily be extended for use with mouse trajectory data and when combined with aggregation techniques, can be effectively used to visualise interaction tasks where the temporal component is important such as usability studies.

Evaluating the functionality of VizAnalysisTools is significant both in terms of performance and usability. Currently, the tool supports analysis of a single user session but as the user sessions increase, performance issues on the Web can arise. Techniques have been explored to apply best practice in order to make systematic approaches to successfully develop and maintain high-quality Web applications [11]. Evaluation metrics for usability are also being examined in this regard. [21] have proposed models for quantifying and assessing usability within the area of HCI. Assessing the benefits which visual analysis of mouse movements brings to map personalisation can be achieved by comparing this approach with that of a pure data mining technique similar to that proposed in [18]. These considerations will be taken into account in order to identify the strengths and weaknesses in the next steps of the design and development of

VizAnalysisTools. While the tool described here represents a specialised case for visually analysing mouse movements, with slight modification, the tool can be used for visualising generic trajectory data to increase the benefits of the system.

6   Conclusions

This paper has described a Web-based tool to support map personalisation by geovisual analysis of mouse movement data. The paper focuses on describing the features of the tool for analysing the data associated with mouse interactions of users interacting with a Web-based GIS. The trends, behaviours and usage patterns identified through the geovisual analysis of this interaction data can be used within map personalisation in order to reduce information overload. The tool uses Web 2.0 technology to deliver an interactive interface which provides analysts with functionality to identify patterns in mouse movement data and discover important areas of the map. Additionally summary reports outline the key usage statistics for a particular map user. The tool is developed with the intention to make it dynamically accessible via the Web, making it interoperable and compliant with existing standards, and to equip it with a range of spatio-temporal analysis tools as recommended by a recent research study in the field of visual analytics [1]. Through the addition of aggregation and clustering components, the tool can be scaled for analysing the movement data from multiple entities which will add significantly to the power of this geovisual analysis tool.

Acknowledgements

Research presented in this paper was funded by a Strategic Research Cluster grant (07/SRC/I1168) by Science Foundation Ireland under the National Development Plan. The authors gratefully acknowledge this support.

References

1. Andrienko, G., Andrienko, N., Demsar, U., Dranschc, D., Dykesd, J., Fabrikante, S., Jernf, M., Kraakg, M., Schumannh, H., Tominskih, C.: Space, Time and Visual Analytics. International Journal of Geographical Information Science 24(10), 1577–1600 (2010)
2. Andrienko, G., Andrienko, N., Wrobel, S.: Visual Analytics Tools for Analysis of Movement Data. ACM SIGKDD Explorations Newsletter 9(2), 38–46 (2007)
3. Andrienko, N., Andrienko, G., Voss, H., Bernardo, F., Hipolito, J., Kretchmer, U.: Testing the Usability of Interactive Maps in CommonGIS. Cartography and Geographic Information Science 29(4), 325–343 (2002)
4. Arroyo, E., Selker, T., Wei, W.: Usability Tool for Analysis of Web Designs Using Mouse Tracks. In: CHI 2006 Extended Abstracts on Human Factors in Computing Systems, pp. 484–489. ACM, New York (2006)
5. Ballatore, A., McArdle, G., Kelly, C., Bertolotto, M.: RecoMap: An Interactive and Adaptive Map-based Recommender. In: Proceedings of the 2010 ACM Symposium on Applied Computing, pp. 887–891. ACM, New York (2010)

6. Chen, C.: Information Visualization. Wiley Interdisciplinary Reviews: Computational Statistics 2(4), 387–403 (2010)
7. Claypool, M., Le, P., Wased, M., Brown, D.: Implicit Interest Indicators. In: Proceedings of the 6th International Conference on Intelligent User Interfaces, pp. 33–40. ACM, New York (2001)
8. Curbera, F., Duftler, M., Khalaf, R., Nagy, W., Mukhi, N., Weerawarana, S.: Unraveling the Web Services Web: An Introduction to SOAP, WSDL, and UDDI. IEEE Internet Computing 6(2), 86–93 (2002)
9. Demsar, U., Virrantaus, K.: Space-time Density of Trajectories: Exploring Spatiotemporal Patterns in Movement Data. International Journal of Geographical Information Science 24(10), 1527–1542 (2010)
10. Fisher, D.: Hotmap: Looking at Geographic Attention. IEEE Transactions on Visualization and Computer Graphics 13(6), 1184–1191 (2007)
11. Ginige, A., Murugesan, S.: Web Engineering: An Introduction. IEEE Multimedia 8(1), 14–18 (2002)
12. Hagerstrand, T.: What About People in Regional Science? Papers in Regional Science 24(1), 6–21 (1970)
13. Haklay, M., Singleton, A., Parker, C.: Web Mapping 2.0: The Neogeography of the Geoweb. Geography Compass 2(6), 2011–2039 (2008)
14. Keim, D.: Information Visualization and Visual Data Mining. IEEE Transactions on Visualization and Computer Graphics 8(1), 1–8 (2002)
15. Mac Aoidh, E., Bertolotto, M., Wilson, D.: Analysis of Implicit Interest Indicators for Spatial Data. In: Proceedings of the 15th Annual ACM International Symposium on Advances in Geographic Information Systems, p. 47. ACM, New York (2007)
16. Mac Aoidh, E., Bertolotto, M., Wilson, D.: Understanding Geospatial Interests by Visualizing Map Interaction Behavior. Information Visualization 7(3), 275–286 (2008)
17. Mac Aoidh, E., McArdle, G., Petit, M., Ray, C., Bertolotto, M., Claramunt, C., Wilson, D.: Personalization in Adaptive and Interactive GIS. Annals of GIS 15(1), 23–33 (2009)
18. McArdle, G., Ballatore, A., Tahir, A., Bertolotto, M.: An Open-Source Web Architecture for Adaptive Location Based Services. In: The International Archives of the Photogrammetry, Remote Sensing and Spatial Information Sciences, Hong Kong, vol. 38(2), pp. 296–301 (2010)
19. Mueller, F., Lockerd, A.: Cheese: Tracking Mouse Movement Activity on Websites, A Tool for User Modeling. In: CHI 2001 Extended Abstracts on Human Factors in Computing Systems, pp. 279–280. ACM, New York (2001)
20. O'reilly, T.: What is Web 2.0. Design Patterns and Business Models for the Next Generation of Software 1, 17 (2007), available at SSRN: http://ssrn.com/abstract=1008839
21. Seffah, A., Donyaee, M., Kline, R., Padda, H.: Usability Measurement and Metrics: A Consolidated Model. Software Quality Journal 14(2), 159–178 (2006)
22. Tahir, A., McArdle, G., Ballatore, A., Bertolotto, M.: Collaborative Filtering- A Group Profiling Algorithm for Personalisation in a Spatial Recommender System. In: Proceedings Geoinformatik, Kiel, Germany, pp. 44–50 (2010)
23. Wu, W., Noble, W.: Genomic Data Visualization on the Web. Bioinformatics, 1541 (2004)

On the Requirements for User-Centric Spatial Data Warehousing and SOLAP

Ganesh Viswanathan and Markus Schneider

Department of Computer & Information Science & Engineering
University of Florida
Gainesville, FL 32611, USA
{gv1,mschneid}@cise.ufl.edu

Abstract. Data warehouses and OLAP systems help to analyze complex multidimensional data and provide decision support. With the availability of large amounts of spatial data in recent years, several new models have been proposed to enable the integration of spatial data in data warehouses and to help analyze such data. This is often achieved by a combination of GIS and spatial analysis tools with OLAP and database systems, with the primary goal of supporting *spatial analysis dimensions*, *spatial measures* and *spatial aggregation operations*. However, this poses several new challenges related to spatial data modeling in a multidimensional context, such as the need for new spatial aggregation operations and ensuring consistent and valid results. In this paper, we review the existing modeling strategies for spatial data warehouses and SOLAP in all three levels: conceptual, logical and implementation. While studying these models, we gather the most essential requirements for handling spatial data in data warehouses and use insights from spatial databases to provide a "meta-framework" for modeling spatial data warehouses. This strategy keeps the *user as the focal point* and achieves a *clear abstraction* of the data for all stakeholders in the system. Our goal is to make analysis more user-friendly and pave the way for a clear conceptual model that defines new multidimensional *abstract data types* (ADTs) and operations to support spatial data in data warehouses.

1 Introduction

For more than a decade, data warehouses have been at the forefront of information technology applications as a way for organizations to effectively use information for business planning and decision making. They contain large repositories of analytical and subject-oriented data, integrated from several heterogeneous sources over a historical time-line [1,2]. The technique of performing complex analysis over the information stored in the data warehouse is popularly called Online Analytical Processing (OLAP). The large increase in the availability of spatial data in recent years has lead to increased challenges in storing such information and analyzing them. Data warehouses provide an effective way to manage spatial information by providing large-scale storage, multidimensional data management and OLAP querying capabilities together in one system.

J. Xu et al. (Eds.): DASFAA Workshops 2011, LNCS 6637, pp. 144–155, 2011.

Spatial data warehouses (SDWs) are large, subject-oriented, repositories of data integrated from a variety sources of a long timeline with native support for spatial objects and advanced operations on them. These operations on the spatial objects can include basic querying operations, such as *"find the city with the largest sales volume for iPads in the state of Florida in 2010,"* or map generalization operations such as *"find all states where the top five school districts out-performed all others (within that state) between 2005 and 2010 in terms of student grades,"* or spatial analysis operations such as convex hull: *"find the smallest convex region containing all the college towns where more than 2500 units of Kinect were sold in 2010"*. Many other interesting spatial aggregation queries are possible when spatial data is fully integrated into data cubes and an effective approach for multidimensional querying is available on them.

OLAP operations are often categorized as distributive, algebraic and holistic [3,4], depending on whether the measures of high level cells can be easily computed from their low level counterparts, without accessing base tuples residing at the finest level. For example, in the classic *sales(location,time,product)* data, the total sales of an item at [Florida, 2010] can be calculated by adding up the total sales of [Florida, January 2010] ... [Florida, December 2010], without looking at base data points such as [Florida, 20 March 2010], which means that SUM is a distributive measure. In comparison, AVG is often cited as an algebraic or semi-distributive measure, in that AVG can be derived from two distributive measures: SUM and COUNT, i.e., algebraic measures are functions of distributive measures. Holistic measures such as standard deviation require data at the specific requisite level for all computations. Similarly, spatial querying and aggregation operations such as spatial roll-up, drill-down and selection also involve several levels of data manipulation. For example, consider a drill down operation from Country (region) to county (maps) to cities (string labels for points). This complex navigation operator can be very useful in mining several levels of spatial information such as geo-spatial and video data.

Upon reviewing existing modeling approaches for spatial data warehousing (Section 2) we found that one of the major shortcomings of existing models is the heavy focus on direct ad-hoc implementation strategies such as a combination of OLAP tools or GIS mapping clients with databases to create a pseudo spatial data warehouse. However, for effective multidimensional data modeling and analysis what is needed is a refined data warehouse architecture that keeps the *user as the focal point* and achieves a *clear abstraction* of the data for all stakeholders in the system. Hence our proposal is for a sound conceptual model built on abstract data types (ADTs) and using the cube metaphor for OLAP analysis while natively supporting spatial data along the data dimensions and as measures for aggregation. The user view is created by using a generic textual analysis language such as an extension of MDX that helps to write SOLAP queries. Finally, a set of transformation rules from the conceptual model to logical design strategies such as relational OLAP (ROLAP) [2], multidimensional OLAP (MOLAP) [1] and hybrid OLAP (HOLAP) [5] is also needed to help complete the design of the spatial data warehouse. Overall, this paper provides a new insight into the fundamental requirements for designing a user-friendly spatial data warehouse model by providing an objective analysis of the essential requirements for it.

The rest of this paper is organized as follows. Section 2 provides a review of existing literature regarding data warehouse models and OLAP, spatial data modeling, spatial data warehouse models, and user interface tools used to design these systems. Section 3 discusses the essential requirements for spatial data warehouses and OLAP. Section 4 presents our meta-framework as a path for developing an enhanced conceptual model based on the cube metaphor that is capable of natively supporting spatial data and aggregations on them. Section 5 concludes the paper and mentions topics for further research.

2   Related Work

In this section, we review existing research on data warehousing and OLAP tools, spatial data modeling and associated implementation strategies, leading to the list of essential requirements for spatial data warehousing (in Section 3). Figure 1 illustrates the various domains that need to be considered for deciding the architecture of a spatial data warehouse (Section 4). A survey of the state-of-the-art in each of these domains is the topic of the current section.

Over the past decade several approaches have been proposed for modeling data warehouses. Now, we present a study of the best available conceptual and logical models for data warehousing. Existing conceptual modeling approaches can be broadly classified into *Extensions of Entity Relationship (E/R) models* ([6,7,8,9,10]), *Extensions of Unified Modeling Language (UML)* ([11,12,13]) and *Ad-hoc* ([14,15,16,17]) design models. Several different logical models have also been proposed to model multidimensional data in the past few years. The data cube operator was formally introduced in [4] in an attempt to extend the relational model to suit multidimensional analysis. A complete survey of the properties of several earlier logical design models can be found in the works of Blaschka et.al. [18], Vassiliadis et.al. [19] and Pedersen et.al. [20]. Though many of these models aid in the relational representation of aggregate data, contributions like the ALL operator and hierarchies are essential even from a multidimensional perspective. One of the earliest approaches for multidimensional modeling was introduced by Kimball in [21]. This *dimensional modeling* approach proposes an informal methodology to derive the multidimensional schema and provides a way to develop a relational implementation in the form of the star schema. Dimensional modeling imposes some rules on the modeling but results in a data model that has the access methods defined clearly by virtue of the relationships [21,1]. Users are also better able to relate to the "see measure by dimensional value(s)" paradigm rather than a simple "collection of values". The approach involves discovering the data-marts for the data-warehouse space, listing all dimensions for each data-mart, using an ad-hoc matrix to capture user requirements, and then designing a fact table with measures added to each grain of detail along the dimension levels. The model presented by Agrawal et.al. in [22] is a logical data model for multidimensional databases. The cube is defined as a set of *dimensions* (each associated with a domain) and a set of *elements* (measures). A mapping is provided between the dimensions and the set of elements. The elements of the cube can be 0,1 (the Boolean Cube) or a n-tuple of elements. This model does not require the dimensions to have a ranked, discrete domain. Instead the mapping function can be used to provide a symmetric treatment between measures and dimensions. An algebra is also defined over the model with operations such as *push* and *pull* (to transform a dimension into measure and vice-versa), *destroy dimension*, *restriction* (to constraint member values), and *join* (to combine two cubes). Several other operations like *cartesian product*, *natural join*, and *associate* are also mentioned. However, this model does not discuss the handling of explicit multiple hierarchies among dimensions or the problem of imprecision due to double counting during data aggregation.

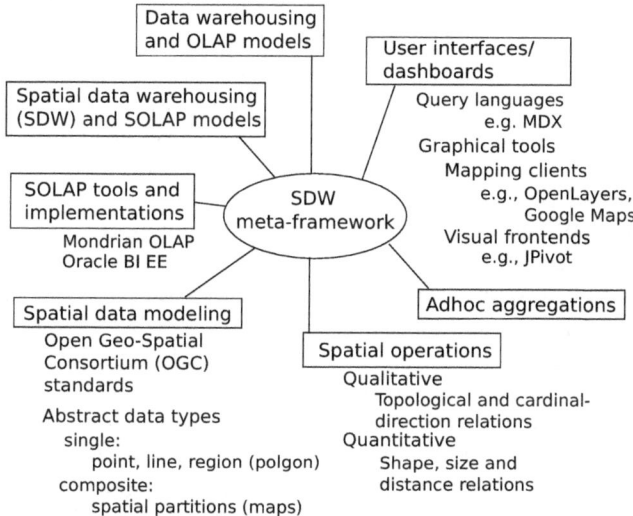

Fig. 1. An illustration of the various domains considered during the design of the spatial data warehouse meta-framework

Spatial data warehousing (SDW) has become a topic of growing interest in recent years. This is primarily due to the explosion in the amount of spatial information available from various sources such as GPS receivers, communication media, online social networks and other geo-spatial applications. Consequently some spatial OLAP tools are now available to help model and analyze such data.

An early approach to spatial online analytical processing (SOLAP) is [23], which mentions essential SOLAP features classified into three areas of requirements. The first is to enable *data visualization* via cartographic (maps) and non-cartographic displays (e.g., 2D tables), numeric data representation and the visualization of context data. Second, *data exploration* requires multidimensional navigation on both cartographic and non-cartographic displays, filtering on data dimensions (members) and support for calculated measures. The third area discussed involves the *structure of the data*, for example, the support for spatial and mixed data dimensions, support for storage of geometric data over an extended time period, etc. The conceptual design models for spatial data warehouses are extensions of E/R and UML diagrams or ad-hoc design approaches. Among extensions of E/R models, [24] presents a clear integration of spatial data for OLAP by extending the MultiDimER and MADS approaches. Among other ad-hoc design approaches, [25] presents a formal framework to integrate spatial and multidimensional databases by using a *full containment* relationship between the hierarchy levels. In [26], the formal model from [20] is extended to support spatially overlapping hierarchies by exploiting the partial containment relations among data levels, thus leading to a more flexible modeling strategy. Bimonte et.al. [27,28] present the GeoCube model for spatial datawarehouse design, based on a formal schema and instance definition for cube elements. GeoCube also extends conventional SOLAP operations with five new operations namely, *classify, specialize, permute, OLAP-buffer* and *OLAP-overlay*. However, one of the shortcomings of this approach is the use of n-n mappings between data-dimensions and facts. Since each cell of the data cube is a unique catesian product of the associated data dimensions, this n-n mapping weakens the cube structure and makes it difficult to apply constraints and dynamic schema changes during OLAP operations.

The logical SDW design models aim to provide support for spatial data dimensions [29], spatial measures [30,31,32,33] and spatial aggregations [34]. The concept of spatial measures (with a specific geometric part) is either defined as references to spatial objects [35,23] or as the results of topological, distance or metric operations [24,23], or as values associated with a spatial data dimension in the data cube [31,36]. In addition to supporting spatial objects, most GIS models use both geometric (e.g., the extent of fire spread is shown as a *polygon*) and thematic or descriptive attributes (e.g., state name) to help qualify geometric data objects [37]. This is a very useful feature for supporting spatial aggregation operations and map generalizations (such as moving from *state* level to *country* level in the *location* hierarchy). A discussion of spatial hierarchies and topological operators in a conceptual SDW model is presented in [38]. Sekhar et.al. [32] extend the MapCube operator to support spatial data and aggregations, but the model is rather constrained and not easily extendable for user-defined queries.

The major implementations of SOLAP tools can be broadly classified as *OLAP dominant, GIS dominant,* or *integrated OLAP and GIS* solutions [33]. OLAP approaches provide means for aggregation of data, while GIS approaches focus on geometric operations and visual data selections while limiting multidimensional data analysis. Another approach is the integration of OLAP and GIS systems [31,29,27]. The GeoMondrian [39] project aims to develop an opensource implementation of a SOLAP analysis server. Currently, it provides a spatially enabled version of the Mondrian OLAP server [40]. However, it is unclear if GeoMondrian has a clear underlying spatial data model with SOLAP operators. It seems to be essentially built ad-hoc, by using a combination of the Java

Topology Suite [41] (which provides spatial operations according to OGC standards) and Mondrian (which provides the OLAP operations on thematic attributes) with PostGIS (which provides the spatial data types). These together create a functional spatial data analysis toolkit supporting the integration of spatial data and operations in an OLAP server.

For modeling spatial data there are now several established approaches in the database community. An introduction to basic spatial data types is given in [42]. The ROSE algebra [43,44] provides a more robust discussion of spatial data types by introducing types such as *point, line* and *region* for simple and complex spatial objects and describes the associated spatial algebra. Composite spatial objects (collections of points, lines and regions) are presented as *spatial partitions* or *map* objects. Similarly, the Open GIS Consortium also provides a Reference Model [45] as a standard for a representing geo-spatial information. Qualitative spatial operations include topological relations [46] such as *disjoint, meet, overlap, equal, inside, contains, covers* and *coveredBy*, and cardinal direction relations. Quantitative relations on spatial objects include metric operations based on the size, shape and metric distances between objects or their components. All these operations can be used to query and analyze spatial data in the data warehouse.

3   Requirements for User-Centric Spatial OLAP

For a data warehouse model to be effective in modeling, storing and querying data there are several essential requirements. Blaschka et.al. [18] provide a list of requirements for multidimensional modeling for OLAP applications. In [20], Pedersen et.al. present eleven requirements for a multidimensional model, using a clinical data warehousing application as an example and then present a formal model that accommodates those requirements. By studying the popular models for multidimensional data (Section 2) and several new applications, we now compile a list of basic requirements for an effective user-centric spatial data warehouse model.

The basic requirements for an effective multidimensional model for data warehouse design can be summarized as follows:

1) *Simple user view*: The user view of the data warehouse must be simple and intuitive, yet capable of capturing the full dimensionality of the data. This user view should be independent of implementation aspects. Thus, for e.g., the use of facts and dimension *tables* (thereby exposing ROLAP implementation) should be avoided in the user view. Instead an abstract view, like a *data cube* with available OLAP querying operations on it would make it easier for users to navigate across data dimensions and perform analysis.

2) *Implementation independent conceptual design*: The conceptual model for the SDW should also be free from implementation aspects (such as star schema design) to be able effectively model data for the analyst's needs.

3) *Separation of structure and values*: There should be an explicit separation of schema and instances, i.e., structure and values. A formal conceptual data model should allow for independence between specification and implementation,

and the ability to alter implementations without affecting the user's view of the OLAP system.

4) *Explicit hierarchies*: Hierarchies (with several levels of member or measure categories) should be supported explicitly in the *data dimensions* and *facts*.

5) *Multiple hierarchies*: Multiple hierarchies along the data dimensions and even measure values should be supported.

6) *Descriptive attributes*: Thematic or descriptive attributes for members and measures (geometric or otherwise) should be supported for enabling selection, navigation and aggregation queries over them during analysis.

7) *Support for attribute aggregation*: The model must provide good support for aggregation on both geometric and thematic attributes apart from metric computations.

8) *Complex measures*: The model should support multiple (composite) and complex members and measures. Example, a cell in the cube can include several measures such as sales quantity and profit. Location can be a complex object such as a *polygon* representing Italy with the Vatican as a *hole* object inside it.

9) *Handling different levels of granularity*: The model should be able to handle data with multiple levels of granularity (dynamic multi-level hierarchies.)

10) *Support for irregular hierarchies*: There must be support for non-conformant (non-onto, non-strict and ragged) hierarchies and generalization/ specialization (is-a) relationships.

11) *User-defined aggregates*: User defined aggregation functions should be supported. These may even include ad-hoc operations such as *ratio* (metric) and multi-level buffer (geometric) operations.

12) *Handling data imprecision/ summarizability*: The model should be able to handle data imprecision so that double-counting of data is avoided, and non-additive data are not summarized.

13) *Handling uncertainty*: The model should also be able to handle the uncertainty in the data using techniques such as *data lineage tracking*.

14) *Handling change over time*: The model should be able to handle updates and deletions over time. Any re-calculations of measure values should be consistent and correct.

15) *Multiple facts/ cube schema*: The model should allow for multiple facts, data dimensions and data cubes to be present in the schema.

16) *Drill-across capability*: The model should support drilling across dimensions, i.e., sharing of dimensions among different fact cubes.

17) *Drill-through capability*: The model should support drilling through capability to be able to query the base level (raw) data.

18) *Dimensionless aggregation*: There should be support for aggregations along attributes that are not part of the data dimensions or hierarchies themselves, such as thematic attributes.

19) *Measureless aggregation*: Aggregation on thematic attributes of facts (measures) should be supported.

20) *Online aggregation*: The model should allow for multiple levels of *online* aggregation, i.e., dynamic, multi-level query design.

21) *Support for spatial hierarchies*: The model should support generalization and specialization hierarchies on spatial objects. This would, for example, enable roll-up operations from *cities* to *states* and *country* level in the *location* hierarchy. The linking of spatial hierarchies to thematic attribute hierarchies should also be allowed.

22) *Support for spatial dimensions*: Data dimensions and hierarchies should natively support spatial data types and operations on them such as roll-up and drill-down on spatial hierarchies (for example, from *state* to *city* level).

23) *Support for spatial measures*: The data cube should be capable of storing and managing spatial measures as both simple, complex and a composite (map) spatial objects.

24) *Support for spatial aggregations*: The model should explicitly support aggregation operations on the spatial measures and members. For e.g., "a *convex hull* on cities having the top-k highest sales of iPads in every state in 2010".

25) *Support for adhoc geo-spatial aggregations*: The model should support adhoc, user-defined geo-spatial operations on both spatial measures, members and their thematic attributes.

4   A Meta-Framework for Spatial Data Warehouse Design

After reviewing the existing data warehouse and SOLAP modeling approaches and generating the list of essential requirements for an effective spatial data warehouse model, we now provide a broad insight into how a spatial data warehouse architecture should be constructed for supporting user-centric OLAP.

For providing user-friendly spatial data analysis it is essential to use an abstract data model to design and construct the data warehouse. This can be provided by a conceptual design view that fully abstracts over the underlying implementation details. To allow users to interact with the conceptual cube an user view (query language or visual map interface) can be used to expose the set of data types and operations for OLAP analysis. At each level, explicit support for spatial data must be provided using spatial data types which can be single objects such as point, line or region, or a combination of these in terms of spatial partitions or map objects. Figure 2 illustrates such a meta-framework that we propose for spatial data warehouse design.

The conceptual model should provide built-in support for spatial objects by using abstract data types (ADTs) or by extending multidimensional data types such as perspectives (data dimensions) and subjects (facts) to include spatial values. Later, additive, semi-additive and holistic classes of aggregation operations can be defined over them [3,4,47]. For example, in Table 1, we show examples of possible spatial aggregation operators (the second line in each of the three categories). Such a conceptual design view can be an extension of the $\mathcal{BigCube}$ model [17] which provides ADTs arranged over different levels to create the conceptual cube or one of the other conceptual modeling approaches supporting spatial data analysis [26,24,34,28].

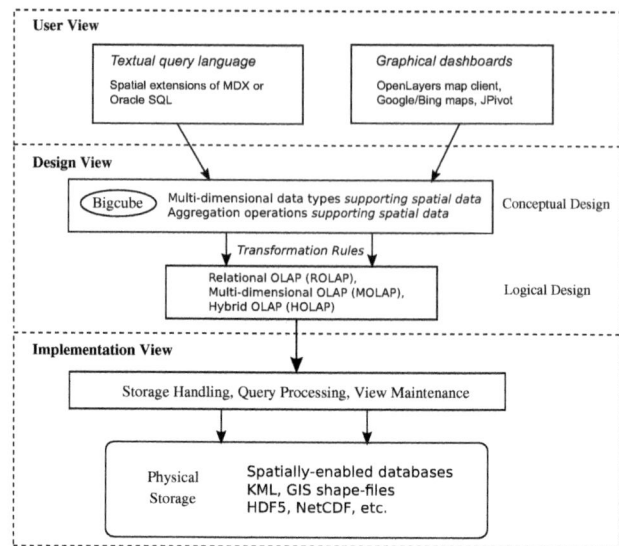

Fig. 2. A meta-framework for spatial data warehouse design illustrating the distinct conceptual and logical design levels and the user view for OLAP analysis

Table 1. Examples of non-spatial and spatial aggregation operators

Type	*BigCube* Aggregation Operator
Additive	*Sum, Count, Max or Apex, Min or Base, Concatenate, Convex Hull, Spatial Union, Spatial Intersection*
Semi-Additive	*Average, Variance, Standard Deviation, MaxN, MinN, Centroid, Center of Gravity, Center of Mass*
Non-Additive	*Median, MostFrequent, Rank, LastNonNullValue, FirstNonNullValue, Minimum Bounding Box, Nearest Neighbor, Equi-Partition*

A set of transformation rules are needed from the conceptual model to the logical design level. The logical design can be done in one of three ways. Data warehouse star, snowflake or galaxy schema can be constructed and the corresponding relational tables are stored in a database linked by foreign keys and other functional dependencies. This is called Relational OLAP or ROLAP. In multidimensional OLAP design, data cubes can be constructed in memory to store and operate over the data warehouse. This is very similar to the cube model used for conceptual design. However though multidimensional querying is often faster in comparison to relational querying, this approach can lead to increased memory and storage requirements. A balance between these two approaches is achieved in Hybrid OLAP by using a combination of relational and multidimensional design strategies. For example, in-memory multidimensional arrays can be used for constructing the materialized views that enable faster query processing on frequently accessed measures and data dimensions, while base level data (at highest granularity) is still stored in relation datasets. A drill-through operation can be used to retrieve the raw data when required.

The user view can include generic textual query languages, a visual graphical dashboard of map clients such as OpenLayers [48], Google or Bing maps or tabular representations using tools such as JPivot. A combination of these tools is often required for effective data visualization and user-friendly analysis to design multiple levels of queries.

This meta-framework can be used to represent the exact semantics of spatial aggregation operations on different view levels. For example, a query such as *"find all adjacent states where more than 5000 iPhone units where sold in 2010"* leads to a selection on thematic attributes followed by a test for the *meet* topological relation on the spatial partitions to generate the required results.

5   Conclusions and Future Work

In this paper, we present a broad overview of existing conceptual, logical and implementation strategies for spatial data warehouses (SDWs). By studying these models we arrive at a set of essential requirements for incorporating spatial data in data warehouses. These are used to propose a "meta-framework" for modeling spatial data warehouses. This framework consists of a user-friendly conceptual cube model that abstracts over logical design details such as star or snowflake schemas and implementation details such as the maintenance of materialized views. Further, user-friendly views are proposed for the SDW by means of a generic textual query language like a spatial extension to MDX, and graphical dashboards or cartographic mapping tools such as JPivot, OpenLayers or Google map visualizations. Overall, this provides a comprehensive view of the existing state-of-the-art in spatial data warehouse modeling and lays the foundation for describing the exact semantics of SOLAP operations on data cubes and to develop a complete spatial data warehousing solution.

References

1. Kimball, R., Ross, M.: The Data Warehouse Toolkit: The Complete Guide to Dimensional Modeling, 2nd edn. (2002)
2. Inmon, W.: Building the data warehouse. Wiley, Chichester (2005)
3. Han, J., Kamber, M.: Data mining: concepts and techniques
4. Gray, J., Bosworth, A., Layman, A., Pirahesh, H.: Data Cube: A Relational Aggregation Operator Generalizing Group-By, Cross-Tab, and Sub-Totals. In: Int. Conf. on Data Engineering, p. 152 (1996)
5. Pedersen, T., Jensen, C.: Multidimensional database technology. Computer 34(12), 40–46 (2002)
6. Franconi, E., Kamble, A.: A data warehouse conceptual data model. In: Proc. of Scientific and Statistical Database Management, pp. 435–436 (2004)
7. Kamble, A.: A conceptual model for multidimensional data. In: 5th Asia-Pacific Conf. on Conceptual Modelling, vol. 79, pp. 29–38 (2008)

8. Sapia, C., Blaschka, M., Höfling, G., Dinter, B.: Extending the E/R Model for the Multidimensional Paradigm. In: ER 1998: Workshops on Data Warehousing and Data Mining, pp. 105–116. Springer, Heidelberg (1999)
9. Malinowski, E., Zimányi, E.: Hierarchies in a multidimensional model: from conceptual modeling to logical representation. Data Knowledge Engineering 59(2), 348–377 (2006)
10. Tryfona, N., Busborg, F., Christiansen, J.: starER: A conceptual model for data warehouse design. In: Proc. of ACM 2nd Int. Workshop on Data Warehousing and OLAP, pp. 3–8 (1999)
11. Abelló, A., Samos, J., Saltor, F.: YAM^2: a multidimensional conceptual model extending UML. Information Systems 31 (2006)
12. Luján-Mora, S., Trujillo, J., Song, I.: A UML profile for multidimensional modeling in data warehouses. Data Knowledge Engineering 59(3), 725–769 (2006)
13. Prat, N., Akoka, J., Wattiau, I.: A UML-based data warehouse design method. Decision Support Systems 42(3), 1449–1473 (2006)
14. Golfarelli, M., Maio, D., Rizzi, S.: The Dimensional Fact Model: a Conceptual Model for Data Warehouses. Int. Journal of Cooperative Information Systems 7, 215–247 (1998)
15. Hüsemann, B., Lechtenbörger, J., Vossen, G.: Conceptual Data Warehouse Design. In: Workshop on Design and Management of Data Warehouses, pp. 3–9 (2000)
16. Zepeda, L., Celma, M., Zatarain, R.: A Mixed Approach for Data Warehouse Conceptual Design with MDA. In: Int. Conf. on Computational Science and Its Applications, pp. 1204–1217 (2008)
17. Viswanathan, G., Schneider, M.: BigCube: A MetaModel for Managing Multidimensional Data. In: Proceedings of the 19th Int. Conf. on Software Engineering and Data Engineering (SEDE), pp. 237–242 (2010)
18. Blaschka, M., Sapia, C., Höflng, G., Dinter, B.: Finding Your Way through Multidimensional Data Models. In: 9th Int. Workshop on Database and Expert Systems Applications, p. 198 (1998)
19. Vassiliadis, P., Sellis, T.: A survey of logical models for OLAP databases. SIGMOD Record 28(4), 64–69 (1999)
20. Pedersen, T., Jensen, C., Dyreson, C.: A foundation for capturing and querying complex multidimensional data. Information Systems 26(5), 383–423 (2001)
21. Kimball, R.: A dimensional modeling manifesto. DBMS Magazine 10(9), 58–70 (1997)
22. Agrawal, R., Gupta, A., Sarawagi, S.: Modeling Multidimensional Databases. In: Proceedings of the 13th Int. Conf. on Data Engineering, pp. 232–243 (1997)
23. Rivest, S., Bedard, Y., Marchand, P.: Toward better support for spatial decision making: defining the characteristics of spatial on-line analytical processing (SOLAP). Geomatica-Ottawa 55(4), 539–555 (2001)
24. Malinowski, E., Zimányi, E.: Representing spatiality in a conceptual multidimensional model. In: Proceedings of the 12th Annual ACM Int. Workshop on Geographic Information Systems, pp. 12–22. ACM, New York (2004)
25. Ferri, F., Pourabbas, E., Rafanelli, M., Ricci, F.: Extending geographic databases for a query language to support queries involving statistical data. In: Int. Conf. on Scientific and Statistical Database Management, pp. 220–230. IEEE, Los Alamitos (2002)
26. Jensen, C., Kligys, A., Pedersen, T., Timko, I.: Multidimensional data modeling for location-based services. The Int. Journal on Very Large Data Bases (VLDBJ) 13(1), 1–21 (2004)
27. Bimonte, S., Tchounikine, A., Miquel, M.: Geocube, a multidimensional model and navigation operators handling complex measures: Application in spatial olap. In: Advances in Information Systems, pp. 100–109 (2006)

28. Bimonte, S., Miquel, M.: When spatial analysis meets olap: Multidimensional model and operators. IJDWM 6(4), 33–60 (2010)
29. Scotch, M., Parmanto, B.: SOVAT: Spatial OLAP visualization and analysis tool. In: Proceedings of the 38th Annual Hawaii Int. Conf. on System Sciences (HICSS), p. 142b. IEEE, Los Alamitos (2005)
30. Han, J., Koperski, K., Stefanovic, N.: GeoMiner: a system prototype for spatial data mining. In: Proceedings of the 1997 ACM SIGMOD International Conference on Management of Data, pp. 553–556. ACM, New York (1997)
31. Marchand, P., Brisebois, A., Bédard, Y., Edwards, G.: Implementation and evaluation of a hypercube-based method for spatiotemporal exploration and analysis. ISPRS Journal of Photogrammetry and Remote Sensing 59(1-2), 6–20 (2004)
32. Shekhar, S., Lu, C., Tan, X., Chawla, S., Vatsavai, R.: MapCube: A visualization tool for spatial data warehouses. Geographic Data Mining and Knowledge Discovery, 73 (2001)
33. Rivest, S., Bédard, Y., Proulx, M., Nadeau, M., Hubert, F., Pastor, J.: SOLAP technology: Merging business intelligence with geospatial technology for interactive spatio-temporal exploration and analysis of data. ISPRS Journal of Photogrammetry and Remote Sensing 60(1), 17–33 (2005)
34. Gomez, L., Haesevoets, S., Kuijpers, B., Vaisman, A.: Spatial aggregation: Data model and implementation. Information Systems 34(6), 551–576 (2009)
35. Stefanovic, N., Han, J., Koperski, K.: Object-based selective materialization for efficient implementation of spatial data cubes. IEEE Transactions on Knowledge and Data Engineering 12(6), 938–958 (2002)
36. Han, J., Stefanovic, N., Koperski, K.: Selective materialization: An efficient method for spatial data cube construction. In: Wu, X., Kotagiri, R., Korb, K.B. (eds.) PAKDD 1998. LNCS, vol. 1394, pp. 144–158. Springer, Heidelberg (1998)
37. Rigaux, P., Scholl, M., Voisard, A.: Introduction to spatial databases: with application to GIS. Morgan Kaufmann, San Francisco (2002)
38. Malinowski, E., Zimányi, E.: Spatial hierarchies and topological relationships in the spatial multiDimER model. In: Jackson, M., Nelson, D., Stirk, S. (eds.) BNCOD 2005. LNCS, vol. 3567, pp. 17–28. Springer, Heidelberg (2005)
39. GeoMondrian Project (December 2010),
    http://www.spatialytics.org/projects/geomondrian/
40. Pentaho Analysis Services: Mondrian Project (December 2010),
    http://mondrian.pentaho.org/
41. Java Topology Suite (JTS) (December 2010),
    http://www.vividsolutions.com/jts/
42. Shekhar, S., Chawla, S.: Spatial databases: a tour. Prentice-Hall, Englewood Cliffs (2003)
43. Guting, R., Schneider, M.: Realm-based spatial data types: the ROSE algebra. The VLDB Journal 4(2), 243–286 (1995)
44. Guting, R., De Ridder, T., Schneider, M.: Implementation of the ROSE algebra: Efficient algorithms for realm-based spatial data types. In: Egenhofer, M.J., Herring, J.R. (eds.) SSD 1995. LNCS, vol. 951, pp. 216–239. Springer, Heidelberg (1995)
45. Open GIS Consortium: Reference Model (December 2010),
    http://openlayers.org
46. Schneider, M., Behr, T.: Topological relationships between complex spatial objects. ACM Transactions on Database Systems (TODS) 31(1), 39–81 (2006)
47. Ruiz, C., Times, V.: A taxonomy of solap operators. In: XXIV Simpósio Brasileiro de Banco de Dados, Fortaleza, CE (2009)
48. OpenLayers mapping client (December 2010), http://openlayers.org

Optimal Bandwidth Selection for Density-Based Clustering

Hong Jin[1], Shuliang Wang[1,2,*], Qian Zhou[2], and Ying Li[3]

[1] State Key Laboratory of Software Engineering, Wuhan University, Wuhan 430079, China
slwang2005@whu.edu.cn
[2] International School of Software, Wuhan University, Wuhan 430079, China
[3] School of Mathematics and Statistics, Wuhan University, Wuhan 430079, China

Abstract. Cluster analysis has long played an important role in a wide variety of data applications. When the clusters are irregular or intertwined, density-based clustering is proved to be much more efficient. The quality of clustering result depends on an adequate choice of the parameters. However, without enough domain knowledge the parameter setting is somewhat limited in its operability. In this paper, a new method is proposed to automatically find out the optimal parameter value of the bandwidth. It is to infer the most suitable parameter value by the constructed model on parameter estimation. Based on the Bayesian Theorem, from which the most probability value for the bandwidth can be acquired in accordance with the inherent distribution characteristics of the original data set. Clusters can then be identified by the determined parameter values. The results of the experiment show that the proposed method has complementary advantages in the density-based clustering algorithm.

Keywords: Density-based clustering, Bayesian posterior probability estimation, Optimal bandwidth selection.

1 Introduction

The rapid advance in spatial data acquisition, transmission and storage results in the growth of vast computerized datasets at unprecedented rates. For numerous data–based applications, efficient methods of data analysis can make use of the information implicitly contained in the data [1]. As a primary means of data analysis, cluster analysis helps to understand the natural grouping and structure in a dataset [2]. The clustering algorithms can be regarded as an approach to get insight into the distribution of a data set. According to the different criteria of similarity measurement and clustering evaluation, the commonly used clustering algorithms may be based on partition, hierarchy, density and grid [3]. The density-based algorithms are to discover the clusters of arbitrary shape and it is easy to be extended. Each cluster corresponds to a relatively dense area of data distribution, by looking for low-density regions separated by the connectivity of high-density area [4].

---

* Corresponding author.

J. Xu et al. (Eds.): DASFAA Workshops 2011, LNCS 6637, pp. 156–167, 2011.

Meanwhile, it is not sensitive to the existence noise. However, the quality of its clustering result mainly depends on the input parameters. DENCLUE is such a representative.

DENCLUE (DENsity based CLUstEring) is a generic clustering algorithm based on kernel density estimation. By means of adjusting the bandwidth of the kernel function, the density-based clustering algorithm is able to efficiently get insight into the distribution of a data set. Since the effectiveness of kernel density estimation depends on the selection of bandwidth, the algorithm is supposed to optimize the selection of the bandwidth in order to improve the accuracy. In this paper a new approach is proposed to optimize the bandwidth selection by using Bayesian inference. So the appropriate clustering results can be acquired more quickly in accordance with the inherent distributed characteristics of the original data set. Theoretical analysis and experimental results show that the approach has good clustering quality and computing performance, and the parameter selection is more objective with good robustness.

The rest of the paper is organized as follows. In section 2, the related principles are introduced such as kernel density estimation and Bayesian inference. And it illustrates how the parameter estimation model can be constructed with respect to the above mentioned principles. In section 3, it is the process of the proposed algorithm that includes its theoretical foundations such as Bayesian posterior density estimation and MCMC (Markov Chain Monte Carlo) method as well as the rationality of the parameter setting method. In section 4, an experimental evaluation is provided. For the experiments, analog data is used as the related paper commonly used. The results are concluded in section 5 along with some issues for future work.

2 Related Principles

Density-based clustering is to model the distribution density of dataset as the sum of the influences of individual data objects by using the functions under kernel density estimation [5]. For kernel density estimation, the contribution of each point to the overall density function is expressed by an influence or kernel function [6]. The overall density function is simply the sum of the influence functions associated with each data point.

2.1 Basic Idea of Density Based Clustering Algorithm

DENCLUE is a clustering algorithm on a group of density distribution functions [7]. Given a space Ω containing dataset D=$\{x_1, x_2, ..., x_n \}$ in d -dimensional space, the basic idea of the algorithm is followed.

(1) The kernel density estimator of the overall density function
Assume that the probability distribution associated with each observed data point uniformly distributes in different dimensions. $\forall x \in \Omega$, the probability density can be estimated as equation (1).

$$\hat{f}^{D}(x) = \frac{1}{nh^{d}} \sum_{i=1}^{n} K\left(\frac{x - x_i}{h} \right) \tag{1}$$

Where, K(x) is the kernel function in terms of product kernel that will be explained in the subsequent part. It generally chooses a symmetric density function that has single peak at the origin such as square wave function and Gaussian function. Constant h is called the bandwidth of the kernel function [8]. In accordance with the above hypothesis, the bandwidth value in different dimensions can be viewed as the same.

(2) Center-Defined Cluster
Given a density-attractor x^*, if there exists $C \subseteq D$ satisfying the condition that $\forall x \in C$, x is density attracted by $x*$ and $f^D(x^*) \geq \xi$ where ξ is the preset parameter noise threshold, and C is called the cluster centered with x^*.

(3) Arbitrary-Shape Cluster.
An arbitrary-shape cluster for the set of density-attractors X is a subset $C \subseteq D$ where

① $\forall x \in C$, $\exists x^* \in X$: $f^D(x^*) \geq \xi$, x is density-attracted to x^*, and
② $\forall x_i^*, x_j^* \in X(i \neq j)$: $\exists$ a path $P \subset Q$ from x_i^* to x_j^* with $\forall y \in P$: $f^D(y) \geq \xi$.

Obviously, there are two important preset parameters in the algorithm such as the bandwidth and the noise threshold. The bandwidth affects the efficiency of the overall density function estimator as well as the number of the density-attractors or the clusters. Let h_{max} represents the maximum of the bandwidth under the condition that the density function $f^D(x)$ has only one density-attractor. While h_{min} represents the minimum of the bandwidth under the condition that the density function $f^D(x)$ has n density-attractors. Each value in the interval $[h_{min}, h_{max}]$ corresponds to an appropriate clustering result about the dataset [8]. Consequently, the value of the bandwidth can be selected from the interval $[h_{min}, h_{max}]$ in order to naturally acquire hierarchy clustering result. Considering the optimal bandwidth, it is acknowledged that the bandwidth value in the maximal interval $I \subset [h_{min}, h_{max}]$ which keep the number of density-attractors remain constant corresponds to an appropriate clustering result. When the bandwidth h is ready, the clustering result can be determined by the noise threshold ξ.

2.2 Parameter Estimation Model

It is important to set the parameter for the density-based clustering algorithm. Here is the bandwidth to be estimated under Bayesian Theorem. First, the kernel density estimation is used to equationte the likelihood function. Then, the parameter estimation is modeled by choosing an empirical prior density function, along with MCMC (Markov Chain Monte Carlo) method to sample the parameter space.

(1) Bayesian Inference
Bayesians views unknown parameter values as random quantities using probability distributions to represent its uncertainty [9].

Let D represent the observed data and θ represent the model parameters. The joint probability distribution P(D, θ) over all random quantities is equation (2), in which θ is able to be multi-dimensional [10].

$$P(D, \theta) = P(\theta)P(D|\theta) \tag{2}$$

Where we call P(θ) the prior density and P(D|θ) the likelihood function. Once given the observed data D, the posterior distribution of the parameter θ can be acquired as equation (3) according to Bayes Theorem.

$$P(\theta|D) = \frac{P(\theta)P(D|\theta)}{\int P(\theta)P(D|\theta)d_\theta} \tag{3}$$

It represents the distribution of θ condition on the observed data D. Since the denominator of equation (3) is not relevant to θ, it can be simplified as being proportional to the prior times the likelihood and formalized as equation (4).

$$P(\theta|D) \propto P(\theta)P(D|\theta) = P(\theta)L(\theta;D) \tag{4}$$

Seen from equation (4), the posterior is a conditional distribution for the model parameters given the observed data.

(2) Parameter Space Sampling

By obtaining samples x_t(t=0,1,…,n) from the distribution $P(x)$, various features of the distribution $P(x)$ can be calculated. For a Bayesian, x is comprised of model parameters and $P(x)$ is called a posterior distribution [9]. From equation (3), with MCMC it has to know the distribution of x up to the constant of the normalization [11]. The notation t expresses an ordering or sequence to the random variables in MCMC. When x_t are independent, the approximation can be made as accurate as needed by increasing n. Under the condition that x_t are not independent, it doesn't limit its usefulness as long as they are sampled from the entire domain of $P(x)$ in correct proportions [12]. By means of constructing a Markov Chain taken $P(x)$ as its stationary distribution, this can be resolved.

In the MCMC methods, the key is how to construct chains that the stationary distribution is the interested one. In this paper, the random-walk metropolis- hastings sampler are chosen to construct the Markov Chain when generating a sequence samples of the target distribution referred to as the posterior distribution on the model parameter.

3   Density-Based Clustering Algorithm Using the Optimal Bandwidth Selection

The effectiveness of the density-based algorithm depends on the subjective preset of the two parameters bandwidth and noise threshold. The choice of the bandwidth has significant impact on the estimation result of the overall density function causing difference with respect to the number and the pattern of the clusters. If the choice of the bandwidth is closer to the original distribution of the data set, the natural clustering results and the number of the categories can be acquired. Suitable value of the noise threshold makes the algorithm focusing on the calculation of high-density area in order to decrease the computing time.

3.1   The Structure of the Algorithm

Regarding the bandwidth as the parameter to be estimated, the parameter estimation is modeled by using Bayesian method and MCMC sampling. Such the estimated bandwidth may make the overall density function better fit the inherent distribution of the original data set. When the data space is multi-dimensional, it is easy to be further extended [13]. According to the estimated bandwidth, the noise threshold can be subjectively preset before starting the clustering algorithm. With the estimated bandwidth and the existing clustering results, the noise threshold can be further adjusted to acquire a more accurate clustering pattern. Besides these, during the process of searching density-attractors, it uses conjugate gradient hill-climbing method instead of the gradient hill-climbing method to accelerate the convergence speed. The structure of the proposed approach is as Fig.1 shows.

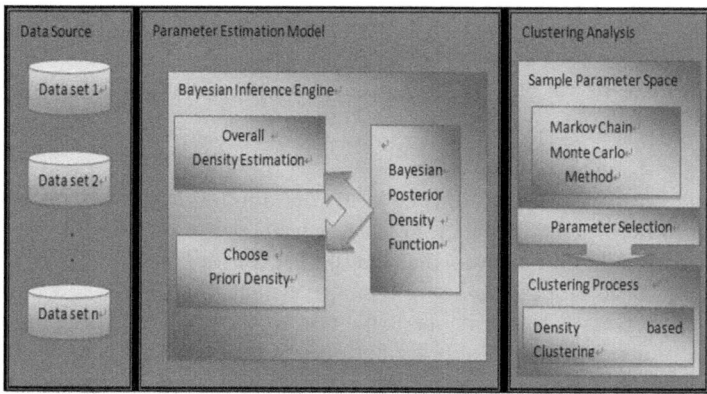

Fig. 1. Density-based Clustering Algorithm Using the Optimal Bandwidth Selection

3.2 Optimal Bandwidth Selection Model

The key step of the approach is how to select the optimal bandwidth value. In this section, the modeling process of the parameter estimation is illustrated by Bayesian Theorem and MCMC method. It first discusses the typically calculated form logarithm of the likelihood function for the parameter to be estimated. Then, with the assumed parameter prior density function, the Bayesian posterior density function of the parameter can be constructed. Using the MCMC simulations to sample the parameter space, the expected value of the parameter can be obtained.

Note that the bandwidth matrix can be restricted to a class of positive definite diagonal matrix with the corresponding kernel function known as a product kernel [14]. When choosing a full bandwidth matrix, it is identical to pre-rotating the original data with an optimal amount and then still using a diagonal bandwidth matrix. Consequently, the general form of kernel density estimator can be transformed to be as equation (5) shows.

$$\hat{f}_H(x) = \frac{1}{n} \sum_{i=1}^{n} \prod_{j=1}^{d} \frac{1}{h_j} K\left(\frac{x - X_{ij}}{h_j}\right) \tag{5}$$

In particular, $K(\cdot)$ is univariate kernel density function associated with product kernel, and h_j represents the different bandwidth value in each dimension.

According to the above kernel density estimator of $f(x)$, the log pseudo-likelihood function for the bandwidth matrix H can be got as equation (6).

$$L(x_1, x_2, ..., x_n | H) = \sum_{i=1}^{n} \log \hat{f}_{H,i}(x_i) \tag{6}$$

Where the leave-one-out estimator is as equation (7):

$$\hat{f}_{H,i}(x_i) = \frac{1}{(n-1)} \sum_{\substack{j=1 \\ j \neq i}}^{n} \pi \prod_{m=1}^{d} \frac{1}{h_m} K\left(\frac{x_i - X_{jm}}{h_m}\right) \tag{7}$$

Regarding the non-zero elements of the bandwidth matrix as parameters, the posterior density of the parameters based on the log pseudo-likelihood function can be obtained according to equation (4).

Assume that the prior density of each non-zero component of H is as probability distribution function (8) shows:

$$P(h_j) \propto \frac{1}{1 + h_j^2} \tag{8}$$

It is proved to be effective that the above priors can put low probability on the region of the parameter space where the likelihood function is flat [15]. We can get the joint prior of all elements of H in the product form of these marginal priors. Then, using Bayes Theorem, the logarithmic posterior of H is as equation (9) shows.

$$P(H|D) \propto \sum_{j=1}^{d} \log P(h_j) + \sum_{i=1}^{n} \log \hat{f}_{H,i}(x_i) \tag{9}$$

In case of a diagonal bandwidth matrix, all elements of H can be sampled through the Metropolis-Hastings algorithm with the acceptance probability computed through (9). Meanwhile, the corresponding kernel function known as a product kernel.

4 Case Study

To demonstrate the effectiveness and efficiency of the proposed method, an experiment is performed using synthetic data. In this section, it starts the algorithm described in section 3 via several bivariate data sets. Given a dataset generated from simulation, we sample the diagonal bandwidth matrix from its corresponding posterior density defined in equation (9) using the random-walk metropolis-hastings algorithm.

After the sample paths of H for each dataset are obtained, the posterior mean acts as an estimation of optimal bandwidth are calculated. With the estimated bandwidth, the density based clustering algorithm is initialized. And for another parameter the noise threshold is subjectively set to a certain value which can be adjusted by the existed clustering result.

4.1 The Procedure of Optimal Bandwidth Selection

Taken two-dimensional dataset as an example, the process of optimal bandwidth selection can be instantiated as follows. According to the above information, the parameter estimation model by Bayesian method and MCMC sampling can be constructed as equation (9). By means of simulation, it provides three data sets that are commonly used in the relative papers. In the experiment, the three synthetic sample databases depicted in Fig.2 are used.

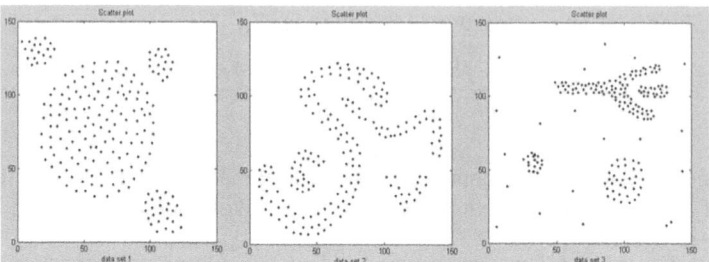

Fig. 2. The Original Data Sets

Therefore, the accuracy of the proposed algorithm is evaluated by visual inspection. Judging from the morphological, for sample dataset 1 there are four ball-shaped clusters with significantly different sizes. For sample dataset 2 it contains four clusters of non-convex shape. While in sample dataset 3 it has four clusters of different shape and size with additional random noise. In order to clearly distinguish the different clusters in the clustering results, it visualizes each cluster found by different color.

As for each dataset the optimal bandwidth are calculated by the parameter estimation model. With respect to the corresponding data set, the optimal bandwidth can be acquired by the expected value of the sample points in the generated Markov Chains.

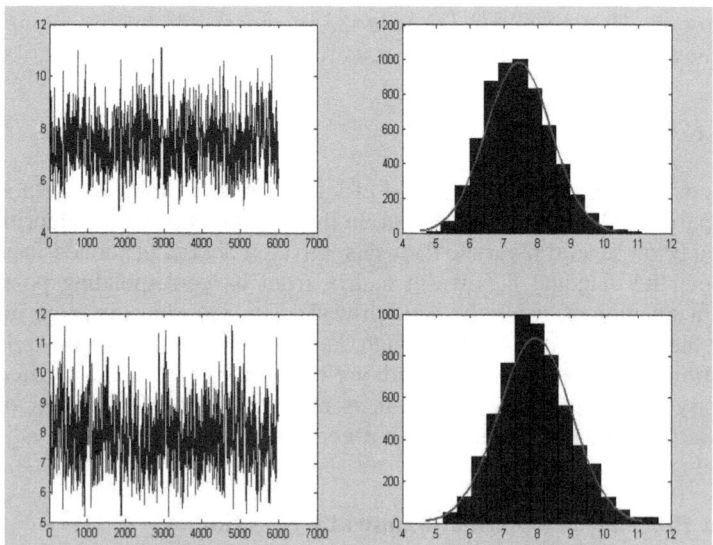

Fig. 3. The Markov chain and Statistic histogram of dataset 1

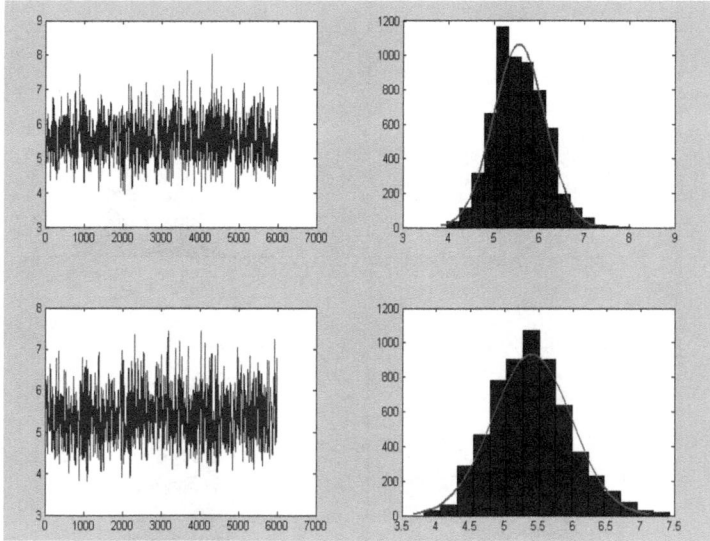

Fig. 4. The Markov chain and statistic histogram of dataset 2

For dataset 1, the left panel in Fig.3 shows the Markov Chains in two dimensions, while the right panel represents the Statistic Histogram to the relevant dimension.

On the basis of the calculation result, the optimal bandwidth for data set1 is equal to 7.9. As shown in Fig.3, it indicates that the bandwidth values approximately the same in different dimensions. The sample values for the bandwidth in both dimensions centralized in the interval [6, 10, 16]. The corresponding Statistic Histograms reflect the most likely values for the bandwidth.

Taking the arbitrary shape clustering into consideration, an experiment on dataset 2 is also given. The same as stated above, in Fig.4 the simulated result is shown. It can be seen that the bandwidth values in different dimensions are fairly close to each other.

In accordance with the calculated result, the optimal bandwidth for data set2 is equal to 5.4. Seen from the corresponding statistic histogram, the values around 5.5 is the most frequently values sampled in the parameter space. And the possible values for the bandwidth in both dimensions distribute in interval [4, 6].

Especially, in the third case a specific realization of data set3 with random noise is provided. The simulated result is as Fig.5. Obviously, the bandwidth values in different dimensions exactly not distribute in the approximate interval which is differ from the above two cases. It just reflects that the overall density function of this dataset is affected by the random noise.

According to the simulation results, the bandwidth values in different dimension are respectively equal to 13.2 and 5.8. It demonstrates the inherent distribution of the original data. Here the related statistic histograms represent the most probability values of the bandwidth in different dimensions. During the process of clustering analysis, a discussion on both dimensions will be given.

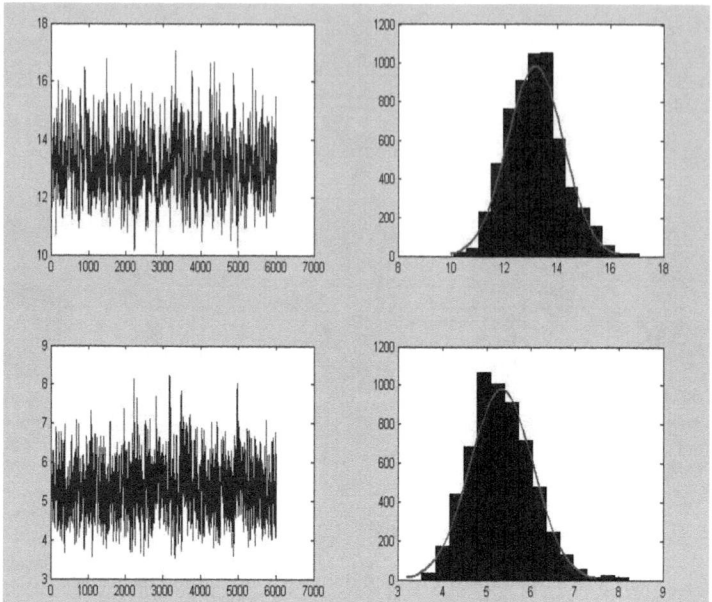

Fig. 5. The Markov chain and statistic histogram of dataset 3

4.2 Clustering Analysis

Density-based clustering algorithm needs two parameters such as the noise threshold and the bandwidth. Together with the optimal bandwidth acquired by the above simulation, the preset noise threshold value may be used to initialize the density-based clustering algorithm. Comparing to the selection of bandwidth, the selection of noise threshold is less important when determining the clustering results.

The calculated bandwidth value for each dataset can be used to start the clustering process. For data set1, the noise threshold $\xi=2$ and the optimal bandwidth $\sigma=7.9$, the clustering result of which is in Fig.6. Obviously, the clusters found by this approach

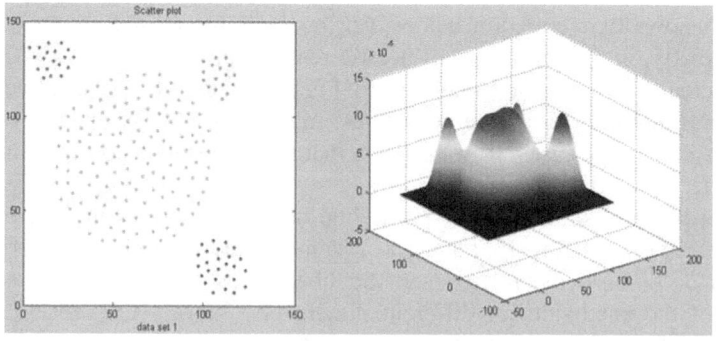

Fig. 6. The clustering result of dataset 1

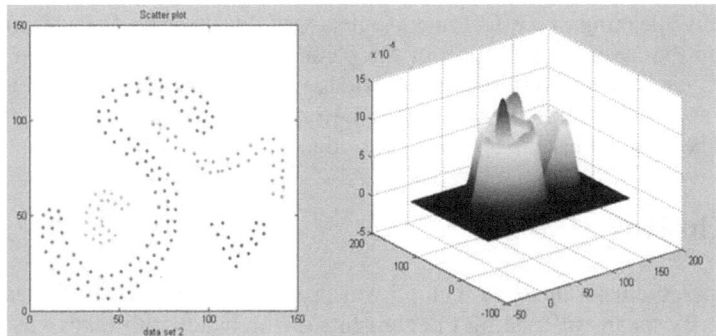

Fig. 7. The clustering result of dataset 2

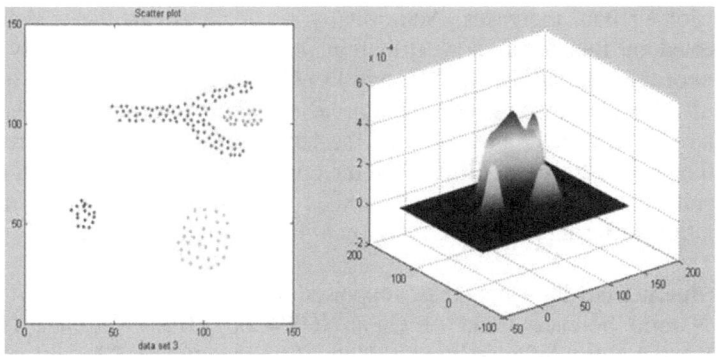

Fig. 8. The clustering result of dataset 3

well reflect the distribution of the original data set. Meanwhile, the density function on the basis of kernel estimator for dataset 1 is also given.

The experiment process on dataset 2 mainly reflects the applicability of arbitrary shape clusters. According to the above bandwidth value calculated for dataset 2, the clustering result can be acquired as Fig.7 shows where the preset parameter $\xi=2$ and $\sigma=5.6$. For visualization, the density function based on kernel estimator is given simultaneously.

When there is noise, the experiment on data set3 is given. Based on the above analysis, the bandwidth values in different dimensions are not the same. One is 5.3 and the other is 13.2, which indicates that the distributed characteristic of the original dataset in two dimensions are heterogeneous. In consistent with the basic idea of density-based clustering, the bandwidth in two dimensions is regarded as the same. Therefore, two cases are respectively considered in simulation. Under the parameter values $\xi=2$ and $\sigma=13.2$, it can be seen that the original dataset is divided into four clusters as Fig.8 shows. While under the conditions that $\xi=2$ and $\sigma=5.3$, the original dataset is divided into several clusters affected by the noise which are not given here. To maintain consistency, the density function of dataset 3 is given too.

Generally speaking, the effect and efficiency of density-based clustering algorithm depends on the carefully selection of the parameter value. The choice of the related parameters has reliance on domain knowledge or subjectivity. Nevertheless, for the proposed approach in this paper, one of the important parameters such as the bandwidth can be automatically adjusted in accordance with the original dataset.

5 Conclusions

In this paper, a cluster analysis method was proposed on the density-based clustering algorithm. By means of treating the elements of the bandwidth matrix as parameters to be estimated, it constructed a parameter estimation model by Bayes Theorem. It provides MCMC algorithms to sample the parameter space. Numerical simulations show that the resulting bandwidths are superior and it has no increased difficulty as the dimension of data increases. Additionally, its main advantage is that the bandwidth selected by this parameter estimation model can more accurately reflect the distribution of the original dataset. Though Denclue is not fundamentally a grid-based technique, it does employ a grid-based approach to improve efficiency. The length for the grid-base is amount to the value of the bandwidth. In the further research, the length of the grid-base can be different in the corresponding dimension on the basis of the bandwidth matrix. Obviously, under this condition it will be more efficient and accurate in the clustering result in consistent with the original dataset.

Acknowledgements. This paper is supported by National 973 (2007CB310804), National Natural Science Fund of China (6074300), Best National Thesis Fund (2005047), and Natural Science Fund of Hubei Province (CDB132).

References

1. Ankerst, M., Breuing, M.M., Kriegel, H.P.: OPTICS: ordering points to identify the clustering structure. In: Proc. of the 1999 ACM SIGMOD International Conference on Management of Data, pp. 49–60. ACM Press, New York (1999)
2. Hinneburg, A., Keim, D.A.: An efficient approach to clustering in large multimedia databases with noise. In: Proc of the 4th International Conference on Knowledge Discovery and Data mining, pp. 58–65. AAAI Press, Menlo Park (1998)
3. George, K., Han, E.H., Kumar, V.: CHAMELEON: a hierarchical clustering algorithm using dynamic modeling. IEEE Computer 27(3), 329–341 (1999)
4. Ester, M., Kriegel, H.P., Sander, J.: A density-based algorithm for discovering clusters in large spatial databases with noise. In: Proc.of the 2nd International Conference on Knowledge Discovery and Data Mining, pp. 226–231. AAAI Press, Menlo Park (1996)
5. Tan, P.N., Steinbach, M., Kumar, V.: Introduction to Data Mining. Pearson Education, London (2006)
6. Gentle, J.E.: Computational Statistics. Springer, New York (2001)
7. Han, J., Kamber, M.: Data Mining: Concepts and Techniques. Morgan Kaufmann, San Francisco (2000)
8. Gan, W.Y., Li, D.Y.: Hierarchical Clustering based on Kernel Density Estimation. Journal of System Simulation 16(2), 302–309 (2004)

9. Dellaportas, P., Forster, J.J., Ntzourfras, I.: On Bayesian model and variable selection using MCMC. Statistic and Computing 12(2), 27–36 (2002)
10. Gelman, A., Carlin, J.B., Stern, H.S., Rubin, D.B.: Bayesian Data Analysis, 2nd edn. Chapman&Hall, London (2004)
11. Chen, M.H., Shao, Q.M., Ibrahim, J.G.: Monte Carlo Methods in Bayesian Computation. Springer, New York (2000)
12. Gilks, W.R., Richardson, S., Spiegelhalter, D.J.: Introducing Markov chain Monte Carlo. In: Gilks, W.R., Richardson, S., Spiegelhalter, D.T. (eds.) Markov Chain Monte Carlo in Practice, pp. 1–19. Chapman and Hall, London (1996a)
13. Terrell, G.R., Scott, D.W.: Variable kernel density estimation. Annals of Statistics (20), 1236–1265 (1992)
14. Duong, T., Hazelton, M.L.: Plug-in Bandwidth Selectors for Bivariate Kernel Density Estimation. Journal of Nonparametric Statistics (15), 17–30 (2003)
15. Scott, D.W.: Multivariate Density Estimation: Theory, Practice, Visualization. Wiley, New York (1992)
16. Fang, M., Wang, S.L., Jin, H.: Spatial Neighborhood Clustering Based on Data Field. In: Cao, L., Feng, Y., Zhong, J. (eds.) ADMA 2010, Part I. LNCS, vol. 6440, pp. 262–269. Springer, Heidelberg (2010)

Developing an Oracle-Based Spatio-Temporal Information Management System

Lei Zhao, Peiquan Jin, Lanlan Zhang, Huaishuai Wang, and Sheng Lin

School of Computer Science and Technology,
University of Science and Technology of China, 230027, Hefei, China
jpq@ustc.edu.cn

Abstract. In this paper, we present an extension of Oracle, named STOC (Spatio-Temporal Object Cartridge), to support spatio-temporal data management in a practical way. The extension is developed as a PL/SQL package and can be integrated into Oracle to offer spatio-temporal data types as well as spatio-temporal operations for various applications. Users are allowed to use standard SQL to access spatio-temporal data and functions. After an overview of the general features of STOC, we discuss the architecture and implementation of STOC. And finally, a case study of STOC is presented, which shows that STOC is effective to represent and query spatio-temporal data on the basis of Oracle.

1 Introduction

Nowadays, many applications show the demand on moving objects management. However, traditional database systems, i.e., the relational DBMSs, can not deal with spatio-temporal data efficiently. Therefore, it has been one of the most important goals in recent research on spatio-temporal databases to design and implement a practical spatio-temporal DBMS.

Previous research on spatio-temporal databases were mainly focused on spatio-temporal semantics[1-2], spatio-temporal data models [3-4], spatio-temporal indexes [5] and spatio-temporal query processing [6-7], whereas little work has been done in the implementation of practical spatio-temporal DBMSs owing to the complexity of spatio-temporal data models and the lack of effective implemental techniques. Although some commercial or open-source DBMSs [8-10] offer support for handling special data types, it is still not feasible to use them to support spatio-temporal data. Recently, the object-relational database technology has been paid a lot of attention in spatio-temporal data model [11] and system implementation, due to its extensibility on new data types and functions.

In this paper, we present an extension of Oracle, named STOC (Spatio-Temporal Object Cartridge), to support spatio-temporal data management in a practical way. The unique features of STOC can be summarized as follows:

(1) It is SQL-compatible and built on a widely-used commercial DBMS (*see Section 2*), namely Oracle. Thus it can be easily used in real spatio-temporal database applications and provides a practical solution for spatio-temporal data management under current database architecture.

J. Xu et al. (Eds.): DASFAA Workshops 2011, LNCS 6637, pp. 168–176, 2011.
© Springer-Verlag Berlin Heidelberg 2011

(2) It supports various spatio-temporal data types (*see Section 3.1*), such as *moving number*, *moving bool*, *moving string*, *moving point*, *moving line*, and *moving region*. Combined with the ten types of spatio-temporal operations (*See Section 3.2*) supported by those new data types, users can represent many types of spatio-temporal data involved in different spatio-temporal applications and query different spatio-temporal scenarios.

(3) It can support real spatio-temporal applications (*See Section 4*). A case study concerning moving taxi cars shows that the STOC system is able to store spatio-temporal data captured by GPS and GIS technologies, and also able to represent different types of spatio-temporal queries.

2 Overview of STOC

STOC (Spatio-Temporal Object Cartridge) is a spatio-temporal cartridge based on Oracle and Oracle Spatial. The detailed implemental architecture of STOC is shown in Fig.1. STOC extends spatio-temporal data types and operations using the PL/SQL scripting language. Once installed, STOC becomes an integral part of Oracle, and no external modules are necessary. Users can use standard SQL to gain spatio-temporal support from Oracle. No external work imposes on users.

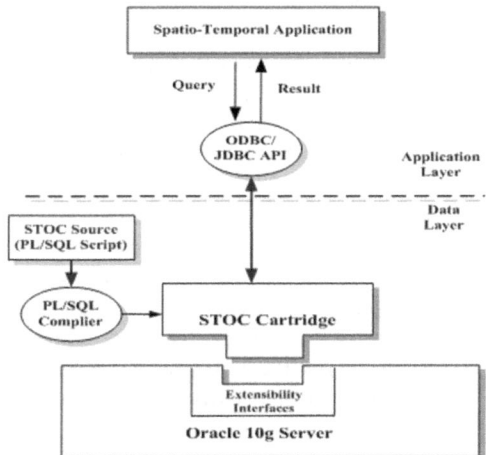

Fig. 1. The Architecture of STOC

We design the spatio-temporal data type model in STOC in order to represent the spatio-temporal objects and complex spatio-temporal changes. STOC extends two categories of new data types into Oracle, namely spatio-temporal data types and temporal data types (as shown in Fig.2). As Oracle has already supported spatial data management from its eighth version, which is known as Oracle Spatial, we build STOC on the basis of Oracle Spatial so as to utilize its mature technologies in spatial data management. The spatio-temporal data types contain moving spatial types and moving base types. The former refers to moving spatial objects in real world, while the latter refers to those numeric, Boolean, or string values changing with time.

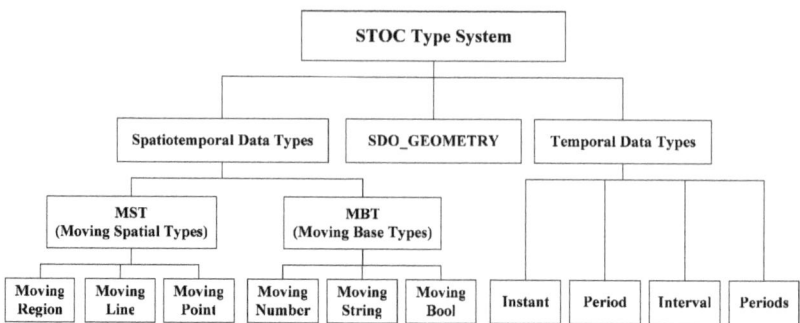

Fig. 2. The Type System of STOC

3   Implementation of STOC

3.1   Moving Data Types in STOC

3.1.1   Moving Base Types (MBT)

MBT represents the base data types changing with time, including *Moving Bool, Moving String* and *Moving Number*, which are the essentials for describing spatio-temporal changes.

Moving Bool and *Moving String* represent the discretely-changing Boolean values and String values respectively. It consists of a set of *period snapshots* ordered by time. A period snapshot is a <*value, [from, to)*> pair. The value of object is constant at the period *[from, to)*. We use the VARRAY data type to organize the set of period snapshots, which is the same as follows.

Moving Number represents the changing numerical values with time. Moving number is a set of *moving number period snapshots* ordered by time. Moving number period snapshot is a quadruple <*t_type, num, chR, [from, to)*>. The *t_type* is the change type of moving number snapshot, when it equals to *0* meaning discrete change, or equals to *1* meaning continuous change. The *num* is the value at the beginning of this period. The *[from, to)* is the valid period about this snapshot. The *chR* represent the change rate of value in this period. If the change of moving number is discrete, *chR* is always *0*.

3.1.2   Moving Spatial Types (MST)

MST represent the spatial data types changing with time, including *Moving Point, Moving Line* and *Moving Region*, which are the cores of moving objects. MST object is a three-dimension object with spatial space (two-dimension) and temporal space (one-dimension).

Figure 3 shows the definition for moving point using PL/SQL. The *t_type* is the change type of moving point, when it equals to *0* meaning discrete change, or equals to *1* meaning continuous change. The *srid* is the coordinate system id for moving point. STOC supports the coordinate systems for Oracle Spatial, such as *NULL* (user-defined coordinate system), *8307* (WGS-84 coordinate system). Those coordinate systems are described in MDSYS.CS_SRS system table. *M_point_units* is a set of

```
CREATE OR REPLACE TYPE M_POINT_U AS OBJECT
  (point mdsys.SDO_GEOMETRY,
   cRX NUMBER,--changeRateofX
   cRY NUMBER,--changeRateofY
   period T_period) ;
CREATE OR REPLACE TYPE M_POINT_U_ARRAY IS VARRAY (102400) OF M_POINT_U;
CREATE OR REPLACE TYPE M_POINT AS OBJECT
   (t_type NUMBER,
    srid NUMBER,
    m_point_units M_POINT_U_ARRAY) ;
```

Fig. 3. Defining Moving Point using PL/SQL

moving point period units ordered by time. *Moving point period unit* is a quadruple *<point, cRX, cRY, [from, to)>*. The *point* is the start position at the beginning of the period for the moving point. The *cRX* and *cRY* represent the change rates for x-axis position and y-axis position respectively. The *[from, to)* is the valid period for this moving point unit.

Since moving line only supports the discrete change, similar to moving bool and moving string, moving line is a set of *moving line period units* ordered by time. Each moving line period unit indicates the position for moving line and its valid period.

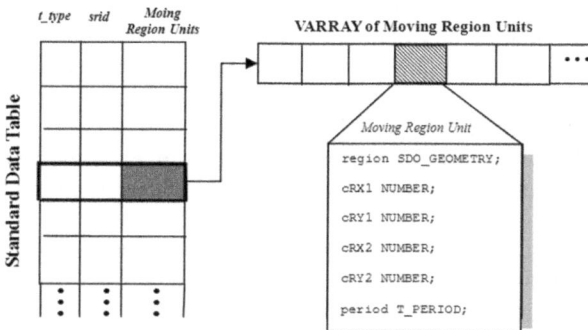

Fig. 4. Data Structure of Moving Region

The data structure of moving region is shown in Fig.4. In this data structure, the *t_type* is the change type of moving region; the *srid* is the coordinate system id; the *m_region_units* is a set of *moving region period units* ordered by time. The moving region period unit is a six-tuple. The *period* is the valid period. On the one hand, if moving region changes discretely, the *region* stores the location information of moving region in this period. On the other hand, if the change is continuous, the *region* stores the *MBR* of moving region at the beginning of this period. The *cRX1*, *cRY1*, *cRX2*, *cRY2* represent the bottom-left and top-right points' change rates for x-axis position and y-axis position respectively.

3.2   Spatio-Temporal Operations in STOC

STOC provides the following types of spatio-temporal operations to support various spatio-temporal queries. All the operations are implemented by PL/SQL and as member functions of spatio-temporal data types.

Object data management operations: for moving object, add or delete the specific period unit to manage the data of moving object.

Object attribute operations: get the attributes of moving object. For example, get the max/min value of moving number; get the change type, coordinate system and speed for moving point.

Temporal dimension project operations: get the temporal information of moving object, such as valid periods or start time about moving object.

Value dimension project operations: project the moving object to value dimension to get the value range of moving object. For moving bool/ moving string/moving number, return the set of its values. For moving point, if it changes discretely, return the set of points, if it changes continuously, return the segments of trajectory. For moving line, return the set of lines. For moving region, if it changes discretely, return the set of regions, or return the region where the moving MBR passed.

Temporal selection operations: return the value of moving object at specific *instant*, *period* or *periods*.

Quantification operations: return whether or not the moving bool/moving string always/sometimes equals to the specific bool/string in valid time.

Moving Boolean operations: return the logical relation between two moving bool object. It is used to support the complex logical query. For *not* operation, the result is false when the moving bool is true in one period and vice versa. For *and* operation, the result is false when at least one moving bool is false in one period, or the result is true when both moving bool objects are true in the same period. For *or* operation, the result is true when at least one moving bool is true in one period, or the result is false when both moving bool objects are false in the same period.

Temporal relation operations: return the temporal relations between each period unit of moving object and specific *instant*, *period*. The temporal relations are defined in [12].

Object relation operations: return the relations between moving object with moving object or non-moving object which may change with time. The data type of return value is moving string. For moving bool/moving string, return equal or not equal. For moving number, return greater than, less than or equal. For moving spatial type, return the spatial relations defined in [13].

Distance operations: return the distance between moving object with moving object or non-moving object which may change with time. The data type of return value is moving number.

4   Case Study: A Traffic Information System

To demonstrate the functions of STOC, we use a traffic dataset to study the effectiveness of STOC. The dataset was also used in BerlinMOD [14].

4.1   Create BerlinMOD Database

The schema for BerlinMOD database is shown in Fig.5. The information about moving cars is stored in table *dataScar*. Table QueryPoints /QueryRegions /QueryInstants /QueryPeriods / QueryLicences record the query conditions for position of point

```
dataScar (Licence: Varchar2,
          Model: Varchar2,Type:Varchar2, Trip: M_Point);
QueryPoints    (Id: NUMBER, Pos: SDO_GEOMETRY);
QueryRegions   (Id: NUMBER, Region: SDO_GEOMETRY);
QueryInstants (Id: NUMBER, Instant: T_Instant);
QueryPeriods   (Id: NUMBER, Period: T_Period);
QueryLicences (Id: NUMBER, Licence: Varchar2);
```

Fig. 5. Database Schema of BerlinMOD

(SDO_GEOMETRY data type), position of region (SDO_GEOMETRY data type), instant (T_Instant data type), period (T_Period data type), and the license of car (Varchar2 data type). The *id* is the prime key for each query table.

4.2 Spatio-Temporal Queries

STOC supports the various kinds of spatio-temporal queries referred by [11, 14]. According to the query conditions and types, those can be divided into six types of spatio-temporal, which are (1) temporal range query, (2) spatial range query, (3) spatio-temporal range query, (4) spatio-temporal distance query, (5) spatio-temporal topology query and (6) spatio-temporal aggregate query.

Q1 (Temporal Range Query): Where have the vehicles whose licenses are from QueryLicences been at each instant in QueryInstants?

```
SELECT LL.Licence AS Licence, II.Instant AS Instant, C.Trip.at_instant
(II.Instant) AS Pos
   FROM dataScar C, QueryLicences LL, QueryInstants II
   WHERE  C.Licence = LL.Licence AND C.Trip.at_instant (II.Instant) IS NOT
NULL;
```

Q2 (Spatial Range Query): Which license numbers belong to vehicles that have passed the points from QueryPoints?

```
SELECT PP. Pos AS Pos, C.Licence AS Licence
   FROM dataScar C, QueryPoints PP
   WHERE     sdo_geom.relate(C.Trip.get_trajectory),     'ANYINTERACT'     ,
PP.Pos,0.005)='TRUE';
```

Q3 (Spatio-Temporal Range Query): Which vehicles traveled within one of the regions from QueryRegions during the periods from QueryPeriods.

```
SELECT RR. Region AS Region, PP. Period AS Period ,C.Licence AS Licence
   FROM dataScar C, QueryRegions RR, QueryPeriods PP
   WHERE     C.Trip.at_period(PP.Period) IS NOT NULL AND sdo_geom.relate
(C.Trip.at_period(PP.Period).get_trajectory()
   ,'ANYINTERACT',RR.region,0.005)='TRUE';
```

Q4 (Spatio-Temporal Distance Query): What are the pairs of licence numbers of "trucks", that have ever been as close as 10m or less to each other?

```
SELECT V1.Licence AS Licence1, V2.Licence AS Licence2
   FROM dataScar V1, dataScar V2
   WHERE  V1.Licence < V2.Licence AND V1.Type = 'truck' AND V2.Type =
'truck' AND V1.Trip.distance_m_point(V2.Trip,0.005).get_min_number()<=10;
```

Q5 (Spatio-Temporal Topology Query): What are the pairs of licence numbers of "trucks", that have meet each other?

```
SELECT V1.Licence AS Licence1, V2.Licence AS Licence2
FROM dataScar V1, dataScar V2
WHERE  V1.Licence  <  V2.Licence AND V1.Type = 'truck' AND V2.Type =
'truck'      AND      V1.Trip.relation_m_Point      (V2.Trip,0.005).sometimes
('EQUAL')='TRUE';
```

Q6 (Spatio-Temporal Aggregate Query): Which points from QueryPoints have been visited by a maximum number of different vehicles?

```
CREATE VIEW visited_car(pp_id ,licence) AS
    SELECT PP.ID AS Pos , c.licence
    FROM QueryPoints PP, dataScar C
    WHERE sdo_geom.relate(C.Trip. get_trajectory(), 'ANYINTERACT',
    PP.Pos,0.05)='TRUE';
CREATE VIEW PosCount AS
    SELECT vc.pp_id AS Pos_id , COUNT(*) AS Hits
    FROM visited_car vc
    GROUP BY vc.pp_id ;
SELECT N.Pos_id, N.Hits
FROM PosCount  N
WHERE N.Hits = (SELECT MAX(Hits) FROM PosCount );
```

5 Related Work

Early work in implementing spatio-temporal database systems employed a layered approach [16, 17], which aimed at implementing an additional spatio-temporal layer on top of standard relational DBMS. Since the relational DBMS has little built-in spatio-temporal support, generated query may become very complex and potentially difficult to optimize for the underlying representation of spatio-temporal data. The object-relational approach, which refers to extending an extensible DBMS with spatio-temporal plug-ins, has been the most focused approach in recent years, since this makes it feasible to implement a spatio-temporal DBMS completely. Now major database vendors have provided techniques for this extensibility, including Oracle and IBM. And some extensions have been developed for spatial/temporal/spatio-temporal data management. The typical examples are Informix Timeseries Datablade [18] and Informix Geodetic Datablade [19]. The Timeseries Datablade is designed for capturing single attribute that changes over time, in which the temporal data types are very restricted. The Geodetic DataBlade is a spatio-temporal extension provided by Informix, but it is only designed for specific GIS applications.

There are also some research prototypes developed by other researchers. SECONDO [20] realizes the abstract moving object data types and operators defined in [11] and supports spatio-temporal queries and analysis by employing Berkeley DB as data storage. Nonetheless, SECONDO is unable to combine with existing DBMSs and is not compatible with SQL. SpADE [21] is built on MySQL, with the extension of a moving object module and an index module. However it only supports a limited set of spatio-temporal queries, namely the range query and the nearest neighbors query. Hermes [22] is based on the trajectory database of Oracle in support of range query, nearest neighbors query and similar trajectory query of moving point. But it does not support moving regions and spatio-temporal analysis query. TrajStore [23] only focused on spatio-temporal range query on massive datasets, thus can not suit for different spatio-temporal applications.

6   Conclusions

In this paper, we present a spatiotemporal extension of Oracle called STOC, aiming at providing practical support of spatiotemporal data management for various commercial applications. STOC is developed using the cartridge technology provided by Oracle, which enables us to add new data types as well as functions and indexing methods into the kernel of Oracle. The most valued feature of STOC is that it is SQL-compatible and allows users to develop spatio-temporal applications upon Oracle.

Acknowledgements

This work is supported by the National High Technology Research and Development Program ("863" Program) of China (No. 2009AA12Z204), the National Science Foundation of China (no. 60776801), the Open Projects Program of National Laboratory of Pattern Recognition (20090029), the Key Laboratory of Advanced Information Science and Network Technology of Beijing (xdxx1005), and the USTC Youth Innovation Foundation.

References

1. Sistla, P., Wolfson, O., et al.: Modeling and Querying Moving Objects. In: ICDE, Birmingham, UK (1997)
2. Bohlen, M., Jensen, C., Skjellaug, B.: Spatio-Temporal Database Support for Legacy. In: 1998 ACM Symposium on Applied Computing, Georgia (1998)
3. Erwig, M., Güting, R.H., et al.: Spatio-Temporal Data Types: An Approach to Modeling and Querying Moving Objects in Databases. GeoInformatica 3(3), 265–291 (1999)
4. Güting, R.H., Bohlen, M.H., et al.: A foundation for representing and quering moving objects. ACM Transactions Database Systems 25(1), 1–42 (2000)
5. Pfoser, D., Jensen, C.S., Theodoridis, Y.: Novel Approaches to the Indexing of Moving Object Trajectories. In: Proceedings of VLDB (2000)
6. Benetis, R., Jensen, C.S., et al.: Nearest neighbor and reverse nearest neighbor queries for moving objects. In: IDEAS, pp. 44–53 (2002)
7. Lema, J.A.C., Forlizzi, L., et al.: Algorithms for moving objects databases. The Computer Journal 46(6), 680–712 (2003)
8. Oracle Corp. Oracle Spatial User's Guide and Reference, http:// www.oracle.com /technology/products/spatial/spatial_doc_index.html
9. SQL Server Spatial Data, http://www.microsoft.com/sqlserver/ 2008/en/us/spatial-data.aspx
10. PostGIS, http://postgis.refractions.net/
11. Forlizzi, L., Güting, R.H., et al.: A Data Model and Data Structures for Moving Objects Databases. In: SIGMOD, pp. 319–330 (2000)

12. Allen, J.F.: Maintaining knowledge about temporal intervals. Communication of ACM 26, 832–843 (1983)
13. Egenhofer, M., Franzosa, R.: Point-Set Topological Spatial Relations. International Journal of Geographical Information Systems 5(2), 161–174 (1991)
14. Düntgen, C., Behr, T., Güting, R.H.: BerlinMOD: a benchmark for moving object databases. VLDB J. 18(6), 1335–1368 (2009)
15. Wolfson, O., Sistla, P., et al.: DOMINO: Databases for Moving Objects Tracking. In: SIGMOD, Philadelphia, PA, pp. 547–549 (1999)
16. Torp, K., Jensen, C.S., Bohlen, M.H.: Layered Implementation of Temporal DBMSs-Concepts and Techniques. In: DASFAA, Melbourne, Australia, pp. 371–380 (1997)
17. Torp, K., Jensen, C.S., Snodgrass, R.T.: Stratum Approaches to Temporal DBMS Implementation. In: IDEAS, Cardiff, Wales, pp. 4–13 (1998)
18. Informix Corp. Informix Timeseries DataBlade Module User's Guide, Version 3.1 (1997)
19. Informix Corp. Informix Geodetic DataBlade Module User's Guide, Version 2.1 (1997)
20. Güting, R.H., de Almeida, et al.: Secondo: Anextensible DBMS platform for research prototyping and teaching. In: ICDE, pp. 1115–1116 (2005)
21. Ooi, B.C., Huang, Z., et al.: Adapting Relational Database Engine to Accommodate Moving Objects in SpADE. In: ICDE, Istanbul, pp. 1505–1506 (2007)
22. Pelekis, N., Frentzos, E., Giatrakos, N., Theodoridis, Y.: HERMES:Aggregative LBS via a trajectory DB engine. In: ISIGMOD, pp. 1255–1258 (2008)
23. Cudre-Mauroux, P., Wu, E., Madden, S.: TrajStore: An Adaptive Storage System for Very Large Trajectory Data Sets. In: ICDE (2010)

Some Research Directions in FlashDB[*]

Sang-Won Lee

School of Information and Communication Engineering,
Sungkyunkwan University,
Chunchun 300, Jangan, Suwon, Korea
wonlee@ece.skku.ac.kr

Abstract. Flash memory based SSDs (flash SSDs) are becoming popular as an alternative storage to harddisk, and it is not unrealistic to witness in the foreseeable future that flash SSDs replace harddisks as the main secondary storage in enterprise databases. In fact, Oracle has already started to use flash SSDs as its main storage in performing TPC-C benchmark [1]. In this talk, we will outline some personal research directions in flash memory database (in short, FlashDB) under way. First of all, we will show the multipurpose uses of the log in flash memory, which has been mainly regarded as a write performance booster in flash memory [2][3]. As one of specific examples, we will explain how the concept of log in the in-page logging scheme can be extended to support multiversions and fast recovery in flash memory in a very effective way with a modest overhead. Second, we are investigating on a hybrid architecture of flash memory and phase change ram (i.e. PRAM). Although some advocates of non-volatile memory have predicted that flash memory will give way to non-volatile memory soon (e.g. by the year 2012), the performance of PRAM is far lagging behind its promise. For this reason, we believe that they will co-exist, complementing each other, for a while. As a hybrid architecture of flash memory and PRAM, we suggest to use PRAM as the log area in in-page logging [4], report a preliminary performance result, and explain its several architectural advantages. Third, we are exploring how to leverage flash SSDs as cache in memory hierarchy. As an alternative design, we suggest FlashCache scheme and report its preliminary performance result.

Keywords: Flash Memory, SSD, Database.

References

1. Transaction Processing Council, http://www.tpc.org
2. Lee, S.-W., Moon, B.: Transactional In-Page Logging for Multiversion Read Consistency and Recovery. In: 27th IEEE International Conference on Data Engineering. IEEE Computer Society, Los Alamitos (2011)
3. Ouyangyz, X., Nellansy, D., Wipfely, R., Flynny, D., Panda, D.K.: Beyond Block I/O: Rethinking Traditional Storage Primitives. In: 17th IEEE International Symposium on High Performance Computer Architecture. IEEE Computer Society, Los Alamitos (2011)
4. Lee, S.-W., Moon, B., Park, C., Hwang, J.-Y., Kim, K.: Accelerating In-Page Logging with Non-Volatile Memory. Bulletin of the IEEE Computer Society Technical Committee on Data Engineering 33(4), 41–47 (2010)

---

[*] This research was supported by MKE ITRC, Korea.

J. Xu et al. (Eds.): DASFAA Workshops 2011, LNCS 6637, p. 177, 2011.
© Springer-Verlag Berlin Heidelberg 2011

Page-Level Log Mapping: From Many-to-Many Mapping to One-to-One Mapping

Jing Xu, Fang Xie, and Jianhua Feng

Department of Computer Science and Technology, Tsinghua University
{j-xu08,xie-f03}@mails.tsinghua.edu.cn, fengjh@tsinghua.edu.cn

Abstract. Flash memory has been widely used as secondary storage in many systems, such as mobile devices, portable computers and enterprise servers. However, due to the unique characteristics of flash memory, the optimization of flash-based systems for exploiting the superior properties as well as overcoming the limitations of flash memory becomes an important and challenging problem. In this paper, we propose *page-level log mapping* to address this problem. It adopts *backward link* technique to optimize the logical-to-physical page mapping, which can improve the read and write performance of flash-based systems. It also incorporates flash-optimized policies for buffer management, free page allocation and garbage collection. Experimental results show that our approach achieves high efficiency across a wide range of workloads, flash types and memory constraints, and significantly outperforms state-of-the-art methods.

1 Introduction

As a non-volatile storage media, flash memory has been increasingly used in a wide spectrum of systems, including mobile devices, portable computers and enterprise servers. This mainly owes to the superior properties of flash memory compared with magnetic disks, such as lower access latency, lower power consumption, higher density, lighter weight and better shock resistance. However, conventional systems are not likely to yield the potential optimal performance on flash memory. Because they do not take into account the unique characteristics of flash memory, such as asymmetric read and write costs. Moreover, traditional *in-place update* is prohibitively expensive for flash memory due to the *erase-before-write* limitation. Therefore, the optimization of flash-based systems for exploiting the superior properties as well as overcoming the limitations of flash memory is a significantly important and challenging problem.

Many recent studies have been focused on improving the write performance of flash-based systems. One approach is based on *out-of-place update*, which is usually provided by a middle layer called *Flash Translation Layer* (FTL) [1]. This approach has the disadvantage that updating even a small part of a page will lead to invaliding the original page and writing a new page entirely. Therefore, it is not well-suited to the workloads containing a large number of random fine-grained writes, such as *online transaction processing* (OLTP) and metadata updates. To address this problem, *log-based* approaches [11,13] record the changes made to

J. Xu et al. (Eds.): DASFAA Workshops 2011, LNCS 6637, pp. 178–189, 2011.
© Springer-Verlag Berlin Heidelberg 2011

a page as logs. Multiple updates can be applied at once with a log page write to reduce the amortized update cost. For a read operation, the original data page and corresponding logs need to be fetched to re-create the up-to-date page. There are two methods for the retrieval of logs. One is to store logs in the specific regions of flash memory. For example, *in-page logging* (IPL) [11] co-locates a data page and its logs in the same block, thus the logs can be found by scanning the in-block log region. Since each read operation will cause a log region scan, the overhead might out-weight the saving from the reduced update cost. The other method is to maintain a mapping between pages and their logs. As the logs belonging to a page may be scattered over several log pages, and a log page may contain the logs of multiple pages, it is a complicated *many-to-many* mapping. To maintain such a mapping could cause large memory overhead.

In this paper, we propose a new approach called *page-level log mapping* (PLM) for the optimization of flash-based systems. The key idea is to link each data page and its logs together with *backward links* . Essentially, PLM replaces the complicated many-to-many mapping with a simple one-to-one mapping. The main contributions of our work are summarized as follows.

- We propose the PLM approach which adopts the backward links technique to optimize the logical-to-physical page mapping of flash-based systems.
- We develop two implementations of PLM incorporating flash-optimized strategies for buffer management, free page allocation and garbage collection.
- We conduct extensive experiments for performance evaluation. The results show that our approach significantly outperforms state-of-the-art methods.

The rest of this paper is organized as follows. Section 2 introduces the design overview of PLM. In Section 3, we describe the two implementations of PLM. Section 4 presents performance evaluations. Finally, we survey related work in Section 5, and summarize this paper in Section 6.

2   Design Overview

2.1   Basic Concepts

To re-create the latest version of a logical page, its *related pages*, i.e., the original data page (*B-page*) and corresponding log pages (*L-pages*), need to be read. As discussed in Section 1, existing methods either result in large memory overhead to maintain the mapping, or expensive I/O cost for log region scan. The essential issue is that the mapping is a complicated many-to-many mapping, as illustrated in Figure 1(a). For example, Logical page 1 is mapped to three related pages, i.e., B-page$_1$, L-page$_{1,2}$ and L-page$_{1,3}$, where L-page$_{1,2}$ (L-page$_{1,3}$) denotes the L-page that keeps the logs of Logical Page 1 and 2 (Logical Page 1 and 3). To address the issue, we propose to link the related pages of a logical page together with *backward links* , i.e, newer related pages point to older ones. Only the mapping between logical pages and their latest related pages, called *page-level log mapping* (PLM), needs to be maintained. For example, in Figure 1(b),

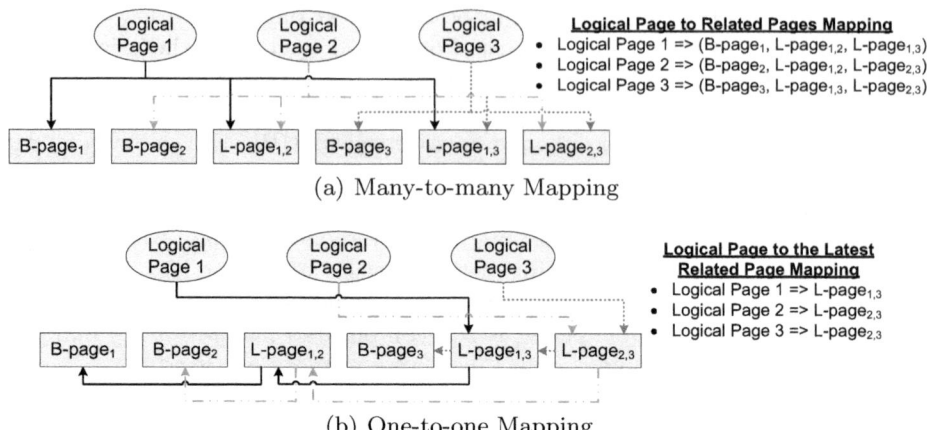

(a) Many-to-many Mapping

(b) One-to-one Mapping

Fig. 1. Mappings between logical pages and their related pages

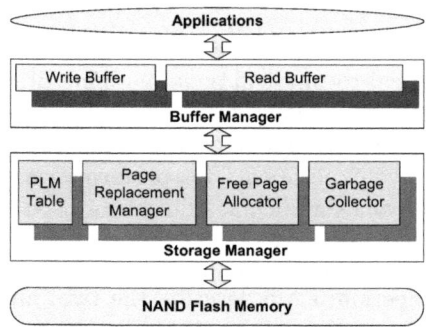

Fig. 2. The system architecture of PLM

Logical page 1 is mapped to L-page$_{1,3}$. Following the backward links started from L-page$_{1,3}$, other related pages, i.e., L-page$_{1,2}$ and B-page$_1$, can be efficiently retrieved. Since the mapping becomes a simple one-to-one mapping, it can reduce the memory overhead and avoid expensive log region scan.

2.2 System Architecture

As shown in Figure 2, our optimizations focus on the buffer manager and the storage manager. Here we briefly describe the main components, and the details are presented in Section 3. The buffer manager contains two independent buffers—a read buffer and a write buffer—with different flash-optimized eviction policies. In storage manager, the *PLM table* is used to maintain the mapping between logical pages and their latest related pages. The *page replacement manager* is incorporated to write the newest version of a logical page as a new B-page at the optimal time. Thus reading a frequently updated page will not lead to a large number of L-page accesses. The *free page allocator* can dynamically adjust the layout of B-pages and L-pages by using a free page allocation policy. The *garbage collector* is responsible for storage reclamation which involves choosing appropriate blocks, writing up-to-date pages and erasing these blocks for reuse.

3   The Implementations of the PLM Approach

In this section, we present two implementations of PLM, called *block associative log mapping* (BALM) and *fully associative log mapping* (FALM).

3.1   The Block Associative Log Mapping

To facilitate storage reclamation, block associative log mapping (BALM) co-locates the B-page and L-pages belonging to a logical page in the same block. BALM differs from IPL [11] as follows. Firstly, IPL preserves a log region in each block, while BALM writes pages in a block sequentially[1], therefore B-pages and L-pages are stored in a mixed manner. Secondly, unlike IPL has identical and static ratio between B-pages and L-pages in each block, BALM dynamically adjusts the ratio for individual blocks. Thirdly, when no free space is available in a block for newly generated logs, BALM does not merge the block immediately as IPL. Instead, it writes the up-to-date page as a new B-page to a different block. In the following, we describe the details of BALM.

Read Operations: To re-create the up-to-date page, we propose to merge logs when traversing along the backward links, then apply them to the B-page together (Algorithm 1). Once part of the page is reconstructed from the logs, it has already been the newest version. The subsequent (older) logs that update the same part can be discarded immediately (Algorithm 1 Line 4).

Page Replacement: There are two cases that BALM will replace the related pages of a logical page with a new B-page. One case called *force replacement* occurs when no free pages are available in a block. We implement force replacement in the write operation (Algorithm 2). While the logs of a logical page are to be flushed out, BALM packs them with the logs of other logical pages in the same block as an L-page. If the block is not full, BALM writes the L-page to the first free page in the block. Otherwise, it stores the current version of the logical page as a new B-page to a different block. The other case is called *active replacement* . When the number of L-pages belonging to a logical page exceeds the pre-specified *Max_Log_Size*, BALM writes the latest version of the page as a new B-page in the same block. For example, as illustrated in Figure 3(a), Logical Page 3 is mapped to B-Page$_3$ after active replacement. Compared with Figure 1(b), fewer pages need to be fetched for reading Logical Page 3.

Next, we discuss when is the optimal moment to apply active replacement. Given a sequence of page accesses, as shown in Figure 3(b). Without loss of generality, consider Logical Page 1. W_{P1} denotes the B-page write, E_{P1} indicates the time when the in-memory copy is evicted from the read buffer, and

---

[1] For MLC flash memory, the pages in a block should be written sequentially in order to avoid disturbing data on previously-written pages [16].

Input: the logical page address *addr*
Output: the current version of the logical page P

1 initialize an empty page L_Σ;
2 $P_R \leftarrow$ read page PLMTable $[addr]$; // fetch the latest related page
3 **while** P_R *is not a B-page* **do**
   // merge logs in a non-overwrite manner
4   merge the logs belonging to *addr* in P_R to L_Σ;
5      $P_R \leftarrow$ read page $P_R.backwardLink[addr]$; // follow the backward link
6 $P \leftarrow$ apply L_Σ to P_R; // finally apply the logs to the B-page
7 **return** P;

Algorithm 1. The Read Operation

Input: the logical page address *addr*, the L-page to be written P_L
Output: the return code

1 $B \leftarrow$ get the information of the block that PLMTable $[addr]$ belongs to;
2 **if** B *is not full* **then**
3   write P_L to the first free page of B;
4   update PLMTable;
5   **return** WRITE_ALL;
6 $P_N \leftarrow$ Allocate(); // allocate a new page in another block
7 write the current version of logical page *addr* to P_N; // force replacement
8 update PLMTable;
9 **return** WRITE_ONE;

Algorithm 2. The Write Operation

R_{P1} denotes the read operation. N_1 denotes the number of L-pages written between W_{P1} and R_{P1}. N_2 and N_3 are similar to N_1. We consider three cases, *no-replacement*, *replace-after-read* and *replace-before-evict*. The total costs of these cases are summarized in Table 1, where C_w is B-page write cost, C_u is L-page write cost, C_r is page read cost. The cost of replace-before-evict is smaller than replace-after-read, as $N_2 > N_1$. The essential reason is replace-before-evict applies active replacement in a lazy manner. Compared with no-replacement, replace-before-evict has lower cost when $N_2 > C_w/C_r$. Therefore we propose to set Max_Log_Size as C_w/C_r. When a logical page is evicted from the read buffer and the number of its L-pages is larger than Max_Log_Size, BALM applies active replacement. Since the value of C_w/C_r depends on the I/O characteristics of flash memory, and the value of Max_Log_Size directly impacts the ratio between B-pages and L-pages in each block, by setting Max_Log_Size as C_w/C_r, BALM in fact adjusts the ratio according to the I/O characteristics of flash memory.

Free Page Allocation: BALM allocates free pages on demand. For an L-page, it allocates the next free page if the block is not full. Otherwise, it applies force replacement. We introduce the concept of *force replacement loss* which represents, compared to active replacement, the number of L-pages that have not been

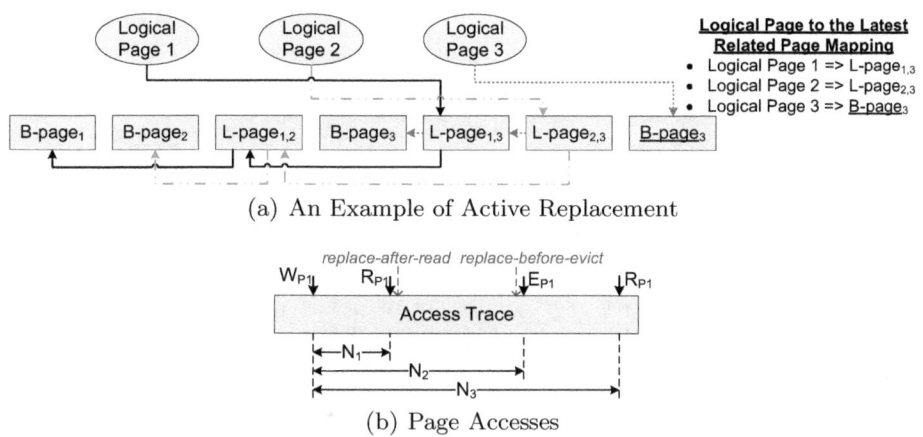

(a) An Example of Active Replacement

(b) Page Accesses

Fig. 3. The active replacement in BALM

Table 1. The total cost of applying active replacement

Case	Cost
no-replacement	$C_w + N_3 \cdot C_u + (N_1 + 1 + N_3 + 1) \cdot C_r$
replace-after-read	$C_w + N_3 \cdot C_u + (N_1 + 1) \cdot C_r + C_w + (N_3 - N_1 + 1) \cdot C_r$
replace-before-evict	$C_w + N_3 \cdot C_u + (N_1 + 1) \cdot C_r + C_w + (N_3 - N_2 + 1) \cdot C_r$

written due to force replacement. Our goal is to find the block with the minimum force replacement loss. Given a block B, we use E_B to evaluate the expectation of force replacement loss: $E_B = (Max\_Log\_Size \cdot N - \sum (Max\_Log\_Size - L_i))/N_e$, where N is the number of pages in a block, N_e is the number of empty pages in B, and L_i is the number of L-pages belonging to a logical page $P_i (\forall P_i \in B)$. The block with the minimum E_B is chosen for B-page allocation. From the above formula, we can see that it always prefers the block containing more logical pages and less L-pages. In other words, the block containing pages, which are not updated frequently, is more likely to be selected. It also prefers the block having more empty pages. Since blocks that are unlikely to become full soon are always chosen for page allocation, frequent force replacement can be avoided.

Garbage Collection: BALM calls garbage collection when empty pages in each block are less than a pre-defined threshold and there comes a B-page allocation request. BALM selects the block containing the fewest valid pages to reclaim first. The up-to-date pages of the block are re-created, then written to other blocks. The allocation of these newly generated B-pages follows the free page allocation policy described above. Finally the PLM table is updated and the block is erased. Since BALM dispenses these new B-pages to other blocks, it actually re-balances the distribution of B-pages during garbage collection.

Buffer Management: BALM has two independent buffers   a read buffer used to keep several logical pages and a write buffer for log write caching. The read buffer is divided into two regions. The *working region* contains recently used

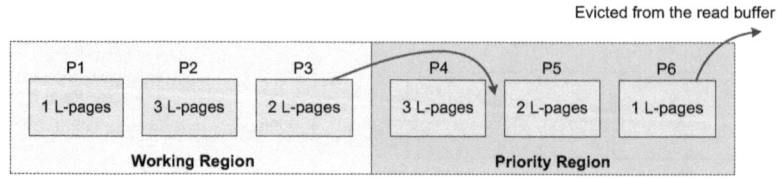

Fig. 4. The read buffer

logical pages. The *priority region* consists of pages which are candidates for eviction and ordered by the number of L-pages. When a page needs to be evicted from the read buffer, the least recently used page in the working region is moved into the priority region, then the page in the priority region that has the least number of L-pages is selected for eviction. For example, in Figure 4, $P3$ is added between $P4$ and $P5$, and $P6$ is evicted. If the evicted page is marked as needing force replacement or active replacement, it will be written out as a new B-page. Otherwise, it can be discarded immediately, since all the changes made to it have been kept in the write buffer. This policy has the advantage that it takes into account both the hit rate and read cost. The working region works like a traditional LRU list, so it can generally achieve high hit rate. The priority region always keeps the pages with more L-pages. As these pages have relatively high read costs, keeping them in the read buffer can reduce the total read cost.

The write buffer employs an LRU-like eviction strategy. When it becomes full, the logs of the least recently updated page are written to flash memory. The main idea here is that the recently updated pages are likely to be updated again in the near future. Keeping their logs in the buffer is particularly beneficial for write-coalescing. Thus the logs will not be scattered in many L-pages.

3.2   The Fully Associative Log Mapping

The fully associative log mapping (FALM) is another implementation of PLM. It stores B-pages and L-pages in different blocks, i.e., *data blocks* and *log blocks*. FALM can store the logs belonging to a logical page in any pages of any log blocks. We describe the differences between FALM and BALM as follows.

Page Replacement: Since L-pages are not stored with the B-page in the same block, FALM never triggers force replacement. But it still applies active replacement to improve the overall performance.

Free Page Allocation: FALM allocates data (log) blocks on demand, and writes pages in each block sequentially. The newly generated B-page (L-page) is written to the first free page of the data (log) block. When the data (log) block becomes full, an empty block is allocated as the new data (log) block.

Garbage Collection: FALM calls garbage collection when empty blocks are less than a pre-specified threshold. For data blocks, it selects the block with the fewest valid pages. For log blocks, the oldest log block is reclaimed first. After writing up-to-date pages to new locations, the PLM table is updated and these blocks are erased. There is no need to modify the backward links pointing to the reclaimed blocks, because relying on the updated PLM table we will access the new B-pages and never traverse along the old backward links.

Buffer Management: FALM employs similar buffer policies as BALM, we outline the differences here. Firstly, FALM does not consider force replacement when a logical page is evicted from the read buffer. Secondly, FALM is able to pack the logs belonging to multiple blocks into an L-page. Recall the log packing of BALM must be performed in the same block. Therefore, FALM generally obtains better write performance than BALM.

In summary, the strategies for page replacement and free page allocation of FALM are simpler than BALM. However, since FALM locates B-pages and L-pages in different blocks, the system initialization and garbage collection are more complicated and expensive.

4   Experimental Evaluation

We implemented the proposed PLM approach, i.e., BALM and FALM, the out-of-place update (OPU) approach and in-page logging (IPL) [11]. OPU is similar to FTL using page-level mapping [4], but implemented in the buffer and storage managers to ensure a fair comparison. We employed a trace-driven simulation environment, and evaluated the total time (including the time taken for garbage collection) under each workload to measure the performance.

Workloads: A workload consists of a sequence of read and update operations. We used two real workloads called *Index* and *Dirik*. The Index workload was generated by running a MySQL server under TPC-C benchmark. We only kept the index accesses, as it represents the access pattern containing a large number of random fine-grained updates. The Index workload consists of 19.9 million operations. The Dirik workload was collected from a laptop running real user applications [5]. It is a read-intensive workload, and the update granularity is larger than the Index workload. About 262 thousand operations are contained in this workload. In order to investigate the performance when varying the granularity and percentage of update operations, we also used two synthetic workloads called *Random* and *Normal*. The access pattern of the Random workload was randomly generated, while the Normal workload follows a normal distribution. Each of them comprises 4.19 million operations. By default, the average update size is 512B, and the percentage of update operations is 50%.

Table 2. The characteristics of flash memory

Type	Capacity	Page Size	Block Size	Page Read	Page Write	Block Erase
MLC				$284\mu s$	$1,833\mu s$	$31,950\mu s$
SLC$_1$	$4GB$	$(4K+128)B$	$(256+8)KB$	$80\mu s$	$200\mu s$	$1,500\mu s$
SLC$_2$				$130.9\mu s$	$405.9\mu s$	$2,000\mu s$

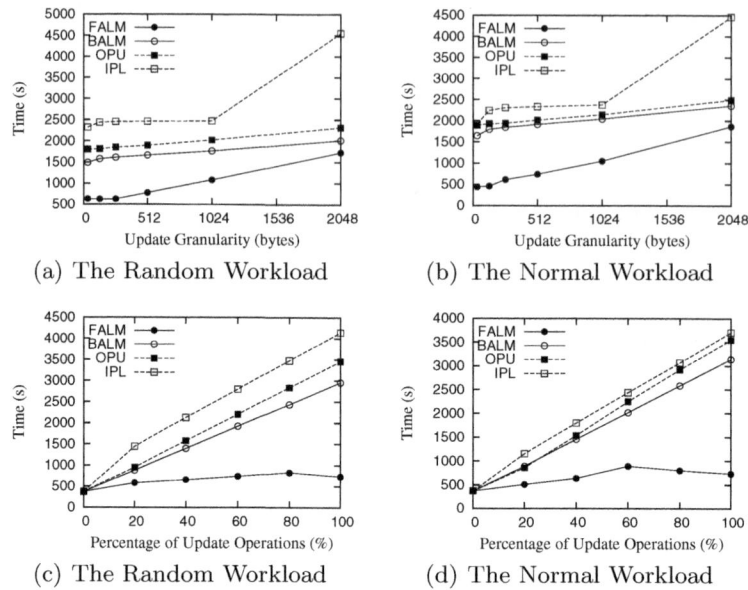

(a) The Random Workload   (b) The Normal Workload

(c) The Random Workload   (d) The Normal Workload

Fig. 5. The performance when varying update granularity and rate

Setup: Table 2 summarizes the characteristics of flash memory used in our experiments. Unless stated otherwise, the MLC flash memory was used. We set the logical page size as 4KB, the same as the physical page size. We gave 128MB and 1.25GB flash memory to each approach under the Index and the Dirik workload, 4GB under the Random and the Normal workload. By default, 128KB and 10MB RAM was given for in-memory buffer under the Index and the Dirik workload, 2MB under the Random and the Normal workload. In BALM and FALM, the ratio between the read and the write buffer size is 7 : 1.

Varying update granularity: First we discuss the impact of update granularity. As shown in Figure 5(a) and Figure 5(b), the performance of IPL decreases dramatically when increasing the average update size from 1KB to 2KB. The reason is that larger update granularity causes the in-block log regions of IPL become full quickly, and leads to frequent merge operations. Since OPU always rewrites the whole page, increasing update granularity has little impact on its performance. As for our methods, BALM, unlike IPL, can dynamically adjust the ratio between B-pages and L-pages in each block, while FALM inherently triggers no force replacement and can pack logs from multiple blocks. Therefore, they achieve higher performance than other approaches for different update sizes.

Varying update rate: Figure 5(c) and Figure 5(d) show the performance when the percentage of update operations is varied from 0% (i.e., read-only) to 100% (i.e., update-only). Our methods generally outperform other approaches. This is because, by using the proposed flash-optimized policies, our methods achieve a better tradeoff between the read and write costs, and can obtain higher overall

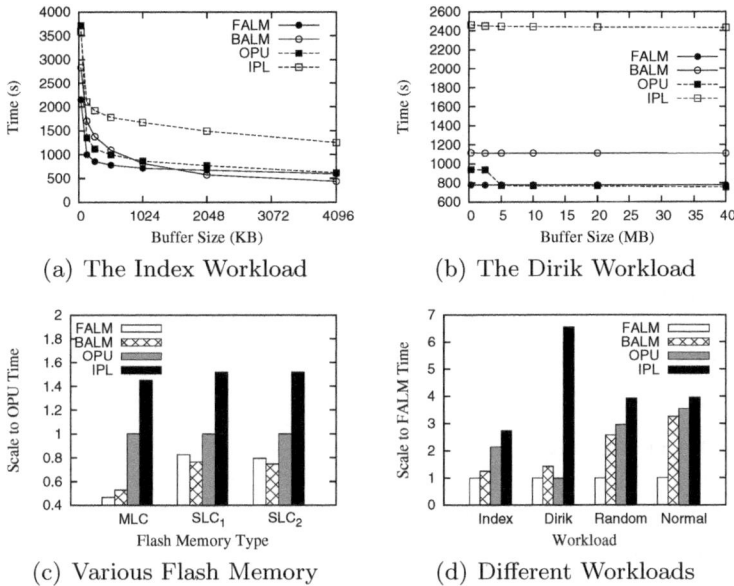

Fig. 6. The performance with different buffer sizes, workloads and flash memory

performance for different update rates. In particular, FALM exhibits excellent performance even under update-intensive workloads. This owes to that FALM packs the logs belonging to multiple blocks, which usually enables better effect of log packing and significantly reduces the amortized update cost.

Varying buffer size: We illustrate the performance under different memory constraints in Figure 6(a) and Figure 6(b). OPU and IPL are generally less efficient than our methods, and the performance gaps become larger when smaller buffer is used. This trend indicates that our methods could perform well even with small RAM. The reason is OPU and IPL employ the traditional LRU, while our methods use buffer management policies optimized for flash memory. As these policies consider not only the asymmetric costs of page read and write, but also the different write costs of pages having different numbers of logs, our methods can take full advantages of available RAM to improve performance.

Various flash memory: We conducted several experiments on different flash memory (Table 2). As presented in Figure 6(c), all values are scaled to the time taken by OPU for comparison convenience. It can be seen that our methods outperform other approaches on all types of flash memory. In particular, compared with the best of other approaches, our methods improve the performance by about 50% on MLC flash memory, and by 20%-30% on SLC flash memory. This is due to that our methods adjust the ratio between B-pages and L-pages based on the I/O characteristics of underlying flash memory, therefore, are able to adapt to various flash memory.

Different workloads: Here we investigate the performance under different workloads. As shown in Figure 6(d), for comparison convenience, all values are scaled to the time taken by FALM. The results demonstrate that our methods can generally obtain higher performance than other approaches across different types of workloads. The main reasons are stated as follows. Unlike IPL will suffer from frequent merge opertions caused by log overflow (note that IPL performs much worse than other approaches under the Dirik workload), our methods dynamically adjust the ratio and distribution of B-pages and L-pages to reduce the number of page writes and block erasures. Compared with OPU, our methods use logs to decrease the total amount of data to write. Therefore, under workloads containing many random fine-grained updates, which OPU is not well-suited to, our methods still can achieve high efficiency.

5 Related Work

FTL [1] and its varieties [3,4,12,8,6,10] were proposed to improve the write performance of flash-based systems. The main idea is to maintain a physical-to-logical address mapping and apply out-of-place update to overcome the erase-before-write limitation. The work of Lee et al. [11] focused on the design of flash-based DBMS called IPL, which records changes as logs on the per-page basis and writes them to in-block log regions. When a log region becomes full, IPL merges the block by writing up-to-date pages to an empty block. Since all log regions are static, IPL might suffer from frequent merge operations caused by log overflow. Na et al. [13] addressed this problem by allocating a log region dynamically. However, their work concentrated on the implementation of B^+-tree on flash memory. A recently proposed approach is called page-differential logging (PDL) [9]. When a page write request is issued to flash memory, PDL compares the original page stored in the flash memory with the page to write. Then PDL only writes the difference between these two pages. This approach is implemented in flash memory driver. It might be inefficient to implement PDL at upper layers of systems as our proposed methods, as each update operation will lead to a read operation for fetching the original page.

Flash-aware buffer management policy has been extensively studied in many literatures. Most proposed approaches, such as BPLRU [7], CFLRU [15] and CFDC [14] take into consideration that flash memory has asymmetric costs of page read and write. However, they are not best suited to log-based systems where different pages can have different read costs. To address this problem, Cesana et al. [2] proposed the Multi-Buffer Manager, which is customized for databases using log-based approach such as IPL [11]. Multi-Buffer Manager divides the buffer into a set of local buffers, each of which contains pages of the same cost. By adjusting the maximum size of local buffers, Multi-Buffer Manager can reduce the total cost of reads and writes in flash memory.

6 Conclusion

This paper has proposed a new approach called page-level log mapping (PLM) to optimize flash-based systems. It adopts backward link technique to support efficient reads and writes, and can yield the optimal overall performance. Two implementations of PLM were developed, and flash-optimized policies were devised for buffer management, free page allocation and garbage collection. We have conducted extensive experiments to evaluate the performance. The results experimentally proved the practicality and efficiency of our proposed approach.

References

1. Intel Corporation. Understanding the Flash Translation Layer (FTL) Specification. Technical Note (1998)
2. Cesana, U., He, Z.: Multi-buffer manager: Energy-efficient buffer manager for databases on flash memory. ACM Trans. Embedded Comput. Syst. 9(3) (2010)
3. Choudhuri, S., Givargis, T.: Performance improvement of block based NAND flash translation layer. In: CODES+ISSS (2007)
4. Gupta, A., Kim, Y., Urgaonkar, B.: DFTL: a flash translation layer employing demand-based selective caching of page-level address mappings. In: ASPLOS (2009)
5. Hsu, W., Smith, A.J.: Characteristics of I/O traffic in personal computer and server workloads. IBM Systems Journal 2(2) (2003)
6. Kang, J.-U., Jo, H., Kim, J., Lee, J.: A superblock-based flash translation layer for NAND flash memory. In: EMSOFT (2006)
7. Kim, H., Ahn, S.: BPLRU: A buffer management scheme for improving random writes in flash storage. In: FAST (2008)
8. Kim, J., Kim, J.M., Noh, S., Min, S.L., Cho, Y.: A space-efficient flash translation layer for compactflash systems. IEEE Transactions on Consumer Electronics 48(2) (2002)
9. Kim, Y.-R., Whang, K.-Y., Song, I.-Y.: Page-differential logging: an efficient and DBMS-independent approach for storing data into flash memory. In: SIGMOD Conference (2010)
10. Lee, S., Shin, D., Kim, Y.-J., Kim, J.: LAST: locality-aware sector translation for NAND flash memory-based storage systems. Operating Systems Review 42(6) (2008)
11. Lee, S.-W., Moon, B.: Design of flash-based DBMS: an in-page logging approach. In: SIGMOD Conference (2007)
12. Lee, S.-W., Park, D.-J., Chung, T.-S., Lee, D.-H., Park, S., Song, H.-J.: A log buffer-based flash translation layer using fully-associative sector translation. ACM Trans. Embedded Comput. Syst. 6(3) (2007)
13. Na, G.-J., Lee, S.-W., Moon, B.: Dynamic in-page logging for flash-aware B-tree index. In: CIKM (2009)
14. Ou, Y., Härder, T., Jin, P.: CFDC: a flash-aware replacement policy for database buffer management. In: DaMoN (2009)
15. Park, S.-Y., Jung, D., Kang, J.-U., Kim, J., Lee, J.: CFLRU: a replacement algorithm for flash memory. In: CASES (2006)
16. Peter, D.: Empirical evaluation of NAND flash memory performance. SIGOPS Oper. Syst. Rev. 44(1) (2010)

A Novel Method to Extend Flash Memory Lifetime in Flash-Based DBMS*

Zhichao Liang, Yulei Fan, and Xiaofeng Meng

School of Information, Renmin University of China, Beijing, China
{zhichaoliang,fyl815,xfmeng}@ruc.edu.cn

Abstract. Over the past decades, flash memory has been widely used in hand-held devices, such as PDA, digital camera, cell phone and USB stick. Moreover, as the capacity increases and the price drops gradually, flash memory is becoming the promising replacement of hard disk, even in the enterprise application. As a novel storage medium that is totally different from magnetic disk, flash memory enjoys faster access speed, smaller size, lighter weight, less noise and better shock resistance. However, flash memory suffers from erase-before-write and limited write-erase cycles on the other side, which means the abuse of write, especially small and random write, will wear a flash block out quickly. In this paper, we analyze the free space management in traditional DBMS and point out its disadvantage when used on flash device. Based on this, we propose a new solution involving free space management and buffer management, in which we replace the traditional free space management method employed in most disk-based DBMS, such as space map or free list, with the Append Only(AO) to avoid useless search and use a stand-alone write buffer to reduce the number of small writes to underlying flash device. Evaluation experiments based on four different trace files show that, in comparison with the traditional strategy, our solution reduces 74.5% of page writes in average, and accordingly succeed in extending the lifetime of flash device.

Keywords: Flash-based DBMS, Free space management, Buffer management.

1 Introduction

Over the past decades, flash memory has been used in most of the mobile and embedded systems(e.g.PDA, digital camera, cell phone and USB stick) as a main storage medium for storing and managing personal and multimedia data. Moveover, as the development of flash design technology, the capacity of flash memory increases so fast even beyond the Moore's law(e.g.8GB in 2004 and 16GB in 2005 for NAND-type flash) while the price drops contiguously. Compared with magnetic disk, flash memory promises faster access speed, higher

* This research was partially supported by the grants from the Natural Science Foundation of China (No.60833005, 61070055); the National High-Tech Research and Development Plan of China (No.2009AA011904).

J. Xu et al. (Eds.): DASFAA Workshops 2011, LNCS 6637, pp. 190–201, 2011.

shock resistance while being much smaller and less power hungry, therefore it has begun to step into general applications like DBMS and OS running on laptops, desktops, or even servers in large data centers[1].Solid-states drive(SSD), known as a package consists of flash chips and a flash translation layer(FTL), is becoming the next generation data storage in place of magnetic disk.

However, flash memory is not almighty God. Because of the erase-before-write, out-place update and page-write, there are two situations in which flash memory can't work gracefully. The first is frequent random write, which will lead to more erase operations and make flash memory reach its erase-write limit earlier. The second is frequent small write, which will consume the available pages quickly and shrink the lifetime of flash memory indirectly. As we all know, general-purpose workloads like OS or DBMS are more stressful than their counterparts on handheld devices, so it will be very challenging when we just drop flash memory in these workloads. As far as we know, the problem of lifetime is serious.

According to Micro's data sheet[2], under some particular workloads, a 60GB SSD only has write-lifetime of 42 TB, which is a reduction in write-lifetime by a factor of 7. Here we use the word "write-lifetime", as its definition in[3], which means the total number of writes that can be issued to the device over its lifetime(e.g. an SSD with 80GB of NAND flash with 10,000 erase-cycles per block owns a maximum write-life time of 800TB). Therefore, concerning the price of flash memory, the poor lifetime hinders its widespread use in general-purpose application. In this paper, we mainly focus on how to extend the flash memory lifetime in DBMS.

It is well known that, in order to maximize storage utilization, traditional disk-based DBMS adopts free space management to solve the problem of choosing a disk page to hold a newly allocated data record. As far as we know, the most commonly used methods to keep track of free space are space map and free list[4]. The former records the summary information about the amount free space in each page while the latter maintains a free list, which is a list of pages that are likely to be able to hold a new record. Although these two strategies employ different data structures and different algorithms, they provide the same function: choosing a page with fit free space to satisfy the record write request and they do play an important role in the storage management. However, we must be clear that both space map and free list are based on the assumption the in-place update is supported on the underlying device, but it is impossible on flash memory. Consequently, when we just simply migrate the traditional DBMS on the flash memory, we got ourselves in an embarrassment: all the work done in free space management are in vain cause we can't overwrite the returned pages with free space physically. In order to solve this problem, we propose to replace the complicated methods with Append Only(AO), which allocates new empty pages for write requests and appends them to the tail of original file. Although AO avoids the useless searches in space map or free list, it dose no help for extending the lifetime of flash memory, thus we create a stand-alone write buffer in main memory to reduce the number of random and small writes. In the write buffer, we collect all modifications to the data record(e.g. insert, update) and they are not flushed out by the write operation until a buffer page is filled up. Combined the AO and write buffer, we mitigate the bad effect caused by the small and random writes and the results of experiment demonstrate our method reduces 74.5% page write in average and achieves the target of extending the flash memory lifetime.

The rest of this paper is organized as follows. Section 2 discusses the characteristics of flash memory and their impact on application designing. Section 3 analyzes the methods used in traditional free space management and the problem when they are used on flash memory. Our method is described in Section 4, including the AO, write buffer and merge operation. Section 5 presents the results of our experiment evaluation. Section 6 surveys some excellent related work and Section 7 summarizes all this paper.

2   Characteristics of Flash Memory

There are two types of flash available in the current market: NOR and NAND. NOR flash is designed as the replacement of EEPROM and is mainly used to store the programs. NAND flash, however is designed as the mass storage devices and also is the focus of this paper. Moreover, according to the number of bits stored on each cell, NAND flash can be categorized into single-level cell(SLC) and multi-level cell(MLC). As a novel storage medium that is totally different from magnetic disk, flash has many special characteristics.

- No mechanical latency. As an electronic device, flash memory stores data as charge trapped on a floating gate between the control gate and the channel of a CMOS transistor, so it has no mechanical components like magnetic head in disk. Hence, flash memory has no mechanical latency(seeking time and rotation time) and enjoys efficient random access.
- Two level hierarchical structure. A flash chip contains a set of blocks and a block(typically 128KB) consists of many pages. Page(typically 512bytes for SLC and 2KB for MLC) is the basic unit of data access.
- Asymmetric read, write and erase speed. There are three basic data operations in flash memory: read, write and erase, which present different operation performance correspondingly. Read and write are page-granularity operations and erase is a block-granularity one. As shown in Table 1, we can know that read is the fastest operation, write is much more time-consuming than read and the erase is the most costly among all the three operations.
- Out place update. For magnetic disk, we can overwrite the contents in the original address physically when we want to update a data item, which we describe as in-place update. Unfortunately, it is impossible for flash memory. Flash memory requires erase operation before overwriting, which means you have to erase the whole block even if you only want to update a small data item stored in one of its pages. As we have seen, the erase operation is very costly, frequent in-place updates will degrade the system performance significantly. Consequently, flash memory choose to invalidate the original

Table 1. Performance:Flash Disk

	Page Read(2KB)	Page Write(2KB)	Block Erase(128KB)
Hard Disk [1]	14.1ms	14.1ms	N/A
Flash Chip [2]	25μs	200μs	1.5ms

page and write all the valid data and the updated data to another new empty one, which we describe as out-place-update.

- Limited write-erase cycles. Only a finite number of erasures are allowed on flash memory before the bit error rate of the device becomes unacceptably high. Typically, SLC flash supports 100K erasures per flash block, and the MLC flash only supports about 10K erasures because of the higher bit density. Actually the error rate for MLC deveices increases dramatically with wear, and its non-zero even for brand-new devices[5]. Hence, as SSD technology moves towards MLC, we can expect the lifetime problem be more serious.

Obviously, we have to take all of these characteristic in our consideration when we design the flash-based system.

3   The Methods Used in Traditional Free Space Management

As we mentioned above, the most commonly used methods for keeping track of free space are space maps and free lists. Space maps are usually placed at well known positions of a file that contains summary information about the amount of free space in a set of pages. Taking the space maps implemented in PostgreSQL as an example, we can have a better understanding. As of PostgreSQL 8.4 each relation has its own, extensible free space map stored in a separate "fork" of its relation[7]. In order to keep the map small, PostgreSQL only record the free space at a granularity of 1/256th of a page, that is to say, it is that the stored value is the free space divided by pagesize/256(rounding down). To assist in fast searching, just like the upper structure shown in Fig.1, the map isn't simply an array of per-page entries, but has a binary tree structure above these underlying heap pages. To search for a page with X amount of free space, traverse down the tree along a path where n≥X, until you hit the bottom. If both children of a node satisfy the condition, you can pick either one arbitrarily.

Compared with space map, the structure of free list is much easier. The free list manages all the pages containing free space left for future records or created from deletion by linking them together as a doubly linked list(the lower structure in Fig.1). When a new record write request comes, we check the free list first to see if any page with enough free space available. Any page with new free space

---

[1] Hard Disk: Hitach HDP725025GLA380,250G,7200rpm.
[2] Flash Chip: Samsung K9XXG08UXA, 4G Nand Flash.

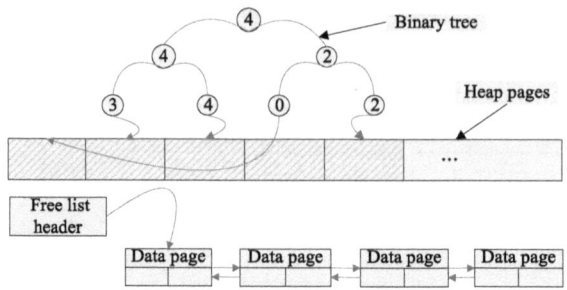

Fig. 1. Space Map and Free List

will be linked to the beginning of the free list and any page will not be delinked until the free space is used up.

Apparently, both space maps and free lists, to some extent, maximize the storage utilization in disk-based DBMS. However, they are not suitable for flash memory. The free space in the pages returned by these methods can't be utilized because of the out-place update characteristic. We still have to allocate a new empty page to accommodate the modified and unmodified data and leave the original page to be erased, hence all the searches done in space map or free list are in vain.

4   Our Solution

This section presents our proposed solution. An overview is described in Section 4.1. Section 4.2 and Section 4.3 give this solution in details, including the algorithms and data structures used. Finally, Section 4.4 explains a separate module to eliminate the side-effect created by this novel solution.

4.1   Overview

Taking the characteristics of flash memory into consideration, we propose to replace all the stuff in the traditional free space management with Append Only(AO), namely, we allocate new empty pages as soon as a write request comes and append them to the tail of original files. In this simple and fast way, we avoid the useless searching for a page with free space. However, AO does no help with reducing the frequent random and small writes, hence we propose to allocate a stand-alone write buffer in main memory to collect the writes and flush them to flash deveice later. Fig.3 shows the architecture of our solution, in which we read the data records from the DBMS storage into shared memory and add the updated or inserted records into write buffer temporarily and finally flush the pages to DBMS storage following the AO pattern.

Discussion. This method seems to be questioned for a limitation: data consistency loss becomes possible concerning the volatility of main memory, but it is only "seems". For the data consistency loss, we must make it clear before

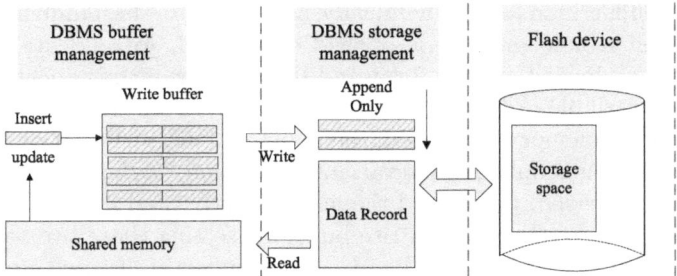

Fig. 2. The Overview of Our Method

our further discussion. Our solution for extending the lifetime of flash memory has nothing to do with anything about logging and we just focus on how to handle the data records, so the logging just goes on. Consequently, our solution would lead to more logging data to read when the system crashes but not consistency loss.

4.2 Free Space Management

As we pointed out in Section 3, the methods or algorithms expecting to use the free space in the allocated pages do not work on flash memory, only resulting in complicated data structures and useless search, hence we propose to use the simplest and fastest algorithm, Append Only(AO, see Algorithm 1).

Input: WriteRequest(Page *P*, Relation *Rel*)
Output: NULL
1: ASSERT(*Rel* exists);
2: find the file *F* where the *Rel* is stored;
3: allocate a page *NP* and append it to the tail of *F*;
4: write the contents of *P* to *NP*;
5: return;

Algorithm 1. Append Only Write

About how to store the relations, different tactics are employed in storage management, such as storing different relations in separate files in PostgreSQL and mixing all relations in a single series of files in MySQL. However, we don't care here, we just append the pages to the corresponding files. To some extent, we convert the random write into sequential write in this way, while the latter enjoys high efficiency.

4.3 Write Buffer

Write buffer is a stand-alone area in main memory. Here we use the adjective "stand-alone", which indicates that write buffer will not be affected by the normal buffer replacement strategy such as RAND, FIFO, LRU and CLOCK. The size of this buffer increases dynamically according to the number of relations opened. When a new relation is opened or created, we allocate a particular number of pages in write buffer for it and the number of the allocated pages is a configuration variable, *PagesForRelation*. However, in case the write buffer grows beyond the main memory can afford, the number of distinct relations maintained is limited by another configuration variable, *RelationsForBuffer*. When this number would be exceeded, we discard the least recently used relation. Accordingly, we can easily control the size of write buffer by setting these two variables.

There are two operations handled by write buffer: insert and update. When a new record is to be inserted to a relation, we add this record into the pages that belong to the relation directly. When a record is to be updated, we read the original record into the shared memory firstly, and then we also add the updated record, including the modified columns and unmodified columns, into the corresponding pages. Those pages in write buffer will not be flushed out until they are filled up or a checkpoint occurs. This method enjoys several advantages. Firstly, write buffer delays the data commit and reduces the number of small write. Secondly, write buffer only stores and flushes the inserted and updated records, so it reduces the amount of data to write and the number of write operation to execute. Thirdly, write buffer reorganizes the records from different pages into a single page, so the number of random write declines. And lastly, concerning the data locality, a record just updated maybe need to be modified again immediately, thus we can quickly update it if this record is still queuing and waiting for its write operation in write buffer.

Apparently, write buffer and AO disperse all the records to the entire relation file and might destroy the data distribution. However, as shown in[8], the clustering by key values is not that beneficial. Since flash memory enjoys outstanding random read speed, the non-clustering method outperform its clustering counterpart in most cases, thus it is more important to reduce the number of write operations than to maintain the data distribution.

4.4   Merge Operation

According to the analysis above, our solution reduces the number of write, but it abuses the storage space in a different way. Using the traditional method, if we want to update some records, we just copy all the records from the original page to main memory, modify the target records, flush all these records to underlying flash device and the FTL will allocate a new page to accommodate these records, leaving the original page to be erased. In contrast, since our method only flushes the updated records out, the other valid records are not moved. Fig.4 demonstrates the difference. When the update occurs, the traditional method allocates a new page and leaves the updated page to be erased while our method allocates a new page to collect the updated records and leave the updated page with obsolete records to be still valid. Obviously, this is a nightmare of storage utilization, especially for the update-intensive applications, hence we design a periodical merge operation to eliminate this side effect.

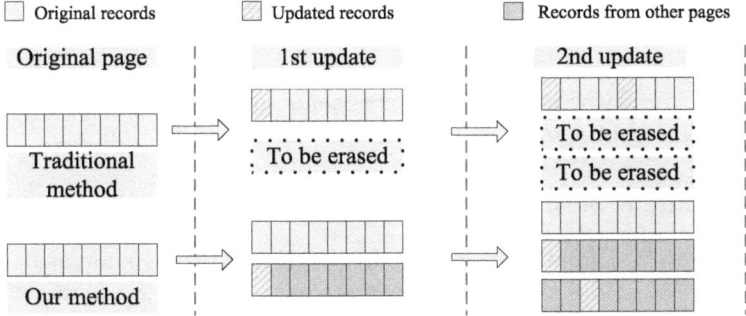

Fig. 3. Updates in Traditional and Our Methods

In order to merge the pages with "holes", we need to keep track of their usage, so we introduce Obsolete Address Log(OAL). OAL has the same mechanism with the normal logging except the stuff it logs. When a record is updated or deleted, a corresponding OAL log is created. The OAL log has a layout like(PageId, Offset, Length), which describes the accurate location of the record in the original page and we flush the logs out with the normal ones. In this way we get all the information about the storage utilization. When a merge operation is incurred, we scan the OAL and count the obsolete records in every page, if the number is over the threshold, we start to merge the page(see Algorithm 2). The OAL will be destroyed for keeping itself slim and reconstructed for storing the remaining and incoming OAL logs when the merge is done.

Input: MergeRequest(ObsoleteAddressLogFile L)
Output: NULL
 1: ASSERT(L exists);
 2: scan the records in L and build a hash index for them with the *PageId* as the hash key;
 3: **for** each hash key in the index **do**
 4:   count the number of records n in its buckets;
 5:   **if** (n > the threshold) **then**
 6:     load the page P into main memory;
 7:     **for** each record in P **do**
 8:       **if** (this record can not be found in the index) **then**
 9:         add this record to the write buffer;
10:       **else**
11:         delete the record in the index;
12:       **end if**
13:     **end for**
14:     delete P;
15:   **end if**
16: **end for**
17: delete L;
18: create L;
19: write the remaining records in the index to L;
20: free the index;
21: return;

Algorithm 2. Merge Operation

5   Evaluation Experiments

This section presents the experiment environment and the results. Section 5.1 describes the simulator we used in experiments and how the experiments were performed. The results and some analysis are shown in Section 5.2.

5.1   Experiment Setup

We used a flexible simulator[14] for our experiment. This simulator not only simulates the behaviors of different kinds of flash memory and different kinds of FTL algorithms, but also provides functions that measure the number of read, write and erase. The settings in the experiment are as follows: the size of page is 2KB, the size of block is 128KB and the FTL uses a very simple page-level mapping theme. For comparison, we implemented the traditional method and our method in the simulator. Moreover, we collected four kinds of trace files with varying ratio of read to write(2/8, 4/6, 6/4 and 8/2 respectively) and each of these trace files includes 1000,000 I/O requests.

The experiment was performed like this: for both the traditional method and our method, the simulator accessed the four trace files and processed read/wrie requests archived in them. The performance was measured by counts of read, write and the overall latency. Since our methods delay the erase operation intrinsically and different FTL algorithms employ different garbage collection strategies, we don't take the count of erase into consideration to avoid the effect caused by the underlying FTL. In the simulator, the overall latency is calculated by applying a weight to each operation: weight of 25 for a read operation, 200 for a write operation and 1500 for an erase operation.

5.2   Performance Results and Analysis

Fig.4 presents the counts of page write. It is seen that our method enjoys a 74.5% decrease averagely. Since we only collect the updated records to the write buffer and flush them to the underlying devices following the AO pattern, we reduce the number of small writes and maximize the storage utilization effectively. Fig.5 shows the counts of page read. We can see that, compared with the traditional method, ours has to read more pages to satisfy the read requests. Concerning we disperse the updated records into the pages different from their original ones, it is sensible. On average, our method increases 53.5% page read. The results of overall latency are shown in Fig.6. From the figure we can tell our method enhances the performance by 67.75% on average, which mainly owe to the decline of write I/O. Although our method increases the number of read operations, it dose cut down the number of write operations, and it improves the overall latency consequently. Accordingly, we can conclude that the traditional method do better in terms of the number of read operations while our method do better in terms of the number of write operations. As we have analyzed above, the frequent small write and random write can be poison for the lifetime of flash memory. Our method reduces the number of small write and random write with

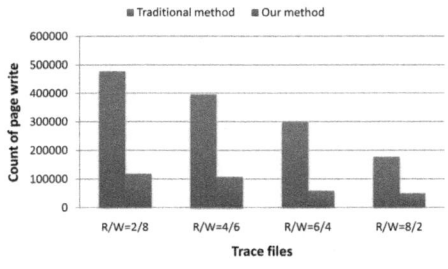

Fig. 4. Write Count for Comparison

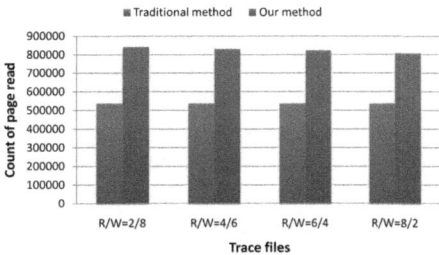

Fig. 5. Read Count for Comparison

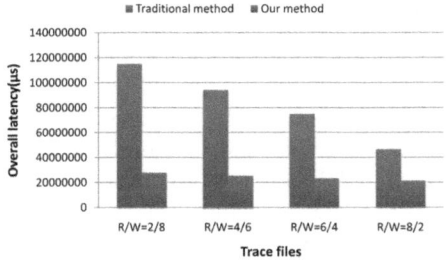

Fig. 6. Overall Latency for Comparison

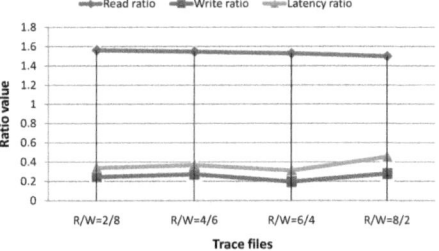

Fig. 7. Tendency of Ratio

AO and write buffer, hence it protects the lifetime of flash memory from being destroyed by the abuse of write.

In addition, Fig.7 summarizes the tendency of read, write and latency ratio of our method to traditional method across the four trace files. From that, we can find that our method achieves its best at the trace file with the ratio 6/4, which implies that our method doesn't support "the more writes the better effectiveness". The possible reason is that more writes lead to more merge operations, which lead to more writes in turn. Moreover, when we want to get all the newest data records from a specific page, we may need an extra read operation even only one record is updated in the page, so a high ratio of read to write will not relax this situation significantly, and that is why the counts of read operations keep high across the four trace files as shown in Fig.7.

6 Related Work

As a novel storage medium that is totally different from magnetic disk, flash memory is getting more and more research attention in recent years. We can simply categorize all the excellent research work into several aspects. Some work focus on the measurements on flash memory. Bouganim et al.[12] proposed uFlip, a benchmark for measuring the response time of flash IO patterns, to help the researchers to fully understand the performance characteristic of flash devices.

Chen *et al.*[13]also conducted intensive experiments and measurements on different types of SSDs, including the low, middle and high ends. Some work focus on how to adjust the traditional methods in DBMS to take fully advantage of the unique characteristics of flash memory. Chen[9]proposed to use the USB flash drives for synchronous logging performance. Nath *et al.*[10]proposed FlashDB, which uses a novel self-tuning index that dynamically adapts its storage structure to workload and underlying storage device. Tsirogiannis *et al.*[11] proposed the PAX layout and corresponding algorithms for scan and join to improve the query processing on SSD. Some work focus on how to make full use of different types of flash devices from the point of hierarchy. Yim[6] designed a novel memory hierarchy including the volatile RAM, non-volatile RAM and SLC/MLC flash chips to achieve high-speed I/O. Moreover, some nice work on FTL[15], [18], [16] also contributed good ideas for the research on flash memoty.

7   Conclusions

The contribution we make in this paper are as follows: First, we analyze the free space management methods in the traditional DBMS and point out they are useless in flash-based DBMS, cause we can't update any data on flash memory in space. Second, we propose to use Append Only(AO) algorithm served as the free space management method, avoiding the useless searches in the DBMS pages and random write pattern. Third, in order to reduce the number of small writes, we propose to use a stand-alone write buffer. Lastly, we conduct a performance evaluation of our method based on four different trace files. The results demonstrate our method reduce 74.5% page write and achieve the target of extending the flash memory lifetime.

References

1. Caulfield, A.M., Grupp, L.M., Swanson, S.: Gordon: Using Flash Memory to Build Fast, Power-efficient Clusters for Data-intensive Applications. In: Proceedings of the 14th International Conference on Architectural Support for Programming Languages and Operating Systems (ASPLOS), pp. 217–228 (2009)
2. Micron: C200 1.8-Inch SATA NAND Flash SSD, http://download.micron.com/pdf/datasheets/realssd/realssd_c200_1_8.pdf
3. Soundararajan, G., Prabhakaran, V., Balakrishnan, M., Wobber, T.: Extending SSD Lifetimes with Disk-Based Write Caches. In: Proceeding of 8th USENIX Conference on File and Storage Technologies (FAST), pp. 101–114 (2010)
4. McAiliffe, M.L., Carey, M.J., Solomon, M.H.: Towards Effective and Efficient Free Space Management. In: Proceedings of the 1996 ACM SIGMOD International Conference on Management of Data (SIGMOD), pp. 389–400 (1996)
5. Grupp, L.M., Caulfield, A.M., Coburn, J., Swanson, S., Yaakobi, E., Siegel, P.H., Wolf, J.K.: Characterizing Flash Memory: Anomalies, Observations, and Applications. In: Proceeding of the 42st Annual IEEE/ACM International Symposium on Microarchitecture (MICRO), pp. 24–33 (2009)
6. Yim, K.S.: A Novel Memory Hierarchy for Flash Memory Based Storage Systems. Journal of Semiconductor Technology and Science 5(4), 262–269 (2005)

7. PostgreSQL: Source Code Download, `http://www.postgresql.org/ftp/source/v8.4.3/postgresql-8.4.3.tar.bz2`
8. Bae, D.-H., Chang, J.-W., Kim, S.-W.: Clustering and Non-clustering Effects in Flash Memory Databases. In: Proceeding of 20th International Workshop on Database and Expert Systems Application (DEXA), pp. 4–8 (2009)
9. Chen, S.: FlashLogging: Exploiting Flash Devices for Synchronous Logging Performance. In: Proceedings of the 2009 ACM SIGMOD International Conference on Management of Data (SIGMOD), pp. 73–86 (2009)
10. Nath, S., Kansal, A.: FlashDB: Dynamic Self-tuning Database for NAND Flash. In: Proceedings of the 6th International Conference on Information Processing in Sensor Networks (IPSN), pp. 410–419 (2007)
11. Tsirogiannis, D., Harizopoulos, S., Shah, M.A., Wiener, J.L., Graefe, G.: Query Processing Techniques for Solid State Drives. In: Proceedings of the 2009 ACM SIGMOD International Conference on Management of Data (SIGMOD), pp. 59–72 (2009)
12. Bouganim, L., Jónsson, B.T., Bonnet, P.: uFLIP: Understanding Flash IO Patterns. In: Fourth Biennial Conference on Innovative Data Systems Research, CIDR (2009)
13. Chen, F., Koufaty, D.A., Zhang, X.: Understanding intrinsic characteristics and system implications of flash memory based solid state drives. In: Proceedings of the 11th International Joint Conference on Measurement and Modeling of Computer Systems (SIGMETRICS), pp. 181–192 (2009)
14. Jin, P., Su, X., Li, Z., Yue, L.: A flexible simulation environment for flash-aware algorithms. In: Proceedings of the 18th ACM Conference on Information and Knowledge Management (CIKM), pp. 2093–2094 (2009)
15. Chung, T.-S., Park, D.-J., Park, S., Lee, D.-H., Lee, S.-W., Song, H.-J.: System Software for Flash Memory: A Survey. In: Sha, E., Han, S.-K., Xu, C.-Z., Kim, M.-H., Yang, L.T., Xiao, B. (eds.) EUC 2006. LNCS, vol. 4096, pp. 394–404. Springer, Heidelberg (2006)
16. Kim, J., Kim, J.M., Noh, S.H., Min, S.L., Cho, Y.: A Space-Efficient Flash Translation Layer for Compactflash Systems. IEEE Transactions on Comsumer Electronics 48(2), 366–375 (2002)
17. Kang, J.-U., Jo, H., Kim, J., Lee, J.: A Superblock-based Flash Translation Layer for NAND Flash Memory. In: Proceedings of the 6th ACM & IEEE International Conference on Embedded Software (EMSOFT), pp. 161–170 (2006)
18. Lee, S.-W., Park, D.-J., Chung, T.-S., Lee, D.-H., Park, S., Song, H.-J.: A Log Buffer-based Flash Translation Layer Using Fully Associative Sector Translation. IEEE Transactions on Embedded COmputing Systems 6(3) (2007)

Log-Compact R-Tree: An Efficient Spatial Index for SSD

Yanfei Lv, Jing Li, Bin Cui, and Xuexuan Chen

School of Electronics Engineering and Computer Science, Peking University
Key Lab of High Confidence Software Technologies (Ministry of Education),
Peking University
{lvyf,leaking,bin.cui,xuexuan}@pku.edu.cn

Abstract. R-Tree structure is widely adopted as a general spatial index with the assumption that the deployed system is equipped with magnetic hard disk. While the application of SSD becomes more and more popular, traditional optimization of R-Tree structure on SSD is much less desirable than that on magnetic hard disk. Existing flash-aware index approaches employ log mechanism to reduce random writes at a cost of large amount of read. A novel index named Log Compact R-Tree (LCR-tree) is proposed in this paper. Distinguished from previous attempts, compacted log is introduced to combine newly arrival log with origin ones on the same node, which renders great decrement of random writes with at most one additional read for each node access. Extensive experiments illustrate that the proposed LCR-Tree can achieve up to 3 times benefit against existing approaches.

1 Introduction

Flash memory devices are widely spread in various applications from small embedded device to large data center as an ideal alternative for hard disk. The ubiquitous use of flash memory benefits from its outstanding characteristics: high I/O performance, low power consumption and high reliability. Flash memory based disk, i.e., Solid State Drive SSD, has been applied widely as a substitute of hard disk. SSD has asymmetric I/O performance: the read speed is usually much faster than write, especially random write. On the other hand, the unique characteristics of flash memory based devices also make approaches designed for hard disk no longer work well for flash memory. Thus, flash-based algorithms and data structures have become a hot research field in recent years. As a key component of database system, lots of flash-specific index structures are introduced.

R-Tree [4] is a general index structure for multi-dimensional data and has been integrated into many DBMSes such as PostgreSQL [18], MySQL [13]. An example of R-Tree is given in Figure 1. The updates on tree-based indexes are often small and scattered over the whole tree structure. Those small sized random writes severely deteriorate the overall index performance on flash memory. A lot of methods on B-Tree were proposed to overcome this obstacle [19,1,12]. As a

J. Xu et al. (Eds.): DASFAA Workshops 2011, LNCS 6637, pp. 202–213, 2011.

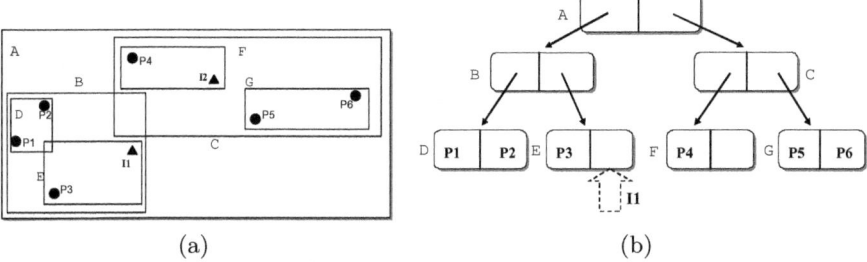

Fig. 1. An example of R-Tree structure

variant of B-Tree structure, such problem also exists in R-Tree. Nevertheless, From the I/O aspect, R-tree has its unique features compared with B-Tree, which render those optimizations unsuitable for the R-Tree.

- Overlapped read. There generally exist overlaps between the Minimum Bounding Rectangle (MBR) of nodes on the same level. Consequently, searching a node in R-Tree may have to access multiple nodes on the same level, so the number of page read is likely to exceed the height of R-Tree, where lies the first significant difference between B-Tree and R-Tree.
- MBR Cascaded update. The update of leaf node may lead to update MBR of the nodes from the root to this leaf, as the boundary of the parent node may be enlarged by the insertion. Thus, the update to the higher level node is much more frequent than that in B-Tree.

Majority of existing works compact the random updates into sequential "log write" which forms one log chain for each node. The original data will not be overwritten until the length of log exceeds a certain threshold. The major disadvantage of log chain methods is that the number of read operations sharply increases with the length of log chain. Nowadays, the improvement of both flash chip and design technology decreases the gap between read and write performance of SSD [2]. Consequently, the read also takes a crucial part in the overall performance especially in R-Tree design. How to diminish random write operation with small influence on the read performance is the key problem for R-Tree design on the SSD.

To the best of our knowledge, RFTL proposed by Wu [20] is the only work on flash-aware R-tree design, which is transferred from a similar method on B-Tree. It combines the small random updates and stores them as sequential log. Each node maintains a log list, to facilitate log searching. However, the higher a node is, the longer the log chain is, because update frequency of higher level is relatively high in R-Tree. In addition, higher level nodes are read intensive, so this strategy incurs too much reads. LA-Tree (Lazy-Adaptive) [1], the state-of-the-art approach on B-tree, adopts layered buffer to batch updates and defers overwriting leaf nodes. Considering that access to a lower level node must be preceded by loading buffers of higher level nodes, huge numbers of the updates in the higher level buffer may be irrelevant to the leaf node we required. Those extra read operations significantly burden the I/O task of LA-Tree.

In this paper, we propose a novel tree index structure named LCR-Tree (Log Compacted R-Tree). In the LCR-Tree, a new log to a node is compacted with its former ones, and the logs for different nodes are compacted into one page if possible. By doing this, it effectively decreases the number of random write operations without huge increment of reads, and thus strikes a balance between read and write performance. The designed LCR-Tree has the following key features. 1) Random writes are compacted into sequential writes. 2) The read operations are at most twice as much as the original R-Tree. 3) The method is easy to implement and only requests a little modification on the existing R-Tree.

The rest of this paper is organized as follows: Section 2 introduces background knowledge about flash memory and related work. In Section 3, a new index LCR-Tree is presented in detail. Experimental results are reported in Section 4, and finally we conclude the paper in Section 5.

2   Preliminaries

2.1   Introduction to SSD

There are various kinds of SSDs, and are with significant performance difference among them. We list some of the I/O bandwidths in Table 1 including sequential read, random read, sequential write and random write. The bandwidth of ADATA is tested on a PC with the environment introduced in the experiment part, while other data are referenced from paper [11].

Table 1. Bandwidths of SSDs with 2KB access unit (MB/sec)

Device	Seq Read	Rnd Read	Seq Write	Rnd Write
ADATA	14	8.8	13	0.1
Samsung	8.2	6.5	6	0.1
Mtron	20	17	23	0.3
Intel	36	14	26	2.5

As shown in the table, although the bandwidths vary among different SSDs, they have stable trend with access patterns for a certain SSD. 1) The sequential read has the similar speed with sequential write. Furthermore, random read speed is comparable with sequential access. 2) Random write is one or two magnitude slower than other patterns. Though, in practice, the real workload is often semi-random, the performance gap still remains several times. Therefore, the key problem for index structure design in flash-based devices is to prevent random write at the expense of not many extra read operations.

2.2   Related Work

Flash based database system has become a fruitful research field [7,9,3]. In-page logging [8] and page-differential logging [6] are two general page deployment strategies to deal with the random update in database system. In-page logging stores the data and its log together in a page, but unfortunately it is difficult for the current SSD to deal with such small-grained update. Page-differential logging adopts the byte difference between the up to date and original data as the log. However, there are some troubles when it is adopted in index, e.g., data deletion in a node may cause a large difference.

A lot of works have been done on flash-aware index design [22,21,10]. μ-Tree [5] is a novel index structure for one-dimensional data. It integrates the nodes from the root to leaf into one page to facilitate update. However, it has low scalability for large amount of data as the node size is too small. Most of the tree structures adopt the logging technology. There are two types of logging methods: direct log which associates the log direct with its original data, and layered log which stores the log to a sub-tree into the root of that sub-tree. Typical direct logging methods are BFTL [19] and RFTL [20], which integrate small updates on different nodes together as sequential write and maintain a log address list for each node in the main memory. The read pressure of these methods is very high. In-page logging strategy has been applied in B-Tree structure optimization [15] and [14], and they also have the problem of in-page logging which was discussed previously. FD-Tree [12] and LA-Tree [1] are two flash-based indexes using Layered log. The log stored in a node contains all the updates to the sub-tree with the root of that node. The updates will be flushed to child nodes layer by layer. However, in case of reading one node, some updates to the sub-tree must also be accessed. If it was applied on the R-Tree, the extra read operations will further deteriorate the overall performance. Flash DB [16] reduces the overhead of reading a log chain by dynamically switch between Log-Mode and Disk-Mode. This is a combination of the log method in essence and does not fundamentally solve the problem of log chain.

3   The LCR-Tree

3.1   Overview of LCR-Tree

In this section, we introduce the proposed novel index named Log-Compact R-Tree (LCR-tree). The key idea underlying is that all previous logs on one node are merged together to the same page named compact log, which ensures at most only one additional read on the page when accessed. In addition, the compact logs for different nodes are merged into one page if possible, which remarkably decreases the write operations.

In order to simplify the log management and accelerate the log writing, we organize all the logs into a separate section, a sequentially written area on SSD. Thus, the overall LCR-Tree contains two components, named Tree part and Log part respectively. The overview of LCR-Tree is given in Figure 2(b) with R-tree implementation in Figure 2(a) for comparison.

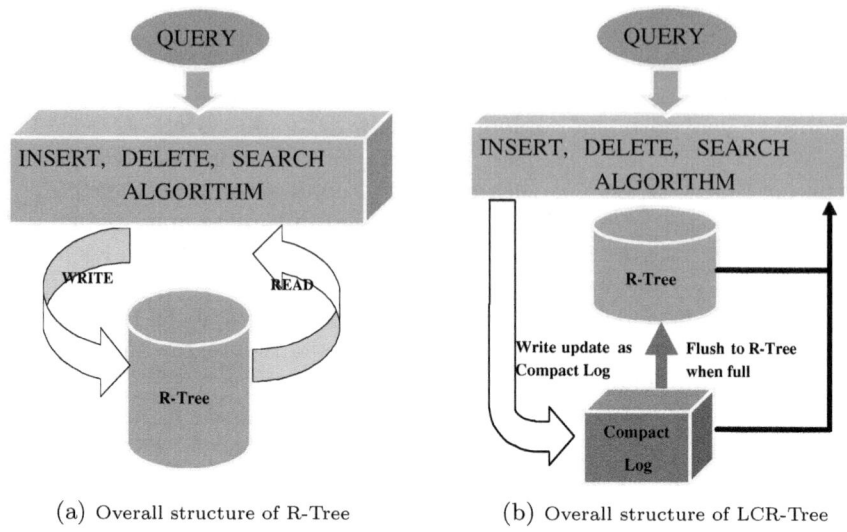

(a) Overall structure of R-Tree   (b) Overall structure of LCR-Tree

Fig. 2. The LCR-Tree structure compared with R-Tree

It can be observed that the Tree part is an ordinary R-Tree, which stores the original data of R-Tree. The Log part is employed to store the update on original R-tree, aiming to reduce the random write operation on the structure. The log area sequentially records the difference between the current R-tree with its origin version after each modification. The logs would be merged back to the original tree when full. In order to facilitate the log finding, we also maintain a mapping table in the main memory. Once a new update on a node arrives, we also read the previous log on this node out and merge them together named compact log. While adding newly compact log, LCR-Tree will try to integrate this log with previous compact logs on other nodes into one page so that writes on the Log section are processed in the page unit. The data structure of the Compact Log are carefully designed, and hence is enabled to store significant modification logs in a given area, which ultimately decreases the write operations, especially the random write operations.

Note that, the modifications are stored in Compact Log rather than the origin R-Tree, we need to check both Compact Log and origin R-Tree to get the up-to-date R-Tree nodes when dealing with read operations on R-tree. LCR-Tree does not modify the R-Tree search algorithm except the access to log and can ensure the final tree is as same as what is supposed to be. The compact log design ensures that we can easily get the up-to-date R-Tree with at most one additional read on the Log part for each node.

Another benefit of the separate Log part design is that it can decrease the coupling between two components as low as possible and guarantee the maximum reuse of the R-Tree code. Hence, our design has good flexibility and robustness. In fact, LCR-Tree only modifies the write and read interface of search algorithm, which renders it easy to implement and can be applied to other variants of

---

Algorithm 1. ReadNode subroutine

Input: the node address to read denoted as D
Output: the node is returned
1 $Node_{origin}$ = read original node from R-Tree ;
2 Log_m = find the related log in main memory;
3 $Entry$ = search the log address in the mapping table using D;
4 **if** *Entry is not NULL* **then**
5 | Log_f = load the related log on SSD;
6 **end**
7 $Node_{return}$ = Apply Log_m and Log_f to $Node_{origin}$;
8 return $Node_{return}$;

R-Tree such as R^+-Tree and R^*-Tree conveniently without much modification to the original source code.

3.2   Design Details of LCR-Tree

In this part we introduce the design details of LCR-Tree. We first present the data structures of implementation, followed by the key algorithms. The compact log for one page is a list of cells, each of which stores one update. There are three types of cells: insert, update and delete, the first two bits of the cell indicate the type. Following the type is the branch ID to process and the new value. This structure ensures the maximum log length of one node is limited to one page. To accelerate the log searching, a mapping table between original node address and its log address is maintained in main memory. Each entry of the map also stores the length of log list.

The process of reading a node should also be adjusted to adapt LCR-Tree data structure. The revised ReadNode function has been listed in the Algorithm 1. Note that the node in R-Tree is probably NULL (line 1), this corresponds to the case that a node is newly inserted and not merged into R-Tree yet which is explained in the insert procedure. In this case, the node can be obtained using logs only. We can observe that the read operation is at most twice as the original R-Tree. The searching process in LCR-Tree is straightforward with the help of this ReadNode routine and thus omitted here.

The key technique of LCR-Tree is the insert strategy which is illustrated in Algorithm 2. With the help of ReadNode procedure, we can locate the node to insert into and finish the original R-Tree insertion process. The difference is that we record all the nodes updated and the newly generated nodes, if some nodes are split, during the insert operation (line 2). For the updated nodes, we combine the log with their previous ones and record the compacted log together (lines 3-7). For the new nodes, the trick here is to write the nodes as logs into the Log section directly (lines 8-11). This is correct because the parent node has been updated and the log can be obtained in the mapping table between R-Tree Node and log. In case that the log is full, the log will be merged with the original R-Tree (lines 12-14), which reclaims the Log part. In the worst case, the number of writes is the same as the original R-Tree. However, in practice the

---

Algorithm 2. LCR-Tree_Insert

 Input: objects in the same node to insert
 Output: objects are inserted into LCR-Tree
 1 find the node to insert using the new ReadNode subroutine;
 /* Invoke original R-Tree insert and record the nodes updated
    and generated   */
 2 $< NodeSet_u, NodeSet_g >$ = original R-Tree insert algorithm;
 3 **foreach** *Node in* $NodeSet_u$ **do**
 4 | merge the new log with previous one;
 5 | write back to the compact log;
 6 | update mapping table in memory;
 7 **end**
 8 **foreach** *Node in* $NodeSet_g$ **do**
 9 | write *Node* as compact log;
 10 | generate a entry in mapping;
 11 **end**
 12 **if** *the Log part is full* **then**
 13 | mergeLog() ;   /* Merge Log part with R-Tree */
 14 **end**

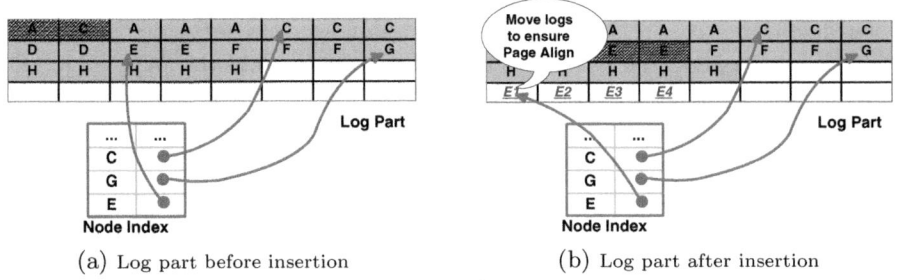

(a) Log part before insertion   (b) Log part after insertion

Fig. 3. An update example in LCR-Tree

logs of different nodes can be integrated into a page and written back as a batch. Modifications on the same node only result in one write operation when flushed. In addition, most of the writes in LCR-Tree are sequential writes. Consequently, the performance of LCR-Tree is expected to be largely improved.

An example of compact log write for insertion is illustrated in Figure 3. Reminding the graph and original R-Tree in Figure 1, an entry named I1 is supposed to be inserted into node E. The Log part before insertion is shown in Figure 3(a). Each line in the table represents a page in Log Part while each cell denotes one update for a certain node. Slashed cell means expired log and blank one stands for free page. The node index resides in main memory to help log searching. The Log Part after insertion is given in Figure 3(b). Newly arrival log E3 and E4 is the update for node E. First they are merged with E1 and E2. The compacted log is written to a new free page with the node index updated.

When the system crashes because of power failure or run-time error, the mapping table in main memory will vaporize. The crash recovery is not considered in the current version of LCR-Tree. However, it is easy to extend the strategy to support crash recovery. We can record the page address with its corresponding log, then after the system reboot, we can reconstruct the R-tree by scanning the original R-tree and inversely scanning the log part.

4   Experimental Results

We compare our LCR-Tree with standard R-Tree and RFTL [20] in this section to show the superiority of the proposed index. The program is developed in Visual Studio 2008 environment using C++. All experiments are performed on a Windows 7 PC with 2.0 GHz Intel Core CPU and 1 GB of physical memory. The SSD we used is an ADATA SSD with 30GB capacity which has been introduced in the second section. We first conduct the experiment on synthetic data sets and then on real geographic data. Table 2 gives the parameters used in our experiments. The *LogRatio* means the percentage of log size to tree size. Unless stated explicitly, the default parameter values, given in bold, are used.

Table 2. Experimental parameters

Parameter	Value
Node Size	1KB, 2KB, **4KB**, 8KB
Total Entry Number	10K, 20K, 50K, **100K**, 200K
Log Ratio	1% 2%, 4%, 8%, 10%, **20%**,50%, 100%
Insert Ratio	0%, 20%, **40%**, 60%, 80%, 100%

4.1   Experiments on Synthetic Data Sets

In this part, we generate a set of two-dimensional points uniformly distributed in the 10000*10000 area and insert them into spatial index. Then different queries are performed on the indexes to validate the performance of indexes. The execution time is used as the criterion for all the experiments.

First we tune the log size and node size of LCR-Tree. The result is given in Figure 4. We observe the index constructing process with 100K entries. The size of log part ranges from 1% to 100% of the tree part size. The execution time of constructing R-Tree is illustrated in Figure 4(a). The time first decreases and then slightly increases with the enlargement of log size. When the log size is small, more merges between log and original data are performed. Thus, the number of update operations grow and performance degrades. In addition, a larger log size causes more read operation and the performance also deteriorates. The optimal log size is around 20% and taken as the default value.

We also test the run time against various node size, the results are given in Figure 4(b). The best node size is around 4KB, and the performance degrades for node size either larger or smaller than 4KB. This is reasonable as SSD has no

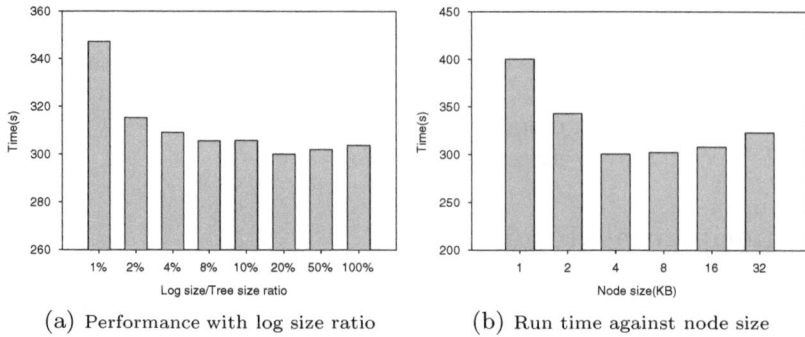

(a) Performance with log size ratio   (b) Run time against node size

Fig. 4. Parameters tuning

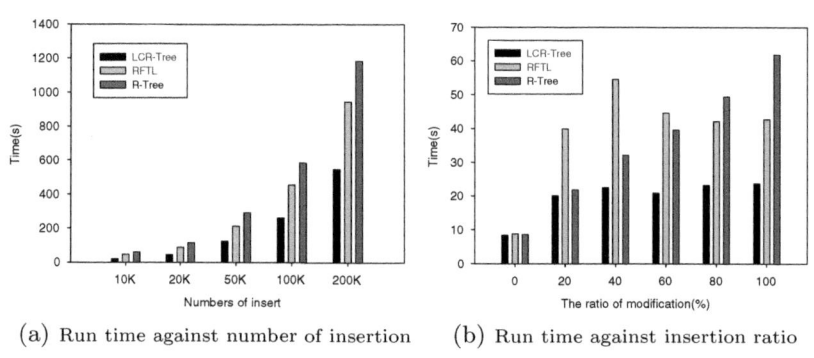

(a) Run time against number of insertion   (b) Run time against insertion ratio

Fig. 5. Evaluation of LCR-Tree on synthetic data sets

advantage at reading small-sized block while large node incurs more cost when accessing and updating one node. In the other experiments, unless explicitly denoted, we use 4KB as the default node size.

In this experiment, we evaluate the performance of LCR-Tree by comparing with RFTL and traditional R-Tree. The log size is fixed to 100 tree nodes size both for LCR-Tree and RFTL, and we vary the number of entries to be inserted into the index. The results are given in Figure 5 (a). With the increment of entries to be inserted, execution time of all the indexes grows quickly. Our strategy reduces the execution time by 58% and 46% on average compared with traditional R-tree and RFTL respectively.

We also generate a workload mixed with insert and search operations which is performed on an index with 100K entries and record the execution time. The number of total requests is 10K and insert ratio varies from 0% to 100%. The results are shown in Figure 5 (b). The execution time of R-Tree rises almost linearly with the increment of insert ratio, while LCR-Tree has no obvious change after 40% insert ratio. RFTL has the worst performance when insert ratio is 40%. The reason is that reading a node is likely to access more paths through the tree due to the overlapped read characteristic of R-Tree, while most insertions only need to go through one path. However, reading in RFTL is the costliest among three indexes, so RFTL exhibits worst performance. When the insert ratio is low, RFTL has shorter log chain and lower read cost and thus the overall performance is better. Furthermore, when insert ratio is high, although the log chain is longer, RFTL has small amount of read operations. Consequently, it takes less execution time than the case with 40% insert ratio, and performs better than R-tree.

4.2 Experiments on Real Spatial Data Sets

Real data sets are used to evaluate LCR-Tree which is downloaded from R-Tree Portal web site [17]. The descriptions of these data sets are listed in Table 3.

We perform the tree construction process and record the execution time on each data set. The environment remains the same as described above. Figure 6 gives the results. As shown in the figure, the execution time of our algorithm

Table 3. Description of real spatial data sets

Name	MBR Number	Description
CAS	98,451	the streams of California
LAS	131,461	LA streets
LARR	128,971	LA rivers and railways

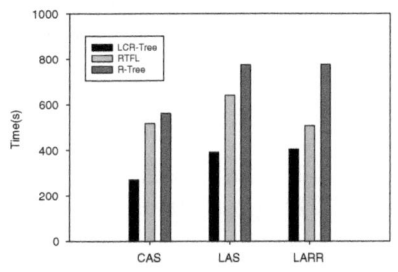

(a) Run time comparison

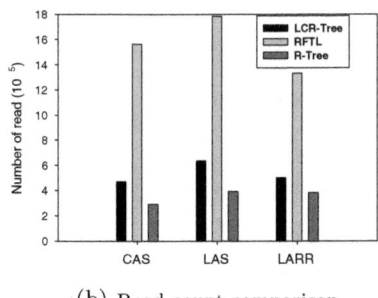

(b) Read count comparison

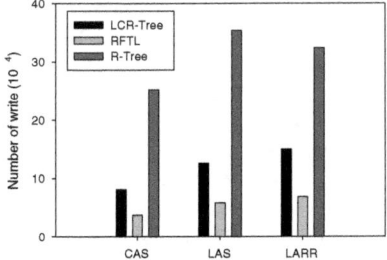

(c) Write count comparison

Fig. 6. Evaluation of LCR-Tree on real spatial data sets

is only 50% of R-Tree 65% of RFTL on average, which illustrates the proposed LCR-Tree is efficient for real data sets. The detailed read and write counts are given in Figure 6(b) and (c). Both RFTL and LCR-Tree exhibit less writes but more reads compared with R-Tree, so both of them are trade-offs of I/O operations. RFTL always has the least number of writes, however, at the cost of huge increment of reads, which counteracts the effect of decrement of writes and hence accounts for its poor improvement over R-Tree as shown in experiments. The proposed LCR-Tree achieves an effective balance between I/O performance with respect to the characteristics of flash-based devices.

5   Conclusion and Future Work

The ubiquitous use of flash devices poses great challenge to traditional index structures. In this paper, we have presented a novel flash-aware variant of R-Tree, named LCR-Tree. The LCR-tree deploys the compact log mechanism to improve the write performance significantly with only a little increment of read overhead. The LCR-Tree has good feasibility and is easy to implement. The experimental study on both synthetic and real data sets shows that the LCR-Tree can achieve up to 3× gains over the R-Tree and existing flash-based index RFTL.

In ongoing work, we plan to enhance our LCR-Tree to make full use of the various I/O cost of different brand of SSDs. It is also necessary to adjust the log merging time automatically according to the evolution of the work load. Furthermore, it is interesting to combine the adjacent pages as an I/O batch to further improve the overall performance of LCR-Tree.

Acknowledgement

This research was supported by the grants of Natural Science Foundation of China (No. 60873063) and MIIT grant 2010ZX01042-001-001-04.

References

1. Agrawal, D., Ganesan, D., Sitaraman, R.K., Diao, Y., Singh, S.: Lazy-Adaptive Tree. An optimized index structure for flash devices. PVLDB 2(1), 361–372 (2009)
2. Bouganim, L., Jónsson, B.T., Bonnet, P.: uFLIP: Understanding flash io patterns. In: CIDR (2009)
3. Chen, F., Koufaty, D.A., Zhang, X.: Understanding intrinsic characteristics and system implications of flash memory based solid state drives. In: SIGMET-RICS/Performance, pp. 181–192 (2009)
4. Guttman, A.: R-trees: A dynamic index structure for spatial searching. In: SIGMOD Conference, pp. 47–57 (1984)
5. Kang, D., Jung, D., Kang, J.U., Kim, J.S.: mu-tree: an ordered index structure for nand flash memory. In: EMSOFT, pp. 144–153 (2007)
6. Kim, Y.R., Whang, K.Y., Song, I.Y.: Page-differential logging: an efficient and dbms-independent approach for storing data into flash memory. In: SIGMOD Conference, pp. 363–374 (2010)

7. Koltsidas, I., Viglas, S.: Flashing up the storage layer. PVLDB 1(1), 514–525 (2008)
8. Lee, S.W., Moon, B.: Design of flash-based DBMS: an in-page logging approach. In: SIGMOD Conference, pp. 55–66 (2007)
9. Lee, S.W., Moon, B., Park, C., Kim, J.M., Kim, S.W.: A case for flash memory ssd in enterprise database applications. In: SIGMOD Conference, pp. 1075–1086 (2008)
10. Li, X., Zhou, D., Meng, X.: A new dynamic hash index for flash-based storage. In: WAIM, pp. 93–98 (2008)
11. Li, Y., He, B., Yang, J., Luo, Q., Yi, K.: Tree indexing on solid state drives. PVLDB 3(1), 1195–1206 (2010)
12. Li, Y., He, B., Luo, Q., Yi, K.: Tree indexing on flash disks. In: ICDE, pp. 1303–1306 (2009)
13. MySQL: Creating Spatial Indexes, http://dev.mysql.com/doc/refman/5.0/en/creating-spatial-indexes.html
14. Na, G.J., Lee, S.W., Moon, B.: Dynamic in-page logging for flash-aware b-tree index. In: CIKM, pp. 1485–1488 (2009)
15. Na, G.J., Moon, B., Lee, S.W.: In-page logging B-tree for flash memory. In: Zhou, X., Yokota, H., Deng, K., Liu, Q. (eds.) DASFAA 2009. LNCS, vol. 5463, pp. 755–758. Springer, Heidelberg (2009)
16. Nath, S., Kansal, A.: FlashDB: dynamic self-tuning database for nand flash. In: IPSN, pp. 410–419 (2007)
17. Portal, R.T.: Spatial (geographical) Datasets, http://www.rtreeportal.org
18. PostgreSQL: PostgreSQL Index, http://www.postgresql.org/docs/8.1/static/indexes-types.html
19. Wu, C.-H., Chang, L.-P., Kuo, T.-W.: An efficient B-tree layer for flash-memory storage systems. In: Chen, J., Hong, S. (eds.) RTCSA 2003. LNCS, vol. 2968, pp. 409–430. Springer, Heidelberg (2004)
20. Wu, C.H., Chang, L.P., Kuo, T.W.: An efficient r-tree implementation over flash-memory storage systems. In: GIS, pp. 17–24 (2003)
21. Yin, S., Pucheral, P., Meng, X.: Pbfilter: indexing flash-resident data through partitioned summaries. In: CIKM, pp. 1333–1334 (2008)
22. Zeinalipour-Yazti, D., Lin, S., Kalogeraki, V., Gunopulos, D., Najjar, W.A.: MicroHash: An efficient index structure for flash-based sensor devices. In: FAST (2005)

An FTL-Agnostic Layer to Improve Random Write on Flash Memory

Brice Chardin[1,2], Olivier Pasteur[1], and Jean-Marc Petit[2]

[1] EDF R&D, France
[2] Université de Lyon, CNRS,
INSA-Lyon, LIRIS, UMR5205, F-69621, France

Abstract. Flash memories are considered a competitive alternative to rotating disks as non-volatile data storage for database management systems. However, even if the Flash Translation Layer – or FTL – allows both technologies to share the same block interface, they have different preferred access patterns. Database management systems could potentially benefit from flash memories as they provide fast random access for read operations although random writes are generally not as efficient as sequential writes. In this paper, we propose a simple data placement algorithm designed for flash memories, to reorganize inefficient random writes in a quasi-sequential access pattern. This access pattern is first established encouraging for a subset of flash devices by identifying a strong correlation between spatial locality and write performances, with a distance being defined to quantify this effect. This design is then validated by a formalization with a mathematical model, along with experimental results. With this optimization, random write potentially become as efficient as sequential write, improving random write speed by up to two orders of magnitude.

1 Introduction

For the sake of interchangeability, many flash memories include a Flash Translation Layer – abbreviated as FTL – to comply with the block interface, a rotating disk legacy. In addition to providing block write and read operations, the FTL manages flash chips complex writing mechanism. However, this layer is implemented with proprietary and undocumented software, which makes flash devices appear as "black boxes" from a system's point of view [3].

Advantageously, this FTL allows a straightforward substitution between both storage technologies. Yet, most database management systems include rotating disks-oriented optimizations, which are not relevant for flash memories. Even if both technologies use the same block interface, they have different preferred access patterns. Database management systems could potentially benefit from flash memories as they provide fast random access for read operations. Still, for FTL-based devices, random writes are generally not as efficient as sequential writes [5] and most optimization techniques for flash memories relate to this specific issue.

J. Xu et al. (Eds.): DASFAA Workshops 2011, LNCS 6637, pp. 214–225, 2011.

In this paper, we identify a strong correlation between write performances and spatial locality for a subset of FTL-based devices; and define a distance to quantify this effect. From this property, we propose a simple data placement algorithm, which trades flash memory space for random write performances. Its efficiency is validated by a formalization with a mathematical model, along with experimental results. With this optimization, random write potentially become as efficient as sequential write, improving random write speed by up to two orders of magnitude.

The rest of this paper is organized as follow. Section 2 introduces NAND flash memories and different types of mapping used in the FTL. Section 3 emphasizes the importance of locality on these devices for write performances and defines a distance between consecutive writes to quantify this effect. In section 4, we derive from this property an optimization technique for random writes, using an indirection layer to minimize this distance, thus avoiding scattered writes. In section 5, we present an approximate model for this algorithm. The results of both our experiments and model are reported in section 6. Related works are described in section 7. Then, section 8 summarizes the contributions of this paper.

2   NAND Flash Memories

NAND flash memory is a non-volatile storage technology, which allows three low-level data-access operations: read, write (or *program*) and erase. Still, erasing is performed at a different granularity than reading or writing: NAND flash chips are divided into blocks that can be erased independently, each block containing a fixed number of pages, each of which being individually accessible for reading or writing. As overwrites are not allowed, a full block must be erased prior to writing on one of its already used page. Additionally, pages within a block must be written sequentially.

To handle this complex writing mechanism, most flash memories include a Flash Translation Layer (FTL) that redirects writes on available (erased) pages and stores the associations between the logical sector identifier and its physical location in an address translation table.

In most cases, this translation operates on a page-level basis or on a block-level basis [6]. With a page mapping FTL, each logical page has its associated physical page. After an overwrite, the translation table is updated with the new physical location and the old physical location is marked as obsolete to be reclaimed by a garbage collection mechanism. With a block mapping FTL, each logical block has its associated physical block and an additional logging area, which consists of log blocks. When a page is overwritten, new data are appended to the last log block. Garbage collection merges a data block with all its associated log blocks by copying every valid page on a new (erased) data block and updating the translation table.

Each mapping granularity has its own drawbacks. Page mapping has a higher memory overhead because of its larger address translation table, while block mapping performances are highly dependent on empty blocks availability, to serve as log blocks.

3   Write Spatial Locality for FTL-Based Devices

As FTL enclosed in flash devices are usually proprietary and undocumented, studies have been conducted to identify preferred write access patterns for such devices. In [2], Birrell et al. identify a strong correlation between the average latency of a write operation and the gap between writes, as long as this gap is less than the size of two flash blocks. They conclude that write performance varies with the likelihood that multiple writes will fall into the same flash block, which is a manifestation of an underlying block or hybrid-mapping FTL. As a result, a fine-grained mapping is mandatory for high performance flash memories, but we believe that such a mapping can be efficiently provided by an additional layer, distinct from the FTL. Indeed, Wang et al. study in [12] the effectiveness of log-structured file systems for flash-based DBMS, since these file systems tend to write large data blocs in sequence. Their experiments validate potential benefits as they achieve up to x6.6 performance improvement.

uFLIP [3] is a component benchmark designed to quantify the behavior of flash-memories when confronted to defined I/O patterns. Some of these patterns relate to locality and increments between consecutive writes. Their results confirm that localizing random writes greatly improve efficiency and large increments lead to performances which could be even worse than random writes.

We propose a similar approach to quantify the effect of spatial locality on FTL-based devices, by introducing a notion of distance between consecutive writes. In our experiments, the average write duration for each distance d is evaluated by skipping $|d|-1$ sectors between consecutive writes. This metric can be negative to discriminate between increasing and decreasing address values. From the results of previous works, we conjecture a usual behavior where, up to a distance d_{max}, the average cost of a write operation $cost(d)$ is approximately proportional to d.

To validate this assumption, we measured the effect of distance on a variety of flash devices. Although individual write durations are erratic, their average value converge when this access pattern is sustained. Figure 1 shows that our assumption is verified for a flash-based SSD[1] and a USB flash drive[2].

Scattered writes (ie. $d \geq d_{max}$) are typically 20 to 100 times slower than sequential writes for flash memories with a block-mapping FTL[3]. Consequently, and because of this proportional performance pattern, reducing the average distance between consecutive writes can significantly improve efficiency, even if strict sequential access (d=1) is not achieved. The optimization described in the following section focuses on this access pattern, skipping as little sectors as possible.

---

[1] SSD Mtron MSD SATA3035-032, sector size 4 KiB.
[2] Flash chip HYNIX HY27UG088G5B with an ALCOR AU6983HL controller, sector size 4 KiB.

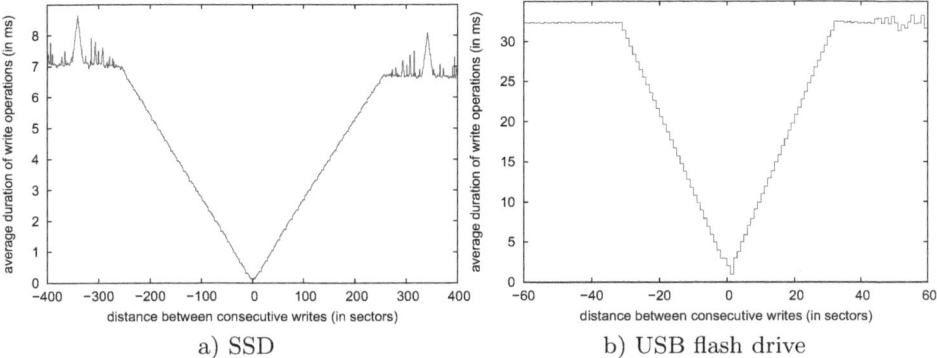

a) SSD   b) USB flash drive

Fig. 1. Influence of distance on write duration

4   Gathering Random Writes

Online transaction processing usually have a part of its workload constituted of small random writes. The optimization described in this section converts these random writes into sequential writes, while skipping sectors containing valid data, as this access pattern should increase performances on flash memories. With this optimization, sectors containing valid data (used sectors) and free (unused) sectors are mixed on the device. An additional indirection layer is used to redirect logical writes to unused sectors by minimizing the distance between consecutive writes.

To allow data retrieval, correspondences between physical and logical locations are stored in an address translation table, with every logical sector being associated with a physical sector. Unused physical sectors – not associated with any logical sectors in the address translation table – do not contain any useful data, and therefore constitute a pool of free sectors available for writing.

To overwrite a logical sector, data are assigned to a pool sector adjacent to the previous write. Then the logical-physical association stored in the address translation table is updated, the previously associated sector therefore being freed and added to the pool. Figure 2 illustrates how logical writes are assigned to physical locations, when writing successively on logical sectors 0, 3 and 0.

This optimization does not require garbage collection, as the size of the pool remains constant: physical sectors containing obsolete data are immediately added to the pool, and can be overwritten. Yet, as an independent and internal mechanism, the FTL might still use garbage collection to handle flash erasures.

Any logical access pattern, whether sequential or random, will lead to a quasi-sequential physical access pattern. Consequently, the average distance (and thus write efficiency) is determined exclusively by the proportion of pool sectors. As increasing pool size requires additional non-volatile memory space, this characteristic can be adjusted to obtain an expected efficiency. As a downside, sequential reads are also transformed into random reads. However, this behavior is not an issue for flash devices, as random reads are as efficient as sequential reads [3].

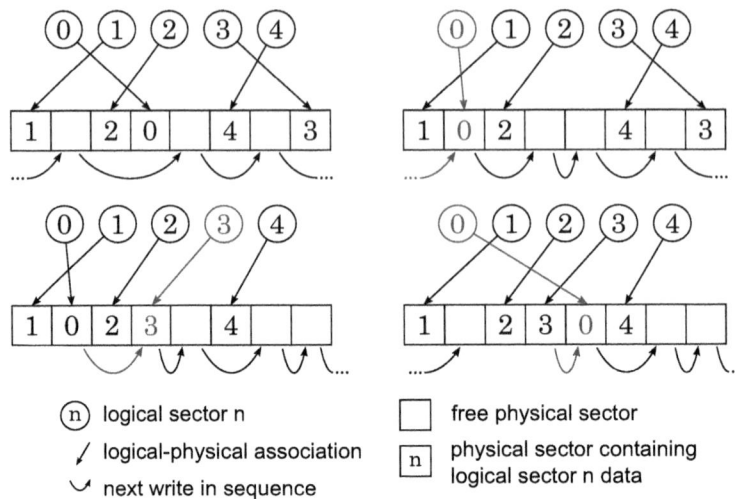

Fig. 2. Optimization overview

Read operations only induce lookups in the address translation table, which is a negligible overhead.

To prevent revisiting regions of the memory recently accessed, where pool sectors should have been exhausted, only positive distances are considered in this optimization. Additionally, the addressable space is assumed to be circular, in order to avoid handling edges differently.

The first data structure used by the redirection algorithm is the address translation table. This table – named T – binds every logical sector to a flash sector. As this optimization aims at minimizing the average distance between consecutive writes, the most recently written sector is referred to as $f_\frown$. A simple version of the redirection algorithm involves the following operations when writing a logical sector l:

1: $f \leftarrow$ closest pool sector from $f_\frown$
2: write data on f
3: $T(l) \leftarrow f$ {update the translation table}
4: $f_\frown \leftarrow f$ {keep the reference of the most recently written sector}

Operation (1) – searching the pool sector closest to the previous write – has to be implemented carefully with an adequate data structure. In our implementation, to hasten lookups of this sector, we keep references of every sector in the pool in an ordered list P, where sectors are arranged by increasing distances from $f_\frown$. $P(0)$ is thus the closest free sector from $f_\frown$, followed by $P(1)$, etc.

Including this ordered list of pool sectors allows efficient retrievals of closest pool sectors. Nevertheless, this list has to be updated with each newly freed sector, whenever the translation table is altered. With this additional data structure, writing on a logical sector l implies the following operations:

1: $f \leftarrow T(l)$
2: write data on $P(0)$
3: $T(l) \leftarrow P(0)$ {update the translation table}
4: remove $P(0)$ from P {update the list of pool sectors}
5: add f in P

Operations (4) and (5) can be done asynchronously (ie. during the subsequent write in a write-intensive environment) without any consistency issue, as the list of pool sectors P can be rebuild from the translation table T. Consequently, P might not be up-to-date for each write request, which results in a slight increase of the average distance between consecutive writes if the closest pool sector from $f_\frown$ is not yet referenced in this list. However, this case appears infrequently for large pools and can be neglected.

5 Model

To estimate write speed improvement provided by this algorithm, we propose to model its behavior by evaluating the average cost of a write operation. This model is based on the simplifying assumption that pool sectors are uniformly distributed within flash sectors. This state is also supposed to be stable with occurring writes. Additionally, writing cost is expected to be determined exclusively by its physical distance from the previous write.

Under these approximations, the overall speed improvement can be evaluated given the probability to obtain each possible distance, and their associated costs. For this model, the following notations are used :

 - $\mathbb{F}$ is the set of sectors accessible from the device (flash sectors),
 - $\mathcal{D}(a, b)$ is the distance between two sectors a and b,
 - $\mathbb{L}$ is the set of logical sectors,
 - $\mathbb{P}$ is the set of pool sectors; by definition, $|\mathbb{P}| = |\mathbb{F}| - |\mathbb{L}|$.

Definition 1. *Let $p(d)$ be the probability that the sector $f_i \in \mathbb{P}$ which minimizes $\mathcal{D}(f_\frown, f_i)$ also verifies $\mathcal{D}(f_\frown, f_i) = d$, namely having a distance d between two consecutive writes.*

With the uniform distribution assumption, the probability $p(d)$ to get a distance d between two consecutive writes can be estimated as the ratio between favorable and possible distributions:

$$p(d) = \frac{\binom{|\mathbb{F}|-d-1}{|\mathbb{P}|-1}}{\binom{|\mathbb{F}|-1}{|\mathbb{P}|}} \tag{1}$$

The cost of a write operation conditioned by its distance from preceding write, $cost(d)$, can be approximated but also measured from the device, as shown in Fig. 1. For our evaluations, the latter is believed to be more accurate.

Given these two parameters, $p(d)$ and $cost(d)$, the average cost of a write operation, named $cost_{\text{avg}}$, amounts to:

$$cost_{\mathrm{avg}} = \sum_{d=1}^{|\mathbb{L}|} p(d) \times cost(d) \qquad (2)$$

Estimations from this model are reported in Sect. 6, together with experimental results. In addition to theoretical performance gains, resource usage can be quantified as this optimization trades server CPU and RAM, as well as flash memory space for writing speed.

CPU overheads occur when handling the translation table and the list of pool sectors during a write operation. These overheads relate to the following operations:

- search for the closest pool sector, which is $\mathcal{O}(1)$ when pool sectors references are stored in an ordered list,
- update the translation table, which is $\mathcal{O}(1)$,
- update the list of pool sectors, which is $\mathcal{O}(\log |\mathbb{P}|)$ with optimized data structures, such as skip-lists.

Updating the list of pool sectors is the only operation with significant CPU cost. However, as stated in Sect. 4, this update can be asynchronous. For read operations, looking up correspondences between logical and flash sectors in the address translation table results in constant CPU overhead, which is negligible compared to a flash sector read duration.

Server RAM overheads are caused by the translation table and the list of pool sectors maintained in main memory. These overheads amounts to $\mathcal{O}(|\mathbb{L}| \times \log |\mathbb{F}|)$ for the translation table, and $\mathcal{O}(|\mathbb{P}| \times \log |\mathbb{F}|)$ for the pool. Total RAM overhead adds up to $\mathcal{O}(|\mathbb{F}| \times \log |\mathbb{F}|)$.

As pool sectors are stored on the device, and do not hold any useful data, flash memory space overhead amounts to $|\mathbb{P}|$ sectors.

The last significant trade-off involves sequential writes. Since, with this algorithm, performances do not depend on the access pattern, logical sequential writes have the same performances as logical random writes. Existing attempts to sequentialize accesses would not bring any additional performance gain and should be discarded.

6   Results

To validate this optimization together with the model detailed in the previous section, the data placement algorithm is tested on both devices mentioned in section 3. These tests consist in evaluating the average cost of logical random writes for varying sizes of the pool.

Figure 3 shows experimental results and the model expectations for the USB flash drive. To compare with conventional access patterns, random write and sequential write iops (respectively 30 and 1060) are also reported on this figure.

To obtain performances equivalent to sequential writes, consequent sacrifices have to be made in terms of flash memory space. In our experiment, 95% of

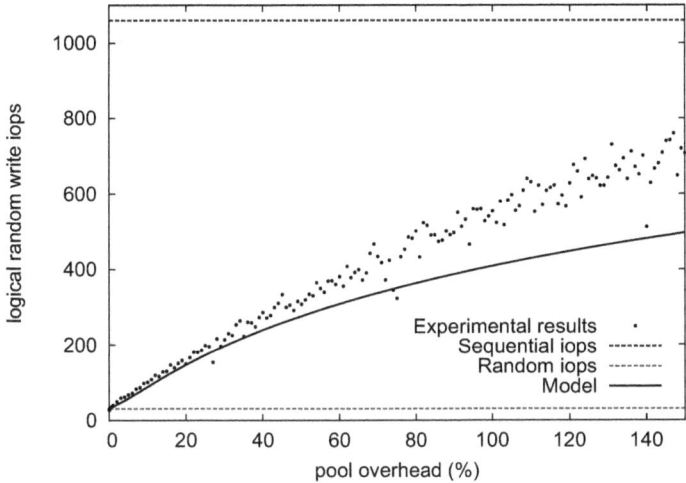

Fig. 3. Logical random write performance for 100,000 logical sectors

sequential write efficiency is achieved when the pool is about three times larger than the logical address space. However, we achieved significant improvements over random writes with acceptable trade-offs, as we have a ten times improvement with 50% flash memory space overhead.

Contrastingly, writing speed on the SSD is improved with distances below $d_{\max} = 256$, instead of $d_{\max} = 32$ for the USB flash drive. As a result, notable improvements are achieved on the SSD with relatively lower sacrifices. Experimental results for this device are reported on Fig. 4.

Another noticeable difference was suggested by measurements obtained in Sect. 3: Fig. 5 focuses on small distance values. Remarkably, a quasi-sequential access pattern with a distance of 4 sectors between consecutive writes shows relatively good performances. Highest iops are achieved with a pool size of about 29,000 sectors, which results in an average – but still random – distance of 4. This property allows optimal performances to be achieved with much less overhead. Indeed, our optimization reaches 7796 average iops for 4KB logical random writes at a cost of only 687 KB of RAM, and 14.5% flash overhead for 800 MB of usable data space. Compared to physical random writes 134 average iops, performances are improved by ×58.

Determining this optimal pool size is not straightforward, and depends on the sector size. With 16 KiB sectors, experiments give an optimum distance of 2; and about 1.6 for 32 KiB sectors. A possible explanation for this behavior is that interleaving favors non-zero sized skips to access multiple internal flash chips in parallel [13].

Unfortunately, this "peak" behavior might not be representative of flash solid-state drives. Among the twelve SSD with uFLIP results available, only one (by the same manufacturer as ours) expose the same characteristic. This singularity is only a facultative additional benefit as it was not part of our initial

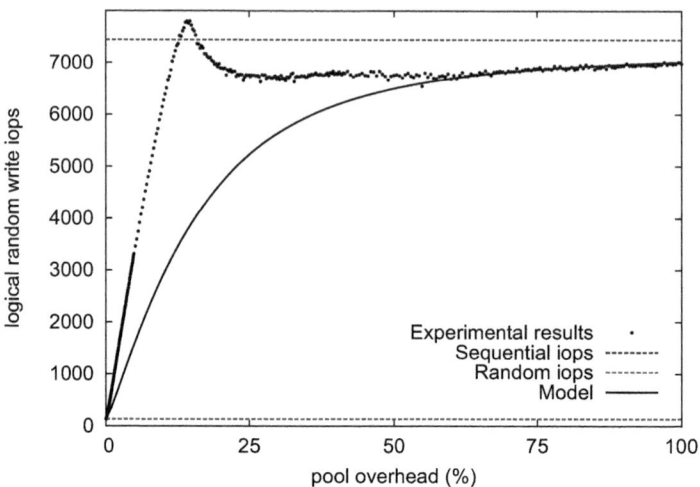

Fig. 4. Logical random write performance for 200,000 logical sectors

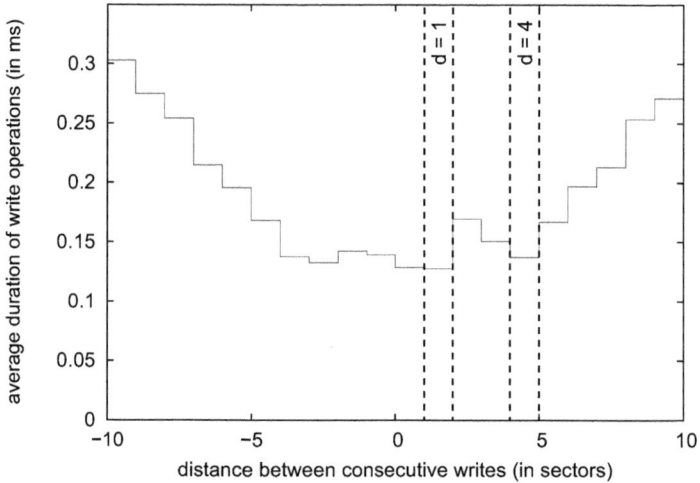

Fig. 5. Small variations of gaps for quasi-sequential writes

assumptions. Our model might be closer of what we would expect with a regular SSD, and still shows considerable performances gains, such as a ×40 improvement over random writes with 25% flash memory overhead.

Still, these results reveal some limitations of our model. One of its simplifying assumptions is that writing cost is determined exclusively by its physical distance from the previous write, which might not be accurate.

Moreover, experiments outperform our model as memory is accessed in only one direction – increasing physical addresses – to prevent revisiting regions of the memory where pool sectors should have been exhausted. This improvement over our theoretical uniform distribution reduces the average distance between consecutive writes.

7   Related Works

Many optimizations have been conceived with flash chips characteristics in mind. A frequent design avoids in place updates with log-based methods.

The In-Page Logging Approach [8] allocates a portion of each bloc to write updates of its pages. This optimization improves writing speed, as updates are written sequentially inside the erase unit, at the expense of more read operations. Garbage collection consists in merging data pages with their log sectors on a new empty block.

Page-Differential Logging [7] uses a similar approach, except page differential are logged. Writing is improved as differentials of multiple pages can be combined to fit in a single page. Also, differentials are recomputed from the original page for each overwrite, which implies that reading a logical page involves at most two physical read (the original page and its last differential). The garbage collection mechanism is also improved, as merging a page with its current differential is not required (both can be copied separately).

Log-structured file systems, with a distinction between file systems designed for raw flash chips (without FTL) – like YAFFS, LogFS, JFFS – and those designed for block devices – like LFS – use methods comparable to the in-page logging approach, and therefore provide similar benefits and drawbacks. Additionally, I/O patterns of log-structured file systems for block devices when accessing multiple files tend to be of small size and scattered.

Regarding more specific use cases, B-File [10] is an abstraction layer for self-expiring items on flash memories. Depending on their expiration date, items are written sequentially in appropriate erase units to avoid copying valid data on deletion. Another approach defines an Append and Pack Layout [11], which divide the database in two separate datasets, respectively write-hot and write-cold. These datasets are written sequentially in multiples of the erase block size, with space reclamation when the memory is full.

The main differences between these approaches and our optimization are the necessity of a garbage collection mechanism and decreased read performances as a logical read rely on multiple physical reads. In contrast with these works, we optimize data access exclusively over the FTL. As a result, our approach is not applicable to raw flash chips.

RS-Wrapper [13] is a simple conversion between random writes and sequential writes for FTL-based devices. When random writes are adequately dense, their experiments show that reading the missing pages to overwrite sequentially the entire data range outperforms overwriting exclusively modified pages. However, reorganizing these random writes in a quasi-sequential access pattern has not been tested.

FlashLogging [4] is an efficient mechanism for synchronous logging on multiple low capacity flash devices. While the use case differs from our proposition, this approach could be used to address the non-volatile issue of our current optimization. Indeed, data written to the device are volatile, since the address translation table is stored in RAM, and is needed to rebuild the database. Logging its modifications on additional flash devices could provide an efficient solution. This issue could also be managed by writing logical addresses together with data, similarly to the FTL internal functioning.

On a different but related subject, enterprise class SSD can provide better random write performance at the cost of additional RAM, processing power and spare blocks (not accessible from the host) [1,9]. However, these designs focus on random write and provide invariably good performances for the entire device. Most database applications mix random and sequential accesses and do not require such homogeneous random write efficiency. By adding a software layer, our optimization permit using less expensive personal-class SSD with good, yet spatially limited, random write performances. This is also applicable to removable flash media, which have lessened hardware capabilities.

8   Conclusion

In this paper, we first introduced a notion of distance, and described its impact on flash memories write performances. Based on this property, we proposed a data placement algorithm, which significantly improves random write performances. Our contributions emphasize the importance of locality for these FTL-based devices, and we believe quasi-sequential access patterns to be of use for future data placement optimizations.

Compared to native write operations over the FTL, our optimization benefit from the host available RAM and processing power to improve random write efficiency on portions of the device. This method support localized performances adjustment, while flash memories offer homogeneous behaviors.

For the SSD used in our experiments, we achieved an improvement of up to ×58 at a cost of only 14.5% flash overhead. In this best case scenario, this technique even caused random write to perform slightly better than sequential write, by 3.5%. This optimization is, to some extent, also applicable on flash memories with less capacity, as results with a USB flash drive show a ×10 improvement with 50% flash overhead. Conjointly, we proposed a model to predict write performances, however, future works are needed to enhance its accuracy and help tradeoffs adjustments.

Another perspective relate to data volatility, which might be addressed in future works with solutions proposed in Sect. 7. Still, this optimization is already applicable for indexation or temporary tables, where volatility is acceptable.

References

1. Agrawal, N., Prabhakaran, V., Wobber, T., Davis, J.D., Manasse, M., Panigrahy, R.: Design Tradeoffs for SSD Performance. In: 2008 USENIX Annual Technical Conference, pp. 57–70. USENIX Association (2008)
2. Birrell, A., Isard, M., Thacker, C., Wobber, T.: A Design for High-Performance Flash Disks. SIGOPS Operating Systems Review 41(2), 88–93 (2007)

3. Bouganim, L., Jónsson, B.T., Bonnet, P.: uFLIP: Understanding Flash IO Patterns. In: 4th Biennial Conference on Innovative Data Systems Research (2009)
4. Chen, S.: FlashLogging: Exploiting Flash Devices for Synchronous Logging Performance. In: 35th International Conference on Management of Data, pp. 73–86. ACM, New York (2009)
5. Gray, J., Fitzgerald, B.: Flash Disk Opportunity for Server Applications. Queue 6(4), 18–23 (2008)
6. Kim, J., Oh, Y., Kim, E., Choi, J., Lee, D., Noh, S.H.: Disk Schedulers for Solid State Drives. In: 7th International Conference on Embedded Software, pp. 295–304. ACM, New York (2009)
7. Kim, Y.R., Whang, K.Y., Song, I.Y.: Page-Differential Logging: An Efficient and DBMS-independent Approach for Storing Data into Flash Memory. In: 36th International Conference on Management of Data, pp. 363–374. ACM, New York (2010)
8. Lee, S.W., Moon, B.: Design of Flash-Based DBMS: An In-Page Logging Approach. In: 33th International Conference on Management of Data, pp. 55–66. ACM, New York (2007)
9. Lee, S.W., Moon, B., Park, C.: Advances in Flash Memory SSD Technology for Enterprise Database Applications. In: 35th International Conference on Management of Data, pp. 863–870. ACM, New York (2009)
10. Nath, S., Gibbons, P.B.: Online Maintenance of Very Large Random Samples on Flash Storage. PVLDB 1(1), 970–983 (2008)
11. Stoica, R., Athanassoulis, M., Johnson, R., Ailamaki, A.: Evaluating and Repairing Write Performance on Flash Devices. In: 5th International Workshop on Data Management on New Hardware, pp. 9–14. ACM, New York (2009)
12. Wang, Y., Goda, K., Kitsuregawa, M.: Evaluating Non-In-Place Update Techniques for Flash-Based Transaction Processing Systems. In: Bhowmick, S.S., Küng, J., Wagner, R. (eds.) DEXA 2009. LNCS, vol. 5690, pp. 777–791. Springer, Heidelberg (2009)
13. Zhou, D., Meng, X.: RS-Wrapper: Random Write Optimization for Solid State Drive. In: 18th Conference on Information and Knowledge Management, pp. 1457–1460. ACM, New York (2009)

Energy Efficiency Is Not Enough, Energy Proportionality Is Needed!

Theo Härder, Volker Hudlet, Yi Ou, and Daniel Schall

Databases and Information Systems Group
University of Kaiserslautern, Germany
{haerder,hudlet,ou,schall}@cs.uni-kl.de

Abstract. Due to the *energy consumption/resource utilization* characteristics of todays centralized DB servers, the fastest configuration is also the most energy-efficient one. Extensive use of SSDs alone cannot enable a fundamental change of this overall picture, because the storage-related energy consumption is typically only a little fraction of the overall energy budget. Even, when this storage-related share is (almost) completely reduced by optimized flash-aware buffer management, the saving effect achieved may be limited by less than ∼10%. Therefore, we have designed a cluster of wimpy computing nodes called WattDB, where the individual nodes are dynamically attached and detached to the cluster on demand – depending on the current workload needs –, thereby aiming at energy-proportional DB management.

1 Introduction

In recent years, *data management on new hardware* evolved into an active and attractive field of research. Among the topics considered for data management are multi-core processors, vectorized query processing, flash memory, (column-oriented) main-memory databases, OLAP- or OLTP-specific optimizations, etc. Most efforts exclusively aim at performance enhancements for specific database applications – highlighted by the statement "One Size Does Not Fit All!". Representative approaches include MonetDB [1], Greenplum [2], VoltDB [3], SanssouciDB [4], and others.

A few years ago, we started to explore the use of flash memory and the integration of SSDs (solid state disks) into the DB I/O architecture. Unlike most other projects, we primarily addressed energy efficiency for all tasks of database management, but also considered potential performance gains. Our strong hope was to improve both goals – "Keeping Performance While Saving Energy?" [5], which was supported by our initial experiments.

The remainder of this paper is organized as follows: Sect. 2 reviews indicative research results concerning our experiments of exploiting SSDs for DB management, whereas our findings concerning flash-aware DB buffers are summarized in Sect. 3. Sect. 4 highlights the need of energy proportionality. How such a goal can be approximated in DB applications, is sketched in Sect. 5, where we discuss design issues for our *WattDB* system. Concluding remarks and future challenges are outlined in Sect. 6.

J. Xu et al. (Eds.): DASFAA Workshops 2011, LNCS 6637, pp. 226–239, 2011.

2 Experimental Results and Critical Observations

SSD $\neq$ *SSD* – this fact was drastically demonstrated in [6]. It is not only true when using a collection of different device types, but also for instances of the same device type. Reasons are continuous improvements of the SSD types as well as refinements of the flash translation layer (FTL) to optimize measures for wear leveling. All these optimization efforts are proprietary and kept as "business secrets" of the device manufacturers – resulting in a black-box view of SSD devices, often exhibiting surprising behavior. Table 1 gives an overview of the impressive success these efforts gained for almost all characteristics chosen.

Table 1. Storage device characteristics

Device	Model	IOPS		Idle (W)	Active (W)		EUR/GB
		Read	Write		Read	Write	
HDD1	WD 7.2K RPM	70		5.3	6.3		0.38
HDD2	WD 10K RPM	210		4.2	5.7		0.77
HDD3	Fujitsu 15K RPM	500		8.4	9.9		0.76
SSD1	SuperTalent 32GB	2,700	50	1.3	1.7	2.0	9.00
SSD2	MTRON MSP	12,000	130	1.2	2.0	2.0	18.00
SSD3	Intel X-25-M G1	35,000	3,300	0.6	1.3	2.4	2.40
SSD4	Intel X-25-M G2	35,000	8,600	0.9	1.2	4.1	2.44
SSD5	Crucial RealSSD	60,000	45,000	0.8	1.1	1.7	2.01

Strong variations at the instance level are caused by the read/write/erase asymmetry which is amplified by the prevailing conditions or housekeeping tasks of an individual device, e. g., device "aging", state-dependent garbage collection, frequency of erasures as reactions to load patterns, etc.

All these reasons lead to the conclusion that current, state-of-the-art SSDs cannot be easily integrated and used for database management. In any case, we have to cope in DB environments with a spectrum of heterogeneous SSDs with differing and instable access behavior. Furthermore, the data sheets of the device manufacturers often report performance figures which can never be met by typical DB workloads. To support our conclusion, we want to repeat here some indicative, empirical results of previous publications [6,7].

2.1 SSD Performance Measurements

Our initial experiments focused on DB-specific read and write performance of SSDs, for which the essential characteristics – taken from the data sheets of the manufacturers – are listed in Table 1. To stress all devices with the same access patterns, we developed a tool (similar to uFlip and IOmeter[1]) that allows us to perform benchmarks on the devices. The tool supports adjustable page sizes and

---

[1] http://uflip.inria.fr/~uFLIP/, http://www.iometer.org

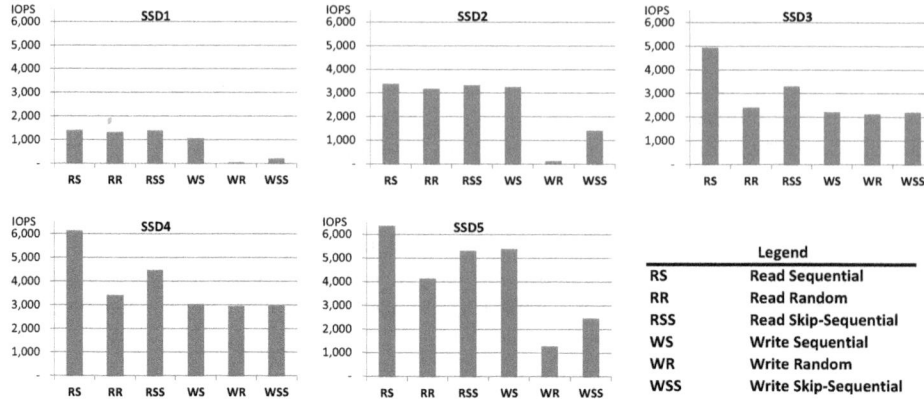

Fig. 1. Performance measurements using different SSD types

is able to read and write different access patterns from/to the devices; we used 32KB page sizes in all experiments[2].

Our first test pattern is sequentially accessing n pages simulating DB scan operations. A second pattern is randomly accessing all pages of the test file, where each page is fetched only once. Finally, the *skip-sequential* pattern accesses pages sequentially, but skips randomly over some pages. Hence, we mimic kind of index scans via (sorted) reference lists. Illustrated in Fig. 1, the results for the SSDs considered are briefly commented in the following:

- SSD1: Our results clearly show slow performance under all tested patterns as well as heavily degraded write performance. But, random read is as fast as sequential read.
- SSD2 shows improved performance to SSD1. Still, random writing is tremendously slower than other access methods.
- SSD3, as the first Intel generation, is really good at sequential reading, while random reading is comparatively slow. All write patterns are performing equally well – a significant operational difference w. r. t. SSD1 and SSD2.
- SSD4, as next Intel generation, came with the additional feature *TRIM support*[3]. It reveals improved overall performance for all patterns. Nevertheless, the disk is showing the same challenges as the first generation.
- While reading on SSD5 is faster than on all other SSDs we tested, random writing stresses this device remarkably. Although the data sheet promises 60,000 IOPS for random read, we could not get even close to this number (a fact we experienced in most other tests).

2.2   Result Interpretation

Using these performance results, we examine common assumptions, expert wisdom, and rules of thumb regarding SSDs [9] and show that not all of them are

---

[2] According to [8], 32 KB is the preferable page size for SSDs.
[3] http://t13.org/Documents/MinutesDefault.aspx?keyword=trim

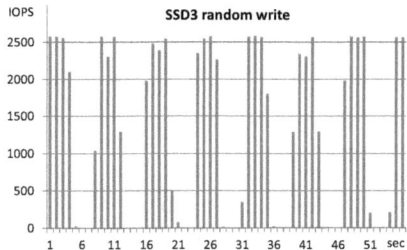

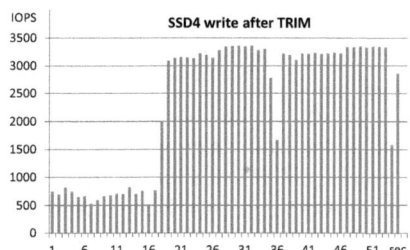

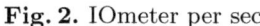

Fig. 2. IOmeter per sec   **Fig. 3.** TRIM effects

true. Given that we cannot design a DBMS tailor-made to a single SSD of a specific type and that varying workloads are present, we may critically observe the following issues.

Random access. SSDs should not suffer from random access, in fact, they do. Random access may be substantially slower than sequential access.

Therefore, sequential accesses should still be preferred over random accesses, although it is not as vital as on hard disks. For hard disks, a rule of thumb recommends an index-based scan only if the selectivity factor of the predicate to be evaluated is below ∼1–3%, otherwise a sequential scan of the whole table is advised. On SSDs, the selectivity factor can be shifted to higher percentages. Because of the different SSDs' performance characteristics, it is not possible to spot a clear break-even point.

Database query optimizers can decide between random and sequential access based on configurable disk parameters.[4] Simple rules of thumb, however, do not work anymore [10], especially if the performance characteristics fluctuate over time. Therefore, optimizing algorithms under wrong assumptions or device models can make overall performance even worse. Furthermore, developers for flash-aware buffer algorithms have to consider that device-specific tweaks might be obsolete in no time.

Unstable and fluctuating behavior. Using a tweaked IOmeter version, we were able to get more detailed performance data from our devices. Fig. 2 visualizes the write performance of SSD3 in pages/second on a per-second basis. As illustrated, every 4 to 5 seconds, performance is heavily degraded for about 3 seconds. We conclude, the drive is performing internal reorganization like freeing up flash blocks or searching for another writable block.

While benchmarking SSD4, we had a look at the TRIM command introduced for this model and observed an interesting behavior. Fig. 3 depicts our write performance measurement right after deleting ∼130 GB of files on the drive and issuing corresponding TRIM commands to the drive. In this graph, a heavily degraded performance in the first half of the measurement is evident. Apparently, the SSD tries to free up flash blocks while we were simultaneously

---

[4] E.g.: http://publib.boulder.ibm.com/infocenter/db2luw/v9r7/index.jsp?
topic=/com.ibm.db2.luw.admin.perf.doc/doc/c0005051.html

applying a write load to it. The proprietary FTL mapping particularly concerns device caching, block allocation, and garbage collection. All these mechanisms are software controlled and entirely hidden to the upper software layers. Hence, optimization decisions in the OS or DBMS may be counterproductive and sometimes even worsen the time-consuming house-keeping activities. As inferred from Fig. 2, write latency may extremely vary. While less than ~400 μs in the best case, we have observed outliers of more than some hundred ms, that is a device-dependent variance of more than ~200–500. In contrast, disks with a device-dependent variance of ~2–5 exhibit quite stable access behavior and lend themselves to reliable optimizer decisions.

Another aspect is a kind of heterogeneity among the SSD types present in a DBMS environment, where several heterogeneous SSDs may coexist in an application (or they may be dynamically exchanged). As a consequence, tailor-made algorithms for specific SSD types, e. g., concerning indexing or buffer management, are not very useful. The same arguments apply for specific workload optimizations (pure OLTP or OLAP processing, mixed workloads with varying degrees of reads/writes). A continuous adjustment or exchange of algorithms affected is not very practical in productive DBMS applications.

Read/write asymmetry. It does not seem to be true in general. SSD1 and SSD2, for example, do not exhibit degraded performance for (sequential) write, they are equally fast as sequential read. On all other SSDs, an asymmetry is measurable, but still not as bad as advertised. Exploited for buffer management, this observation may make at least some prevalent assumptions obsolete.

Slower when full? As a consequence of the erasures, overwriting some blocks on a full disk should be much slower than writing to an empty disk. We verified this assumption by filling all drives with random data and repeating our tests afterwards. No significant differences were measurable.

Impact of queue depth. To gain more insights, we repeatedly measured various queue depths. By using a random read pattern, we give the FTLs a fair chance to optimize the queue. As reported in [6], the only significant improvement was observed between QD 1 and QD 2. Beyond this point, extending the QD did not improve data throughput. We did not expect this result, because manufacturers use even higher queue depths for their performance measurements.

Energy consumption. Fig. 4a shows the absolute power consumption of the SSDs we tested. For this test, a sequential read pattern is used. Write patterns might consume even more energy. Obviously, the drives do consume energy when being idle; therefore, they are not as energy saving as expected. Power consumption ranges from ~4–6 Watt for consumer disks to ~9–14 Watt for disks of enterprise server quality. The SSDs' power profiles are substantially different to those of hard disks; their power consumption for idle states and peak loads is considerably lower, for example, up to a factor of ~15 and ~8 for SSD3 and HDD3, respectively (see Table 1). Fig. 4b shows how many pages can be read by each SSD consuming one Joule of energy. As illustrated, *pages/Joule* are constantly

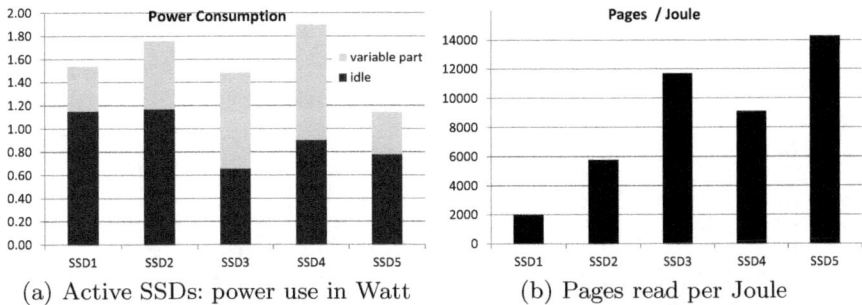

(a) Active SSDs: power use in Watt   (b) Pages read per Joule

Fig. 4. Energy-related SSD measurements

rising, thus newer SSDs are getting more energy efficient. On conventional disks we measured only ~600–1,800 *pages/Joule*. Anyway, a more differentiated comparison is cumbersome, because of the different performance characteristics and their implications on energy efficiency.

The critical question is whether these considerable differences of energy consumption at the device level cause perceptible energy saving at the system level. For this reason, we compare disk- and SSD-based DBMS buffer management in the next section.

3   Findings in DBMS Buffer Management

Because of the read/write/erase asymmetry, buffer management tailor-made to the SSD characteristics is a key issue for flash-aware DBMS optimizations. The fact "whether a page is read only or modified" is an even more performance-critical criterion for the replacement decision [11] than in disk-based DBMSs.

3.1   Objectives of Flash-Aware Replacement Algorithms

Even with SSDs, maintaining a high hit ratio – the primary goal of disk-based buffer algorithms – is still important, because the bandwidth of main memory is at least an order of magnitude higher than the interface bandwidth of storage devices. To exploit the SSD characteristics to the extent possible, flash-aware buffer management should observe the following basic principles [7]:

P1. Minimize the number of physical writes.
P2. Address write patterns to improve the write efficiency.
P3. Keep a relatively high hit ratio.

We have designed the CFDC (Clean-First Dirty-Clustered) algorithm [12] which perfectly addresses all these key issues. As our competitors, quite a number of flash-aware replacement algorithms were proposed trying to approximate these principles more or less successfully. Here, we cross-compare CFDC to the general-purpose algorithm LRU [11] and the algorithms CFLRU [13], LRU-WSR [14], and REF [15], which are tailor-made for SSD use.

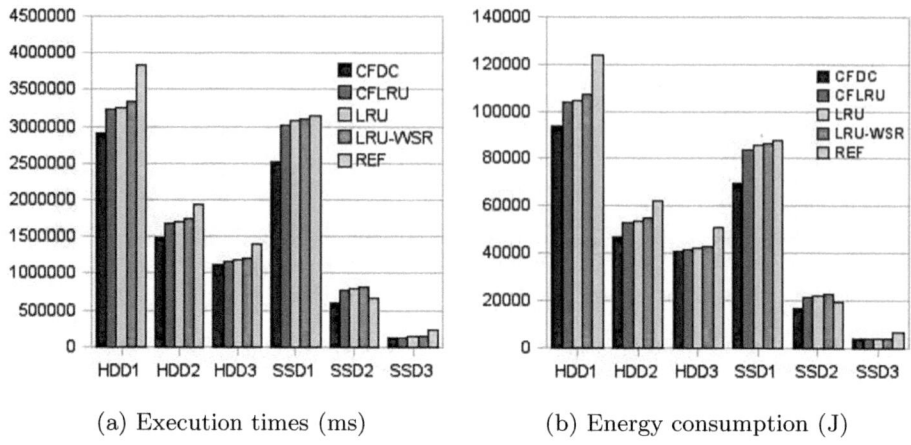

(a) Execution times (ms)   (b) Energy consumption (J)

Fig. 5. Performance and energy consumption running the TPC-C trace

Because DBMSs often use heterogeneous storage devices of mixed types, an important question needs to be answered: how well do these algorithms perform on conventional magnetic disks? Furthermore, how much energy is used for buffer management in either case, i. e., how energy efficient are these algorithms in differing environments? Here, we want to repeat these answers given in [7], thereby preparing our arguments for energy-proportional DBMS management.

3.2 Experiments

Our test environment consists of an Intel Core2 Duo processor and 2 GB of main memory. Both the OS (Ubuntu Linux with kernel version 2.6.31) and the DB engine are installed on an IDE magnetic disk (system disk). The test data (as a DB file) resides on a separate magnetic disk/SSD (data disk, denoted as SATA). The data disks, as listed in Table 1, are chosen to represent low-end (HDD1/SSD1), middle-class (HDD2/SSD2), and high-end (HDD3/SSD3) devices. They are connected to the system one at a time.

Using a relational DBMS, we recorded an OLTP trace (a buffer reference string) of a 20-minutes TPC-C workload with a scaling factor of 50 warehouses. We ran the trace for each of the five algorithms under identical conditions and repeated it for each of the devices under test. Using a buffer size of 8000 pages (64 MB), we minimized the influence of the device caches. But, not the absolute but the relative differences are most expressive.

The illustration of the recorded execution times and energy consumptions in Fig. 5 is considered to be indicative for what we can expect as typical behavior and optimization potential under the various storage device settings. The difference between the execution times of the algorithms becomes smaller on SSD3 (see Fig. 5) due to two reasons: 1. The I/O cost on SSD3 is much smaller than on other devices, yielding the buffer layer optimization less significant; 2. This device has supposedly the largest device cache, since it is the newest product among the devices tested.

Performance gain and energy saving are impressive, if we compare the results among devices of the same class. CFDC turns out to be the best performing algorithm on all devices. Because the time needed to run the trace is – under continuous and hardly varying utilization of the computer system – proportional to the energy consumption, CFDC is also the most energy-efficient one. This observation is in accord with the general thesis [16] that "within a single node intended for use in scale-out (shared-nothing) architectures, the most energy-efficient configuration is typically the highest performing one".

4   Energy-Proportional Computing

The key effect observed in Fig. 5 is further explained by Fig. 6, which contains a break-down of the average working power of major hardware components of interest, compared with their *idle* power values. The figures shown for the configurations HDD3 and SSD3 are indicative for all configurations; they are similar for the other device types, because IDE and ATX – consuming the lion's share of the energy – remain unchanged.

Ideally, the power consumption of a component (and the system) should be determined by its utilization. However, for both configurations, there is no significant power variation when the system state changes from idle to working or even to its full utilization. Furthermore, no clear difference can be observed between the various algorithms, although they have different complexities and, in fact, also generate different I/O patterns. This is due to the fact that, independent of the workload, the processor and the other units of the mainboard consume most of the power (the ATX part in the figure) and these components are not

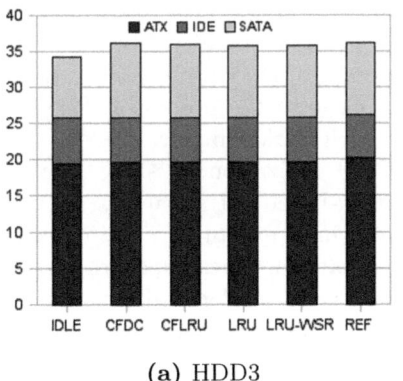

(a) HDD3

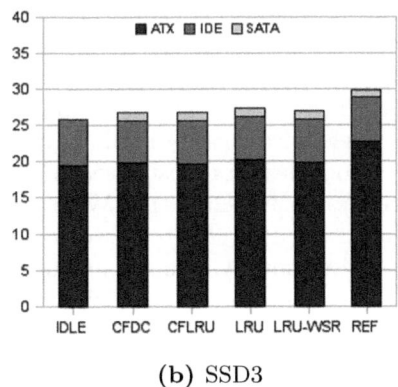

(b) SSD3

Fig. 6. Break-down of average power (W)

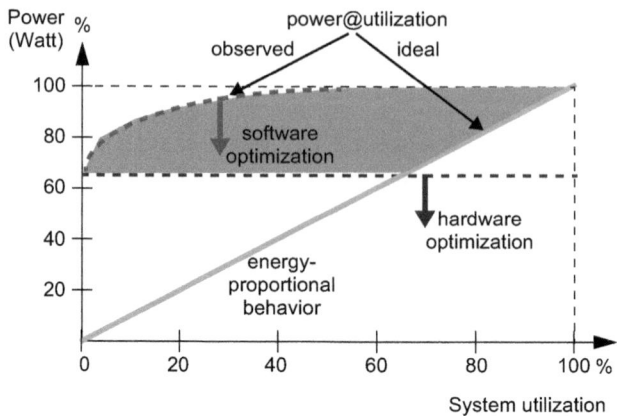

Fig. 7. Power use over system utilization: single computing node

energy-proportional, i. e., their power use is not proportional to the system utilization caused by the workload[5].

The break-down of the average working power in Fig. 6 reflects the average system utilization obtained for individual trace executions. If we evaluate how energy consumption depends on system utilization, we roughly get for our configuration – with a single computing node – the characteristics sketched in Fig. 7. Main-memory power usage is more or less independent of system utilization and increases linearly with the memory size, i. e., the number of RAM modules. In our case, we only had a single 2GB module. Increasing the main-memory size by adding more RAM modules would rapidly shift in our scenario the relative power use close to 100%, even in the idle case.[6] As indicated in Fig. 7, the scope for optimizing the relative energy efficiency by software means is limited and would almost disappear when a large memory is present. But this scope could be widened, if hardware optimizations could be invented (e. g., reduction of RAM's energy consumption). Using a *single computing node*, we would never come close to the ideal characteristics of energy proportionality. Note, we cannot just switch off RAM chips, especially in the course of DBMS processing, because they have to keep large portions of DB data close to the processor. Preserving this reference locality is the key objective of each DBMS buffer.

Google servers mostly reach an average CPU utilization of ~30%, but often even less than ~20% [17]. With our current flash-based optimizations, we do not obtain any noticeable effect on overall energy saving – except for continuous peak load situations. Given normal load patterns and arrival times, an average request is processed more efficiently, i. e., system resources are allocated for shorter intervals, thereby reducing system utilization even further. What kind of system architecture enables a substantial approximation of ideal energy proportionality?

---

[5] The elapsed time T of the workload almost completely determines its energy consumption E (note, $E = \bar{P} \cdot T$, where $\bar{P}$ is the average power measured).

[6] Memory of enterprise server quality consumes ~10 Watt per 4GB DIMMS [16].

5   Design Considerations of WattDB

Current approaches to *data management on new hardware* almost exclusively focus on high performance for continuous peak loads in specific application areas and – to achieve this goal – primarily rely on extremely large main memories. But from a "green perspective", it is unreasonable to build systems, e. g., main-memory DBMSs for OLAP applications, which have a much lower average utilization and a power usage profile as sketched in Fig. 7[7].

For this reason, we have started the WattDB project, where a cluster of wimpy, *shared-nothing* computing nodes replaces the powerful DB server machine. The cluster core consists of a *single node*[8] and can attach further nodes without interrupting DB processing. In this way, the cluster can scale up to n nodes and is able to smoothly grow and shrink dynamically – depending on the current workload needs. Apparently, due to this dynamic node attaching/detaching, WattDB as a cluster will stepwise approximate the ideal course of power usage, i. e., its behavior is becoming energy proportional. Note, the cluster dynamics, i. e., the time span [18] where low-utilized nodes are disconnected from the cluster and deactivated or where overload situations are resolved by reactivating switched-off nodes, is a key question to be answered by the project.

Each of the individual computing nodes must be able to access the entire database. As a consequence, we need to build an I/O architecture, where – at each point in time and each cluster configuration – all external storage devices (SSDs or HDDs) can be dynamically shared by all attached nodes, i. e., the *shared-nothing* processing architecture of the cluster has to be supported by a *shared-disk* I/O architecture.

As a consequence of dynamic node fluctuation, DB cluster coordination becomes a frequent task to optimally support DB processing and maintenance as well as concurrency control and logging/recovery, etc. Static task assignment to specific computing nodes creates *single points of failures* and may quickly lead to unbalanced system behavior [19]. Therefore, static and physical partitioning of storage structures and runtime responsibilities is impractical. Hence, new partitioning schemes and procedures based on logical predicates have to be developed. Instead of allocating physical partitions, flexible physiological DB partitioning is needed – a new outstanding challenge to make WattDB work.

5.1   Architecture Overview

Our cluster currently consists of ten nodes with identical processors and main memory. Two nodes are equipped with four hard disks each to serve as DB storage. All nodes are interconnected by a Gigabit-Ethernet as depicted in Fig. 8. To minimize the energy footprint of each node, *Intel Atom D510* light-weight CPUs are used. In combination with the installed 2 GB of main memory and

---

[7] Such an extreme main-memory DB server would steadily consume 2.5 KW for each TB of main memory installed, no matter whether it is idle or working.

[8] In this case, all coordination, query processing, and storage-related tasks have to be performed by this node.

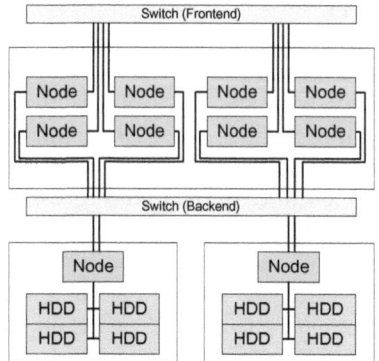

Fig. 8. Physical cluster layout

Fig. 9. Photo of the cluster

Table 2. Energy consumption of a single node

	off	suspend	idle	100% CPU 1 disk	100% CPU 2 disks	100% CPU 4 disks
Power Consumption [W]	3	3	29	32	35	41
% of peak	7%	8%	71%	78%	86%	100%

the mainboard, each node consumes less than 30 W in idle mode. The disks are designed for mobile use, so each disk consumes only 3 W. Our hardware is *Amdahl-balanced*, hence, processing power and data throughput are matched [20]. Still, a single node is not energy proportional as Table 2 shows. When idle, about 70% of the node's peak power is consumed (see also Fig. 6). Although the single nodes are not energy proportional, the desired energy proportionality is approximated by load-dependent deactivation/reactivation of cluster nodes.

Dynamically powering nodes impacts all layers of database software. As a first step, we have analyzed the impact of node fluctuations for *Storage Mapping*, *Query Processing* and *Cluster Coordination*.

5.2   Storage Mapping and Partitioning

Hard disks are one of the reasons, why common servers cannot be energy proportional, as explained in Sect. 2. They continuously consume energy to keep their platters spinning. Therefore, it would be a great opportunity for saving power, if unused disks could be switched off. If the related storage capacity is not needed right now, entire nodes could be powered down. Dynamically partitioning data by their access frequencies may gain some improvements [21], but it is insufficient, because data on cold disks would have high access cost. Instead, we propose a mechanism for dynamically consolidating data while keeping I/O performance agreements. The system's storage can be scaled from energy efficiency requirements by consolidating data to as few disks as possible to high-performance needs by distributing data to more disks for faster parallel access.

Fig. 10 shows examples of partitioning schemes. *Table A* is separated into two partitions that reside on a single disk. This is the most energy-efficient storage solution, because only a single disk needs to be powered. In turn, access performance is limited, as one disk can only achieve a certain number of IOPS. In contrast, *table B* is partitioned, too, but the partitions are distributed to separate disks. Therefore, access to table B is up to two times faster than that to table A. At the same time, random IOPS should double compared to the first scheme, but energy consumption doubles as well. Finally, *table C* shows an even more distributed case, where data is distributed to four disks. While this raises the relative energy consumption of the data, access bandwidth and IOPS again are increased.

In these examples, the storage mapping resulted in a balanced tree of partitions. WattDB will support even more flexible partitioning schemes. Hence, high-traffic areas of a table can be divided into finer grained partitions than rather cold areas – causing unbalanced partition trees. Management of a partition subtree, e.g., either table C, partition C.1, or partition C.1.1 in Fig. 10, is delegated to a single node. By distributing a table to more nodes, the effective main-memory buffer for this table is increased.

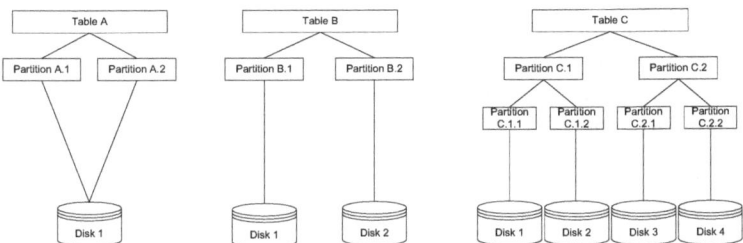

Fig. 10. Storage partitioning schemes

5.3 Query Processing

By providing such an an energy-aware storage layer, query planning needs to consider the existing data distribution. A resulting query execution plan (QEP) has to reflect the data partitioning schemes and their assignment to nodes. As a consequence, subqueries can be formed to access partitions, process data, and emit intermediate results. Eventually, these results are consumed by a node which combines them to the final output and delivers it to the client. Fig. 11 sketches an example how this work assignment is achieved for a simple query. More sophisticated operations like distributed joins can be executed as well.

Our approach is considered fairly scalable, because partition management including buffering, locking, and recovery, consists of local tasks that scale with the number of nodes in the cluster. Still, a global transaction manager is needed to detect deadlocks in transactions spanning multiple nodes.

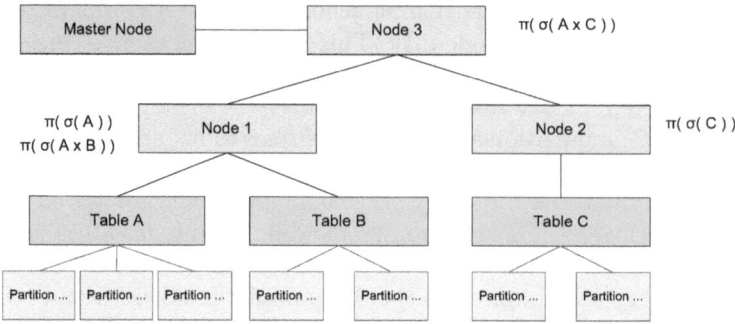

Fig. 11. Node assignment for query processing: an example

5.4   Cluster Coordination

Obviously, a centralized instance, called master, is needed for all coordination tasks. Cluster clients address a dedicated node as an entry point to submit their queries. Furthermore, it has to monitor and tune the performance of the cluster. Moreover, all tasks concerning allocation of new objects, housekeeping and reorganization on depraved storage structures, repartitioning after workload shifts, redistribution of responsibilities as an implication of cluster growth or shrinkage, etc. need centralized control. Therefore, we introduce a master node that will perform all these tasks. This node will also manage global deadlock detection and keep track of the energy consumption.

6   Conclusion and Future Work

We have sketched our main results of recent contributions concerning our flash-related research. At the device level, we have summarized performance behavior and energy efficiency of a spectrum of different SSD types and have discussed important consequences for DBMS processing. At the system level, we have compared performance and energy use of a number of flash-aware buffer management algorithms, when different classes of HDDs and SSDs were used as external storage. We could clearly verify the claim [16] that – within a single shared-nothing computing node – the most energy-efficient configuration is typically the highest performing one.

This observation guided our research endeavor towards energy-proportional computing applied to *data management on new hardware* to seriously observe energy saving – not only for peak loads, but also for low-load situations and even idle times. As a consequence, we designed WattDB, whose core components are currently implemented. In the future, we want to specialize WattDB towards differing directions to provide tailor-made support for the application classes OLTP, OLAP, and MapReduce.

References

1. Boncz, P.A., Manegold, S., Kersten, M.L.: Database Architecture Evolution: Mammals Flourished long before Dinosaurs became Extinct. PVLDB 2(2), 1648–1653 (2009)
2. Greenplum. Driving the Future of Data Warehousing and Analytics (2009), http://www.greenplum.com/
3. VoltDB. Fast, Scalable, Open-Source SQL DBMS with ACID (2010), http://voltdb.com/
4. Plattner, H.: SanssouciDB: An In-Memory Database for Processing Enterprise Workloads. In: Proc. BTW. LNI - P, vol. 180, pp. 2–21 (2011)
5. Härder, T., Schmidt, K., Ou, Y., Bächle, S.: Towards Flash Disk Use in Databases - Keeping Performance While Saving Energy? In: Proc. BTW. LNI - P, vol. 144, pp. 167–186 (2009)
6. Hudlet, V., Schall, D.: SSD!= SSD - An Empirical Study to Identify Common Properties and Type-specific Behavior. In: Proc. BTW. LNI - P, vol. 180, pp. 430–441 (2011)
7. Ou, Y., Härder, T., Schall, D.: Performance and Power Evaluation of Flash-Aware Buffer Algorithms. In: Bringas, P.G., Hameurlain, A., Quirchmayr, G. (eds.) DEXA 2010. LNCS, vol. 6261, pp. 183–197. Springer, Heidelberg (2010)
8. Bouganim, L., Jónsson, B.T., Bonnet, P.: uFLIP: Understanding Flash IO Patterns. In: CIDR (2009)
9. Graefe, G.: The five-minute rule 20 years later (and how flash memory changes the rules). Commun. ACM 52(7), 48–59 (2009)
10. Schall, D., Hudlet, V., Härder, T.: Enhancing Energy Efficiency of Database Applications Using SSDs. In: C3S2E, pp. 1–9 (2010)
11. Effelsberg, W., Härder, T.: Principles of Database Buffer Management. ACM TODS 9(4), 560–595 (1984)
12. Ou, Y., Härder, T., Jin, P.: CFDC: A Flash-Aware Buffer Management Algorithm for Database Systems. In: Catania, B., Ivanović, M., Thalheim, B. (eds.) ADBIS 2010. LNCS, vol. 6295, pp. 435–449. Springer, Heidelberg (2010)
13. Park, S., Jung, D., et al.: CFLRU: a Replacement Algorithm for Flash Memory. In: CASES, pp. 234–241 (2006)
14. Jung, H., Shim, H., et al.: LRU-WSR: Integration of LRU and Writes Sequence Reordering for Flash Memory. Trans. on Cons. Electr. 54(3), 1215–1223 (2008)
15. Seo, D., Shin, D.: Recently-evicted-first Buffer Replacement Policy for Flash Storage Devices. Trans. on Cons. Electr. 54(3), 1228–1235 (2008)
16. Tsirogiannis, D., Harizopoulos, S., Shah, M.A.: Analyzing the Energy Efficiency of a Database Server. In: SIGMOD, pp. 231–242 (2010)
17. Barroso, L.A., Hölzle, U.: The Datacenter as a Computer: An Introduction to the Design of Warehouse-Scale Machines, Synthesis Lectures on Computer Architecture. Morgan & Claypool Publishers (2009)
18. Albers, S.: Energy-efficient Algorithms. Commun. ACM 53(5), 86–96 (2010)
19. Rahm, E.: Evaluation of Closely Coupled Systems for High-Performance Database Processing. In: ICDCS, pp. 301–310 (1993)
20. Szalay, A.S., Bell, G.C., Huang, H.H., Terzis, A., White, A.: Low-power Amdahl-balanced Blades for Data-intensive Computing. SIGOPS Oper. Syst. Rev. 44
21. Li, X., Li, Z., Zhou, Y., Adve, S.: Performance-Directed Energy Management for Main Memory and Disks. Trans. Storage 1, 346–380 (2005)

Flash-Based Database Systems: Experiences from the FlashDB Project

Xiaofeng Meng[1], Lihua Yue[2], and Jianliang Xu[3]

[1] Renmin University of China, Beijing, China
[2] University of Science and Technology of China, Hefei, China
[3] Hong Kong Baptist University, Hong Kong, China

Abstract. The new characteristics of flash memory bring great challenges in optimizing database performance, by using new querying algorithms, indexes, buffer management schemes, and new transaction processing protocols. In this talk, we will first present an overview on the FlashDB project, which was launched in 2009 and supported by the National Natural Science Foundation of China (No. 60833005). The project aims at constructing the fundamental theory and design principles of flash-based database systems including a series of key problems, such as system architecture, storage management and indexing, query processing, transaction processing, buffer management, etc. In particular, we focus on establishing a basis for data management involving flash memory, developing database management systems for flash-based SSDs, and preparing a test bed for flash-based database applications. During the past two years, we have made some achievements in buffer management [1, 2], index structures [3], storage management [4], and SSD simulation platform [5]. After a brief introduction on the current research results in the project, we will discuss some experiences and lessons concluded from the study. We will emphasize several issues that may be open up exciting avenues and influence the direction of the research within the scope of flash-based database systems.

Keywords: Flash Memory, SSD, Flash-based Database.

References

1. Tang, X., Meng, X.: ACR: An Adaptive Cost-Aware Buffer Replacement Algorithm for Flash Storage Devices. In: MDM 2010, pp. 33–42 (2010)
2. Li, Z., Jin, P., Su, X., et al.: CCF-LRU: A New Buffer Replacement Algorithm for Flash Memory. IEEE Trans. on Consumer Electronics 55(3), 1351–1359 (2009)
3. Yin, S., Pucheral, P., Meng, X.: A sequential indexing scheme for flash-based embedded systems. In: EDBT 2009, pp. 588–599 (2009)
4. Zhou, D., Meng, X.: RS-Wrapper: random write optimization for solid state drive. In: CIKM 2009, pp. 1457–1460 (2009)
5. Jin, P., Su, X., et al.: A flexible simulation environment for flash-aware algorithms. In: CIKM 2009, demo (2009)

J. Xu et al. (Eds.): DASFAA Workshops 2011, LNCS 6637, p. 240, 2011.
© Springer-Verlag Berlin Heidelberg 2011

Trading Memory for Performance and Energy

Yi Ou and Theo Härder

University of Kaiserslautern
{ou,haerder}@cs.uni-kl.de

Abstract. Managing extremely large amounts of data with high performance and low power consumption is very difficult. We look at this urgent problem from an architectural perspective and present our prototype design and implementation of a three-layer database storage system, which uses flash-based devices as an intermediate caching layer. The flash-based layer significantly improves the I/O efficiency of the storage system. Therefore, we can reduce the use of energy-inefficient RAM-based memory without compromising the overall system performance. The efficiency of the three-layer storage system is demonstrated by our practical experiments using traces from both standard benchmarks and a real-life application.

1 Introduction

The worldwide data volume is doubling every two years. According to the estimation of IDC, currently 45 GB of data in average exists for each person in the world: that is 281 Billion GB (281 Exabytes) in total. At the same time, IT enterprises are still hungry for data [1]. To cope with the pace of data explosion, the number of installed servers and storage systems is rapidly growing, resulting in a huge amount of energy consumption. One of the greatest challenges for the information management community is to manage extremely large amounts of data in an (both performance and energy) efficient way.

Flash memory is a kind of non-volatile storage media, popularly used in memory cards and USB flash drives. In the desktop PC and server storage markets, solid-state disks based on flash memory (flash SSDs) are also gaining attention, due to their increasing storage capacity and decreasing price. In contrast to traditional hard disk drives (magnetic HDDs), flash SSDs have no mechanical parts and, therefore, allow much faster random accesses. Because flash memory is non-volatile, its active power is much lower (compared on a Watt/GB basis) than that of DRAM, for which a large portion of the active power is consumed to maintain the state of the chip.

Currently, most of the database storage systems follow a classical *two-layer architecture* (2LA) [2], with a RAM-based buffer layer accelerating page requests to and from the persistence layer based on hard disk drives. With an increasing amount of data accommodated at the persistence layer, the capacity of the expensive and energy-inefficient RAM-based buffer typically becomes the performance bottleneck.

J. Xu et al. (Eds.): DASFAA Workshops 2011, LNCS 6637, pp. 241–253, 2011.
© Springer-Verlag Berlin Heidelberg 2011

Table 1. Price and performance of storage devices

Device	Model No.	EUR/GB	Read (ms)	Write (ms)
RAM1	Kingston KVR667D2D8P5/2G	19.00	$\sim$ 10 ns	$\sim$ 10 ns
RAM2	Kingston KHX1600C9D3B1K2/4GX	19.11	$\sim$ 10 ns	$\sim$ 10 ns
RAM3	Kingston KVR1333D3D4R9S/4G	24.70	$\sim$ 10 ns	$\sim$ 10 ns
SSD1	Intel SSDSA2MH160G1GN	2.40	0.029	0.303
SSD2	Intel SSDSA1MH160G2GN	2.44	0.029	0.116
SSD3	Crucial CTFDDAC256MAG-1G1	2.01	0.017	0.022
HDD1	WD WD800AAJS 7200 RPM	0.38	15.000	15.000
HDD2	WD WD1500HLFS 10000 RPM	0.77	4.500	4.500
HDD3	Fujitsu MBA3147RC 15000 RPM	0.76	2.000	2.000

In terms of performance and price and their long-term trend, flash SSDs fit perfectly into the gap between DRAM and magnetic HDDs. Table 1 lists, for each storage media type, the prices and performance figures[1] of three devices (from low-end to high-end). These figures strongly suggest a *three-layer architecture* (3LA), where flash is used as an intermediate caching layer, while conventional and inexpensive HDDs are employed at the bottom layer to accommodate our ever-increasing demand on storage capacity. With such a memory hierarchy, the capacity of the RAM-based layer could be kept relatively small, because a larger amount of pages can be cached on the flash media, which is still much faster than HDDs. To justify the move from 2LA to 3LA, a few questions need to be answered:

Q1. Will the cost of adding the intermediate layer be justified by performance improvements?

Q2. Can we achieve the goal of improving performance while saving energy at the same time?

The major contribution of this paper is giving answers to the questions Q1 and Q2, which are ignored so far in related works. In addition, we contribute in the following aspects:

- We advocate 3LA, the three-layer database storage architecture with flash as the intermediate layer.
- We define the basic interfaces for the three layers and present the prototype of such a storage system.
- Using buffer traces of standard benchmarks and a real-life workload, we accomplish an extensive empirical study, comparing the performance and energy consumption of 2LA and 3LA.

The remainder of this paper is organized as follows: Section 2 discusses related works. Section 3 presents our design of 3LA and related algorithms. Section 4

---

[1] We used the sales prices of Internet stores as of November 2010. Performance figures are derived from the device data sheets, for randomly accessing pages of 4 KB.

reports our empirical study. The concluding remarks and future works are presented in Section 5.

2   Related Work

Multi-level caching has been intensively studied in the past. Zhou et al. [3] characterized second-level buffer access patterns and proposed a set of algorithms for managing the second-level buffer. Those algorithms are not flash-specific, therefore, their major performance metric is the hit ratio. One of them is implemented in our prototype system and included in our experiments.

Koltsidas and Viglas [4] identified three page-flow schemes in a three-level caching hierarchy and proposed flash-specific cost models for those schemes. While addressing both theoretical problems and important implementation issues, their focus is the validation of the cost models and the comparison among those schemes. Energy efficiency and a comparison between 2LA and 3LA are not covered in their work.

Narayanan et al. [5] addressed both complete replacement of disks by SSDs, as well as use of SSDs as an intermediate tier between disks and DRAM. They compare these architectural variants with 2LA using an offline tool, which, given a block-level trace of a workload, suggests the least-cost storage configuration that supports the workload's requirements. They found that replacing disks by SSDs is not a cost-effective option for any of their workloads, due to the higher dollar-per-GB cost of flash SSDs.

Although our goal partially overlaps with that of [5], there are several aspects that distinguish our work fundamentally from theirs: 1. Their traces represent the workload to the disk layer (block level), while our traces represent the workload to the buffer manager (buffer traces). 2. Our observations are quite different from theirs. For example, they found that fewer than 10% of their workloads can benefit from an intermediate layer based on flash, while in our experiments, 3LA is superior to 2LA in most configurations. 3. Our observations are expected to be more accurate, because in their experiments, traces were not executed, but just analyzed by the tool, while our traces are actually run in the real systems.

3   The 3LA Storage System

We consider three layers of software in our storage system, as shown in Figure 1. The RAM layer L_r manages the buffer pool with $|L_r|$ pages in main memory, the flash layer L_f manages the flash-based buffer pool with $|L_f|$ pages, and the disk layer L_d manages the accesses to the magnetic disk or a pool of (possibly inexpensive and redundant) magnetic disks with a total capacity of $|L_d|$ pages.

Considering the relative price and performance ratios of the three types of storage media, e. g., those listed in Table 1, we assume that:

$$|L_r| \leq |L_f| \leq |L_d| \tag{1}$$

Due to these capacity constraints and performance ratios, the hottest pages should be kept in L_r, and L_f should try to keep the hot pages that can not be kept in L_r. As a consequence, replacement policies are required in L_r and L_f.

L_r supports a typical buffer pool interface, e. g., that of the classical fix-use-unfix protocol [6]. Both L_f and L_d provide the interface of reading or writing a page, identified by its logical page number. Each layer only uses the interface provided by the layer directly below it, i. e., there is no cross-layer dependency. In particular, in 3LA, L_r never accesses L_d directly.

However, because L_f and L_d basically have the same interface, L_f can be implemented as an optional layer. When L_f is not present, L_r directly accesses L_d. In that case, 3LA degenerates to 2LA. Such a degeneration is practically used for our experiments in Section 4.

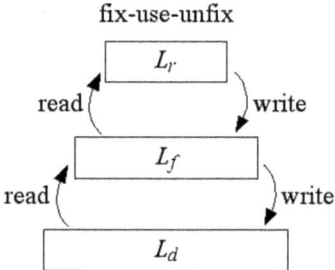

Fig. 1. Inter-layer interfaces in the three-layer architecture

For both architectures, we assume that L_r follows two basic principles: *demand paging* and *write back*. Consequently, we have the following two *invariants*, which are independent of the algorithm and implementation of L_f and valid for both 3LA and 2LA:

I1. L_r calls the read(p) function on the layer directly below it, iff page p is not present in L_r and there is a page request for p to be served by L_r (page fault in L_r).

I2. L_r calls the write(p) function on the layer directly below it, if page p is to be evicted from L_r and p is dirty (modified at least once after entering L_r).

L_r and L_d are basically the same as in the classical two-layer disk-based storage system. For this reason, we only present the replacement algorithms for the management of L_f in the following: the *Local* (LOC) algorithm and the *Global* (GLB) algorithm.

In both algorithms, a list of cache positions L with $|L| = |L_f|$ is maintained in an LRU fashion. A *cache position* identifies a page slot in the flash-based cache and contains a clean/dirty bit. Furthermore, a directory H is maintained, mapping currently cached pages to their corresponding cache positions.

3.1   The LOC Algorithm

In the LOC algorithm, L_f is managed *locally* in an LRU fashion, without re-quiring extra knowledge from L_r. The procedure of reading a page from L_f is shown in Algorithm 1. One difference to an main-memory LRU cache is that flushing a page involves first reading the page from flash and then writing it to the storage. Writing a page p to L_f involves finding its cache position c via H and storing p at c. If p is not found in L_f, it will be written to L_d immediately.

Because $|L_r| \leq |L_f|$ and LOC only has local knowledge, it is possible that some or even all pages in L_r are doubly cached in L_f. However, pages in L_r are not necessarily all in L_f, due to different page reference behaviors at different layers. Note, references to L_f are consequences of buffer faults in L_r.

Algorithm 1. LOC read page from L_f

---

    input   : read request for page p, storage layer L_d
    output : update L and H; return p with content loaded
1 cache position $c \leftarrow$ lookup p in H ;
2 **if** $c \in L_f$ **then**
3 | read p from cache position c ;
4 | move c to MRU position of L ;
5 **else**
6 | victim cache position $v \leftarrow$ LRU position of L ;
7 | page $q \leftarrow$ the page stored at v ;
8 | **if** v *is dirty* **then**
9 | ⌊ read q from cache position v and flush q to L_d ;
10 | read p from L_d and store p at v ;
11 | move v to MRU position of L ;
12   ⌊ update H by replacing entry (q, v) with entry (p, v) ;
13 **return** p;

3.2   The GLB Algorithm

The GLB algorithm is first introduced in [3]. We examine it here in a flash context. The GLB algorithm follows the exclusive scheme [4], i. e., no page is ever cached in L_r and L_f at the same time. For better comprehension, we assume the replacement policy in L_r is also LRU, without loss of generality. Based on this assumption, we can think of a *global* logical LRU list L_g, consisting of the LRU list of L_r at its MRU end, and the LRU list of L_f at its LRU end.

Reading a page p from L_f is requested upon a page fault in L_r (see I1). In case of a cache hit in L_f, p is moved from L_f to L_r (H and L are updated accordingly). In case of a cache miss, p is read directly from L_d to L_r, avoiding doubled caching in L_f. In both cases, a page q is evicted from L_r to L_f. After being read, p becomes the MRU page in L_r (also in L_g).

To "maintain" the logical list L_g, page q currently evicted from L_r should become the LRU page in L_f. Therefore, we have to extend the interface of L_f

(as described in Figure 1) by a new function *evict* called by L_r for passing evicted clean pages to L_f. Note, a write request is called on L_f, only when the evicted page is dirty (see I2). The procedure of processing a write or evict request for page q is the same: flush the page stored at the LRU position v of L if the page is dirty, move v to the MRU position, store q at v, mark v dirty if q is dirty, and update H.

3.3 Discussion

Given the same workload, the global cache hit count (total number of buffer hits in L_r and L_f) of GLB is expected to be higher than that of LOC, because the effective cache size of the latter is smaller, due to doubled caching in L_r. However, in GLB, the number of flash writes equals the number of L_r page evictions. This is OK for a RAM-based second-level buffer, but it is an issue for flash media both in terms of performance (see Section 4) and lifespan [7].

For both algorithms in our current implementation, the dirty pages in L_f (whose cache positions are marked dirty) are flushed to L_d when the system is shutdown, for the sake of consistency. A simple improvement leveraging the non-volatility of flash can be made here: we can just materialize the content of H at shutdown and rebuild H at startup[2], without flushing the "dirty" pages in L_f. This technique not only speeds up the shutdown procedure, but also shortens the warm-up phase of the system, because the hot portion of the pages are likely already in L_f, ready for immediate access. For the LOC algorithm, pages in L_f are up-to-date at restart, iff the dirty pages of L_r are flushed before the shutdown of L_f starts. For the GLB algorithm, page sets L_f and L_r are disjunct, therefore, pages in L_f are automatically up-to-date at restart.

4 Experiment

To answer the questions Q1 and Q2, we did an extensive empirical study based on a fair comparison between 2LA and 3LA, using buffer traces recorded under various workloads. We first present our simulation-based study using TPC-E, TPC-C, and TPC-H traces, before we discuss the experiment ran on real devices using the trace from a real-life application. Our study on energy consumption is based on the following assumption:

A1. The acquisition cost and power consumption of storage media are linear to their capacity in use.

Assumption A1 might not be valid at fine granularity, however, it is reasonable, when observed at a coarser granularity. For example, if the power of a 2-GB DRAM module is 10 W, according to A1, 0.2 GB of DRAM would consume 1 W, which is not valid, because, as long as the module is working, it consumes 10 W, no matter the remaining 1.8 GB are in use or not. But we can safely say that $2n$ GB of DRAM based on the same model consume $10n$ W.

---

[2] The byte size of H is much smaller compared to that of L_r and L_f.

All experiments were done using our prototype implementation of the 3LA storage system, which can also be easily configured to function as a 2LA system, as described in Section 3. For both architectures, our test program only communicates with L_r by sending the logical page requests delivered by the traces to its buffer manager, which manages the L_r buffer pool using the replacement policy LRU. All experiments start with *cold* L_r and L_f buffers. The time used to flush the dirty pages at shutdown is included in the measurements.

In our experiments, we scaled a parameter b (in number of pages) logarithmically. For 2LA, b is the size of the buffer layer, i. e., $|L_r| = b$, while for 3LA, we set $|L_r|$ and $|L_f|$ as follows:

$$|L_f| = n \times b \tag{2}$$

and

$$|L_r| = max(1, \lfloor b - |L_f| \times (C_f/C_r + S_d/S_p) \rfloor) \tag{3}$$

where C_f/C_r is the dollar-per-GB cost ratio of flash to RAM, S_d is the byte size of a directory entry of H, and S_p the page size in bytes. The term $|L_f| \times C_f/C_r$ gives the number of RAM pages that should be reduced to achieve a cost-neutral investment for $|L_f|$ pages of flash memory. The term $|L_f| \times S_d/S_p$ is the number of RAM pages consumed by the directory H for $|L_f|$ pages of flash. We call Formula 2 and 3 the *equi-cost constraints*, because it enforces a fair basis for the comparison among the 2LA and 3LA configurations, i. e., having the same acquisition cost.

The parameter n is used to examine the behavior of 3LA when the size of L_f is scaled. Because the value of $|L_r|$ can not be negative, we have $b - |L_f| \times (C_f/C_r + S_d/S_p) > 0$, which resolves to $n < 1/(C_f/C_r + S_d/S_p)$. Together with the constraint in Formula 1, we have the practical range of n:

$$1 \le n < (C_f/C_r + S_d/S_p)^{-1} \tag{4}$$

If we ignore S_d/S_p, which is relatively small, then we obtain $1 \le n < C_r/C_f$. In our experiments, the page size S_p is 8192 bytes and the directory entry size S_d is 4 bytes. We chose the cost ratio $C_f/C_r = 0.10$, which is very close to the real price ratios according to Table 1. According to Formula 4, the practical range of n is approximately $[1, 10)$. Note n does not have to be an integer. For a given b, the value of n actually controls how much RAM is traded for flash, observing the equi-cost constraints.

4.1   Simulations

For the simulation-based experiments, the *Virtual Execution Time* (T_v) is used as the major performance metrics, defined as:

$$T_v = T_f + T_d \tag{5}$$

Here, T_f and T_d are the simulated device access times elapsed in L_f and in L_d respectively. T_f is defined as:

$$T_f = T_{fr} + T_{fw} = N_{fr} \times C_{fr} + N_{fw} \times C_{fw} \qquad (6)$$

T_{fr} and T_{fw} are the accumulated times reading from and writing to the flash media, N_{fr} is the number of flash reads, C_{fr} the average cost of a flash read, N_{fw} the number flash writes, and C_{fw} the average cost of a flash write. The flash reads and flash writes here refer to the physical reads from and writes to the flash device. They are not to be confused with the read and write requests sent to the L_f software. Similarly, T_d is defined as:

$$T_d = T_{dr} + T_{dw} = N_{dr} \times C_{dr} + N_{dw} \times C_{dw} \qquad (7)$$

The definition of T_v only considers the costs of accessing the storage media and ignores the CPU cost, because all the algorithms involved have a constant complexity. The inter-layer communication costs are ignored as well, because, the dominating cost in the system is the cost of page accessing, not page transferring. In our simulation, we used the average read and write costs close or equal to those of the middle-class devices in Table 1, i.e., $C_{fr} = 0.030$, $C_{fw} = 0.120$, $C_{dr} = 4.5$, and $C_{dr} = 4.5$ (ms).

Figure 2a illustrates the T_v of running a TPC-E trace[3] using 2LA and 3LA. All 3LA configurations tested significantly outperform the 2LA configuration. For better clarity of the chart, we only show the curves for $n = 2$ and $n = 8$. For the $n = 8$ configuration, LOC reduced the virtual execution time by 32% to 35% (for $b = 1000$ to $b = 32000$), compared with 2LA.

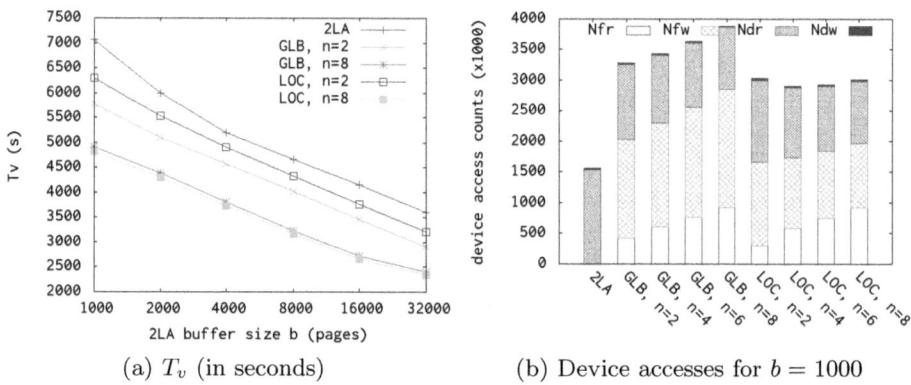

(a) T_v (in seconds)   (b) Device accesses for $b = 1000$

Fig. 2. TPC-E trace performance

The behavior of 3LA is better explained by Figure 2b, where the numbers of device accesses[4] are compared for $b = 1000$. For 2LA, there is no flash device access, while a significant amount of flash device accesses is required for

---

[3] Provided by a leading IT enterprise.
[4] In the simulation, no real device access occurs.

3LA (Figure 2b). For both GLB and LOC, with n scaled from 2 to 8 (thus an increasing $|L_f|$ and decreasing $|L_r|$), the number of flash reads climbs up, indicating a growing number of hits in L_f, and, consequently, the number of disk reads goes down. The latter is equal to the number of *global cache misses* (i. e., a page is neither in L_r nor in L_f). Because of the speed difference of flash to disk, the flash accesses introduced at L_f are paid off in terms of overall performance (Figure 2a).

As shown in Figure 2b, the number of flash writes performed by GLB increases with an increasing $|L_f|$ and a decreasing $|L_r|$, because it depends on the latter, as discussed in Section 3.3. In contrast, the increasing $|L_f|$ reduces the number of flash writes performed by LOC. This is because it reduces the number of L_f cache misses and each cache miss requires a flash write (line 10 of Algorithm 1).

Table 2. Energy consumption of the TPC-E trace for $b = 1000$

Alg.	n	$|L_f|$	$|L_r|$	P_f (mW)	P_r (mW)	$P_f + P_r$ (mW)	T_v (s)	E (J)
2LA		0	1000	0.000	4.121	4.121	7059	29.09
GLB	2	2000	799.02	0.014	3.292	3.307	5776	19.10
GLB	4	4000	598.05	0.029	2.464	2.493	5304	13.22
GLB	6	6000	397.07	0.043	1.636	1.679	5061	8.50
GLB	8	8000	196.09	0.057	0.808	0.865	4905	4.24
LOC	2	2000	799.02	0.014	3.292	3.307	6305	20.85
LOC	4	4000	598.05	0.029	2.464	2.493	5372	13.39
LOC	6	6000	397.07	0.043	1.636	1.679	5024	8.44
LOC	8	8000	196.09	0.057	0.808	0.865	4818	4.17

Table 2 compares the energy efficiency of 2LA and 3LA for $b = 1000$. The $|L_f|$ and $|L_r|$ values in the 3rd and 4th column are calculated according to Formula 2 and 3. Having these values, we can compute the power value of L_r, based on assumption A1, as follows:

$$P_r = |L_r| \times S_p \times P_r^u \tag{8}$$

where P_r^u is the unit power of RAM, having the value 0.503×10^{-9} (W/B) here, derived from the data sheet of RAM2 in Table 1. The power value of L_f, denoted as P_f, is calculated in a similar way, with $P_f^u = 0.873 \times 10^{-12}$ (W/B), derived from the data sheet of SSD2. Having $P_r + P_f$ and the virtual execution times (T_v), we can then calculate the energy consumption values in the last column. Note that the buffer layer of 2LA consumed much more energy than those of 3LA (by a factor of six for $n = 8$). Disk-layer values are not included in the table, because they are of the same size in both architectures.

The results of running the buffer traces of a TPC-C and a TPC-H workload are shown in Figure 3 and Figure 4. In general, these results confirm our observation on the performance advantage of 3LA. For both traces, with b beyond 16000 pages and $n = 8$, the flash cache of 3LA is large enough to accommodate all

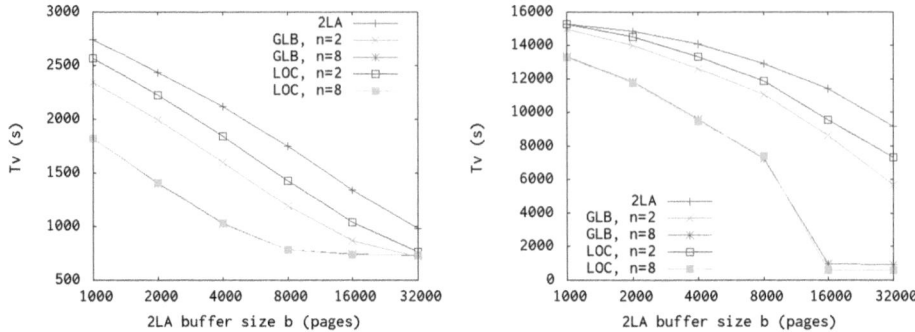

Fig. 3. TPC-C trace performance   **Fig. 4.** TPC-H trace performance

pages of the working sets, which are much smaller than that of the TPC-E trace, therefore, no performance improvement can be observed when b is increased to 32000 pages. The TPC-H trace is highly read intensive, with only 256 page updates out of 6.5 million page requests. That is the reason why the performance of 3LA improves much faster with the growing buffer sizes under the TPC-H workload (Figure 4), compared to the TPC-E and TPC-C cases.

4.2   Running a Real-Life Trace on Real Devices

Complementary to our simulation-based study, we also experimented with a trace from a real-life application on real devices. Our test machine is equipped with an AMD Athlon Dual Core Processor, 1 GB of main memory, and is running Linux (kernel version 2.6.24). HDD2 from Table 1 is used as the storage device in L_d, and SSD2 is used as the flash device in L_f. Both devices are accessed as raw devices, i. e., no file system or OS caching is involved, and our storage system has the control over the access to the devices.

The trace used here is a one-hour page reference string of an OLTP production system of a bank. This trace is well-studied and has been used in [8,9,10,11,12]. It contains 607,390 references to 8-KB pages in a database having a size of 22 GB, addressing 51880 distinct page numbers. About 23% of the requests update the page referenced.

The measured execution times (wall-clock times) are shown in Figure 5. The curves have a shape very similar to that of Figure 3, confirming the accuracy of our simulation. An interesting observation can be made here: for $b = 32000$, the execution time in 3LA increases with n, instead of decreasing with it as in most cases tested. In our case here, the 51880 distinct pages addressed by the trace can be completely accommodated by L_r and L_f, for $n = 2$. Therefore, in such a situation, trading RAM for more flash does not further avoid any access to the disk layer, but reduces the number of buffer hits in L_r and introduces higher numbers of flash accesses, as indicated by Figure 6a, where a breakdown of device I/O is presented, with measured values of T_{fr}, T_{fw}, T_{dr}, and T_{dw}. Nevertheless, the energy consumption decreases with an increasing n, as

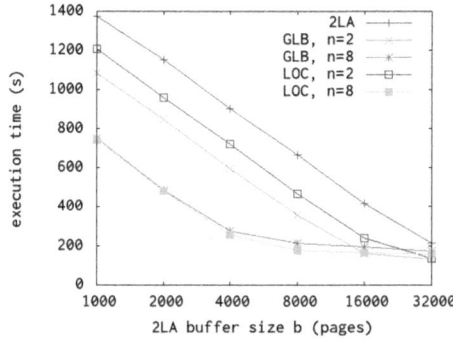

Fig. 5. Execution time (seconds) of the bank trace

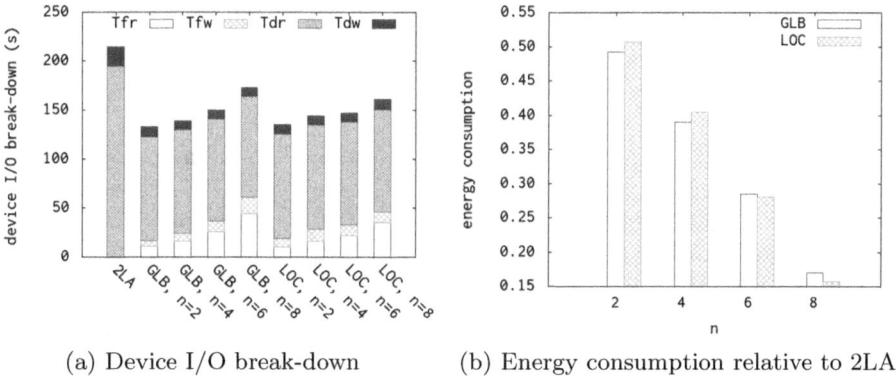

(a) Device I/O break-down   (b) Energy consumption relative to 2LA

Fig. 6. Statistics running the bank trace for $b = 32000$

shown in Figure 6b, which illustrates the energy consumption figures, obtained similarly to those of Table 2, in a relative fashion.

A question arises here: how much RAM should be traded for flash? Or, in our context, what is the break-even point for n? Analytically determining the optimal value for n is a very difficult problem. However, based on our empirical research, we know that for workloads having a small working set that can be kept in the RAM layer, there is no performance benefit of trading RAM for flash, while for workloads with larger working sets that can not fit into the main memory, a larger n generally improves performance as well as energy efficiency. Of course, when n closely approaches C_r/C_f, $|L_r|$ becomes 1 (Formula 3), i.e., the RAM layer has only one page. Such extreme cases should obviously be avoided in system design. Together with Formula 4, our observations can be used as *rules of thumb* in practical applications.

Based on our experiments discussed so far, we can summarize the characteristics of GLB and LOC as follows. For small $|L_f|$, i.e., $|L_f| \sim |L_r|$, GLB achieves higher hit ratios, while for large $|L_f|$, i. e., $|L_f| \gg |L_r|$, LOC is generally better, because GLB's advantage in hit ratios becomes insignificant and it is eaten up by its higher number of flash writes, which is much more expensive than flash reads. A configuration with $|L_f| \gg |L_r|$ is closer to our goal of managing extremely large amounts of data with high performance and low power consumption.

5   Conclusion and Future Work

In this paper, we looked at the problem of using flash as a caching layer between RAM and HDDs from a new perspective: the amount of expensive and energy-inefficient RAM can be reduced due to the support of flash. Our empirical study considered the most important aspects of TCO (Total Cost of Ownership) of a storage system: the acquisition cost and the operating cost (power cost). Our study gives positive answers to the questions Q1 and Q2 and reveals that we can build a 3LA system which is much faster and much more energy efficient than a 2LA system built with the same acquisition cost, meeting the goals of performance and energy efficiency, which are often considered conflicting, at the same time.

In practice, with improved storage system performance, the number of disks, which is sometimes higher than necessary in favor of disk I/O throughput, can generally be reduced, resulting in further operational cost savings due to reduced floor space and cooling requirements.

The performance advantage of 3LA comes from the superior performance/price ratio of flash devices compared with HDDs[5]. This ratio will steadily increase in the next years, while the performance/price ratio of HDDs will remain relatively stable. As a consequence, the performance advantage of 3LA will be even more significant in the future.

LOC and GLB served as the baseline algorithms. No flash-specific optimizations are yet integrated. Techniques such as using different page size at different layers as those discussed in [4] could further improve the performance of 3LA. It could also be interesting to examine hybrid configuration of algorithms, e. g., frequency-based algorithm at one layer and recency-based algorithm at the other layer. As future work, we will also look into such optimizations. However, future improvements expected for performance and energy-efficiency of 3LA do not conflict with our observations made in this paper.

One of the major differences of a flash-based cache to a RAM-based cache is non-volatility. We have discussed a technique leveraging this property to shorten the warm-up phase of the system in Section 3.3, which is not the main focus of this paper and will be empirically evaluated in the future. The non-volatility of the flash layer should be further exploited to speed up processing of transactions, for which durability is required.

---

[5] Similar observations are made in our experiments using the values of HDD3, the high-end HDD in Table 1.

Acknowledgement

We are grateful to IBM (Deutschland and USA) for providing the TPC-E trace and to anonymous referees for valuable comments. This research is partly supported by the German Research Foundation and the Carl Zeiss Foundation.

References

1. Spiegel. Google-chef will noch mehr daten (2010),
   http://www.spiegel.de/netzwelt/netzpolitik/0,1518,716204,00.html
2. Härder, T.: DBMS architecture - the layer model and its evolution. Datenbank-Spektrum 13, 45–57 (2005)
3. Zhou, Y., Chen, Z., et al.: Second-level buffer cache management. IEEE Transactions on Parallel and Distributed Systems 15(6), 505–519 (2004)
4. Koltsidas, I., Viglas, S.D.: The case for flash-aware multi-level caching. Technical Report (2009)
5. Narayanan, D., Thereska, E., et al.: Migrating server storage to SSDs: analysis of tradeoffs. In: EuroSys, pp. 145–158. ACM, New York (2009)
6. Gray, J., Reuter, A.: Transaction Processing: Concepts and Techniques. Morgan Kaufmann, San Francisco (1993)
7. Gal, E., Toledo, S.: Algorithms and data structures for flash memories. ACM Computing Surveys (CSUR) 37(2), 138–163 (2005)
8. O'Neil, E.J., O'Neil, P.E., et al.: The LRU-K page replacement algorithm for database disk buffering. In: SIGMOD, pp. 297–306 (1993)
9. Johnson, T., Shasha, D., et al.: 2Q: a low overhead high performance buffer management replacement algorithm. In: VLDB, pp. 439–450 (1994)
10. Megiddo, N., Modha, D.S.: ARC: A self-tuning, low overhead replacement cache. In: FAST. USENIX (2003)
11. Li, Z., Jin, P., et al.: CCF-LRU: A new buffer replacement algorithm for flash memory. Trans. on Cons. Electr. 55, 1351–1359 (2009)
12. Ou, Y., Härder, T.: Clean first or dirty first? a cost-aware self-adaptive buffer replacement policy. In: IDEAS, Montreal, QC, Canada (2010)

Design of Embedded Database Based on Hybrid Storage of PRAM and NAND Flash Memory

Youngwoo Park, Sung Kyu Park, and Kyu Ho Park

Korea Advanced Institute of Science and Technology (KAIST),
305-701, Guseong-dong, Yuseong-gu, Daejeon, Korea
{ywpark,skpark,kpark}@core.kaist.ac.kr

Abstract. Andorid which is the popular smart phone OS uses a database system to manage its private data storage. Although the database system supports a powerful and lighteweight database engine, its performance is limited by a single storage media, NAND flash memory, and a single file system, YAFFS2. In this paper, we propose a new embedded database system based on hybrid storage of PRAM and NAND flash memory. Using the byte-level and in-place read/write capability of PRAM, we separately manage a journaling process of the database system. It increases the transaction speed and reduces the additional overhead caused by NAND flash memory. We implement our database system using SQLite and dual file systems (YAFFS2 and PRAMFS). Consequently, the proposed database system reduces the response time of the database transaction by 45% compared to the conventional database system. In addition, it mitigates the burden of NAND flash memory management. Moreover, previous database applications can be executed on the proposed system without any modification.

Keywords: Database, NAND flash memory, PRAM, SQLite.

1 Introduction

Recently, various non-volatile memories such as NAND flash memory [16] and PRAM (Phase-change RAM) [17] have been developed very quickly. These non-volatile memories gain acceptance as an alternative storage media.

NAND flash memories have become increasingly popular for the main data storage of the embedded system due to its shock-resistance, low power consumption, and high I/O speed. The characteristics of NAND flash memory are very different from those of traditional magnetic disks. It supports a page-level (512B or 2KB) read and write operation, but the write speed is much slower than the read speed. Once a page is written, the page should be erased first in order to update data on that page. The erase operation performs in terms of block (32KB or 256KB) and takes much longer than page write time. All pages in a block should be unnecessarily erased at the same time. As a result, NAND flash memory cannot be managed by in-place update scheme like a disk. It is very important to optimize the write and erase count for NAND flash memory based system.

J. Xu et al. (Eds.): DASFAA Workshops 2011, LNCS 6637, pp. 254–263, 2011.

Table 1. Characteristics of NAND Flash Memory and PRAM

	PRAM [7]	NAND Flash Memory [16]
Volatility	Non-volatile	Non-volatile
Interface	Byte-addreable	Page based I/O
Capacity	512Mbit	8 Gbit
Read Time	80ns/Word	$60\mu s$/Page$^+$
Write Time	$1\mu s$/Word	$800\mu s$/Page$^+$
Erase Time	N/A	1.5ms/Block*
Endurance	1M write cycles	5K erase cycles

+Page = 2KB, *Block=128KB.

On the other hand, PRAM has begun to gain acceptance as an alternative to the embedded storage because of its high density and performance. PRAM has been developed by key industry manufacturers such as Intel, Samsung, and IBM. Already, Samsung commercialize 32MB PRAM for mobile devices to replace NOR flash memory to store boot-up codes. Unlike NAND flash memory, PRAM supports byte-level read/write and in-place update. These characteristics can complement page-level read/write and out-of-place updates of NAND flash memory in the database system.

Table 1 summarizes the characteristics of NAND flash memory and PRAM. These special features of new storage media impose new challenges to the traditional database system. In order to make the database system adapt to those memories, it is necessary to reconsider the storage architecture and develop new database management algorithms.

Currently, Android, which is the popular smart phone OS releasd by Google, uses database to manage its private data storage[18]. Fig. 1 shows the database system of Android. Android uses SQLite library which is widely used for the embedded system [12]. Because SQLite is a file based database system and NAND flash memory is the main storage of smart phone, YAFFS2 [8] is used for database file management. The database system of Android supports a powerful and lightweight relational database engine available to all applications. However, the storage related management is compeletly assigned to the single file system. It is hard to adopt a new database architecture and develop a specific algorithm of the database system. The performance of Android's database system is limited by YAFFS2 and NAND flash memory.

In this paper, we propose a hybrid storage architecture and transaction method for the embedded database system. In our hybrid storage architecture, NAND flash memory is used for the main database storage. At the same time, phase change random access memory (PRAM) accommodates the temporal database transaction. Because of byte-level read/write capability, PRAM ensures better

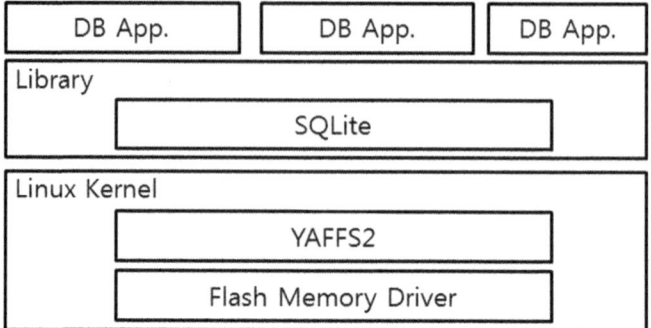

Fig. 1. Database system of Google's Android

performance for small size operations than NAND flash memory. Moreover, it can reduce many writes of NAND flash memory during transaction and decrease the database system overhead for storage management.

In our implementation, we modify SQLite database library to exploit PRAM as an alternative storage. SQLite does make use of temporary files during transactions. We separately store those temporary files to PRAM using a dual file system approach which is YAFFS2 for NAND flash memory and PRAMFS [9] for PRAM. The experimental results show that the proposed embedded database architecture reduces the response time of database transaction by 45% compared to thr traditional database system which consists of single NAND flash memory.

2 Related Work

Many researchers have developed new algorithms for flash-based DBMS. IPL [3] presents in-page logging approach in which the change made to a data page in memory are buffered on the per-page basis instead of writing the page entirely, thus avoiding high latency of write and erase operations. PDL [4] proposes a page-differential logging scheme with three design principles which are writing-difference-only, at-most-one-page writing, and at-most-two-page reading. These principles can guarantee good performance in both read and write operations and prolong the lifetime of flash memory by reducing the number of erase operations.

Some approaches have been tried to scale down DBMS. PicoDBMS [1] is designed for the resource-constrained smartcard. It proposes a new pointer-based storage model with a unique compact data structure and query execution with no RAM consumption, thus matching performance with the smartcard application's requirements. LGeDBMS [2] is designed for mobile systems. It applies the design principle of log-structured file system (LFS) [11]. It can reduce the number of I/Os by writing a log once rather than writing a log for each data change.

However, there is a limitation for designing DBMS only with flash memory. Although the average size of updates in DBMS is about dozens of bytes, the unit of the write operation in flash memory is over 2KB or 4KB. Flash memory also has long erase time, about 1.5ms, but small random requests in DBMS make a large number of erase operations.

Many researches have used non-volatile memories like NOR, FRAM, and PRAM [17] to handle small-sized data. Using the byte-level read/write capability of NOR flash memory, HFFS [10] synchronously stores data as a log in NOR flash memory, thus reducing the write count and providing a long lifespan of NAND flash memory. FRASH [5] proposes a novel file system for FRAM and flash memory, thus resolving long mount time issue. PFFS [6] presents a scalable and efficient flash file system using the combination of flash memory and PRAM. PFFS separates the metadata from the regular data in a file system and saves them into PRAM, thus solving the scalability problem and improving write and garbage collection performance.

We can solve the limitation only with NAND flash memory of the traditional database system by exploring hybrid storage of NAND flash memory and PRAM in this paper. It results in reducing the response time of database transactions and mitigating the burden of NAND flash memory management.

3   Hybrid Storage Architecture

For the embedded database system, we propose a hybrid storage architecture. Fig. 2 compares the proposed architecture with a conventional architecture. As shown in Fig. 2(a), the conventional embedded system uses NOR flash memory as boot-up code storage.

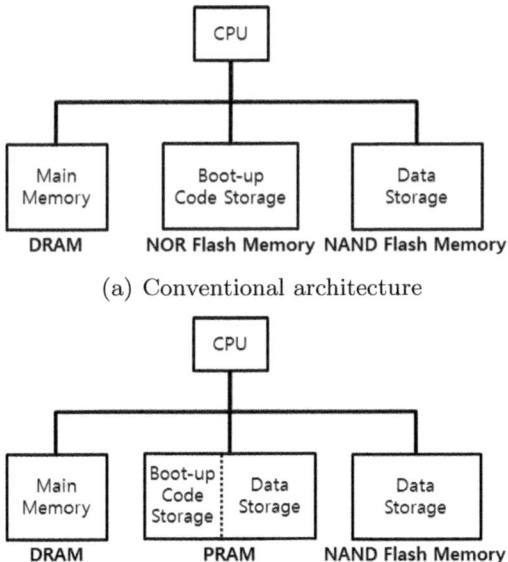

(a) Conventional architecture

(b) Proposed hybrid architecture

Fig. 2. Storage architecture comparison

On the other hand, the proposed hybrid architecture employs PRAM as boot-up code as shown in Fig. 2(b). PRAM supports the byte-level access and XIP (eXecution In Place). There is no problem to replace NOR flash memory to PRAM as boot-up code storage. In addition, the remaining space of PRAM is used as an alternative database storage while NAND flash memory is still main database storage. Because the size of boot-up codes is very small compared to commercialized PRAM size, PRAM can be utilized to enhance database performance. Using PRAM data storage, we can store a part of data generated during the database transaction and reduce unnecessary access to NAND flash memory.

4   Transaction on Hybrid Storage

In our hybrid storage architecture, the data storage of PRAM can be used to optimize several parts of the database system. In this paper, we target to enhance the transaction process. Fig. 3 shows the process of database transaction based on hybrid storage of PRAM and NAND flash memory.

The key of transaction process is to store journal for rollback operation. First, data to be updated are read to a database buffer from NAND flash memory. Then, these original data are updated in the database buffer. After updating data, journal data which are updated parts of data are written to PRAM instead of NAND flash memory. Originally, both data and journal data are written to NAND flash memory. It makes performance degradation because of two reasons. The first one is that the unit of write operation is page whose size is over 2KB,

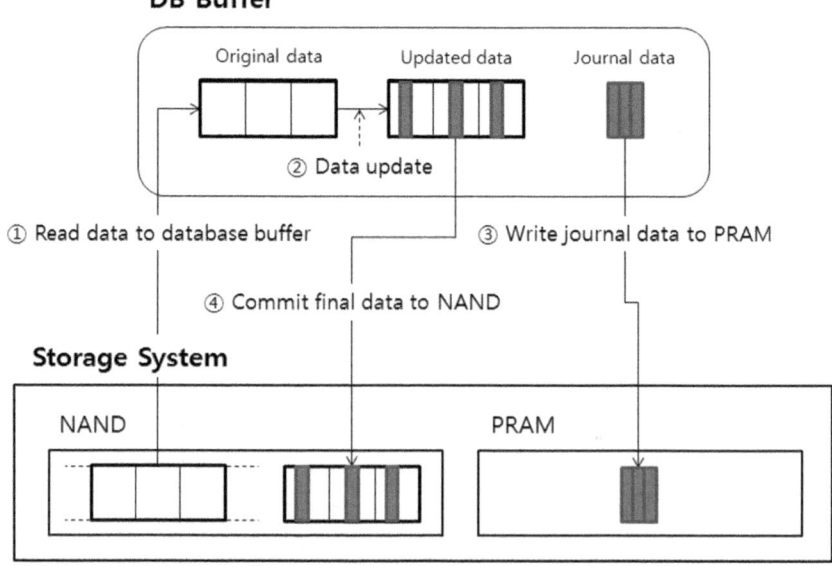

Fig. 3. Transaction Process on hybrid Storage

but the size of journal data is small. Another reason is that write operations in NAND flash memory make additional erase operations due to the characteristic, called out-of-place update, thus it can increase the operation time. Therefore, we can improve performance by writing journal data to PRAM. As whole updated data are written to NAND flash memory, the transaction process is finished.

5   Implementation Issue

We implement our database system using SQLite. SQLite is an in-process library that implements a self-contained, serverless, zero-configuration, and transactional SQL database engine [12]. It is widely used for the embedded system because it is very compact library which can be less 300KB with all features enabled. Currently, Android uses SQLite as its built-in embedded database [13].

If we adopt SQLite library to our hybrid storage architecture, we need a method to access both PRAM and NAND flash memory. Basically, a database in SQLite is a single file. SQLite is implemented on the file system. One way to utilize hybrid storage is to develop a new file system and SQLite transaction algorithm for hybrid storage. However, it causes modification of all database applications and subordinates our database system to specific file system.

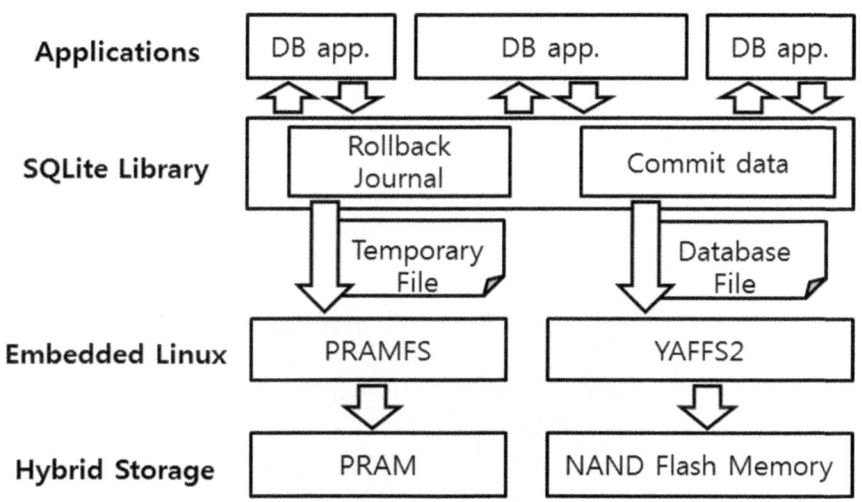

Fig. 4. Software architecture of proposed database system

Instead of file system modification, we use dual file systems and PRAM storage for storing temporal rollback information. As shown in Fig. 4, we use two separate file systems for single database management. YAFFS2, which is used for Googles Android, is NAND flash memory based file system [8]. PRAM is managed by PRAMFS [9] that is a simple file system designed for non-volatile memory. These two file systems are used to store files generated by SQLite database library.

YAFFS2 manages main database storage and PRAMFS stores only temporal file called rollback journal.

Currently, SQLite uses the rollback journal to atomic commit and rollback capabilities in SQLite. The rollback journal is frequently accessed and updated because it stores all information needed to restore the database back to its original state. In addition, the rollback journal is created when a transaction is first started and is deleted when a transaction commits or rolls back [12]. The size of rollback journal is small enough to store in small-sized PRAM storage. Because of non-volatility of PRAM, there is no problem to rollback transaction after sudden system power-off. Therefore, if the rollback journal is managed by PRAMFS, we can reduce a lot of data read/write of NAND flash memory during transaction and increase the database system performance.

Because the rollback journal is originally located in the same directory as the database file, we simply modify the SQLite library to designate the location of rollback journal. We never change a programming interface of SQLite. Previous database applications are compatibly executed on the proposed system. Moreover, there is no file system dependency in the proposed database system. We can use any two file systems to separate rollback journal from database. Even it identically works as conventional SQLite when using single file system.

6   Experiment

6.1   Experimental Environment

In order to evaluate our database system, we use an evaluation board shown in Fig. 5. This evaluation board has a 266MHz ARM processor and 64MB of SDRAM. It can include most of the storage devices that are currently used for embedded systems: NOR flash memory, SLC and MLC NAND flash memory,

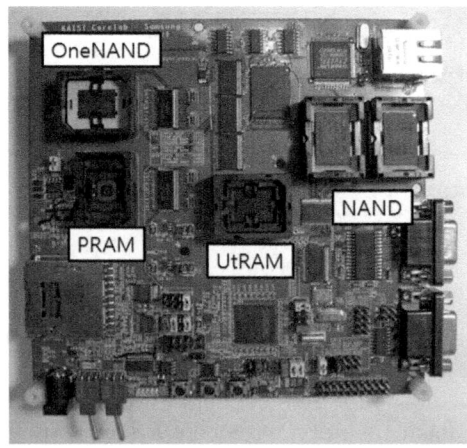

Fig. 5. The evaluation board for experiment

OneNAND, UtRAM, and PRAM. We can organize the proposed hybrid storage architecture on that board. However, PRAM is not currently available, thus the performance evaluation was actually carried out with an emulated PRAM by using UtRAM. We analyze the read/write time of PRAM and emulate that by power backed UtRAM with software delay [6][15]. Our emulation method is enough to evaluate our hybrid storage architecture because PRAM read/write time is deterministic and the data in power backed UtRAM was not volatilized.

6.2   Experimental Result

We implement our database system in a Linux OS with a kernel 2.6.19 version. Latest version of YAFFS2 and PRAMFS are used for file system of NAND flash memory and PRAM. We modify SQLite 3.7.3 vesion to select rollback journal location optionally. The proposed database system is compared to the conventional embedded database system that uses YAFFS2 file system with single NAND flash storage. To evaluate the performance of database, we use DBT-2 benchmark which is a fair usage implementation of the TPC-C benchmark specification [14]. Our results are measured by taking the average of consecutive five tests of DBT-2 benchmark.

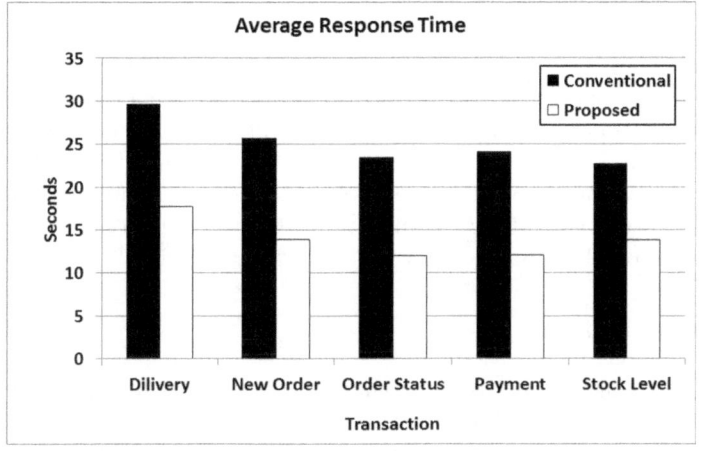

Fig. 6. Average response time of DBT-2 benchmark

Fig. 6 summarizes the average response time of TPC-C transaction. The results show that our proposed database system reduces the response time of all kinds of TPC-C transactions by 45%. The average NOTPM (New Order Transaction Per Minute) value is increased to 0.334 in the proposed database system, while the conventional database system shows only 0.172 NOTPM. This is mainly because the proposed system decreases the overhead of rollback journal. The proposed storage makes it possible to read/write rollback journal by single byte. We can reduce rollback journal update latency.

Table 2. Comparison of Write and Erase Count in NAND Flash Memory between Conventional and Proposed Database

	Conventional Database with Only NAND Flash Memory	Proposed Database with NAND Flash Memory and PRAM
Avg. Write Count	13020	5178
Avg. Erase Count	77	12

In addition, we can exploit in-place update in the proposed system. Table 2 shows the average write and erase count of NAND flash memory made during DBT-2 benchmark execution. Because proposed system updates rollback journal data in PRAM, it reduces many write and erase operations of NAND flash memory and increases the performance of our database system. Moreover, if the write and erase count is reduced, the overhead of NAND flash memory management like garbage collection and wear-leveling may be also decreased significantly.

7   Conclusion

Because of advance of the non-volatile memory technology, we should redesign database system. Each non-volatile memory has very different characteristics. It is important to consider the features of non-volatile memories. In our proposed design, we developed the hybrid storage architecture to use the advantage of PRAM and NAND flash memory. We prove that the embedded database system performance can be increased by simple modification. Our implementation is compatibly used for previous SQLite application. Our idea is further applicable to storage architecture using other non-volatile memories.

8   Future Work

As the further work, we'll separate other temporary files such as master journals, statement journals, TEMP databases, materializations of views and subqueries, transient indices, and transient databases used by VACUUM as well as the rollback journals for writing to PRAM. We can also exploit PRAM for storing frequently updated data for reducing write operations in NAND flash memory. Because PRAM has limited size compared to NAND flash memory, we need to develop allocation, swapping, and buffering schemes in SQLite layer or file system layer according to the data size or type.

References

1. Pucheral, P., Bouganim, L., Valdureiz, P., Bobineau, C.: PicoDBMS: Scaling Down Database Techniques for the Smartcard. In: Proc. the 26th International Conference on Very Large Data Bases (VLDB 2000), pp. 11–20 (2000)

2. Kim, G.-J., Baek, S.-C., Lee, H.-S., Lee, H.-D., Joe, M.J.: LGeDBMS: A Small DBMS for Embedded System with Flash Memory. In: Proc. the 32nd International Conference on Very Large Data Bases (VLDB 2006), pp. 1255–1258 (2006)
3. Lee, S.-W., Moon, B.: Design of Flash-Based DBMS: An In-Page Logging Approach. In: Proc. the 2007 ACM SIGMOD International Conference on Management of Data (SIGMOD 2007), pp. 55–66 (2007)
4. Kim, Y.-R., Whang, K.-Y., Song, I.-Y.: Page-Differential Logging: An Efficient and DBMS-independent Approach for Storing Data into Flash Memory. In: Proc. the 2010 International Conference on Management of Data (SIGMOD 2010), pp. 363–374 (2010)
5. Kim, E.-K., Shin, H., Jeon, B.-G., Han, S., Jung, J., Won, Y.: FRASH: Hierarchical File System for FRAM and Flash. In: Gervasi, O., Gavrilova, M.L. (eds.) ICCSA 2007, Part I. LNCS, vol. 4705, pp. 238–251. Springer, Heidelberg (2007)
6. Park, Y., Lim, S.-H., Lee, C., Park, K.H.: PFFS: A Scalable Flash Memory File System for the Hybrid Architecture of Phase-Change RAM. In: Proc. the 2008 ACM Symposium on Applied Computing (SAC 2008), pp. 1498–1503 (2008)
7. Qureshi, M.K., Karidis, J., Franceschini, M., Srinivasan, V., Lastras, L., Abali, B.: Enhancing Lifetime and Security of PCM-Based Main Memory with Start-Gap with Start-Gap Wear Leveling. In: Proc. the 42nd Annual IEEE/ACM International Symposium on Microarchitecture (MICRO 42), pp. 14–23 (2010)
8. Aleph One Ltd., Yet Another Flash File System, YAFFS (2002), http://www.yaffs.net
9. Protected and Persistent RAM Filesystem, http://pramfs.sourceforge.net
10. Lee, C., Baek, S.H., Park, K.H.: A Hybrid Flash File System Based on NOR and NAND Flash Memories for Embedded Devices. IEEE Transactions on Computers 57(7), 1002–1008 (2008)
11. Rosenblum, M., Ousterhout, J.K.: The Design and Implementation of a Log-Structured File System. In: Proc. the 13th ACM Symposium on Operating Systems Principles (1992)
12. SQLite, http://www.sqlite.org
13. Wikipedia, SQLite, http://en.wikipedia.org/wiki/SQLite
14. Database Test 2 (DBT − 2^{TM}), http://osdldbt.sourceforge.net
15. K1S5616BCM Data Sheet, http://www.samsung.com
16. K9F2G08U0A Data Sheet, http://www.samsung.com
17. Wikipedia, Phase-change memory, http://en.wikipedia.org/wiki/Phase-change_memory
18. What is Android?, http://developer.android.com/guide/basics/what-is-android.html

Hybrid Storage with Disk Based Write Cache

Puyuan Yang, Peiquan Jin, and Lihua Yue

School of Computer Science and Technology,
University of Science and Technology of China, 230027, Hefei, China
yangpuy@mail.ustc.edu.cn

Abstract. Recently, flash-memory-based solid state disks (SSDs) have been considered to be alternatives for traditional magnetic disks. However, it has not come true so far due to some limitations on SSDs, such as high latency of write operation and low reliability in case of unbalanced erasure. Therefore, a practical way is to integrate SSD and magnetic disk and then to obtain a better tradeoff between those two storage medium. In this paper, we investigate the issues of integrating SSD and disk in the storage layer of a database management system. In particular, we propose a new approach to using a magnetic disk as the write cache of an SSD, in which each data page is placed either in disk or in SSD. To find an optimal page placement scheme, we first propose a page migration model, which uses two grains, namely page and block (a set of pages), to perform the migration between SSD and disk. Based on this model, we develop an online approach to determining the optimal places of data pages. We conduct experiments on tailor-made traces to measure the performance of our hybrid storage approach. The results show that our approach ensures most read operations are performed on SSD and most write operations are focused on disk. Meanwhile, our hybrid approach has less runtime than the single-disk-based mechanism.

1 Introduction

Flash memory was commonly used in hand-held devices. As the constantly growing of the capacity of flash memory, flash-memory-based solid state disks (SSDs) are going to be an alternative of traditional magnetic disks and be used as secondary storage devices in modern computer systems.

However, SSD exhibits some different characteristics compared to magnetic disks. Firstly, it has no mechanical latency. Secondly, it has different I/O latencies, i.e., a read operation usually consumes less time that a write operation does. In particular, the random write operation in SSD is considered to be even slower than that in magnetic disk. This has been one of the major problems that influence the performance of SSDs.

To improve the I/O performance of SSD, we propose a hybrid storage approach, which uses both an SSD and a magnetic disk. Moreover, we use the magnetic disk as a write cache of the SSD, which ensures that all writes are first taken place on the magnetic disk and thus can avoid the random writes on the SSD. To take advantage of the high read speed of SSD, we develop an algorithm to migrate read-mostly pages into SSD, if they are first located in the magnetic disk. Those migrated pages are

J. Xu et al. (Eds.): DASFAA Workshops 2011, LNCS 6637, pp. 264–275, 2011.

organized as blocks and all migrations are performed according to a block unit, which aims at making use of the high sequential-write performance of SSD and also reducing the erase times of flash memory.

The rest of this paper is organized as follows. Related work is presented in Section 2. The hybrid storage model is discussed in Section 3. In Section 4, we explain the details of our idea, including the migration algorithm. Experiments are discussed in Section 5. And finally, we conclude the paper in Section 6.

2 Related Work

2.1 Flash Translation Layer

To overcome the limitations of flash memory, SSD employs a software layer called flash translation layer (FTL). FTL is placed between the file system and flash chips, typically stored in a ROM chip, and its main purpose is to provide logical-to-physical address mapping, error recovery, and wear-leveling. Concretely, FTL maintains a mapping of the logical page address to enable a new physical page location on flash memory, and the mapping information is maintained separately in flash memory and main memory for look-up. Besides, FTL always redirects a page write request to an empty area which has been erased earlier.

Many different algorithms have been proposed for the FTL [2]. Their mapping runs at the page level or block level.

Page-mapping FTL deals with the erase-before-write by redirecting a write request to any empty page in the flash memory, and its mapping table records the valid physical page locations. So one erasure can serve x write request when a block contains x pages. However, page-mapping method requires a great number of memories for its information. In some cases, it may take a long time that scanning the whole flash memory to reconstruct the mapping information at start-up.

Block-mapping FTL only maps a logical block address to its physical location, which means that both of the offsets of a page in the physical block and logical block are totally identical. So the memory requirement is much less than page-mapping FTL. However, updating a page requires that the new content of the page must be written to the same offset in another free block and the rest pages of the block have to be copied to the newly allocated block, which results in that updating a page brings up a block erasure and x pages writes.

2.2 Log-Block-Based FTL

Some improvement, called hybrid mapping FTL, have been proposed for the problems of page-mapping FTL and block-mapping FTL. Some representatives of them are called log-block-based approaches [3,4]. In the log-block FTL, a set of flash blocks are called log blocks which the write requests are always directed to. The log-block FTL avoids frequent block erasures by mapping the page addresses in a log block at the page level which means allowing a page to be in any position in a log block. The remaining flash blocks are called data blocks which are managed at the block level, and they generally take up much larger area than log blocks.

There are two kinds of schemes depending on the block association policy: a log block is allocated for only one data block (BAST)[4] and a log block is allocated for multiple data blocks(FAST)[5]. For a write request, firstly data updating takes place in a log block. If there is no free log block available, a log block must be selected as a victim to be freed, and all the valid pages in both the log block and its corresponding data block (or blocks) are migrated into data blocks to make a room for on-going write requests. This step is called block merge.

In the approach of BAST, a log block serves the page writes to only one data block. When merge operation occurs, there are two situations:

(a) If the victim log block does not contain all valid pages of its data block, each valid page—either in the log block or in the corresponding data block—is copied to another empty block. Then the third block become the new data block and the log block and the old data block are erased for later use as log block or data block, which brings up two erasures.

(b) When not only the victim log block contains all valid pages of its data block but also the offsets of pages are same to those of their corresponding pages in the data block, we just simply mark the victim log block as the new data block and only one erasure is necessary for the old data block. This operation is called a switch merge. However, in general BAST shows very low space utilization. Especially for random write patterns with low locality, it shows poor performance because of frequent block merges with log block poorly filled with modified pages.

To resolve the problem of BAST, FAST was proposed. It makes a log block serve write request for multiple data blocks. So it is obvious that the higher space utilization in the log blocks can be achieved and merge operation frequency can be reduced. However, a merge operation may bring more erasures. Fox example, if a victim log block is associated with N data blocks, for each of them, the valid pages are copied to a free block, and then it is necessary that N erasures for the old data blocks and one erasure for the log block.

Log-Block-Based FTL can apparently improve the write performance of SSD, and the write requests is higher locality, the better is the flash write performance, which suggests that the pages written to SSD being organized in block in buffer can make benefit for FTL.

2.3 Hybrid Storage Policy

There have been some good ideas about hybrid storage policy based on SSD and magnetic disk [6, 7]. In article [6], SSD and magnetic disk are regarded at the same level of the memory hierarchy. Three page placement algorithms are discussed deeply which define the proper location for a page. For each page, these algorithms record the total times of the physical read & write and the logical read & write to computer the migration cost (defined by physical I/O cost of flash and magnetic disk). If the I/O benefit of the page in the other device is more than that in the current device and the difference is bigger than migration cost, the page can be migrated to the other device. By recording and analyzing the history I/O information, these algorithms can properly distinguish the read-intensive pages and the write-intensive pages. As experiment shown in article [6], these algorithms are sensitive to the change of the I/O workloads of the page. Inspired by the discussion of these algorithms, we propose a migration model at the block level.

In article [6], the other main policy is the buffer policy. In buffer, except the main queue managed by LRU algorithm, four more queues which store the victim page from the main queue are designed to reduce the total I/O cost: 1) Flash Read Queue stores the clean pages located in flash, 2) Flash Write Queue stores the dirty pages located in flash, 3) Magnetic Read Queue stores the clean pages located in magnetic disk, 4) Magnetic Write Queue stores the dirty pages located in magnetic disk. The pages accessed in the four queues are put back into main queue. According to the order from lowest I/O cost to highest I/O cost, evict operation firstly select the pages in FRQ as victim page, then in MRQ if FRQ is empty, then in MWQ if FRQ and MRQ are empty, then in FWQ if other 3 queues are empty.

SSD and magnetic disk are at the same level, so some page writing certainly take place in flash, and the buffer replacement policy and migration policy are designed at the page level. That may bring some random writes to flash which may result in some non-switch merges as shown in section 2.2. So we propose migrating the data to SSD in block.

In article [6], it may result in some unfit migration that the migration algorithms do not consider the write frequency of a page in local-time. For example, a page located on magnetic disk should be migrated at time t according to the migration algorithm, but from time t there come a series of write requests for the page and the process the page belongs to becomes cold, which means some physical writes inevitably come out. If the page is migrated to flash, these physical writes take place on the flash. After all, the migration algorithm in article [6] can agilely get the read/write-intensive workloads of the pages, but it cannot handle the frequent writes to the pages in flash in local-time.

3   The Hybrid Storage System Model

The hybrid storage system consists of an SSD and a magnetic disk. It is imaginable that such a system may consists of two or SSDs or magnetic disks. However, in this paper we simplify the system model to include only one SSD and one magnetic disk. We remain the issue how to cope with more SSDs and magnetic disks in the hybrid system in the future work.

In the hybrid storage system, SSD can be used to improve the total read performance, while it has a limited lifetime. On the other hand, magnetic disk has better performance in terms of random write. Thus the key issue in the hybrid storage system is how to make use of the high read speed of SSD and the high random write performance of magnetic disk, in order to improve the overall I/O performance of the storage system. Meanwhile, we want to avoid the high-cost random writes on SSD and random read on magnetic disk.

Based on the above consideration, we design different processing strategies to answer write requests and read requests in the hybrid storage system. As shown in Fig.1, for read request, SSD and magnetic disk are used at the same level of the storage hierarchy. To make this model favor for read performance, we propose a migration algorithm to place the pages with read-intensive workloads in SSD. In Fig.1, Rf represents the physical read cost of flash memory, and Rm represents the physical read cost of magnetic disk.

In order to reduce the random writes on SSD, we take the magnetic disk as a write cache for SSD by placing all the updated pages from SSD at the page level in the magnetic disk (as shown in Fig.1 (b)). So the page-grained updating only takes place in the magnetic disk, and then the read-mostly pages in the magnetic disk can be migrated into SSD according to a block unit. The migration algorithm depends on the accessed history of the pages and blocks located in the magnetic disk. As shown in Fig.1 (b), our hybrid storage system employs a two-layer storage hierarchy. In Fig.1 (b), *Wmp* represents the physical write cost to magnetic disk and *Wfb* represents the physical write cost to SSD.

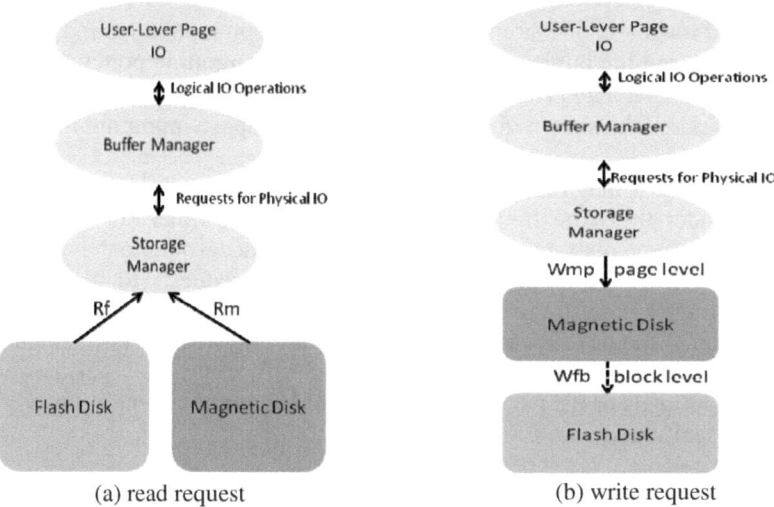

Fig. 1. Different strategies to process read and write requests

The migration unit in our system is different. When a page from SSD is written to the magnetic-disk cache, we just use a general buffer replacement algorithm such as LRU, which does not bring additional system costs. On the other hand, when pages in magnetic disk are migrated into SSD, we use the block unit, i.e., the pages are organized into chunks and then written into SSD. This procedure may introduce more system costs because of the more calling to the I/O interface. So we need to get a proper buffer policy to reduce the migration cost.

4   The Migration Algorithm

4.1   Page Placement

We present an algorithm to deciding the optimal placement of data pages. Whenever the buffer pool is out of space, a page must be selected to be replaced according to our buffer replacement policy. At that point, if the page is fetched from magnetic disk, our migration algorithm can decide whether the page should be placed in SSD along with its corresponding block.

The migration approach is designed as an online algorithm. It only keeps track of the blocks located in magnetic disk and decides when to be migrated. We model the decision process as a two-state system. All pages always change between the two states, saying that either a page is on SSD or on the magnetic disk. As shown in Fig.2, we use different symbols to represent the costs of read and write operation, which are listed as follows:

- *rm*: the cost of reading a random page from the magnetic disk.
- *Wm*: the cost of writing a random page to the magnetic disk.
- *rf*: the cost of reading a random page from SSD.
- *Wfb*: the cost of writing a block to SSD.

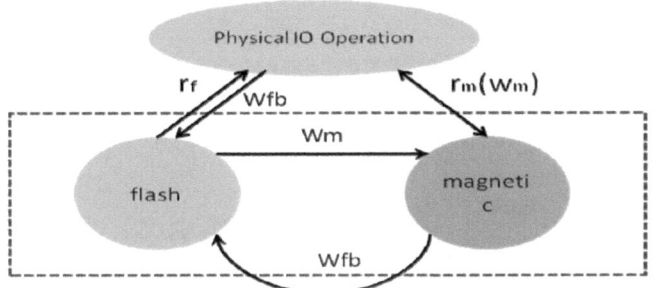

Fig. 2. The migration model

Note that in our system all write operations on SSD are performed at the block unit, so the migration cost from one state to the other is equal to *Wm* or *Wfb*.

Because the magnetic disk is taken as a write cache for SSD, the migration algorithm only focuses on migrating the read-mostly pages in the magnetic disk into SSD. As shown in [7], Read-Threshold Trigger is a good method to get the read-intensive block, but the difference in our system is that the threshold is decided by the physical I/O cost which is inspired by article [6]. On the other hand, the interaction between physical and logical operations is not clear. We need to count the actual I/O and the logical application-level I/O. In article [6], the hybrid algorithm has the best result by counting the actual physical I/O and the possible physical I/O computed by those logical times a probability. We propose a similar algorithm, but the difference in our system is that we only count the read operations and ignore the write operations.

4.2 Block Level Hybrid Algorithm

To exhibit the excellent read performance of SSD, we need to place the read-intensive blocks on SSD. We propose a block level hybrid algorithm to count both the physical and logical operations on data blocks on the magnetic disk. If a page on the magnetic disk gets a read request, we increase the count of its corresponding block. The basic idea is that a physical read operation has more impacts than a logical one and they affect the actual costs.

As shown in Fig.3, we only count the read operations. Line 13, 14 and 15 show that for a block if the read accounting reaches a threshold, the pages of the block on magnetic disk can be migrated to SSD. The threshold is decided by the physical write cost. *Wm* is the cost of the physical writing a page to magnetic disk and *Wfb* is the cost of the physical writing a block to SSD which consist the migration cost, and the formula in Line 15 *Wm x block size* presents the cost of those pages moved back to the magnetic disk later. Line 15 indicates that the migration operation will be carried out when the benefit of read-after-migration is bigger than the migration cost.

The probability that a logical read operation will be realized as a physical one is proportional to the accounting of the read operations. Let n is the number of the total read operations and b is the number of the logical read operations in history. The probability b/n refers to the possibility that a logical read operation will not affect the total read cost. So the formula 1 b/n represents the possibility that a logical read operation will have an impact on the I/O cost.

Algorithm block_hybrid (Page pg)
1. if (pg is a new page)
2.   pg.state ← m.
3.   bk ← pg's corresponding block
4. if (bk is a new block)
5.   bk.lr←0, bk.pr←0.
6. No counting until after the pg's first physical write
7. if (pg.state = m)
8.   After each logical read of pg
9.   bk.lr←bk.lr+1;
10.   After each physical read of pg
11.   bk.pr←bk.pr+1;
12. q ← 1-b/n
13. Cf ← (bk.lr x q + bk.pr) x rf
14. Cm ← (bk.lr x q + bk.pr) x rm
15. if(Cm – Cf > Wm x blocksize + Wfb)
16   for each pg in bk
17.   pg.state ← f
18.   if (no page in bk, pg.state = m)
19.   delete bk node
20.   else
21.   bk.lr←0, bk.pr←0.
22. pg.dirty ← 1.

Fig. 3. The block level hybrid algorithm

In Fig.3, *lr* counts the logical read operations and *pr* counts the physical read operations since the block has at least one page on the magnetic disk. When the page eviction occurs in the buffer, the algorithm determines if the corresponding block of the victim page should be migrated or not.

4.3   Page Level HSLRU-2 Algorithm

The physical write frequency of a page on magnetic disk is measure by the HSLRU-2 algorithm (see Fig.4). HSLRU-2 is similar to LRU-2. The basic idea is that the time is recorded as dirty-last when it is written to the magnetic disk. As shown in article [8], there is also a time threshold to measure the frequency. For a page, if the time interval between the contiguous dirty-last is smaller than a time threshold, the page is in the sequential physical write state. Upon eviction, a page in the sequential physical write state cannot be migrated even if its corresponding block has been proved read-intensive. In other words, it requires that the pages are not in the sequential physical write state and their corresponding block has been proved read-intensive that the pages can be migrated in block to the SSD.

Algorithm HSLRU-2(page pg)
1.   When pg is fetched from disk at time t
2.   if(pg is a new page)
3.   pg.state←m
4.   pg.firstinbuf ← 1
5.   else if(pg.state = m)
6.   pg.firstinbuf←0
7.   else if(pg.state = f)
8.   whenever pg.state←m, pg.firstinbuf← 1
9.   when pg is evicted from buffer at time t'
10.   if(pg.firstinbuf = 0 and pg.state = m and pg.dirty = 1)
11.   if(t' – pg.dirtylast > TimeIntervalThreshold)
12.   pg.dirtylast←t'
13.   pg.dirtyhist(1)←t'
14.   else
15.   pg.dirtyhist(2)←t'
16.   pg.dirtylast←t'
17.   else if(pg.firstinbuf = 1 and pg.state = m)
18.   at next time t0 when pg is fetched
19.   pg.dirtylast & pg.dirtyhist(1,2)←t0
20.   when judge the physical write frequency of pg in buffer
21.   if(pg.dirtyhist(1) < pg.dirtyhist(2))
22.   pg is in the frequent physical write state

Fig. 4. The HSLRU-2 Algorithm

The algorithm as shown in Fig.4 measures the physical write frequency of a page on the magnetic disk. for a page, if the contiguous physical writes to the magnetic disk are proved sequential, the algorithm uses a parameter dirtyhist(2) to record the replacement time, or it uses a parameter dirtyhist(1) to record the replacement time. Besides, it also records the time of each physical write to magnetic disk with a parameter dirtylast. So dirtyhist(1) presents the time of the latest un-sequential physical write while *dirtyhist*(2) represents the time of the latest sequential one. As shown in Line 21, it proves the page is currently in the sequential physical write state that *dirtyhist*(2) is bigger, to which the contrary shows the page is not in that state as shown in Line 24.

In the buffer, the pages which are new on the magnetic disk do not need to be recorded, because those pages are not physically stored on the magnetic disk. The recording starts after their first written to the magnetic disk. The parameter *firstinbuf* shows if we should record the frequency of the page or not.

The time interval threshold should be defined by the frequency of the write requests. Similar to [8], for the multiple processes running, there may frequently comes out some physical write operations which occur to some not hot pages which are always evicted in buffer. So the time interval threshold is better to be designed at the average level of the interval of pages' replacement or it is just decided by the DBA.

In summary, the block level hybrid algorithm keeps track of the information about blocks while the HSLRU-2 algorithm focuses on the pages information. So we can handle the hybrid storage problem at two levels.

5   Performance Evaluation

We have implemented our algorithms in Windows XP to evaluate their performance. Our experiment runs in a simulation environment and uses different types of trace files. We attach a SATA-interfaced SSD to the test computer as our flash device. To reduce the cache effects of the file system, we execute the file operations without using the buffer of the file system.

In order to determine the cost parameter of SSD and magnetic disk, we use IOmeter [11] to measure the I/O costs for each device. We record the average latency of an I/O operation and the IOPS for each device. Table 1 shows the results. The second column is the measured average times, the third column is IOPS, and the fourth column is the costs normalized by the flash random read time and IOPS.

Table 1. I/O costs used in the experiment

Operation	Latency(ms)	IOPS	Costs
Flash random read	0.482	2057	1
Flash random write	235.78	4.24	487
Flash block write	40.128	403553.28	86
Magnetic random read	11.0953	90.11	23
Magnetic random write	5.86	170.7	12

Our trace files include six files created according to the method introduced in [13]. The trace files all contain 300000 write and read requests and 10000 different pages. The features of the trace files are shown in Table 2. The locality means the range of the access, for example 80%/20% presents that the 80% of the accesses focused on the 20% of the pages.

Table 2. Trace files used in the experiment

Trace Type	Requests	Different pages	Read/Write ratio	Locality
T9182	300000	10000	90%/10%	80%/20%
T5582	300000	10000	50%/50%	80%/20%
T1982	300000	10000	10%/90%	80%/20%
T9155	300000	10000	90%/10%	50%/50%
T5555	300000	10000	50%/50%	50%/50%
T1955	300000	10000	10%/90%	50%/50%

In our experiment, we first test the migration result. We use two metrics, namely *Write Hit(WH)* and *Read Hit(RH)*, to measure the effectiveness of our algorithm. Suppose that there are total *N1* read operations and *N2* write operations in the trace file, if *x* read operations are executed on SSD and *y* write operations are focused on magnetic disk, we define *RH* and *WH* as follows:

$$RH = \frac{x}{N1} \qquad WH = \frac{y}{N2} \tag{1}$$

A higher *WH* value indicates that our algorithm moves more write-intensive pages into the magnetic disk, which shows that the cost of write operation is reduced because magnetic disk has better write performance. Similar to *WH*, a higher *RH* value means that our system is able to reduce the additional costs of read operations introduced by magnetic disk. In the experiment, we use the LRU algorithm as the buffer replacement policy and run the trace files using both an SSD and a magnetic disk as storage devices. Besides, we set three different initialization states and measure the migration effectiveness respectively. Those states are labeled as follows:

- HDDP: the pages all on the magnetic disk.
- SSDP: the pages all on the SSD.
- HybridP: half pages on the SSD and the other pages on the magnetic disk.

Table 3 shows the numbers of pages partitioned into SSD and the magnetic disk, under three setups. Table 4 shows the WH and RH values for each initial state.

Table 3. Numbers of pages partitioned into SSD and HDD

Trace	Pages Partitioned on HDD			Pages Partitioned on SSD		
Type	HDDP	SSDP	HbridP	HDDP	SSDP	HybridP
T9182	2227	2423	2645	7773	7576	7355
T5582	6065	6442	4332	3935	3558	5668
T1982	7482	6152	6790	2515	3843	3210
T9155	1992	1997	2024	8008	8003	7976
T5555	5367	4844	4365	4633	5156	5635
T1955	7993	8456	8126	2007	1544	1874

Table 4. The WH and RH values for three initial states

Trace	WH(%)			RH(%)		
Type	HDDP	SSDP	HbridP	HDDP	SSDP	HbridP
T9182	31.42	29.1	31.1	68.84	71.26	69.31
T5582	66.86	63.68	65.73	32.87	36.14	34.29
T1982	92.50	89.86	91.10	7.68	10.27	8.88
T9155	20.66	18.20	20.45	79.50	82.08	79.66
T5555	50.92	47.26	49.35	49.34	53.03	51.10
T1955	87.66	84.67	86.16	12.27	15.15	13.70

In Table 3, we can find out the number of pages partitioned into the magnetic disk and that into SSD is very similar to that of read and writes operations in the trace files, which shows our algorithm can effectively distinguish write-mostly pages from read-mostly ones. Table 4 shows that in the read-intensive trace file, namely T9182 and T9155, we get much higher RH values, while for T1982 and T1955 we get higher WH values. This indicates that our approach can effectively place the write-intensive pages on the magnetic disk and let the read-intensive pages on SSD.

In this paper, our goal is to use the magnetic disk as a write cache to improve the performance of SSD. So we compare the run time of our system with the SSD-only system. Besides, we record the time of the migration to compare the benefit and the additional cost brought by our system. The result is shown in Fig.5.

The line named "Flash" is the run time of different trace files under the SSD-only storage system. The line named "Hybrid" is the run time for the hybrid storage system and the line named "Migration" refers to the time cost for migrate operations. We calculate the runtime for each trace file and get the average value of the execution time. As shown in Fig.5, our system can reduce the total cost for SSD.

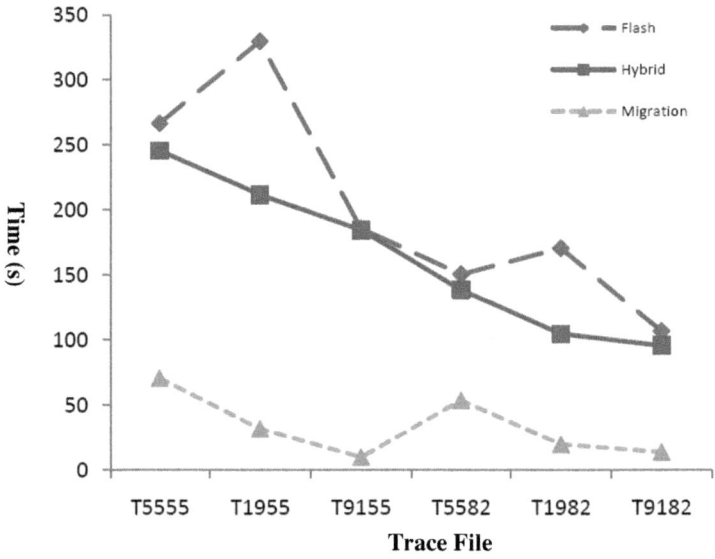

Fig. 5. Comparison of runtime between SSD-only and hybrid storage system

6   Conclusions and Future Work

SSD is regarded as an alternative of traditional magnetic disks. Previous work was usually based on such assumption that SSD will totally replace magnetic disks in a couple of years. However, as time passed, the current situation shows that SSD will not replace magnetic disks, due to some intrinsic limitations of SSD. Therefore, a practical way is to integrate SSD with magnetic disks and build a hybrid storage system. In this paper, we present a hybrid storage model and propose different ways to deal with read and write requests. In particular, we argue to use the magnetic disk as a write cache of SSD, in order to reduce the random write costs of SSD. We also develop a migration algorithm and introduce the new block/page-based migration unit to partition write-mostly pages into magnetic disk while read-mostly pages into SSD. The experimental results have shown that our policy can effectively partition pages according to their access pattern and further reduce the total I/O cost of SSD-based storage system.

In the future, we will investigate the buffer management policy to improve the migration mechanism and further reduce the I/O cost of the hybrid storage system. Besides, we shall consider pulling the energy saving impact into the hybrid storage model. For some high-grade SSD having better performance for random write than HDD and even better random-write than random read, we will research in the migration method by some other standard.

Acknowledgement

This paper is supported by the National Science Foundation of China (No. 60833005 and No. 61073039).

References

1. Kaiputer (2008), http://news.mydrivers.com/1/114/114319.htm
2. Chung, T.-S., et al.: System software for flash memory: A survey. In: Sha, E., Han, S.-K., Xu, C.-Z., Kim, M.-H., Yang, L.T., Xiao, B. (eds.) EUC 2006. LNCS, vol. 4096, pp. 394–404. Springer, Heidelberg (2006)
3. Lee, S.-W., Park, D.-J., Chung, T.-S., Lee, D.-H., Park, S., Song, H.-J.: A log buffer-based flash translation layer using fully-associative sector translation. ACM Trans. on Embedded Computing Systems 6(3) (July 2007)
4. Kim, J., Kim, J.M., Noh, S.H., Min, S.L., Cho, Y.: A space-efficient flash translation layer for Compact Flash systems. Trans. on Consumer Electronics 48(2), 366–375 (2002)
5. Lee, S.-W., Choi, W.-K., Park, D.-J.: FAST: An Efficient Flash Translation Layer for Flash Memory. In: EUC Workshops, pp. 879–887 (2006)
6. Koltsidas, I.: Flashing Up the Storage Layer. In: VLDB 2008. ACM, New York (2008)
7. Soundararajan, G., Prabhakaram, V.: Extending SSD Lifetimes with Disk-Based Write Caches. In: USENIX Conference on File and Storage Technologies. FAST (2010)
8. Neill, E.O.: The Page Replacement Algorithm For Database Disk Buffering. In: SIGMOD 2000, pp. 297–306. ACM, New York (1993)
9. Park, S.-Y., Jung, D., Kang, J.-U., Kim, J.-S., Lee, J.: CFLRU: A Replacement Algorithm for Flash Memory. In: CASES 2006, pp. 234–241. ACM, New York (2006)
10. Ou, Y., Harder, T., Jin, P.: CFDC: a flash-aware replacement policy for database buffer management. In: Science And Technology (DaMoN), pp. 15–20. ACM, New York (2009)
11. Iometer Project, iometer-[user—devel]@lists.sourceforge.net. Iometer Users Guide, http://www.iometer.org
12. Russinovich, M.: v2.01. DiskMon for Windows (2006), http://www.microsoft.com/technet/sysinternals/utilities/disk mon.mspx
13. Yoo, Y., Lee, II., Ryu, Y., et al.: Page Replacement Algorithms for NAND Flash Memory Storages. In: Gervasi, O., Gavrilova, M.L. (eds.) ICCSA 2007, Part I. LNCS, vol. 4705, pp. 201–212. Springer, Heidelberg (2007)

An Analysis of Network Structure and Post Content for Blog Post Recommendation

Wan-Shiou Yang and Yi-Rong Lin

Department of Information Management, National Changhua University of Education,
No. 1, Jin-De Rd., Changhua 50007, Taiwan
{wsyang,yrlin}@cc.ncue.edu.tw

Abstract. The acceleration of Weblogs has increased the perceived information overload for bloggers attempting to find interested or relevant information. Helping bloggers to efficiently locate relevant and high-quality information is imperative. In this research, we therefore propose four approaches that exploit the post citation network, blog-based social network, and post content to facilitate the automatic construction of an authoritative blog post recommender system. The proposed approaches were tested with blog data collected from Baidu Space, and the experimental results revealed that the proposed approaches outperform the content-only approach and the explicit citation approach.

Keywords: Weblog, Recommender System, Information Retrieval, Social Network.

1 Introduction

Over the past few years, the web has experienced an exponential growth in the use of weblogs or blogs. Weblog is a good paradigm of online social network which constitutes web-based regularly updated journals with reverse chronological sequences of dated entries, usually with blogrolls on the sidebars, allowing bloggers link to favorite sites which they are frequently visited [11]. Weblog offers a more open channel of communication, people in the blogosphere read, commentate, cite, socialize and even reach out beyond their social networks, make new connections, and form communities [11]. A blog-based social network thus emerges as a powerful and potentially services-valued form of communication.

There exists a large number of information in the blogosphere, including text-based blog entries (articles) and profile, pictures or figures and multimedia resources [3]. The acceleration of blogs thus has increased the perceived information overload for bloggers attempting to find interested or relevant information [7] [8] [11] [16]. Helping bloggers to efficiently locate relevant and high-quality information is thus imperative and a major focus of the information retrieval and database communities.

The information overload problem can be tackled with by recommender systems [9] [28]. The content-based approach is often used to build the key components of systems for recommending information on the Web [9] [10] [17] [22] [16] [25]. This approach typically employs information retrieval techniques to build personal interest profiles by analyzing the content of articles browsed by the user, and recommends articles whose content is highly consistent with the interest profile of a user.

J. Xu et al. (Eds.): DASFAA Workshops 2011, LNCS 6637, pp. 276–286, 2011.
© Springer-Verlag Berlin Heidelberg 2011

The content-based approach is simple. However, it has a fundamental limitation: Articles are essentially seen as pieces of text. As a result, although articles recommended by the content-based approach are highly consistent with the interest profile of a user, they are not necessarily authoritative. For example, a set of articles might be recommended based on articles about financial investment browsed by a user, but the first 10 hits might be for articles that most people would not consider authoritative.

This research aimed at remedying the above problem of the content-based approach and proposed approaches that exploit the post citation network, blog-based social network, and post content to facilitate the automatic construction of an authoritative blog post recommender system. Specifically, we combined information retrieval techniques with network analysis techniques to find relevant and authoritative blog posts for making recommendation. The proposed approaches were tested with blog data collected from Baidu Space, and the experimental results showed that the proposed approaches outperform the traditional content-only approach and the explicit citation approach.

This paper is organized as follows. Section 2 reviews related works, and Section 3 describes the proposed approaches. The results of evaluating the proposed approaches using empirical data are reported in Section 4, and Section 5 concludes with a summary and a discussion of future research directions.

2 Literature Review

Modern recommendation techniques have their roots in information filtering [9] [28] and aim to filter out information that is irrelevant and uninteresting to a given user. Several approaches have been commonly used for recommending information on the Web. Content-based recommender systems [9] [10] [16] [17] [22] [25] generally employ information retrieval techniques to build a user's interest profile by analyzing the content of items that the user has navigated recently. This approach recommends items that exhibit a high degree of similarity with the interest profile of the target user. Collaborative recommender systems [9] [15] [24] recommend items that similar users have liked. Usage mining has been proposed as an alternative approach for making recommendation [19] [20] [26], and employs data mining techniques to discover usage patterns by analyzing the navigational activities of users. Several recent researches [8] [16] [25] [27] investigated relationships in blog-based social networks and embedded authority and trust models in the proposed systems for recommending Weblogs or bloggers. In this research, we focus on the recommendation of blog posts and combine the analysis of post content and the analysis of network structures for making recommendation.

A number of ranking strategies have been developed to measure the authoritativeness of information on the Web. The two best-known algorithms are HITS [12] and PageRank [21]. The HITS algorithm deduces the hubs and authorities that exist in a subgraph of the Web consisting of both the results of a query and the local neighborhood of these results. The PageRank algorithm precomputes a rank vector that provides a-priori estimates of the importance of all pages on the Web. Several researches develop measures for ranking Weblogs or bloggers based on PageRank and HITS. Adar et al. [1] propose the iRank algorithm that acts on implicit link structure to find those Weblogs that initiate the epidemics, which denote similarity between nodes in content and out-links. Fujimura et al. [5] propose the EigenRumor algorithm to rank bloggers and blog posts at the same time by weighting the hub and authority scores of the bloggers based on eigenvector calculations. Kritikopoulos et al. [13] propose the BlogRank algorithm by using explicit and implicit hyperlinks that inferred from common weblog owners, affiliated authors or commentators, to rank weblogs. Cen et al. [4] analyze people's reading and commenting behaviors in blogspace and propose an algorithm for Weblog ranking. In this research, we infer implicit links in the post citation network and adopt the PageRank strategy to recommend relevant and high-quality blog posts.

The topic-sensitive Web search introduced by Haveliwala [6] is related to our method. The method precomputes topical PageRank vectors prior to query time, which are then combined at query time based on the similarity between topics and the query. This search method has been extended in various ways, such as by computing a topical PageRank vector to each Web page using a probabilistic model [23], by exploring how to personalize PageRank based on features readily available from a Web page and by integrating textual and linkage information based on the textual authority of a Web page [2]. In contrast to these methods, our approaches integrate the textual information from post content and the linkage information from both blog-based social network and post citation network to compute recommendation scores.

3 Proposed Approaches

The PageRank algorithm [21] models the behavior of a random Web surfer who at each time step is at a particular Web page and determines which page to visit next as follows: the probability of the surfer following the hyperlinks on that page is $1-\alpha$, and the probability of the surfer resetting by jumping to a Web page picked at random is α. If the surfer follows the hyperlink structure, then he or she will follow each outgoing link with equal probability. If the surfer resets, the next Web page is picked uniformly at random from all possible pages. Consider the Web as a directed graph, where nodes represent Web pages and edges between nodes represent hyperlinks between Web pages. Let W be the set of all available Web pages, $N=|W|$, F_i be the set of pages that page i links to, and B_i be the set of pages that link to page i. Then the probability $p(j)$ that a surfer is on page j is given by

$$p(j) = (1-\alpha) \cdot \sum_{i \in B_j} \frac{p(i)}{|F_i|} + \alpha \cdot \frac{1}{N} \tag{1}$$

In this research, we adopt the PageRank strategy to analyze the blog post citation network for recommending relevant and authoritative blog posts. A blog post citation network G_I can be formalized as a set of nodes V_I, each of which represents a post, and a set of directed edges $E_I \subseteq V_I^2$, each of which represents a citation by that post to another. The PageRank algorithm, however, can not be directly applied to the post citation network. First, as noted in researches [1] [4] [5] [13], the number of citations to a blog post is generally very small. The scores of blog posts calculated by directly applying PageRank algorithm are too small to permit blog posts to be ranked by importance. Second, the PageRank algorithm will provide an objective global estimate of the importance of each post. A global ranking of a post does not necessarily reflect the importance of that post to a given individual user. For recommending a post to a given individual user, the best one(s) should be those that are most relevant to the interests of the user or of similar users. Therefore, in this paper, we predict how to tackle the citation sparse problem and how to personalize PageRank based on the user's interests so that relevant and high-quality posts can be recommended.

For the citation sparse problem, we propose to infer implicit citation links in this research. Blogs are locations on the Web where bloggers express opinions or experiences by publishing blog posts. The readers are provided with the ability to read the ideas or opinions contained in the blog posts, and to submit their own comments in order to express their agreements or disagreements. Bloggers also frequently provide blogrolls, which is a list of other bloggers, allowing them to link to favorite sites. These reading, commenting, and listing interactions are highly informative [1] [5] and these interaction behaviors can serve as important indications for the purpose of studying the information propagation [4], and bloggers with frequent interactions are apt to exhibit similar information interests. Therefore, in this research, we infer implicit links between posts from the interaction behaviors of bloggers.

Specifically, a blogger interaction network G_2 can be formalized as a set of nodes V_2, each of which represents a blogger, and a set of directed edges $E_2 \subseteq V_2^2$, where E_2 represents interaction links, each of which represents a link by that blogger to another. The implicit links between blog posts are then inferred as follows. Let P_i be the set of all available posts of blogger i and P_j be the set of all available posts of blogger j. If there is an interaction link from blogger i to blogger j in the blogger interaction network G_2, for any pair of posts $p_i \in P_i$ and $p_j \in P_j$, a link $p_i \rightarrow p_j$ is inferred and added to the post citation network G_1 if the content similarity $Similarity(p_i, p_j)$, defined later, is larger than a threshold T.

Currently, the implicit links between posts are inferred using the following approaches:

(1) Listing approach: Listing links between bloggers are used to infer implicit links between posts.
(2) Commenting approach: Commenting links between bloggers are used to infer implicit links between posts.
(3) Reading approach: Reading links between bloggers are used to infer implicit links between posts.
(4) All approach: Listing, commenting, and reading links are all used to infer implicit links between posts.

Above interaction or citation data can be collected from the services provided in most blog sites, such as blogroll, comment, and citation, or automated mechanisms provided in some systems, such as trackbacks (http://www.sixapart.com/pronet/docs/trackback_spec) and reader recording [4].

For personalizing PageRank, we define a surfing pattern of a person guided by her interests as

$$p_a(j) = (1-\alpha) \cdot \sum_{i \in B_j} p_a(i) \cdot p_a(i \to j) + \alpha \cdot p_a^{'}(j) \tag{2}$$

where $p_a(i \to j)$ is the probability that the surfer transitions from post i to post j when she is visiting the post a, and $p_a^{'}(j)$ is the probability that post j is selected at random when performing a random jump given that she is visiting post a. Our method derives the two distributions into Equations 3 and 4 using content similarity:

$$p_a(i \to j) = \frac{Similarity(a, j)}{\sum_{k \in F_i} Similarity(a,k)} \tag{3}$$

$$p_a^{'}(j) = \frac{Similarity(a, j)}{\sum_{k \in W} Similarity(a,k)} \tag{4}$$

We apply a simple information retrieval method [14] to the blog posts. Information retrieval is the task of locating specific pieces of information from a text, which yields useful structured data from unstructured text. Here this specifically involves parsing the raw posts, removing punctuation and prepositions, and grouping stemming words into generalized terms. Our work adopts the bag-of-words model and creates a vector of terms for each post. We use the TF/IDF measure to determine the weight of each term. TF/IDF comprises two factors: (1) *tf* (term frequency), which measures how well that term describes the post contents; and (2) *idf* (inverse document frequency), whose reciprocal measures how often a term appears in the entire collection. A term that appears in many posts of the collection, i.e., with low *idf* value, is not very useful for distinguishing posts in the same collection. Thus, the TF/IDF term-weighting scheme is

$$w_{i,j} = f_{i,j} \times idf_i \tag{5}$$

where $f_{i,j} = \dfrac{freq_{i,j}}{\max_l freq_{l,j}}$ (*freq*$_{i,j}$ is the raw frequency of the ith term on the jth post)

and $idf_i = \log \dfrac{N}{n_i}$ (N is the total number of posts and n_i is the number of posts in which the ith term appears).

The conversion produces vectors that represent the blog posts. The content similarity $Similarity(i, j)$ of two posts i and j is then defined as the cosine of the angle between the vectors of i and j:

$$Similarity(i, j) = cos(i, j) \tag{6}$$

For a user's current interest, for example, given a post a, the top-N posts in the rank vector of a are recommended. For a user's long-term interests, the averaged rank vector of all posts of the user is computed and the top-N posts are recommended.

4 Evaluation

We evaluated the proposed approaches using the blog data collected from Baidu Space (http://hi.baidu.com). The Baidu Space is a Chinese weblog publish platform with millions of users registered now where users can write the blog, upload photos to album, and interact with others by related services. A feature of Baidu Space is that posts in Baidu Space record IDs of newest 8 registered readers. This enables us to trace the reading records and add reading links to the blogger interaction network[1]. We first collected the training data from March 1, 2009 to March 31, 2009. We collected posts which were published in this period, and identified the authors of these posts. Once the bloggers were decided, we inferred the interaction links between them. The listing links were inferred from the blogrolls on the side bar in the blog site of each blogger. The commenting and reading links were inferred by the records of commenting and reading behaviors respectively. We then collected the reading records of each blogger in the training dataset from April 1, 2009 to April 20, 2009. Each blogger thus had a sequence of reading records, and these sequences were used as testing data. Detailed statistics of our training and testing datasets are listed in Table 1.

Table 1. Statistics of the training and testing datasets

	# of posts	1,031,112
	# of citation links	30,933
	# of bloggers	67,778
Training dataset	# of listing links	1,050,226
	# of commenting links	115,023
	# of reading links	1,030,251
Testing dataset	# of sequences	16,945

Also, since the posts were in Chinese and a Chinese sentence contains no delimiters, such as a space, to separate words. We used a well-known Chinese word segmentation technique [18] to segment Chinese words. The TF/IDF measure described in Section 3 was then used to determine the weight of each word. Each post was then represented as a vector of the top-200 Chinese words.

We measured the performances of the proposed approaches by comparing the recommended posts and the reading posts of each blogger in the testing dataset. The precision and recall scheme was adopted as the performance metric for measuring the quality of recommendations, where precision measures the ratio of the number of recommended posts accessed by a blogger to the total number of recommended posts, and recall measures the ratio of the number of recommended posts accessed by a blogger to the total number of posts accessed by the user. The precision (recall) of a recommendation approach is the average precision (recall) of all bloggers in the testing dataset.

---

[1] We may miss some records if a post has more than 8 readers.

We first evaluated the effect of threshold *T*. Figure 1(a) and 1(b) show the precisions and recalls for the number of thresholds (ranging from 0.1 to 0.5 in increments of 0.1) for the top-10 recommendations. In this experiment, α was set to 0.15, as suggested in the original PageRank algorithm. The results of this experiment indicate that the precisions and recalls of the proposed approaches initially increase with the threshold, reaching the maximum at the threshold of 0.3 or 0.4, and then gradually decrease as the threshold increases. This is consistent with our expectation that a smaller threshold (≤ 0.3) implies a greater diversity of post content and a larger threshold (≥ 0.4) results in a less amount of inferred implicit links, and hence affect the performances of the proposed approaches.

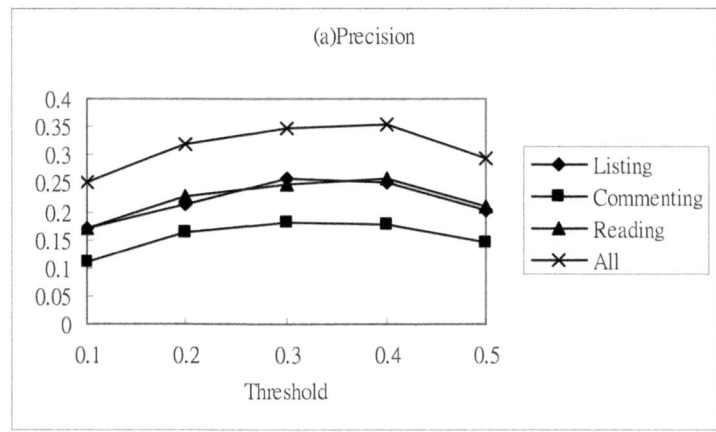

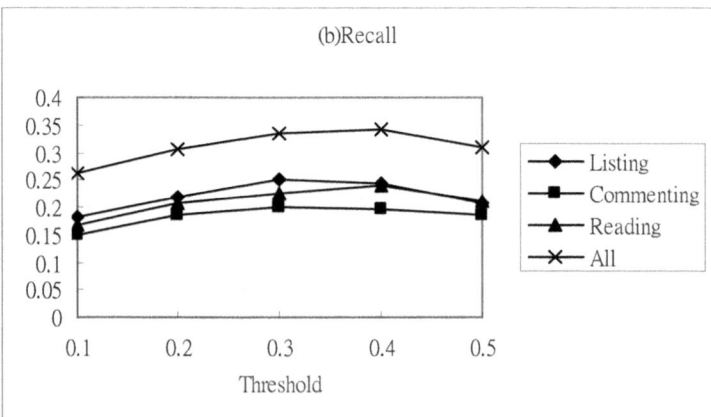

Fig. 1. Precisions (a) and recalls (b) of the recommendation approaches for different thresholds

We then compared the performances of the proposed approaches with the content-only approach and the explicit citation approach. For the content-only approach, we applied the same information retrieval method to the posts. The top-10 posts were then recommended for a given user. For the explicit citation approach, we didn't add implicit links to the post citation network and directly applied the same personalized PageRank approach to the network. In this experiment, α was set to 0.15 and T was set to 0.35. Figure 2(a) and 2(b) show the obtained precisions and recalls respectively.

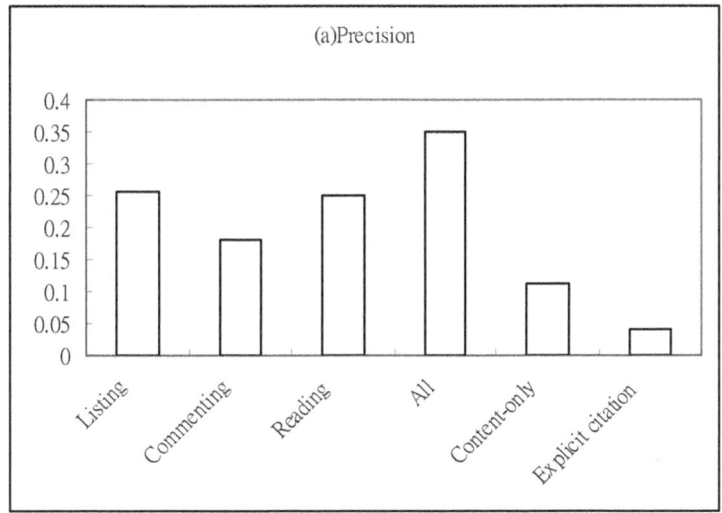

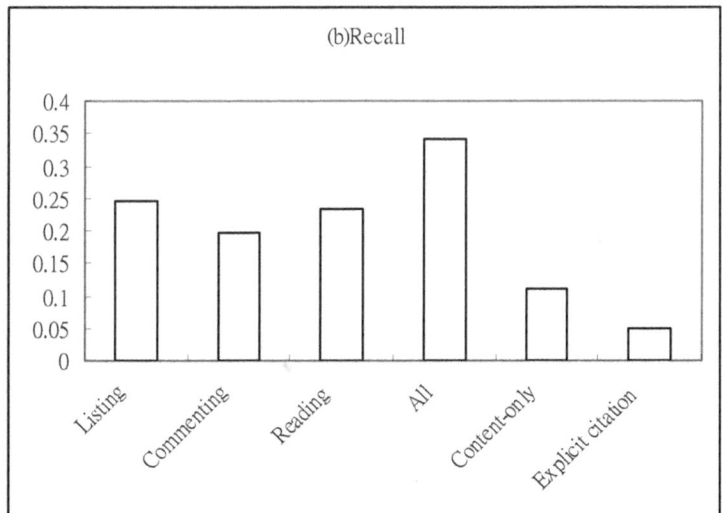

Fig. 2. Comparison of the precisions (a) and recalls (b) of the proposed approaches with the content-only approach and the explicit citation approach

In Figure 2, it can be seen that the All approach has the highest precision and recall values, followed by Listing, Reading, and Commenting in order. The result indicates that the number of inferred implicit links, as listed in Table 2, has significant effect on the performances of the proposed approaches. Also, the precision and recall values of all proposed approaches were higher than the content-only approach. This demonstrates the effectiveness of recommending posts using the post citation network, which is due to the network structure reflecting the authoritativeness of posts. Finally, the proposed approaches outperform the explicit citation approach. It is expected since only 3% (30,933/1,031,112) posts in our training dataset have explicit citations and hence directly analyzing the structure of the post citation network is not useful.

Table 2. The number of implicit links inferred by proposed approaches

	# of inferred implicit links
All approach	12,373,344
Reading approach	7,424,006
Listing approach	8,661,341
Commenting approach	1,856,002

5 Conclusions

This study developed four approaches for recommending blog posts that make use of the post citation network, blog-based social network, and post content. An evaluation of the proposed approaches using the blogging information collected from Baidu Space revealed that the proposed approaches outperform the content-only approach and the explicit citation approach.

This work could be extended in several directions. The present study focused on three kinds of interaction links: listing, commenting, and reading. It would be interesting to extend our work to study more interaction behaviors. Also, the scalability of our approaches could be improved by using data structures such as those proposed in the original PageRank algorithm [21] to reduce the work of calculating rank vectors. Finally, applying our approaches to real applications may further reveal the effectiveness of the proposed approaches.

References

1. Adar, E., Zhang, L., Adamic, L., Lukose, R.: Implicit Structure and the Dynamics of Blogspace. In: Workshop on the Weblogging Ecosystem (2004)
2. Aktas, M., Nacar, M., Menczer, F.: Personalizing PageRank Based on Domain Profiles. In: SIGKDD Workshop on Web Mining and Web Usage Analysis (2004)
3. Cayzer, S.: Semantic Blogging and Decentralized Knowledge Management. Communication of the ACM 47(12), 47–52 (2004)
4. Cen, S., Han, L., Ma, J.: Ranking Weblogs by Analyzing Reading and Commenting Activities. In: International Conference on Web Intelligence and Intelligent Agent Technology (2009)
5. Fujimura, K., Inoue, T., Sugisaki, M.: The EigenRumor Algorithm for Ranking Blogs. In: Workshop on Weblogging Ecosystem (2005)

6. Haveliwala, T.: Topic-sensitive PageRank. In: International Conference on World Wide Web (2002)
7. Hayes, C., Avesani, P., Veeramachaneni, S.: An Analysis of Bloggers and Topics for a Blog Recommender System. In: Workshop on Web Mining (2007)
8. Hsu, W.H., King, A., Paradesi, M., Pydimarri, T., Weninger, T.: Collaborative and Structural Recommendation of Friends Using Weblog-based Social Network Analysis. In: AAAI Spring Symposium on Computational Approaches to Analyzing Weblogs (2006)
9. Huang, Z., Chung, W., Chen, H.: A Graph Model for E-Commerce Recommender Systems. Journal of the American Society for Information Science and Technology 55(3), 259–274 (2004)
10. Joachims, T., Freitag, D., Mitchell, T.: WebWatcher: A Tour Guide for the World Wide Web. In: International Joint conference on Artificial Intelligence (1997)
11. Karger, D.R., Quan, D.: What Would It Mean to Blog on the Semantic Web. Journal of Web Semantics 3(2), 147–157 (2005)
12. Kleinberg, J.: Authoritative Sources in a Hyperlinked Environment. Journal of the ACM 46(5), 604–632 (1999)
13. Kritikopoulos, A., Sideri, M., Varlamis, I.: BlogRank: Ranking Weblogs based on Connectivity and Similarity Features. In: Workshop on Advanced Architectures and Algorithms for Internet Delivery and Applications (2006)
14. Kushmerick, N., Weld, D., Doorenbos, B.: Wrapper Induction for Information Extraction. In: International Joint Conference on Artificial Intelligence (1997)
15. Lashkari, Y.: The WebHound Personalized Document Filtering System (1995), http://rg.media.mit.edu/projects/webhound/
16. Li, Y.-M., Chen, C.-W.: A Synthetical Approach for Blog Recommendation Mechanism: Trust, Social Relation, and Semantic Analysis. In: International Conference on Electronic Business (2007)
17. Lieberman, H.: Letizia: An Agent That Assists Web Browsing. In: International Joint Conference on Artificial Intelligence (1995)
18. Ma, W.-Y., Chen, K.-J.: Introduction to CKIP Chinese Word Segmentation System for the First International Chinese Word Segmentation Bakeoff. In: SIGHAN Workshop on Chinese Language Processing (2003)
19. Mobasher, B., Dai, H., Nakagawa, M., Luo, T.: Discovery and Evaluation of Aggregate Usage Profiles for Web Personalization. Data Mining and Knowledge Discovery 6(1), 61–82 (2002)
20. Nasraoui, O., Petenes, C.: An Intelligent Web Recommendation Engine Based on Fuzzy Approximate Reasoning. In: IEEE International Conference on Fuzzy Systems (2003)
21. Page, L., Brin, S., Motwani, R., Winograd, T.: The PageRank Citation Ranking: Bringing Order to the Web. In: International World Wide Web Conference (1998)
22. Pazzani, M., Billsus, D.: Learning and Revising User Profiles: The Identification of Interesting Web Sites. Machine Learning 27(4), 313–331 (1997)
23. Richardson, M., Domingos, P.: The Intelligent Surfer: Probabilistic Combination of Link and content Information in PageRank. In: Advances in Neural Information Processing Systems, vol. 14, pp. 1441–1448 (2002)
24. Terveen, L., Hill, W., Amento, B., McDonald, D., Creter, J.: PHOAKS: A System for Sharing Recommendations. Communications of the ACM 40(3), 59–62 (1997)
25. Tsai, T.-M., Shih, C.-C., Chou, S.-C.: Personalized Blog Recommendation: Using the Value, Semantic, and Social Model. In: Innovations in Information Technology (2006)

26. Xu, G., Zhang, Y., Zhou, X.: A Latent Usage Approach for clustering Web Transaction and Building User Profile. In: International Conference on Advanced Data Mining and Applications (2005)
27. Wang, J., Li, Q., Chen, Y., Lin, X.: Recommendation in Internet Forums and Blogs. In: Annual Meeting of the Association for Computational Linguistics (2010)
28. Wei, C., Shaw, M.J., Easley, R.F.: A Survey of Recommendation Systems in Electronic Commerce. In: Rust, R.T., Kannan, P.K. (eds.) E-Service: New Directions in Theory and Practice. ME Sharpe Publisher (2002)

Extracting Local Community Structure from Local Cores*

Xianchao Zhang, Liang Wang, Yueting Li, and Wenxin Liang**

School of Software, Dalian University of Technology,
Economy and Technology Development Zone, Dalian 116620, China
xczhang@dlut.edu.cn, L.Wang917@gmail.com, liyueting@mail.dlut.edu.cn,
wxliang@dlut.edu.cn

Abstract. To identify global community structure in networks is a great challenge that requires complete information of graphs, which is not feasible for some large networks, e.g. the World Wide Web. Recently, local algorithms have been proposed to extract communities in nearly linear time, which just require a small part of the graphs. However, their results, largely depending on the starting vertex, are not stable. In this paper, we propose a local modularity method for extracting local communities from local cores instead of random vertices. This approach firstly extracts a large enough local core with a heuristic strategy. Then, it detects the corresponding local community by optimizing local modularity, and finally removes outliers based on introversion. Experiment results indicate that, compared with previous algorithms, our method can extract stable meaningful communities with higher quality.

Keywords: Community structure; Local modularity; Local core.

1 Introduction

Community structure is one of the most relevant features of complex networks, such as social networks [1], the Internet [2], the World Wide Web [3], citation networks [4], biological networks [5]. In order to analyse the structures of complex networks, we often refer to graph theory, organizing a system as a graph with n vertices (entities) and m edges (relationships between vertices). With the growth of interest in the study of networks, especially the Web and online social networks, great efforts have been made on the area of community detection [6–9].

In general, a *community* is a tightly-knit sub-graph in a network, in which the within-group links are stronger or denser than between-group links [1]. Directly optimizing the density inside or the sparseness outside, a number of algorithms have been proposed [10–13], which work well in special cases. Similarly, several

---

* This work was partially supported by NSFC under grant No. 60873180, 61070016, SRF for ROCS, State Education Ministry, and by the Fundamental Research Funds (DUT10JR02) for the Central Universities, China.
** Corresponding author.

J. Xu et al. (Eds.): DASFAA Workshops 2011, LNCS 6637, pp. 287–298, 2011.

approaches came forth based on the concept of flow [14–16]. Recently, researchers have been focusing on algorithms based on the *modularity* metric [17, 6, 18].

However, these techniques require information of the whole graphs, which might be problematic for large graphs like the World Wide Web, as the constraint of running time and memory consumption. To overcome the limitation, local community detecting methods emerged, which just demand a small subset of vertices instead of the whole graphs, such as algorithms [19–23, 8]. Unfortunately, for largely depending on the starting vertex, the results of these technologies are instable. Besides, most of them require several parameters that hard to obtain. Furthermore, there might be the confusion of outliers.

To address these issues, in this paper, we propose an effective and efficient three-phase algorithm to extract local community structure. Firstly, we automatically generate a local core, the densest sub-graph within a community. Then, by maximizing local modularity, all potential members are merged into the community. Finally, a pruning phase is used, during which outliers are removed from the community. Compared with previous methods, our approach attains more stable results by starting from a local core instead of a starting vertex. Besides, our algorithm can stop automatically with the maximum local modularity. Moreover, outliers are removed in the pruning phase, keeping the density of extracted community in a high level.

The rest of the paper is organized as follows. In Section 2, the problem of local community structure detection will be defined and some previous algorithms will be reviewed. Then we will describe our approach and report experiment results in Section 3 and Section 4, respectively, followed by conclusions in Section 5.

2   Preliminaries

Here we firstly define the problem of extracting local community structure in a network, and then review some previous techniques for identifying local communities.

2.1   Local Community

A network[1] is usually represented by a graph $G = (V_G, E_G)$, where V_G is the set of vertices with size $n = |V_G|$, and E_G contains edges of G whose size is $m = |E_G|$. According to previous research, a *local community* is a densely-connected subgraph that is extracted with local information.

Given a graph G, in order to identify a local community C, we should at least have complete knowledge about the connectivity of vertices in the community, i.e. links in C, links outgoing of C, namely, its neighbor set $N = \{v | (u, v) \in E_C, u \in V_C \wedge v \notin V_C\}$. Vertices like u constitute boundary set B. As shown in Fig. 1, the portion of graph G have three parts: local community C, boundary set B and neighbor set N.

---

[1] Here we mainly focus on undirected, unweighted graphs.

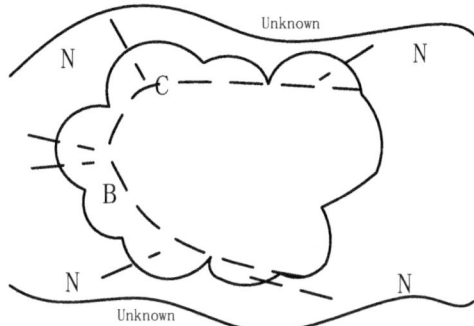

Fig. 1. Portion graph G for identifying local community C

According to the definition, the purpose of local community detection is to find dense vertices set of C with a small number of edges between B and N, namely, a community dense inside (high internal degree) and sparse outside (low external degree).

2.2 Previous Algorithms

Typically, local community detecting techniques randomly start from a vertex v, and gradually merge neighboring vertices one-at-a-time by optimizing a measure metric. Local community detecting algorithms are also applied to obtain global community structure of entire graph by iterative running.

The *modularity* metric has been widely used in global community detection, and is proved to be a very effective metric. It was firstly applied to find local communities by Clauset [19], who proposed a local modularity metric R to measure the "sharpness" of the boundary. However, R method lacks a good stopping criteria, based on predefined parameter k. Similarly, Luo et al. [20] later introduced another modularity metric M. Instead of focusing on the "sharpness" of the boundary, M method measures the radio of internal and external edges. M method would drop community structure when the initial vertex is removed by its pruning strategy.

Approaches based on modularity, R and M, are able to extract more meaningful communities, however, their results usually include many outliers, which would depress the overall quality of community. In view of this issue, Chen et al. [23] put forward a two-phase algorithm based on a new metric L who firstly extracts all possible candidates, and then removes outliers. However, it is too strict to detect integrated community.

Bagrow [21] brought a "outwardness" metric Ω to measure local community structure. Ω method also lacks an appropriate stopping criteria. Andersen [12, 8] advanced a method to find local communities based on bipartite density, which limits to bipartite networks.

All the referred techniques above infer community structure from a random initial vertex, whose consequential community is largely affected by the starting vertex. Therefore, unstable results have always been an intractable issue of these methods.

3   Our Contribution

To tackle the previous drawbacks, we address a three-phase algorithm that takes time polynomial in k (the size of extracted community):

1. Instead of a random starting vertex, we firstly automatically extract a local core as the initial state for local community detection.
2. Starting from the local core, we infer the corresponding community by maximizing local modularity metric, using vertex-at-a-time discovery process. The iterative merging stops when there's no vertex that increases local modularity.
3. This step does pruning to remove outliers.

3.1   Extracting Local Core

A *local core* is a dense sub-graph within a community with high *local density*, generated by a vertex and its neighbors. The core, centric of certain communities, lying in the densest region of the community, is capable of locating the position of the community by rule and line. Instead of starting from a randomly vertex, our approach is able to detect much more stable and meaningful community with initial reliable core.

Given $N(v)$ as the neighbor set of vertex v, local density d_v measures the ratio of the number of edges in $N(v)$ and the number of edges if $N(v)$ is a complete graph:

$$d_v = \frac{2 * Edges(N(v))}{|N(v)|(|N(v)| - 1)} \tag{1}$$

where $Edges(N(v))$ denotes the number of edges that connect vertices in $N(v)$. d_v is in the range of $[0, 1]$.

If vertex v has a high value of d_v, it is very likely to generate a very dense sub-graph with its neighbors, e.g., when $d_v = 1$, the sub-graph is a complete graph. So, if we start finding community from a local core, we have a great initial state of dense and stable structure, which makes it more likely to find perfect community structure.

In order to keep local core with higher local density, we should remove neighbors that have little links to it. Presenting the local core of vertex v as $\mathcal{C}_v$, *vertex density* $d(u, \mathcal{C}_v)$ $(u \in N(v))$ can be defined as:

$$d(u, \mathcal{C}_v) = \frac{Edges(u, N(v))}{|N(v)|} \tag{2}$$

where $Edges(u, N(v))$ is the number of edges that connect u and vertices in $N(v)$.

Neighbors will be included into $\mathcal{C}_v$ if they satisfy:

$$d(u, \mathcal{C}_v) \geq \varepsilon * d_v \tag{3}$$

where ε is a *balance factor* to decide the tightness of $\mathcal{C}_v$. Factor ε can balance local density and the size of $\mathcal{C}_v$: increasing ε would increase local density and decrease community size, vice verse. Note that, local core $\mathcal{C}_v$ should have at least 3 vertices[2].

To extract a local core, we adopt a heuristic approach that checks all vertices in sequence, descending by d_v and edges of v, and finds the vertices set that satisfy Eq. 3. The pseudocode for extracting local core is given in Algorithm 1.

Algorithm 1. Extracting local core

Input: G: the partial graph
Output: $\mathcal{C}_v$: local core generated by vertex v and its neighbors
1: **for** each vertex v in G **do**
2:    Calculate d_v
3: **end for**
4: Sort all the vertices descending by d_v and edges of v. The sorted graph is G'.
5: **for** each $v \in G'$ **do**
6:    Set $\mathcal{C}_v$ to empty.
7:    **for** each $u \in N(v)$ **do**
8:       **if** u satisfies Eq. 3 **then**
9:          Add u to $\mathcal{C}_v$
10:       **end if**
11:    **end for**
12:    **if** $|\mathcal{C}_v| > 2$ **then**
13:       **return** local core $\mathcal{C}_v$
14:    **end if**
15: **end for**

3.2   Merging Vertices

After extracting a local core $\mathcal{C}_v$, we iteratively merge vertices into C by using vertex-at-a-time discovery process subject to maximize local modularity metric R [19]:

$$R = \frac{Edges(B, C)}{Edges(B, C) + Edges(B, N)} \quad (4)$$

The metric R well optimizes the sparseness of the boundary, meanwhile, keeps dense inside the community.

Initially, we place all vertices in $\mathcal{C}_v$ into C, boundary vertices into B and neighboring vertices into N. Then, during each loop, we merge a neighboring vertex u in N into C that results the largest positive change of R. That is, for each $u \in N$, we calculate local modularity R' if we merge u. For the maximum R'_m, if $R'_m > R$, we merge the corresponding u into C and update B and N. This merging process continues until $R'_m <= R$ or N is empty (See Algorithm 2).

---

[2] If there's only two vertices, there's just one edge at most, which is far from dense.

Algorithm 2. Merging vertices

Input: C_v: local core of Step 1 generate by v and its neighbors
        G: the partial graph

Output: C: local community after merging phase

1: Set C, B, N to empty
2: **for** each $u \in C_v$ **do**
3:     Add u to C
4: **end for**
5: Update B and N
6: **while** N is not empty **do**
7:     **for** each $u \in N$ **do**
8:         Calculate R' if u is merged into C
9:     **end for**
10:    Get the maximum value R'_m and current modularity R
11:    **if** $R'_m > R$ **then**
12:       Merge the corresponding vertex of R' to C
13:       Update R, B and N
14:    **else**
15:       Stop merging
16:    **end if**
17: **end while**
18: **return** local community C

Algorithm 3. Pruning phase

Input: C: local community of Step 2
        G: the partial graph

Output: C': the final local community after three phases

1: **for** each $u \in B$ **do**
2:     Calculate $\rho(u, C)$
3:     **if** $\rho(u, C) \leq C$ **then**
4:       Remove u from C
5:       Update B and N
6:     **end if**
7: **end for**
8: **for** each $u \in C \setminus B$ **do**
9:     **if** $deg(u) <= 1$ **then**
10:      Remove u from C
11:    **end if**
12: **end for**
13: **return** final local community C'

3.3 Pruning Phase

With the procedure of Step 2, community structure C is extracted, mixing with outliers. However, in this step, we present a pruning phase to remove outliers weakly connected to C.

In order to remove vertices weakly connected to C, *vertex introversion* $\rho(v, C)$ presented in [26] is utilized, measuring the ratio of internal edges to all edges of v:

$$\rho(v, C) = \frac{Edges(v, C)}{deg(v)} \tag{5}$$

where $deg(v)$ is the degree of v. To optimize internal density, we check all vertices in B and remove those with $\rho(v, C) <= 0.5$. Meanwhile, outliers with only one edge can be removed by checking all inner vertices (vertices in $C \setminus B$). Details of this process are shown in Algorithm 3.

4   Experiment Results

Here we adopt three criteria to evaluate the performance of our algorithm by applying it to three different networks with the ground truth: Zachary's karate club network [27], GN network [11], and NCAA football network[3] [28]. For all the experiments, the default value for balance factor ε is 1.

As the community structure of these three networks are known, we can objectively measure the effectiveness of our algorithm by *precision, recall* and *F-measure*. Given local core C_v, we extract the local community C_v^{alg} and the real community that v belongs to is denoted as C_v^{real}. Then three of these metrics are defined as:

$$Precision_v = \frac{|C_v^{real} \cap C_v^{alg}|}{|C_v^{alg}|} \tag{6}$$

$$Recall_v = \frac{|C_v^{real} \cap C_v^{alg}|}{|C_v^{real}|} \tag{7}$$

$$F - measure_v = \frac{2 \times Precision_v \times Recall_v}{Precision_v + Recall_v} \tag{8}$$

To better illustrate final results, we compare our algorithm with two previous modularity algorithms: R method of Clauset [19], M method of Luo et al. [20], and an algorithm dealing with outliers, L method of Chen et al. [23].

4.1   Zachary's Karate Club Network

Zachary's karate club network [27] is one of the classic studies in social network analysis and has been a benchmark graph for community detection. It shows social interactions between members of a karate club at an American university. The club has 34 members and 2 groups (communities). Fig. 2 illustrates the network and its consensus network structure grouped from Zachary's observations.

Feeding this network into our algorithm and the other three previous algorithms, we extract all the two communities and show average results in Table 1. Our algorithm run once for a local community, while the results of the other three algorithms are the average of multi-run from each vertex as starting vertex. As shown in Table 1, our algorithm gets the highest f measure and a perfect

---

[3] The ground truth of communities (conferences) can be found at http://espn.go.com/college-football/standings/_/year/2006.

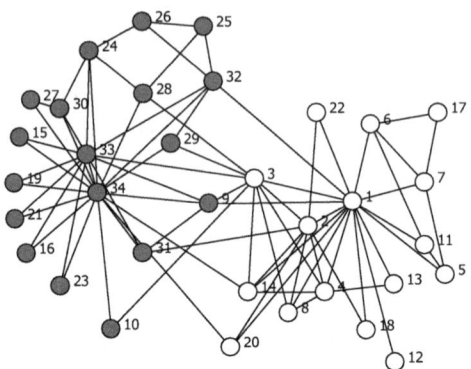

Fig. 2. The network of Zachary's karate club. Vertices of two communities are shown in circle of white and blue, respectively.

precision 1, which indicates that our algorithm identifies a better community structure. R method also gets a high precision, but a low recall, for its stopping criteria based on predefined community size, which is hard to decide for real communities. Both precision and recall of M method are lower, causing by its pruning strategy. The result of L method is even poorer because it's too strict to a community view (reviewed in Section 2.2).

Table 1. Our algorithm's performance on Zachary's karate club network compared with R, M, and L methods

Zachary's karate club network	Algorithms			
	R method	M method	L method	Our method
Precision	0.9715	0.8929	0.9405	**1**
Recall	0.5740	0.6250	0.1896	**0.7049**
F-measure	0.7216	0.7353	0.3156	**0.8269**

4.2   GN Networks

Synthetic networks have been widely used to test community detecting algorithms. We apply our algorithm to a set of computer-generated random graphs with ground truth, GN network [11]. Typically, there're 128 vertices in these graphs, organized into four equal-sized communities. Each of them has a total expected degree $d = 16$, both intra- and inter-community edges included ($d = d_{in} + d_{out}$). By holding the expected degree constant $d = 16$, we change d_{out} to get different graphs with various sharpness of boundaries. Community structure of these graphs becomes vaguer and vaguer as d_{out} increases, here data series for $d_{out} < 8$ are used in our experiments.

   As shown in Fig. 3, we generate 500 networks for each d_{out} and present our average results running once for each network. Selecting each of the 128 vertices as the starting vertex, we get the average precision, recall and f-measure for the

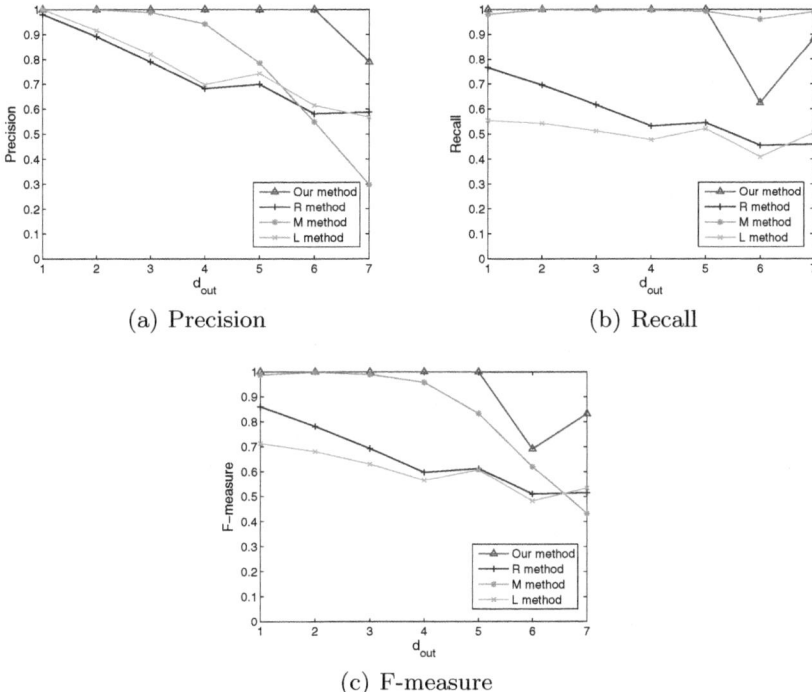

(a) Precision   (b) Recall

(c) F-measure

Fig. 3. (*Color online*)Our algorithm's performance on GN networks compared with R, M, and L methods

other three algorithms for all 500 networks. As shown in Fig. 3(a), our precision is much better than previous algorithms. Though the recall of M method is a little higher when $d_{out} >= 6$ (shown in Fig. 3(b)), our algorithm results the perfect f-measure. More important, our algorithm identify nearly exact real communities for all the networks of $d_{out} <= 5$ that are more approximate to community structure of real networks, which is an outstanding achievement.

4.3   The NCAA Football Network

The first two experiments indicate that our algorithm with local core runs well in both synthetic and real networks, however, there is no notion of outliers. We will apply our algorithm to another network that contains outliers. The network is the schedule for 787 games of the 2006 National Collegiate Athletic Association (NCAA) Football Bowl Subdivision (also known as Division 1-A) [28].

In the NCAA network, there're 115 universities divided into 11 conferences. In addition, there're four independent schools, Notre Dame, Navy, Army and Temple, as well as 61 schools from lower divisions that belong to none of conferences. Each school in a conference plays more games with those in the same conference, so a conference is actually a community. We present the network with a graph containing 180 vertices and 787 edges (115 vertices in 11 communities and 65 outliers).

Table 2. Our algorithm's performance on NCAA football network compared with R, M, and L methods

2006 NCAA football network	Algorithms			
	R method	M method	L method	Our method
Precision	0.6562	0.5735	0.8031	**0.9847**
Recall	0.6396	1	0.6257	**1**
F-measure	0.6478	0.7289	0.7033	**0.9923**

To extract all communities in the NCAA network, we iteratively run our local core algorithm until there's no any more communities. Our algorithm just need run 11 times for all the 11 communities. For the other three algorithms, every vertex has been taken as the starting vertex to extracting corresponding community. Table 2 shows the average precision, recall and f-measure for the four algorithms. Firstly, the disadvantages of metric R and M as we analyse in Section 2.2 have been confirmed by the results. Because they both ignore the confusion of outliers, R and M methods gets a higher recall but a much lower precision, which leads to an unsatisfactory f-measure. L method successfully deals with outliers, but it's too strict to community density that it results lower recall and can not get a good f-measure, either. Obviously, our algorithm with local core gets an excellent results, the highest precision 0.9847 and perfect recall 1, which attains a 0.992 f-measure on average. It proves that our local core algorithm not only perfectly balances the community density and size, but also deals with outliers successfully.

5   Conclusions

A number of local algorithms have been advanced to identify a local community at a time, which cost less on storage and running time. However, previous methods start from a random vertex and their results vary significantly with the choice of starting vertex. What's more, wrongly recognized outliers would decrease community density to some extent. In this paper, we propose a three-phase local modularity approach, extracting local community by starting from a local core, followed by a pruning phase. Our method can automatically stop and do not need the predefined threshold. Experiment results show that our method can extract stable and meaningful community structure with high quality.

As future work, we will focus on the improvement of local core extracting method, as shown in Algorithm 1. Here we use the fixed balance factor ε, which actually should be various for each core. In addition, more effective strategy for outliers detection should be taken into consideration.

Acknowledgements

We'd like to thank Xiaowei Xu for providing the NCAA data and Newman for the Zachary's karate club network.

References

1. Wasserman, S., Faust, K.: Social Network Analysis: Methods and Applications. Cambridge University Press, Cambridge (1994)
2. Faloutsos, M., Faloutsos, P., Faloutsos, C.: On Power-Law Relationships of the Internet Topology. In: Proceedings of Annual Conference of the Special Interest Group on Data Communication (SIGCOMM 1999), pp. 251–262. ACM, New York (1999)
3. Albert, R., Jeong, H., Barabsi, A.L.: Diameter of the World-Wide Web. Nature (401), 130–131 (1999)
4. Hajra, K.B., Sen, P.: Aging in Citation Networks. Physica A, 44–48 (2005)
5. Dorogovtsev, S.N., Mendes, J.F.F.: Evolution of Networks: From Biological Nets to the Internet and WWW. Oxford University Press, New York (2003)
6. Newman, M.E.J.: Modularity and Community Structure in Networks. Proceedings of the National Academy of Sciences 103, 8577–8582 (2006)
7. Dourisboure, Y., Geraci, F., Pellegrini, M.: Extraction and Classification of Dense Communities in the Web. In: Proceedings of the 16th International Conference on World Wide Web, pp. 461–470. ACM, New York (2007)
8. Andersen, R.: A Local Algorithm for Finding Dense Subgraphs. ACM Transactions on Algorithms (TALG) 6, 1–12 (2010)
9. Zhang, X., Li, Y., Liang, W.: C&C: An effective algorithm for extracting web community cores. In: Yoshikawa, M., Meng, X., Yumoto, T., Ma, Q., Sun, L., Watanabe, C. (eds.) DASFAA 2010. LNCS, vol. 6193, pp. 316–326. Springer, Heidelberg (2010)
10. Gibson, D., Kleinberg, J., Raghavan, P.: Inferring Web Communities from Link Topology. In: Proceedings of the 9th ACM Conference on Hypertext and Hypermedia: Links, Objects, Time and Space, pp. 225–234. ACM, New York (1998)
11. Newman, M.E.J., Girvan, M.: Finding and Evaluating Community Structure in Networks. Physical Review E 69, 26113 (2004)
12. Andersen, R., Lang, K.J.: An Algorithm for Improving Graph Partitions. In: Proceedings of the 19th Annual ACM-SIAM Symposium on Discrete Algorithms, pp. 651–660 (2008)
13. Sharan, A., Gupta, S.L.: Identification of Web Communities through Link Based Approaches. In: International Conference on Information Management and Engineering, pp. 703–708 (2009)
14. Flake, G.W., Lawrence, S., Giles, C.L.: Efficient Identification of Web Communities. In: Proceedings of the 6th ACM SIGKDD international conference on Knowledge discovery and data mining, pp. 150–160 (2000)
15. Flake, G.W., Lawrence, S., Giles, C.L., Coetzee, F.M.: Self- organization and Identification of Web Communities. Computer 35, 66–70 (2002)
16. Khandekar, R., Rao, S., Vazirani, U.: Graph Partitioning using Commodity Flows. Journal of the ACM (JACM) 56, 1–15 (2009)
17. Newman, M.E.J.: Fast Algorithm for Detecting Community Structure in Networks. Physical Review E 69, 26133 (2004)
18. Blondel, V.D., Guillaume, J.L., Lambiotte, R., Lefebvre, E.: Fast Unfolding of Communities in Large Networks. Journal of Statistical Mechanics: Theory and Experiment (2008)
19. Clauset, A.: Finding Local Community Structure in Networks. Physical Review E 72(2), 26132 (2005)

20. Luo, F., Wang, J.Z., Promislow, E.: Exploring Local Community Structures in Large Networks. In: Proceedings of the 2006 IEEE/WIC/ACM International Conference on Web Intelligence, pp. 233–239 (2006)
21. Bagrow, J.P.: Evaluating Local Community Methods in Networks. Journal of Statistical Mechanics: Theory and Experiment 2008, P5001 (2008)
22. Andersen, R.: A Local Algorithm for Finding Dense Subgraphs. In: Proceedings of the 19th Annual ACM-SIAM Symposium on Discrete Algorithms, pp. 1003–1009 (2008)
23. Chen, J., Zaizne, O., Goebel, R.: Local Community Identification in Social Networks. In: International Conference on Advances in Social Network Analysis and Mining, pp. 237–242. IEEE, Los Alamitos (2009)
24. Muff, S., Rao, F., Caflisch, A.: Local Modularity Measure for Network Clusterizations. Physical Review E 72(5), 56107 (2005)
25. Hinne, M.: Local Identification of Web Graph Communities. In: Proceedings of the 1st International Conference on Theory of Information Retrieval, pp. 261–278 (2007)
26. Schaeffer, S.E.: Graph Clustering. Computer Science Review 1(1), 27–64 (2007)
27. Zachary, W.W.: An Information Flow Model for Conflict and Fission in Small Groups. Journal of Anthropological Research 33(4), 452–473 (1977)
28. Xu, X., Yuruk, N., Feng, Z., Schweiger, T.A.J.: Scan: A Structural Clustering Algorithm for Networks. In: KDD, pp. 824–833 (2007)

On Summarizing Graph Homogeneously

Zheng Liu and Jeffrey Xu Yu

The Chinese University of Hong Kong
{zliu,yu}@se.cuhk.edu.hk

Abstract. Graph summarization is to obtain a concise representation of a large graph, which is suitable for visualization and analysis. The main idea is to construct a super-graph by grouping similar nodes together. In this paper, we propose a new information-preserving approach for graph summarization, which consists of two parts: a super-graph and a list of probability distribution vectors affiliated to the super-nodes and super-edges. After a carefully analysis of the approximately homogenous grouping, we propose a unified model using information theory to relax all conditions and measure the quality of the summarization. We also develop a new lazy algorithm to compute the exactly homogenous grouping, as well as two algorithms to compute the approximate grouping. We conducted experiments and confirmed that our approaches can efficiently summarize attributed graphs homogeneously and achieve low entropy.

1 Introduction

Researchers make great efforts on mining graph data recently, because of its ability to represent complex relationships among entities in many applicable areas such as Web, social networks, biological networks, telecommunication, etc. In general, it is not an easy task for users to explore a large graph globally in order to find hidden relationships. To solve this problem, graph summarization techniques [6,9,12] have been recently studied. In brief, summarizing is to obtain a concise graph representation, G_S, of a large graph G, where G_S is smaller than G in size for the visualization or analysis. Although specific summarized representations can be various to represent G_S in different approaches, the main idea behind is to construct G_S as a super-graph with super-nodes and super-edges. Here, the nodes in G are partitioned into several groups and each group is represented by a single super-node in G_S. Two super-nodes are connected by a super-edge in G_S if there exist connections of nodes from two corresponding groups in G.

There are two major approaches for super-graph construction. A strict approach [6] requires that a super-edge exists between two super-nodes in G_S only if every pair of nodes residing in the two corresponding super-nodes is connected by an edge in G. A relaxed approach [9, 12] allows two super-nodes to be connected with a super-edge in G_S if there is at least one connected node pairs in G among all the node pairs summarized by the two super-nodes. Here, each super-edge is associated with a participation rate to indicate the percentage of nodes in

J. Xu et al. (Eds.): DASFAA Workshops 2011, LNCS 6637, pp. 299–310, 2011.

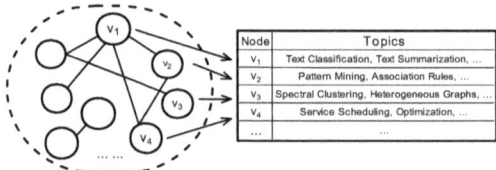

Node	Topics
v_1	Text Classification, Text Summarization, ...
v_2	Pattern Mining, Association Rules, ...
v_3	Spectral Clustering, Heterogeneous Graphs, ...
v_4	Service Scheduling, Optimization, ...
...	...

Fig. 1. DBLP Co-author Network

a super-node that have edges connected to the other super-node. For the strict approach, since only cliques or bipartite cliques can be represented by super-nodes according to the very rigorous requirement, in most cases, the size of the summarized graph cannot be small, even when super-nodes are near-cliques [6]. For the relaxed approach, it is difficult to control the summarization quality. In addition, it requests that the attribute values of nodes in the same group must be the same, which makes it very difficult to handle a multi-attribute graph, in particular, when the number of attributes is large. In this paper, we focus on an information-preserving graph summarization for attribute graphs. An example is shown in Fig. 1, which is a simple DBLP co-author graph to be summarized. Inside the dotted area is a graph where a node represents an author and an edge represents collaboration between two authors. The table associated with shows the main research areas of each author (node), possibly with other information.

The major contributions of this paper are summarized below. First, we focus on how to obtain an optimized approximately homogeneous grouping on which a graph summarization can be constructed by relaxing all three conditions. We propose a unified model using information theory which unifies both attribute information and structural information. Second, we propose a new lazy algorithm to compute the exactly homogeneous grouping, and two new approximate algorithms. Third, we conduct testing using real datasets to confirm the quality and efficiency of our proposed approaches.

The remainder of this paper is organized as follows. Section 2 starts a careful analysis of the graph summarization problem and Section 3 presents our concept of approximately homogenous grouping based on information theory. We introduce the summarization framework in Section 4. Experimental results are reported in Section 5 and related works are discussed in Section 6. Finally, Section 7 concludes this paper.

2   Problem Statement

We focus on an attribute graph $G(V, E, T)$, where V and E represent a set of nodes and a set of edges, respectively. T is a mapping function. Given a set of attributes $(A_1, A_2, \cdots, A_d)$, for each node $v_i \in V$, $T(v_i) = (a_1, a_2, \cdots, a_d)$ represents the attribute vector of v_i, where a_i is a value of A_i. In this work, we concentrate on categorical attributes. For a categorical attribute A_i with n distinctive values, we can represent an attribute value using a n-bitmap, where all bits are zero except for the bit which corresponds to the attribute value. Following the similar definitions in [9], a node grouping $\Gamma = \{V_1, V_2, \cdots, V_k\}$ of a graph $G(V, E, T)$ is to partition all nodes of G into non-overlapping k groups, where V_i represents a non-empty subset of nodes V. Given a graph node grouping Γ, the set of neighbor super-nodes for node $v_i \in G$ is denoted as $NG(v_i) = \{\Gamma(v_j)|(v_i, v_j) \in E\}$ where $\Gamma(v_j)$ represents the super-node that uniquely contains v_j. For $V_j \in NG(v_i)$, let $|V_j|_{v_i}$ denote the number of edges from v_i to any nodes in V_j. An exactly homogeneous grouping is defined as follows.

Definition 1. (Exactly Homogeneous Grouping) [9]
Given a graph $G = (V, E, T)$, an exactly homogeneous grouping Γ satisfies the following conditions: $\forall v_i, v_j \in V$, if $\Gamma(v_i) = \Gamma(v_j)$, then (1) $T(v_i) = T(v_j)$, (2) $NG(v_i) = NG(v_j)$, and (3) $|V_k|_{v_i} = |V_k|_{v_j}$, $\forall V_k \in (NG(v_i) \cup NG(v_j))$.

Based on the exactly homogeneous grouping Γ, a graph summarization G_S can be constructed as follows. A super-node S_i represents a group V_i, for all groups in Γ, and all nodes of G summarized by a super-node in G_S have the same attribute values. The super-edges among super-nodes in G_S imply that every node of G summarized by a super-node has the same pattern of connecting nodes to other nodes summarized by other super-nodes. For example, suppose that S_i has super-edges to S_j, S_k, and S_l. It shows that every node of G summarized by S_i has edges to some nodes of G summarized by S_j, S_k, and S_l. Below, we use V_i and S_i interchangeably.

Exactly homogeneous grouping shows the best summarization of a large graph G in terms of homogeneity criterion. However, the size of G_S based on the exactly homogeneous grouping is too large as a graph summarization, which can be almost as large as G. The problem to be studied in this paper is how to obtain a better small graph summarization G_S by relaxing the conditions given in Definition 1. Tian et al. in [15] relaxed only the condition (2). In other words, the nodes in the same group must have the same attribute vector, but the neighbor relationships are allowed to be similar. Apparently, the relaxation has problems when the number of attributes is large. When the number of attributes is large, it becomes impossible to find a small number of groupings, say k, such that all nodes in the same group have the same attributes. Also, it is questionable if it is sufficient to relax the condition (2) only. In this paper we propose to relax all the three conditions. Our approximately homogeneous grouping is defined as follows.

Definition 2. (Approximately Homogeneous Grouping)
Given a graph $G = (V, E, T)$, a number k, a graph node grouping Γ is called approximately homogeneous grouping, if it satisfies $\forall v_i, v_j \in V$, if $\Gamma(v_i) = \Gamma(v_j)$, (1) $dist_1(T(i), T(j)) \leq \varepsilon_1$, (2) $dist_2(NG(v_i), NG(v_j)) \leq \varepsilon_2$, and (3) $dist_3(|V_k|_{v_i}, |V_k|_{v_j}) \leq \varepsilon_3$, $\forall V_k \in (NG(v_i) \cup NG(v_j))$. Here, $dist_1(\cdot)$, $dist_2(\cdot)$, $dist_3(\cdot)$ are three distance functions, and ϵ_1, ϵ_2, and ϵ_3 are three thresholds.

In an approximately homogeneous grouping, nodes in the same group are considered as homogeneous so long as their attributes and neighbor information are similar. Apparently, once the distance functions and thresholds are determined, there might be many approximately homogenous groupings which satisfy the above definition. We focus on finding the best one among all these groupings. Suppose $F(\cdot)$ is a score function to measure the quality of approximately homogeneous groupings. The problem of optimized approximately homogenous grouping is formally defined as follows.

Definition 3. (Optimized Approximately Homogeneous Grouping)
Given a graph $G = (V, E, T)$, a number k, a graph node grouping Γ is called Optimized Approximately homogeneous grouping, if $F(\Gamma)$ is minimum.

The ensuing questions are as follows. What distance and score functions should we use? How do we adjust the similarity? We will address these issues in the following sections.

3   An Approximately Homogeneous Grouping Based on Information Theory

In this paper, we propose an information-preserving criterion, based on information theory. We first review some background knowledge, followed by detailed discussions about how to utilize a unified entropy model to measure the quality of the three relaxations in Definition 2 and the quality in Definition 3.

Let x_i be a boolean random binary variable and $p(x_i)$ be its Bernoulli distribution function, $\mathbf{p}(\mathbf{x}) = [p(x_1), \cdots, p(x_d)]$ is a Bernoulli distribution vector over d independent boolean random variables $x_1, \cdots, x_d$ [10]. Let $\mathbf{b_j}$ denote a binary d-element vector. Given a set of binary vectors $D = \{\mathbf{b_1}, \cdots, \mathbf{b_n}\}$, under the assumption of independence, the probability by which they are generated by a distribution vector is estimated as

$$P(D|\theta) = \prod_{\mathbf{b_j} \in D} \prod_{i=1}^{d} p(x_i = \mathbf{b_j^i}), \tag{1}$$

where $\mathbf{b_j^i}$ is the ith element of the binary vector $\mathbf{b_j}$. The best θ, which fits for the model, is $\hat{\theta} = arg\max_\theta \log(P(D|\theta))$. The well-known solution based on the maximum likelihood estimation is

$$p(x_i = 1) = \frac{\sum_{\mathbf{b_j} \in D}^{n} \mathbf{b_j^i}}{|D|}. \tag{2}$$

We use information theory to measure the quality of these distribution vectors. Recall that in information theory, entropy [3] is a measure of the uncertainty (randomness) associated with a random variable X, which is defined as $H(X) = -\sum_{x \in X} p(x) \log_2 p(x)$. Consider a random variable x_i whose value domain is $\{0, 1\}$, the probabilities of x_i equals 0 or 1 is $p(x_i = 0)$ or $p(x_i = 1)$. The entropy of an unknown sample of the random variable x_i is maximized when $p(x_i = 0) = p(x_i = 1) = 1/2$, which is the situation that it is the most difficult to predict the value of a unknown sample. When $p(x_i = 0) \neq p(x_i = 1)$, we know that the value of the unknown sample is more likely to be either 0 or 1 accordingly, which is quantified in a lower entropy. The entropy is zero when $p(x_i = 0) = 1$ or $p(x_i = 1) = 1$. For a Bernoulli distribution vector $\mathbf{p}(\mathbf{x})$, assuming the contained random variables are independent to each other, the total entropy of a Bernoulli distribution vector is

$$H(\mathbf{p}(\mathbf{x})) = -\sum_{i=1}^{d} \sum_{x_i=0}^{1} p(x_i) \log_2 p(x_i) \qquad (3)$$

If binary vectors within the set D are similar to each other, or homogeneous, then for each random variable x_i, most of its values should be similar, resulting in a low $H(\mathbf{p}(\mathbf{x}))$.

Below, we discuss the three relaxations in Definition 2. Based on these observations, we can measure the quality of the three relaxations in an unified model inspired by information theory.

Observation for $dist_1(\cdot)$: For each node $v_i \in V$, $T(v_i) = (a_1, ..., a_d)$ is the attribute vector of v_i, where a_i is the value of attribute A_i. As mentioned, we represent categorical attribute values using bitmaps, so we also use a_i to indicate the bitmap when there is no confusing. For a certain group of nodes, V_j, in an approximately homogenous grouping, the attribute information of each node $v_i \in V_j$ is in form of a binary vector by catenating together these bitmaps, denoted as $\mathbf{a} = (a_1, ..., a_d)$. If these binary vectors are similar, then the attribute information of this group is homogenous. Instead of measuring the distances between pairs of these vectors, a binary Bernoulli distribution vector is learned by Eq. (3) from these vectors. The more similar these vectors are, the less randomness of the distribution vector. The randomness is measure by entropy and low entropy means less randomness.

Observation for $dist_2(\cdot)$: Nodes in the same homogeneous group should have similar neighbors in the super-graph, where $dist_2(\cdot)$ is a distance function based on the neighbor relationship. However, $dist_3(\cdot)$ measures difference between two nodes in the same group in terms of the number of neighbors in their neighbor groups. Apparently, $dist_3(\cdot)$ is a stronger requirement than $dist_2(\cdot)$, so we skip the entropy analysis for $dist_2(\cdot)$ and jump to $dist_3(\cdot)$ in the following.

Observation for $dist_3(\cdot)$: If there is a super-edge between super-nodes S_i and S_j, then nodes in S_i should have similar total number of edges to nodes in S_j. We can keep two histograms for each super-edge (S_i, S_j), namely, S_i-to-S_j and S_j-to-S_i, to record the number of neighbors in S_j (S_i) of nodes in S_i (S_j). We explain it using an example as shown in Fig. 2. There are three groups (super-nodes) in the grouping: S_1, S_2, and S_3. At the upper left corner in Fig. 2, it shows how these super-nodes are connected by super-edges. For example, it indicates that every node in S_2 has 10 neighbors in S_1 on average. The histogram of S_2-to-S_1 is drawn on upper right corner, where The x-axis indicates the number of neighbors in S_1 for a node in S_2. The y-axis indicates the number of nodes in S_2 in corresponding to each value on x-axis. Intuitively, a homogenous group should have a tight spread range on x-axis in the histogram. We use entropy to

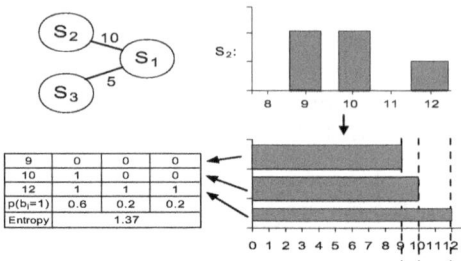

Fig. 2. Connection Strength Example

measure how homogeneous inside each group. To do so, we present the histogram in another way as shown in the bottom right corner. The x-axis still indicates the number of neighbors in S_1 for a node in S_2, while the thickness of each bar indicate the number of nodes in S_1 in corresponding to each value on x-axis. Based on this intuition, we transform each bar in the bottom histogram to a binary vector of all 1's. For example, for bar indicating the number of neighbors is 9, a binary vector of length 9 is constructed. We first catenate 0's at the end of each binary vectors to make them of the same length. Then we remove the common 1's in the suffix of these two vectors. The remaining binary vectors are shown in the bottom left table in Fig. 2. Similar to $dist_1(\cdot)$, a Bernoulli distribution vector is learned from these binary vectors. The more similar these vectors are, the lower entropy of the distribution vector is.

In summary, the homogeneity of a group of nodes can be measured by the concept of entropy of these Bernoulli Vectors. Let $\mathbf{e} = (S_i, S_j)$ denote a super-edge between two super-nodes S_i and S_j, and let v_i denote a node in super-node S_i. The entropy of super-node S_i consists of two parts. Recall that $\mathbf{a}$ is the catenating binary vector of bitmaps. The first is the attribute homogeneity $H(\mathbf{p}(\mathbf{a^m} = 1))$, where $\mathbf{a^m}$ is the mth element in $\mathbf{a}$, and $\mathbf{p}(\mathbf{a^m} = 1)$ is the Bernoulli distribution vector estimated by Eq. (2) based on the vector set of $\mathbf{a}$. Let $\mathbf{b}$ denote the binary vectors for histograms in Fig. 2. The second is the connection strength homogeneity $H(\mathbf{p}(\mathbf{b_j^m} = 1))$, where $\mathbf{b^m}$ is the mth element in $\mathbf{b}$, and $\mathbf{p}(\mathbf{b_j^m} = 1)$ is the Bernoulli distribution vector estimated by Eq. (2) for S_i to S_j. The total entropy for S_i is given below.

$$TotalEntropy(S_i) = \frac{H(\mathbf{p}(\mathbf{a^m} = 1))}{|\mathbf{a}|} + \frac{1}{deg(S_i)} \times \sum_{j=1}^{k} \frac{H(\mathbf{p}(\mathbf{b_j^m} = 1))}{|\mathbf{b_j}|} \qquad (4)$$

Here, $deg(S_i)$ is the degree of super-node S_i in the graph summarization G_S, and k is a user-given parameter for controlling the number of groups in the groping Γ. Now the optimized approximately homogenous grouping is the grouping which minimizes $\sum_{S_i \in \Gamma} TotalEntropy(S_i)$ of the super-graph. What we study next is how to find the optimized approximately homogeneous grouping Γ for a given graph G. Based on Γ, the graph summarization G_S can be constructed.

Algorithm 1. Lazy Exactly Homogeneous Grouping

Input: A graph $G = (V, E, T)$
Output: The exactly homogeneous grouping Γ.

1: Partition the nodes into m initial groups according to distinct attribute vectors **a**;
2: Construct an $n \times m$ node-to-group matrix M;
3: **while** True **do**
4:    Sort rows within each group;
5:    Let $splitflag$ be a all-zero binary vector of length n;
6:    **for** each column in M **do**
7:       **if** $M(i, j) == 0$ and $M(i + 1, j) == 1$ **then**
8:          $splitflag(i) = 1$
9:       **end if**
10:    **end for**
11:    **if** $splitflag$ is all zeros **then**
12:       break;
13:    **end if**
14:    Split each group according to $splitflag$;
15:    Reconstruct the $n \times m'$ node-to-group matrix M;
16: **end while**
17: Output the exact homogeneous grouping Γ.

---

4   Homogeneous Graph Summarization

Our framework for graph summarization contains two steps: the exact homogeneous grouping and the approximate homogeneous grouping.

The Exactly Homogeneous Grouping: Algorithm 1 outlines the procedures to compute the exactly homogeneous grouping based on Definition 1. Recall that **a** is the catenated attribute vector for nodes. Suppose there are m distinct attribute vectors, then nodes in graph G are partitioned into m groups first according to the distinct vectors. Then the algorithm constructs an $n \times m$ node-to-group matrix M, where $M(i, j)$ is the number of v_i's neighbors in S_j. One thing worth noting is that nodes belonging to the same group are stored adjacently in M and the order of groups in rows is the same as the order of groups in columns. At line 6, Algorithm 1 marks the split positions using a binary vector of length n. After having inspected all the groups, Algorithm 1 reconstructs the node-to-group matrix M based the split positions marked. The reason we do not reconstruct M immediately after a split position is that the matrix reconstruction is costly, and that why we called it lazy exactly homogeneous grouping.

It is obvious that the optimized approximately homogeneous algorithm is an NP problem and we present two heuristic algorithms, a bottom-up approximate algorithm and a k-mean approximate algorithm, to solve the approximately homogeneous grouping problem.

The Approximately Homogeneous Grouping: We obtain an approximately homogeneous grouping from an exactly homogeneous grouping Γ. We merge two groups repeatedly in a bottom-up fashion, if the merge can bring minimum entropy increase. The algorithm is presented in Algorithm 2. Algorithm 2 takes

Algorithm 2. The Bottom-Up Approximate Algorithm

Input: The exactly homogeneous grouping $\Gamma = \{V_1, \cdots, V_m\}$; a number k
Output: The approximately homogeneous grouping Γ_A

1: $\Gamma_A = \Gamma$;
2: **while** $|\Gamma_A| > k$ **do**
3:   **for** each group pair V_i and V_j in Γ **do**
4:     $\Gamma_{ij} = \Gamma_A \cup \{V_i \cup V_j\} \setminus \{V_i, V_j\}$;
5:     Compute the $entropy(\Gamma_{ij})$ based on Eq. 4;
6:   **end for**
7:   Let (V_l, V_m) be the pair of groups with the minimum $|entropy(\Gamma_{ij}) - entropy(\Gamma_A)|$;
8:   $\Gamma_A = \Gamma_A \cup \{V_l \cup V_m\} \setminus \{V_l, V_m\}$;
9: **end while**
10: Output the approximate homogeneous grouping Γ_A.

Algorithm 3. The k-Mean Approximate Algorithm

Input: The exactly homogeneous grouping $\Gamma = \{V_1, \cdots, V_m\}$; a number k
Output: The approximately homogeneous grouping Γ_A

1: Random select k the initial groups $C_1, ..., C_k$ from Γ;
2: Evaluate the Bernoulli distribution vectors $\mathbf{a}$ and $\mathbf{b_j}$ for $C_1, ..., C_k$;
3: **repeat**
4:   Evaluate the Bernoulli distribution vectors $\mathbf{a}$ and $\mathbf{b_j}$ for each group $V_i \in \Gamma$;
5:   Catenate $\mathbf{a}$ and $\mathbf{b_j}$ together for V_i;
6:   Assign each group V_i to a new cluster C_i according to the Kullback-Leibler divergence in Eq. 5;
7:   Evaluate the Bernoulli distribution vectors $\mathbf{a}$ and $\mathbf{b_j}$ for $C_1, ..., C_k$;
8:   Catenate $\mathbf{a}$ and $\mathbf{b_j}$ together for $C_1, ..., C_k$;
9: **until** No more changes of the group assignments
10: Output the approximate homogeneous grouping $\Gamma_A = \{C_1, C_2, ..., C_k\}$.

the exact homogeneous grouping Γ as the input. In each iteration from line 2, the algorithm computes the entropy increase introduced by merging possible group pairs, and selects the pair with the minimum extra entropy. There is no need to recompute the total entropy of the whole merged super-graph in order to obtain the entropy increase, because merging two super-nodes (groups) only affects the entropy of their neighbor super-nodes in the corresponding graph summarization G_S. So, in line 5, we only calculate the entropy of the pair of groups to be merged and ones of their neighbor groups. The entropies of other groups are the same as ones in Γ_A.

The k-Mean Approximate Algorithm: In this section, we present a k-mean based approximate algorithm to find the optimized approximately homogeneous grouping using the Kullback-Leibler (KL) divergence. The Kullback-Leibler divergence [10] is a measure of the difference between two distribution vectors $\mathbf{p}$ and $\mathbf{q}$, which is defined as below.

$$KL(\mathbf{p} \parallel \mathbf{q}) = \sum_{i=1}^{d} \sum_{x_i=0}^{1} p(x_i)log\frac{p(x_i)}{q(x_i)} \tag{5}$$

In the view of information theory, KL measures the expected number of extra bits required to encode samples from $\mathbf{p}$ when using a code based on $\mathbf{q}$, rather than using a code based on $\mathbf{p}$. Suppose the Bernoulli distribution vector for a certain node group S_i is $\mathbf{p}$. For each node in group S_i, let $\mathbf{q}$ denote the Bernoulli distribution vector for a node $v_i \in S_i$. We can prove that

$$\sum_{v_i \in S_i} KL(\mathbf{q}(\mathbf{x}) \parallel \mathbf{p}(\mathbf{x})) = -n(S_i) * H(\mathbf{p}(\mathbf{x})). \tag{6}$$

The proof is omitted due to lack of space. So, the optimized approximately homogenous grouping that minimizes $H(\mathbf{p}(\mathbf{x}))$ is the grouping that minimizes $KL(\mathbf{q}(\mathbf{x}) \parallel \mathbf{p}(\mathbf{x}))$, which leads to the following k-mean approximate algorithm, as presented in Algorithm 3.

5   Experimental Results

In this section, we report our experimental results on the real datasets from DBLP Bibliography [1]. We construct a co-author graph with top authors and their co-author relationships, where the authors are from three research areas of database (DB), data mining (DM) and information retrieval (IR). Based on the publication titles of the selected authors, we use a topic modeling approach [4,11] to extract 100 research topics. Each extracted topic consists of a probability distribution of keywords which are most representative of the topic. In the experiments, each author is related to several topics whose probabilities are larger than 5%. By using authors from partial or all areas, we construct four datasets in our experiments. The basic statistics of the four datasets are presented in Table 1. There are total 100 topics and each author can belong to up to five topics. Example of the topics are shown in Table 2, as well as the top keywords in each topic. Fig. 3 presents the frequency distribution of each topic, which is the total number of authors belonging to a topic. All the experiments were conducted on a computer running Windows XP with Intel Core-2 Quad processor and 3GB RAM, but only a single core in the CPU is used for evaluation.

Exactly Homogeneous Grouping: Table 3 presents a comparison between the number of groups and the nodes in the original graphs. The number of

Table 1. The DBLP Bibliography Datasets

	Datasets	# of Nodes	# of Edges	Avg. Degree
D1	DM	1695	2282	1.35
D2	DB	3328	11379	3.42
D3	DB+DM	5023	15262	3.03
D4	D3+IR	6184	18710	3.02

Table 2. The Keywords of The Topics

Topics #	Keywords
32	text, classification, vector, categorization
66	mining, patterns, frequent, sequential...
76	service, scheduling, extending, media
80	clustering, matrix, density, spectral

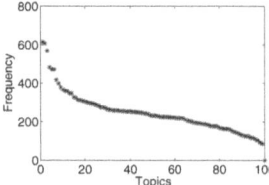

Fig. 3. Topic Frequency in Dataset DB

Table 3. The DBLP Bibliography Datasets

	D1	D2	D3	D4
# of Nodes	1695	3328	5023	6184
# of Attribute Vectors	1492	2931	4401	5409
# of Exact Groups	1604	3219	4829	5912

Fig. 4. Dataset DM

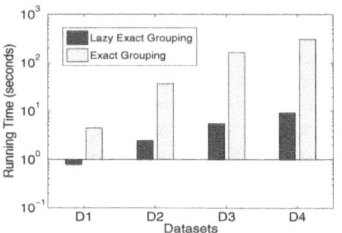

Fig. 5. Exact Algorithms

distinct attribute vectors and the number of exact groups are quite close to the number of nodes. Therefore, the exact homogeneous grouping cannot obtain a graph summary of a reasonable size. Fig. 4 shows the graph structure of the main connected component generated by exact grouping algorithm on dataset DM. In Fig. 5, we compare the running time of our lazy exactly homogenous grouping algorithm with the algorithm, denoted as exact grouping, which reconstructs the matrix M immediately after discovering a split position. The lazy exact grouping algorithm is much faster than the exact grouping algorithm that splits groups immediately.

Approximately Homogeneous Grouping: Due to the high time complexity, we only apply the bottom-up approximate algorithm on the smallest dataset DM. Fig. 6 shows the average entropy of the bottom-up approximate algorithm. Since it works in a bottom-up fashion, we present the average entropy for each possible group number. As the group number shrinks, the average entropy increases. The bottom-up approximate algorithm starts from the exact grouping, so the average entropy at the beginning is 0. Fig. 7 presents the running time of the bottom-up algorithm. Both the x-axis and y-axis are in log scale. The total running time is morn than 10^4 seconds even for the smallest dataset DM. Fig. 8 shows the average entropy of the output approximately homogenous grouping by the k-mean approximate algorithm. The maximum average entropy in Fig. 8 is around 1.1 when k equals to 45, which is quite good since the average entropy is the sum of the average attribute entropy and average neighbor entropy. Note that the maximum possible entropy of a binary random variable is 1. Fig. 9 and Fig. 8 present the running time and the quality for all the four datasets, respectively, for the k-mean approximate algorithm. As we can see in Fig. 9, when the number of desired groups increases, it usually takes longer time to obtain the graph

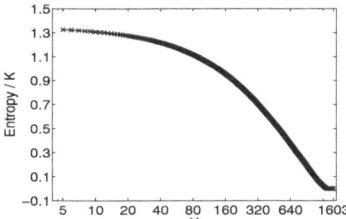

Fig. 6. Quality of Bottom-up Algo.

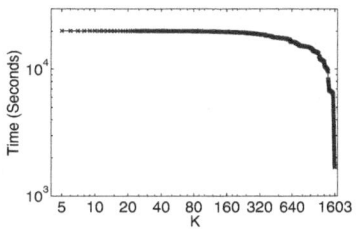

Fig. 7. Timing of Bottom-up Algo.

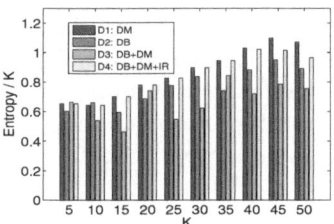

Fig. 8. Quality of k-mean Algo.

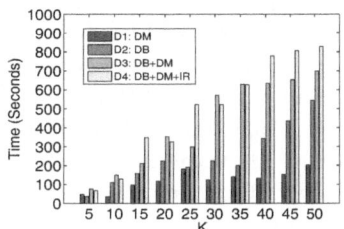

Fig. 9. Timing of k-mean Algo.

summarization. Since k-mean is a heuristic algorithm, the number of iterations is not necessary to be dependent on the number of groups. That explains why sometimes a larger k runs faster than a smaller k. The four datasets are of different sizes in terms of number of nodes. Fig. 9 shows that the running time is in proportional to the dataset size. For dataset DM, k-mean algorithm is almost 100x times faster than the bottom-up algorithm.

6   Related Works

The graph summarization [6,7,9,12] mainly contains researches in two aspects. One of them focuses on visualizing. Recently, Navlakha et al. [6] proposed an approach to summarize graphs without attributes. Their goal is to obtain a compact graph representation with smaller size and can be restored to the original graph with bounded error. They used a single super-node to represent a clique or near-clique, with an addition table recording the missing/added edges. The quality of the summary is its minimum description length. Tian et al. [9] proposed two graph summarization algorithms by aggregating nodes in graphs. Their methods deal with graphs which contain only categorical attributes. The top-down approach first groups together all the node with the same categorical attributes. Then the groups of nodes are repeatedly split until there are k groups. The bottom-up approach first groups together all the node with both the same node attributes and the same edge attributes. Then small groups of nodes are merged into larger groups until there are k groups left. It seems that their methods cannot be applied directly on the graph with many attribute, as discussed.

Zhang et al. [12] extended Tian's approach to deal with numerical attributes by automatically categorizing numerical attribute values in a discovery-driven manner. They also proposed an interestingness measure to identify the most interesting resolutions. Graph compression [2,8,5] is another related area, which focuses on compressing very large graphs in order to store them using smaller disk spaces without losing the ability to answer link queues.

7   Conclusions

In this paper, we study graph summarization using a new information-preserving approach. A graph is summarized by partitioning nodes into groups. We analyzed the approximately grouping criterion and proposed a unified entropy framework to relax all three conditions and measure the summarization quality. We also proposed a lazy exact algorithm, as well as two other approximate algorithms to compute the exactly homogenous grouping and the approximate grouping, respectively. Experiments demonstrate that our methods can summarize attributed graphs efficiently and homogeneously.

Acknowledgment

The work was supported by grants of the Research Grants Council of the Hong Kong SAR, China No. 419008 and 419109.

References

1. Dblp bibliography, http://www.informatik.uni-trier.de/~ley/db/index.html
2. Blandford, D.K., Blelloch, G.E., Kash, I.A.: Compact representations of separable graphs. In: SODA, pp. 679–688 (2003)
3. Cover, T.M., Thomas, J.A.: Elements of information theory. Wiley-Interscience, New York (1991)
4. Hofmann, T.: Probabilistic latent semantic indexing. In: SIGIR, pp. 50–57 (1999)
5. Maserrat, H., Pei, J.: Neighbor query friendly compression of social networks. In: KDD, pp. 533–542 (2010)
6. Navlakha, S., Rastogi, R., Shrivastava, N.: Graph summarization with bounded error. In: SIGMOD Conference, pp. 419–432 (2008)
7. Newman, M.E.J.: The structure and function of complex networks. SIAM Review 45, 167–256 (2003)
8. Raghavan, S., Garcia-Molina, H.: Representing web graphs. In: ICDE, pp. 405–416 (2003)
9. Tian, Y., Hankins, R.A., Patel, J.M.: Efficient aggregation for graph summarization. In: SIGMOD Conference, pp. 567–580 (2008)
10. Yan, X., Cheng, H., Han, J., Xin, D.: Summarizing itemset patterns: a profile-based approach. In: KDD, pp. 314–323 (2005)
11. Zhai, C., Velivelli, A., Yu, B.: A cross-collection mixture model for comparative text mining. In: KDD, pp. 743–748 (2004)
12. Zhang, N., Tian, Y., Patel, J.M.: Discovery-driven graph summarization. In: ICDE, pp. 880–891 (2010)

Expansion Properties of Large Social Graphs

Fragkiskos D. Malliaros[1] and Vasileios Megalooikonomou[1,2]

[1] Computer Engineering and Informatics Department
University of Patras, 26500 Rio, Greece
[2] Data Engineering Laboratory, Center for Information Science and Technology,
Temple University, 1805 N. Broad St., Philadelphia, PA 19122, USA
{malliaro,vasilis}@ceid.upatras.gr

Abstract. Social network analysis has become an extremely popular research area, where the main focus is the understanding of networks' structure. In this paper, we study the expansibility of large social graphs, a structural property based on the notion of expander graphs (i.e. sparse graphs with strong connectivity properties). It is widely believed that social networks have poor expansion properties, due to their community-based organization. Moreover, this was experimentally confirmed on small scale networks and it is considered as a global property of social networks (independent of the graph's size) in many applications. What really happens in large scale social graphs? To address this question, we measure the expansion properties of several large scale social graphs using the measure of subgraph centrality. Our findings show a clear difference on the expansibility between small and large scale social networks, and thus structural differences. Our observations could be utilized in a range of applications which are based on social graphs' structure.

Keywords: Social networks, Expansion, Measurement, Graph Mining.

1 Introduction

Recently, there has been a lot of interest in the study of complex network structures arising in many diverse settings. Characteristic examples are networks from the domain of sociology (e.g. social networks), technological and information networks (e.g. the Internet, the Web, e-mail exchange networks, social interaction networks over social media applications), biological networks (e.g. protein interactions), collaboration and citation networks (e.g. co-authorship networks), and many more [20]. The research interest has mainly focused on understanding the structure, the organization, and the evolution of these networks, and many interesting results have been produced [2].

A better and deeper understanding of network's structure could have multiple benefits in several domains (e.g. better design for graph algorithms and applications). Towards this direction, in this work we study the *expansibility* of large social graphs, a structural property based on concepts from the theory of expander graphs [10]. Our main goal is to explore the expansion properties of large scale social networks and compare it with known results from previous studies

J. Xu et al. (Eds.): DASFAA Workshops 2011, LNCS 6637, pp. 311–322, 2011.

Table 1. Symbols and definitions

Symbol	Definition
G	Graph representation of datasets
V, E	Set of nodes and edges for graph G
$|V|$, $|E|$	Number of nodes and edges
$N(S)$	Neighborhood nodes of node set S
$h(G)$	Expansion factor of graph G
	(or isoperimetric number or Cheeger constant)
$\mathbf{A}$	Adjacency matrix of a graph
a_{ij}	Entry in matrix $\mathbf{A}$
λ_i	i-th largest eigenvalue
u_{ij}	i-th component of j-th eigenvector
$SC(i)$	Subgraph centrality of node i

on small graphs, in order to extract useful conclusions about social networks' structure. Table 1 gives a list of used symbols with their definitions.

Expansion and Expander Graphs

Informally, a graph is a good expander if it is simultaneously sparse and highly connected. More precicely, given a graph $G = (V, E)$, the expansion of any subset of nodes $S \subset V$ with size at most $\frac{|V|}{2}$, is defined as the number of its neighborhood nodes (i.e. those nodes who have one endpoint inside S and the other outside) over the size of the subset S. That is, if $N(S)$ are the neighborhood nodes of S, the expansion factor of the set S is $\frac{|N(S)|}{|S|}$. A graph is considered to have *good expansion properties* if every such subset of nodes has expansion at least $h(G)$, i.e. $h(G) \geq \frac{|N(S)|}{|S|}, \forall S \subset V$ and $|S| < \frac{|V|}{2}$. In other words, the expansion factor of a graph is defined as the minimum expansion over all subsets [10]:

$$h(G) = \min_{\left\{S : |S| \leq \frac{|N|}{2}\right\}} \frac{|N(S)|}{|S|}. \qquad (1)$$

Expansion properties can offer crucial insights into the structure of a graph, and in particular they can inform us about the presence or not of edges which can act as bottlenecks inside the network. This practicaly means that measuring the expansibility of a graph we are able to know to what extent the graph has a modular structure or not. Large expansion factor implies good expansion properties, which means that any subset of nodes will have a relatively large number of edges with one endpoint in this set, and thus poor modularity. In other words, if we think these subsets as cuts of a graph, good expansibility require cuts with large size (i.e. large number of edges crossing the cut). On the other hand, bad expansibility is the opposite behavior. For any subset of nodes it is impossible to satisfy the constraint for a large neighborhood. Hence, such kind of graphs can be easily separated into disconnected subgraphs with the elimination of a small number of edges. It is clear that using notions from the field of sociology and social networks, graphs with poor expansibility correspond to graphs with good community structure.

Contributions and Summary of our Results

In this paper we measure the expansibility of several large social graphs. Based on the above discussion, we expect that social networks will exhibit bad expansion properties, because of the fact that they are organized in communities, i.e. groups of nodes with high density of edges within them, and much lower density between different groups [21]. This structural property was confirmed experimentally from previous studies [7] on *small* social networks, and in several cases is considered as a *global property* of social networks, *independent* of the graph size (e.g., some generative models for social networks are trying to generate synthetic social graphs satisfying this bad expansion property). However, does the same result apply to *large scale* social networks? In other words, how different is the structure of social graphs with a large number of nodes and edges, if any, from that of small graphs?

This is the main question we are trying to answer in this work. We measure the expansion properties of several social graphs with a large number of nodes and edges. In order to do this, we consider the fact that graphs with good expansion properties exhibit large spectral gap between the two largest eigenvalues of the adjacency matrix [10]. Then, utilizing this property together with the measure of subgraph centralilty [6], [7], we characterize the expansibility of these social graphs. Our findings suggest that *large scale social networks, in contrast to small ones, show good expansibility*. This point is particularly significant since it can help us towards a better understanding of large social networks' structure. Furthermore, these observations can be exploited in several domains such as structure-based classification schemes for networks, searching in networks [16] and in applications which may require robustness of the social network over social media applications.

2   Related Work

In this section we review the related work, which can be placed into three main categories: graph structure, applications and spectral graph analysis.

Graph Structure. There is a vast literature on methods for understanding the structure of social networks [21], [11], [18], [15] and generally of complex networks [20]. The key step for these methods is finding properties and laws which the graphs obey. Studying static snapsots of graphs has led to the discovery of interesting properties such as the *power law degree distribution* [9], the *small diameter* [1] and the *triangle power law* [25]. Futhermore, Leskovec et al. [13], [14] showed that time-evolving graphs have diameter which shrinks and stabilizes over time and obey the *densification power law*. For a nice survey one can consult the recent work of Chakrabarti, Faloutsos, and McGlohon [2]. Estrada [7] studied expansion properties of complex networks and showed that social graphs exhibit poor expansibility. However, in contrast with our work, it focuses on small scale networks. On the other hand, we explore large scale social networks and our results suggest a clear difference between their structure (in terms of expansibility) with that of small social graphs.

Applications. The understanding of a network's structure can be exploited in several domains and applications. Generating realistic graphs [2] is such an application, where generators should satisfy the observed properties. Other domains are searching in networks [16], sampling [17] and rumor spreading [4].

Spectral Graph Analysis. Analyzing graphs using spectral techniques (i.e. the eigenvalues and eigenvectors of a matrix representation of the graph (mainly adjacency and Laplacian matrices)) has a long history [3]. More recent related works include spectral algorithms for community detection [23] and spectral counting of triangles in large graphs [25]. As we will see next in this paper, the measure which is used for characterizing the expansibility of social graphs (subgraph centrality) can be computed using the spectrum of the adjacency matrix of the graph.

3 Measuring Expansion Properties

In this section we present the method we used for measuring the expansion properties of social networks, to characterize them as networks with "good" or "bad" expansibility. As we state previously, in order to compute the expansion factor of a graph (which fully characterizes its expansion properties), we need to compute the minimum fraction of neighborhood nodes, over the nodes inside the subset, for all possible subsets of nodes with size at most $\frac{|V|}{2}$. Since this is an NP-hard problem [19], and thus intractable to compute, we need approximation techiniques for the expansion factor of a graph.

Thanks to a very well known result in the field of spectral graph theory, the expansion factor can be approximated using the spectrum of the adjacency matrix $\mathbf{A}$ of the graph, and more precicely the difference between the largest and second largest eigenvalues of $\mathbf{A}$. This difference $\Delta\lambda = \lambda_1 - \lambda_2$ is known as the *spectral gap* of matrix $\mathbf{A}$ and it is related to the expansion factor $h(G)$ through the Alon - Milman inequality[1],

$$\frac{\Delta\lambda}{2} \le h(G) \le \sqrt{2\lambda_1 \Delta\lambda}. \tag{2}$$

Large spectral gap implies big expansion factor and thus a graph with good expansion properties. On the other hand, if λ_2 is close enough to λ_1, the spectral gap will be small and the graph will show poor expansibility.

The above discussion suggests a simple way for characterizing the expansion properties of a graph: compute the spectral gap and if this is large, the graph should have good expansion properties. However, a crucial question in the above

---

[1] This is also known as Cheeger inequality.

claim is *how large* the spectral gap should be for a graph to have good expansibility. As we will see from the experimental study in real-world networks, it is very difficult to measure the quantity of interest solely from the spectral gap of the adjacency matrix.

In this paper we measure the expansibility of a graph using the notion of *subgraph centrality* [6], employing a solution proposed by Estrada [7]. The reason for this decision is twofold: first of all, as we will see in the rest of this paper, the method based on subgraph centrality provides a clear distinction between graphs with different expansion properties. The second reason is that using this method, we can easily compare our results with that of [7], trying to find differences between the structure of large and small scale social graphs.

3.1 Subgraph Centrality

In this section we present the subgraph centrality measure [6] which is the basis for the estimation of the expansion character of a graph. Like other centrality measures in the field of graph theory and network analysis (e.g., degree centrality, betweenness centrality), subgraph centrality determines the importance of a node in the graph taking into consideration all the subgraphs in which the node participates.

More precicely, the subgraph centrality $SC(i)$ of a node $i \in N$ is calculated based on the total number of closed walks in a graph. A closed walk of specific length ℓ is an alternating sequence of nodes and edges starting and ending with a node, $v_1, e_1, v_2, e_2, \ldots, e_{\ell-1}, v_\ell$, where $e_i = (v_i, v_{i+1}) \in E, \forall i = 1, \ldots, \ell - 1$ and $v_1 = v_\ell$. For instance, a closed walk of length three represents a triangle. The subgraph centrality of a node i is defined as the sum of closed walks with different lengths, starting and ending at node i. However, all these walks with different lengths do not contribute equally to the centrality of the node; shorter walks contribute more (this happens because of the fact that in real-world graphs small subgraphs tend to be more interesting (e.g., triangles)). Thus, the subgraph centrality of node i is given by

$$SC(i) = \sum_{\ell=0}^{\infty} \frac{\mathbf{A}_{ii}^\ell}{\ell!}, \tag{3}$$

where the diagonal entry α_{ii} of the matrix $\mathbf{A}^\ell$ contains the number of walks of length ℓ that begin and end at the same node i. Using techniques from spectral graph theory, it can be proved that the subgraph centrality can be obtained from the spectrum of the adjacency matrix $\mathbf{A}$ of the graph. Because of the fact that (3) counts both even and odd length closed walks and more precicely even length walks may be trivial (moving forth and back in the graph), we keep only odd length walks[2] [6]:

---

[2] The graphs used in this study are non-bipartite and thus the number of closed walks of odd length is different from zero.

$$SC(i) = \sum_{j=1}^{N} u_{ij}^2 \sinh(\lambda_j). \tag{4}$$

Now we can write (4) in the form

$$SC(i) = u_{i1}^2 \sinh(\lambda_1) + \sum_{j=2}^{N} u_{ij}^2 \sinh(\lambda_j), \tag{5}$$

where u_{i1} is the i-th component of the principal eigenvector (eigenvector corresponding to the largest eigenvalue λ_1). If the graph has good expansion properties, which means that $\lambda_1 \gg \lambda_2$, then $u_{i1}^2 \sinh(\lambda_1) \gg \sum_{j=2}^{N} u_{ij}^2 \sinh(\lambda_j)$ and relation (5) could be written as

$$SC(i) \approx u_{i1}^2 \sinh(\lambda_1). \tag{6}$$

This means that the principal eigenvector u_{i1} is related to $SC(i)$ as

$$u_{i1} \propto \sinh^{-1/2}(\lambda_1) \, SC(i)^{1/2}. \tag{7}$$

This relation suggests that if the graph has good expansion properties (big spectral gap), u_{i1} will be proportional to $SC(i)$ and a log-log plot of u_{i1} vs. $SC(i)$, $\forall i \in N$ will show a linear fit with slope $1/2$ [7]. Thus, good expansion implies a power-law relationship between the principal eigenvector and the subgraph centrality. On the other hand, graphs with poor expansibility will deviate from this property. Moreover, this behavior can be summarized in the quantity $\xi(G)$, which captures exactly the expansion character of a graph [8]:

$$\xi(G) = \sqrt{\frac{1}{|N|} \sum_{i=1}^{|N|} \left\{ \log(u_{i1}) - \left(\log A + \frac{1}{2}\log(SC(i)) \right) \right\}^2}, \tag{8}$$

where $A = \sinh^{-1/2}(\lambda_1)$. This quantity measures the deviation from the "perfect" linear correlation (in log scale), which occurs when the spectral gap $\lambda_1 - \lambda_2$ is large (and thus the graph has good expansion properties). This is exactly what we propose to use in this paper for measuring the expansion properties of real-world social graphs.

For a better understanding and illustration, we apply this method to two graphs with known expansion properties. The first one is a random graph with 50 nodes produced by the Erdös-Rényi (ER) random graph model [5] with probability $p = 0.3$ (Fig. 1 (a)) and the second one is Newman's collaboration network between 379 researchers in the area of network science (Fig. 1 (c)) [22]. Random graphs are known to have good expansibility [10], and thus we expect linear correlation in log-log scales between the principal eigenvector and subgraph centrality. On the other hand, Newman's collaboration network has bad expansion character because of the fact that nodes form dense modules, with sparse connections between different modules. Hence, we expect deviation from

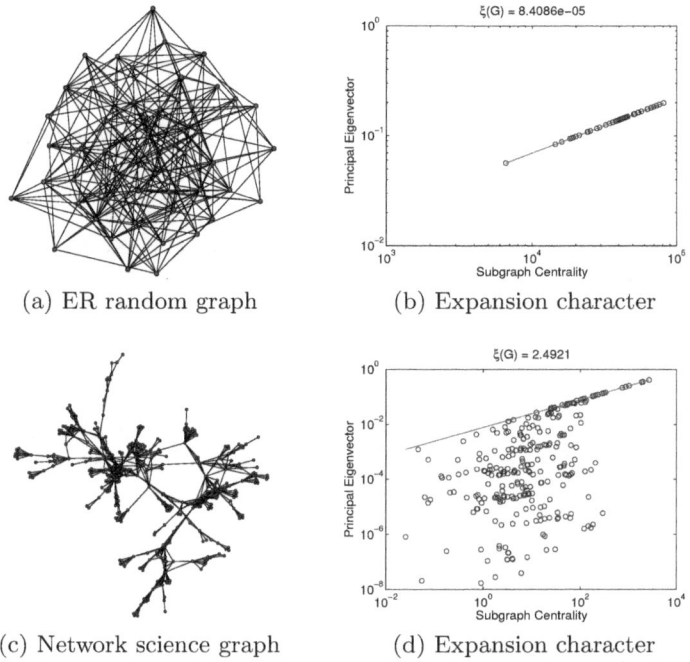

(a) ER random graph   (b) Expansion character

(c) Network science graph   (d) Expansion character

Fig. 1. Two graphs with known expansion properties and the plots of the principal eigenvector vs. subgraph centrality in log-log scale

this "perfect" linear correlation. Figure 1 (b) and (d) depicts these results. Also, we can observe that $\xi(G)$ is much smaller for the ER graph compared with the second one, which agrees with the above discussion.

4   Experimental Results

Equipped with the tools presented in Section 3, we measure the expansion properties of different real-world social graphs shown in Table 2. All these graphs represent social networks with a large number of nodes and edges. The selection of these datasets, except from their large scale, is based on the fact that they were formed under different "rules" and conditions. On the one hand we have networks where edge creation is based on mutual knowledge between individuals (e.g., co-authorship networks). On the other hand, there is a set of social networks, some of which are formed over social media applications, that may not require mutual knowledge (and sometimes confirmation from the other side) for the interaction (e.g. Youtube). In all cases, we consider the graphs as unweighted and undirected. Moreover, we extract the largest connected component and use it as a good representative of the whole graph (this is a standard approach in such kind of studies).

Table 2. Summary of real-world networks used in this study

Network	Nodes	Edges	Description
Epinions [24]	75,877	405,739	Who trusts whom network
Email-EUAll [14]	224,832	340,795	Email network
Slashdot [15]	77,360	546,487	Slashdot social network (Nov. '08)
Wiki-Vote [12]	7,066	100,736	Wikipedia who-votes-on-whom network
Facebook [26]	63,392	816,886	Facebook New Orleans social network
Youtube [18]	1,134,890	2,987,624	Social network from Youtube site
CA-astro-ph [14]	17,903	197,031	Co-authorship network in Astro Physics
CA-gr-qc [14]	4,158	13,428	Co-author. network in General Relativity
CA-hep-th [14]	8,638	24,827	Co-author. network in High Energy Phys.

Figure 2 presents plots of the expansion character of the graphs we examined, together with the values $\xi(G)$. From a first look, it is clear that almost all social graphs (except the last two which we will discuss later) exhibit good expansion properties, showing linear correlation between the principal eigenvector and subgraph centrality in log-log scales. In all plots we have included a red line which represents this ideal behavior in case of graphs with big spectral gap and therefore good expansibility.

The results suggest that social graphs depicted in Fig. 2 (a)-(g), expand very well allowing the selection of arbitrary subsets of nodes with size at most $\frac{|V|}{2}$, such that for every set there is a relatively large number of edges with one endpoint inside the set and the other outside (in other words, every selection of such a subset creates a cut in the graph with a relative large size). Thus, a first conclusion is that these social graphs lack of edges which can act as bottlenecks. Furthermore, this result implies that the nodes inside the networks we examined are not organized based on a clear modular architecture. More precicely, a basic characteristic of the networks' structure is the absence of well defined clusters which can be easily seperated from the whole graph. In other words, the networks lack of clusters (communites) with a clear difference between the number of intra-cluster edges and inter-cluster edges[3].

However, in what degree are the above observations expected? Before trying to answer this question we must repeat that the used datasets correspond to social networks and on-line social networks from social media applications, with a large number of nodes and edges. It is known that the organization of social networks is based on communities (i.e. subgraphs with high intra-community and low inter-community edges). As a result, we expect that social networks will have poor expansion character because of the presence of communities. This means that it is very difficult for all the subsets of nodes to satisfy the constraint for good expansibility. In [7], the author measured the expansibility of a large number of real-world social networks, and showed that almost all of them have bad expansion properties, which is intuitively expected from the above discussion.

---

[3] We must note here that our findings do not imply the absence of communities from social graphs, but the subgraphs which may correspond to communities cannot be easily isolated, since they have a relatively large number of extra-community connections.

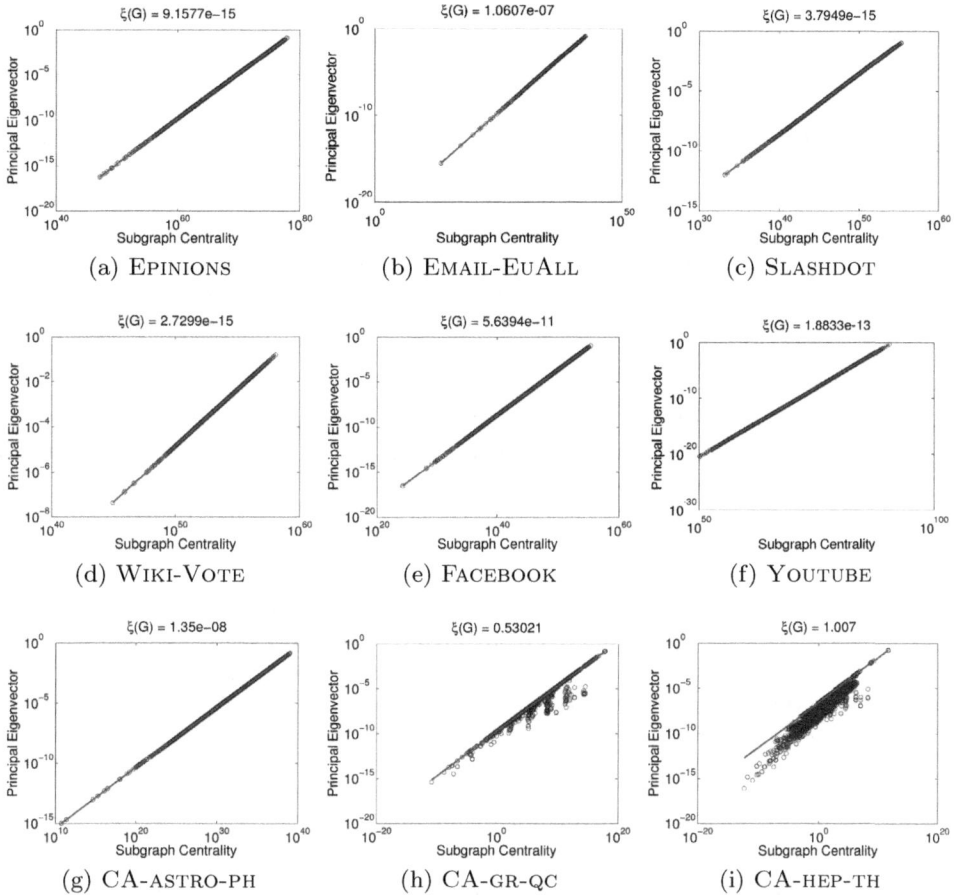

Fig. 2. Expansion properties of large social networks presented in Table 2. All plots depict the principal eigenvector vs. the subgraph centrality in log-log scale, and the $\xi(G)$ value for each graph.

However, it is very important to note that the social graphs studied in [7], have a small number of nodes and edges. Moreover, none of them has arisen from social media applications and generally online social networking, but almost all formed by physical interaction between people.

On the other hand, as our results suggest, the expansion character of large scale social graphs is completely different from that of small scale networks. Almost all studied social networks exhibit good expansibility, which we consider that is mainly due to two reasons. The first one, and the most obvious, is the scale of the network. It seems that, in large scale social graphs it is difficult to find subsets of nodes which can be easily isolated. For example, consider the co-authorship networks CA-ASTRO-PH, CA-GR-QC and CA-HEP-TH. While these networks are formed in a similar way (collaboration between scientists), the first one has about $18K$ nodes and $200K$ edges, while the other two have much smaller order (number of nodes) and size (number of edges) ($4K$ nodes, $13K$ edges and $8K$ nodes, $25K$ edges respectively). Figures 2 (g), (h) and (i) show the expansion properties of these graphs. We can observe their different behavior, where the larger one shows good expansibility with a very small $\xi(G)$ value (1.35×10^{-8}), while the other two do not show this property ($\xi(G) = 0.53021$ and $\xi(G) = 1.007$ respectively).

The second reason we consider for justifying these findings is that most of these networks are created over social networking and social media applications. Thus, because of the fact that the interaction may not require knowledge from both parts, it is easier to be achieved. Of course, something like that is very difficult to happen in social networks which require knowledge of the other part for an interaction.

Application Example. How these findings could be utilized in a real application, such as decentralized search in complex networks? This is a common problem in many applications, where starting from one initial node, we must locate a target node inside the network, without full knowledge of the global network structure (topology). Since the computation of the shortest path to the target node is unable, a strategy is to visit nodes using only local information, in such a way that every subset of visiting nodes has a large neighborhood and thus good expansibility (the goal is to reach the target node, minimizing the number of required steps). Since our findings suggest that large scale social graphs exhibit good expansion properties, the networks tend to be more searchable, making the above searching strategy more efficient.

Computational Issues. While subgraph centrality (4) provides a powerfull tool for measuring the expansion properties of a graph, it requires the computation of all eigenvalue - eigenvector pairs $(\lambda_i, \mathbf{u}_i)$, $\forall i \in N$, of the adjacency matrix $\mathbf{A}$. While this may not be a problem for small graphs, it becomes a computational bottleneck for large scale networks. In order to overcome this, we use the observation of [25], which states that the eigenvalues of real-world graphs are almost symmetric around zero, meaning that their signs tend to alternate. Moreover, because of the fact that the $\sinh(\cdot)$ function keeps the sign of the eigenvalues, we can use only the top strongest eigenvalues and their corresponding eigenvectors to achieve an excellent approximation of the subgraph centrality (in our experiments we keep the first 30 strongest pairs).

5   Conclusions

In this paper we measured the expansion properties of several large scale social graphs, using the measure of subgraph centrality. Our findings show that, in contrast to small social networks, large scale social graphs generally exhibit good expansibility. This is something that has not appreciated previously, and in many cases social graphs were characterized as graphs with poor expansion properties, indepedent of their size. Our observations, except for a better understanding of social networks' structure, could be possibly utilized in several domains and applications such as searching in networks, community discovery and more generally in applications over social networks where the robustness of the underlying structure is a crucial factor. In future work, we plan to further investigate and understand the underlying mechanisms which cause this behavior and how these findings could affect applications which consider the structure of social networks.

Aknowledgements

We would like to thank Alan Mislove for providing some of the network datasets used in this study.

References

1. Albert, R., Jeong, H., Barabasi, A.-L.: Diameter of the world wide web. Nature 401, 130–131 (1999)
2. Chakrabarti, D., Faloutsos, C., McGlohon, M.: Graph mining: Laws and generators. In: Aggarwal, C.C., Wang, H. (eds.) Managing and Mining Graph Data, ch. 3. Springer, Heidelberg (2010)
3. Chung, F.R.K.: Spectral Graph Theory. CBMS, Regional Conference Series in Mathematics, vol. 92. AMS, Providence (1997)
4. Chierichetti, F., Lattanzi, S., Panconesi, A.: Rumour spreading and graph conductance. In: SODA, pp. 1657–1663 (2009)
5. Erdös, P., Renyí, A.: On the evolution of random graphs. Publ. Math. Inst. Hung. Acad. Sci. 5, 17–61 (1960)
6. Estrada, E., Rodríguez-Velázquez, J.A.: Subgraph centrality in complex networks. Phys. Rev. E 71 (2005)
7. Estrada, E.: Spectral scaling and good expansion properties in complex networks. Europhys. Lett. 73(4), 649–655 (2006)
8. Estrada, E.: Network robustness to targeted attacks. The interplay of expansibility and degree distribution. Eur. Phys. J. B 52, 563–574 (2006)
9. Faloutsos, M., Faloutsos, P., Faloutsos, C.: On power-law relationships of the Internet topology. In: SIGCOMM, pp. 251–262 (1999)
10. Hoory, S., Linial, N., Wigderson, A.: Expander graphs and their applications. Bull. Amer. Math. Soc. 43 (2006)
11. Kumar, R., Novak, J., Tomkins, A.: Structure and evolution of online social networks. In: KDD, pp. 611–617 (2006)
12. Leskovec, J., Huttenlocher, D., Kleinberg, J.: Predicting Positive and Negative Links in Online Social Networks. In: WWW, pp. 641–650 (2010)
13. Leskovec, J., Kleinberg, J., Faloutsos, C.: Graphs over Time: Densification Laws, Shrinking Diameters and Possible Explanations. In: KDD, pp. 177–187 (2005)
14. Leskovec, J., Kleinberg, J., Faloutsos, C.: Graph Evolution: Densification and Shrinking Diameters. ACM TKDD 1(1) (2007)
15. Leskovec, J., Lang, K., Dasgupta, A., Mahoney, M.: Community Structure in Large Networks: Natural Cluster Sizes and the Absence of Large Well-Defined Clusters. Internet Mathematics 6(1), 29-123 (2009)
16. Maiya, A.S., Berger-Wolf, T.Y.: Expansion and search in networks. In: CIKM, pp. 239–248 (2010)

17. Maiya, A.S., Berger-Wolf, T.Y.: Sampling Community Structure. In: WWW, pp. 701–710 (2010)
18. Mislove, A., Marcon, M., Gummadi, K.P., Druschel, P., Bhattacharjee, B.: Measurement and Analysis of Online Social Networks. In: IMC, pp. 29–42 (2007)
19. Mohar, B.: Isoperimetric Number of Graphs. J. Comb. Theor. B 47, 274 (1989)
20. Newman, M.E.J.: The structure and function of complex networks. SIAM Review 45, 167–256 (2003)
21. Newman, M.E.J., Park, J.: Why social networks are different from other types of networks. Phys. Rev. E 68, 036122 (2003)
22. Newman, M.E.J.: Finding community structure in networks using the eigenvector of matrices. Phys. Rev. E 74 (2006)
23. Newman, M.E.J.: Modularity and community structure in networks. PNAS 103(23), 8577–8582 (2006)
24. Richardson, M., Agrawal, R., Domingos, P.: Trust management for the semantic web. In: Fensel, D., Sycara, K., Mylopoulos, J. (eds.) ISWC 2003. LNCS, vol. 2870, pp. 351–368. Springer, Heidelberg (2003)
25. Tsourakakis, C.E.: Fast Counting of Triangles in Large Real Networks without Counting: Algorithms and Laws. In: ICDM, pp. 608–617 (2008)
26. Viswanath, B., Mislove, A., Cha, M., Gummadi, K.P.: On the Evolution of User Interaction in Facebook. In: WOSN, pp. 37–42 (2009)

Text Representation Using Dependency Tree Subgraphs for Sentiment Analysis

Alexander Pak and Patrick Paroubek

Université de Paris-Sud, Laboratoire LIMSI-CNRS, Bâtiment 508,
F-91405 Orsay Cedex, France
{alexpak,pap}@limsi.fr

Abstract. A standard approach for supervised sentiment analysis with
n-grams features cannot correctly identify complex sentiment expressions
due to the loss of information when representing a text using the bag-
of-words model. In our research, we propose to use subgraphs from the
dependency tree of a parsed sentence as features for sentiment classification. We represent a text with a feature vector based on extracted
subgraphs and use state of the art SVM classifier to identify the polarity of the given text. Our experimental evaluations on the movie-review
dataset show that using our proposed features outperforms the standard
bag-of-words and n-gram models. In this paper, we work with English,
however most of our techniques can be easily adapted for other languages.

1 Introduction

Bag-of-words is one of the first model of the text representation which is nowadays still often used in sentiment analysis. In this approach, text is usually
represented as a set of unigrams (or bigrams) disregarding their order and relations within the text. Common machine learning techniques such as Naive Bayes
or SVM are then used to perform the sentiment classification of the given text.
Although the accuracy of such approaches can be quite high, especially when
using advanced feature selection techniques and additional lexicons of opinionated texts. We think that this model should be improved or replaced by the one
that can identify more complex sentiment expressions rather than only simple
ones such as 'good movie' or 'bad acting'.

One of the problems of bag-of-words representation is the information loss
when representing a text as a collection of unrelated terms. However, these relations are often very important and may change the degree and the polarity of a
sentiment expressed in the text. We illustrate this problem with a simple example. Let's consider a simple phrase: "This book is **bad**". The sentiment of this
phrase is obviously negative and a standard classifier based on unigrams model
will easily classify this sentence correctly provided a good training dataset. Now
let's make the sentence a little more complex: "This book is **not bad**". In this
case, a simple unigram model will probably fail. However, a bigram model will
still work, capturing 'not bad' as a term with a positive polarity. If we make
the sentence more complex: "This books is **surprisingly not that bad**", both

J. Xu et al. (Eds.): DASFAA Workshops 2011, LNCS 6637, pp. 323–332, 2011.

unigram and bigram models will fail. To make them work, a more sophisticated treatment of negations is needed.

Other than handling negations, n-gram model has problems with capturing long dependencies. A bigram model will capture "I like" as a positive pattern in a sentence such as "**I like** fish", but not in "**I** really **like** fish". If we would advance the task and move to a more refined polarity classification, i.e. identifying not only the polarity of a text (positive or negative), but also the degree of the polarity (low/high or even more precise), the bag-of-words model cannot provide us with the sufficient information.

In order to solve the problems of the bag-of-words text representation, we propose to use parsing dependency tree of a sentence to generate subgraphs that can be used to represent a text. A dependency tree is a graphical representation of a sentence where nodes correspond to words of the sentence and edges represent syntactic relations between them such as 'object', 'subject', 'modifier' etc. Figure 1 depicts a dependency tree of a sentence "I do not like fish very much".

Such a representation of the sentence suits very well for sentiment analysis purposes and even for opinion mining:

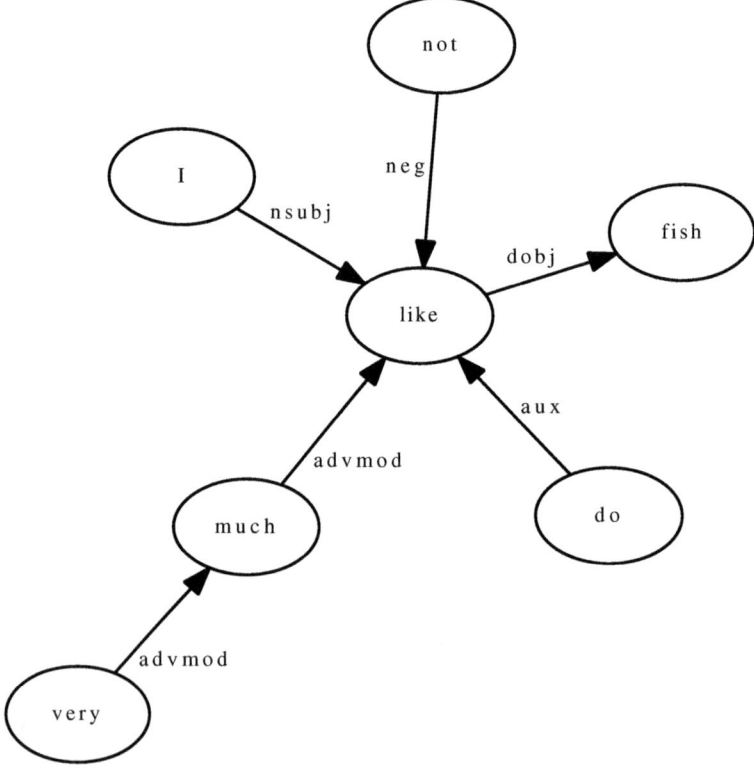

Fig. 1. A dependency tree of a sentence "I do not like fish very much". Nodes represent words, edges represent relations between words.

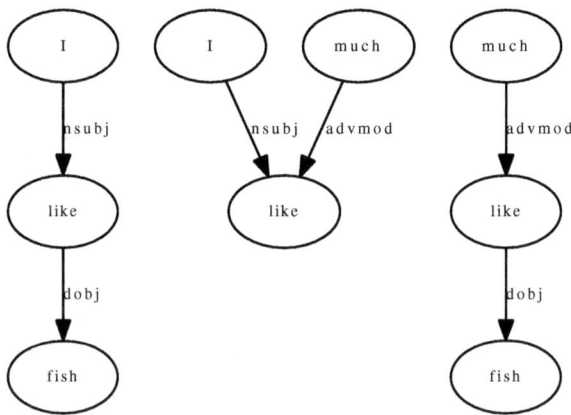

Fig. 2. A representation of a sentence "I like fish much" with subgraphs of size 2

- From the given tree, we can easily identify the negation subgraph "not $\xrightarrow{neg}$ like"
- We can find intensity markers: "very $\xrightarrow{advmod}$ much $\xrightarrow{advmod}$ like"
- Opinion holder: "I $\xrightarrow{nsubj}$ like" and opinion target: 'like $\xrightarrow{dobj}$ fish"

In our approach, we use subgraphs of the dependency tree of the sentences to represent a given text. Similarly to n-grams, we define the size of a subgraph which is equal to the number of edges it contains. Thus, a subgraphs of size 1 contains 1 edge and 2 nodes, a subgraph of size 2 contains 2 edges and 3 nodes etc. For example, the sentence "I like fish much" can be represented by subgraphs of size 2 as depicted on Figure 2.

In the next section, we explain in details how we obtain the dependency tree and generate subgraph representation of a text. In Section 2.2, we show how we use this representation to index movie-review dataset and train an SVM classifier for sentiment polarity classification. We present our experimental evaluation setup and results in Section 3. We discuss the prior research in Section 4 and conclude our work in Section 5

2 Our Method

2.1 Subgraph Representation

We use the output of typed dependencies of Stanford Parser [5] to obtain the dependency tree of a sentence. For the sentence "I do not like fish very much" it produces a list of dependencies from which we can reconstruct the tree:

```
nsubj(like-4, I-1)
aux(like-4, do-2)
neg(like-4, not-3)
dobj(like-4, fish-5)
advmod(much-7, very-6)
advmod(like-4, much-7)
```

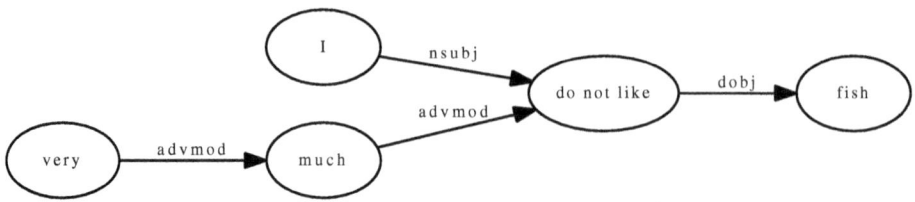

Fig. 3. The dependency tree after combining nodes and pruning edges

However, we would like to obtain a tree where each node has a finite meaning. In our example, nodes as 'not' and 'do' do not have meaning by themselves, and the node 'like' contains only a partial meaning (it lacks the negation). Therefore after we obtain the dependency tree, we combine certain nodes as follows:

- Combine nodes linked with the negation relation ('neg'), such that the edge "not $\xrightarrow{neg}$ like" becomes a single node "not like"
- Combine auxiliary verbs with the main verb ('aux'), such that the edge "do $\xrightarrow{aux}$ like" becomes a single node "do like"
- Combine copulae ('cop')

We also prune the following edges to avoid unnecessary information:

- Determiners, such as "a $\xrightarrow{det}$ book" becomes "book"
- Possesives, such as "my $\xrightarrow{poss}$ book" becomes "book"
- Noun modifiers, such as "dog $\xrightarrow{nn}$ food" becomes "food"

Applying the rules above, the resulting tree for our example would look as in Figure 3.

Finally, the sentence is represented by a set of all possible subgraphs of a size S, where S is equal to the number of edges of subgraphs. In our experiments, we used $S = 1$ and $S = 2$.

For each subgraph we obtain a permutation of subgraphs containing various number (from 0 to $S - 1$) of wildcard nodes. A wildcard node is a node that can match any word. The only exception is that we do not replace verbs and adjectives as they usually possess sentiments. For example, the obtained wildcard subgraphs for "I $\xrightarrow{nsubj}$ like $\xrightarrow{dobj}$ fish" are depicted in Figure 4.

The reason we have add wildcard subgraphs is similar to [1]. Our intension is to be able to match constructions such as "I like fish" and "I like books". In these two constructions only object is different, while the sentiment expression is the same. By introducing wildcard graphs along with specific graphs we are able to capture such phenomena.

2.2   Feature Construction

We represent a given text T as a feature vector $T = \{w_1, w_2, \cdots, w_K\}$, where w_i is a weight of a subgraph i in text T, K is the number of subgraphs in T.

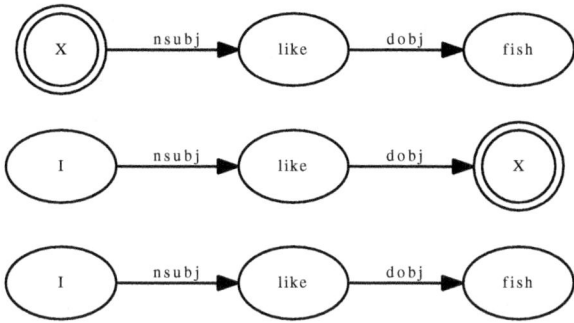

Fig. 4. Obtained wildcard subgraphs for "I like fish"

Similarly to the previous studies in sentiment classification, we experiment with the following types of weighting schemes:

1. Binary, represents whether the graph is presented in the text [11], i.e.:

$$w_i = 1 \ \forall i \in T \tag{1}$$

2. Frequency count:

$$w_i = tf_i \tag{2}$$

where tf_i (term frequency) is the number times a subgraph i occurs in T

3. Smoothed delta TFIDF [9]:

$$w_i = tf_i \cdot \Delta idf_i \tag{3}$$

$$\Delta idf_i = log \frac{N_1 \cdot df_2 + 0.5}{N_2 \cdot df_1 + 0.5} \tag{4}$$

where N_1 and N_2 are total numbers of documents of class 1 and 2, df_1 and df_2 are class frequencies of the graph i (i.e. numbers of documents of classes 1 and 2 in which the graph is occured). In our case, classes 1 and 2 are positive and negative documents.

2.3   Discounting Scheme

The prior research on sentiment analysis showed that review authors tend to express sentiments in the last part of the text [3]. Therefore we decided to capture the position of the sentence in which the subgraph occurs in relation to the whole text. We introduce two strategies on how to add position information when constructing the feature vector. In the first strategy, we divided text into quantiles (3, 4 and 5) and treated subgraphs from different quantiles as independent features. In the second strategy, we added a discounting scheme for term frequency counting:

$$tf_i = \sum_{\forall p_i \in T} f(p_i) \tag{5}$$

where $\{p\}$ is a set of sentence indices where subgraph i occurs and f is a discounting function. p_i is equal to the index of the sentence in which the subgraph occurs divided by the total number of sentences. Thus if the subgraph occurs in the first sentence $p_i = 0$, and if it occurs in the last sentence $p_i = 1$.

As for the discount function we have tried the following:

1. **Uniform**, evenly increases weights of the sentences in the end of the review

$$f(p) = p \tag{6}$$

2. **Sigmoid**, gives more weight to the sentences in the end of the review

$$f(p) = \frac{1}{1 + e^{-10p+5}} \tag{7}$$

3. **Cosine square**, gives more weight to the sentences in the beginning and in the end of the review

$$f(p) = \cos^2(\pi x) \tag{8}$$

4. **Sine square**, gives more weight to the sentences in the middle part of the review

$$f(p) = \sin^2(\pi x) \tag{9}$$

The graphs of the discount functions are given in Figure 5.

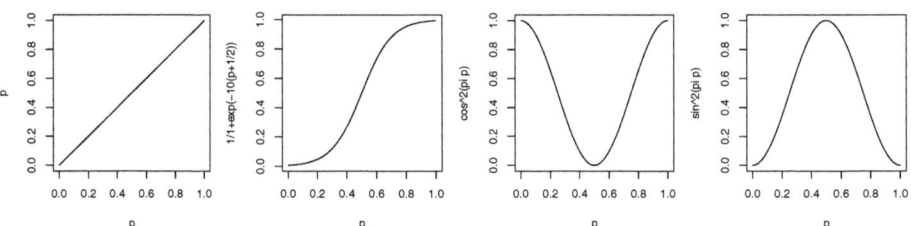

Fig. 5. Discount functions (from the left to the right): uniform, sigmoid, cosine square, sine square

3   Experiments and Results

3.1   Data and Evaluation Setup

We utilized an open source implementation of SVM classifier from the LIBLIN-EAR package [6]. We set default parameters and linear kernel. To evaluate our classifier, we used the movie-review dataset which is often used in sentiment analysis research[1] and initially was used in [11]. The dataset contains 2000 written movie reviews mined from the IMDb (Internet Movie database) web-site[2].

---

[1] http://www.cs.cornell.edu/people/pabo/movie-review-data/otherexperiments.html
[2] http://imdb.com

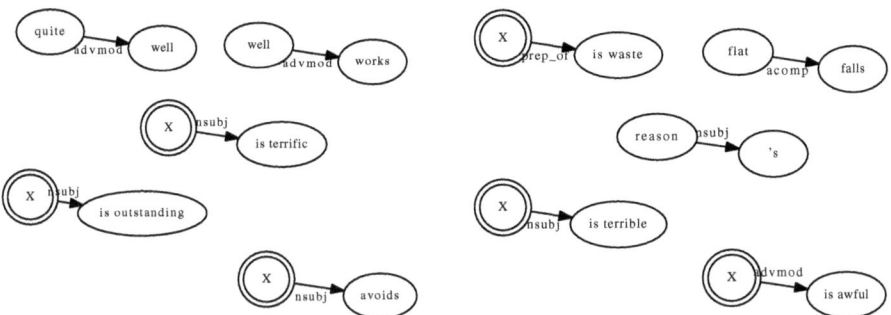

Fig. 6. Positive (on the left) and negative (on the right) subgraphs extracted from the movie-review database

The dataset is evenly split into a positive and a negative set (i.e. 1000 texts in each set). Each document contains a raw text without any HTML formatting. We use our subgraph text representation model with the subgraph of size 1 and also present the results with size 2.

We depict top-5 subgraphs (selected by Δidf score) from positive and negative set in Figure 6 to give an idea of the extracted features.

We apply 10-fold cross validation and measure the classification average accuracy. This way we can compare our results with [11] that we have chosen as the baseline. Pang et al. have obtained the best accuracy of 82.7% using SVM classifier on unigram model with binary features.

3.2   Results

Weighting Scheme. First, we examined which weighting scheme works the best for our subgraph representation model. We have compared binary, frequency and delta TFIDF schemes. The results of the evaluation are presented in Table 1

Table 1. Accuracy comparison for binary, frequency and delta TFIDF weighting schemes

Weighting Scheme	Ave. accuracy (%)	Δ
Baseline	82.7	-
Binary	81.9	-0.6
Frequency (tf)	80.7	-2
Delta TFIDF	**85.1**	+2.4

From the presented results, we can observe that the delta TFIDF scheme yields the highest accuracy of 85.1%. This result outperform the baseline on 2.4%. Binary weighting scheme (81.9%) performs better than frequency counting (80.7%). Similar results were obtained in previous researches [11][9]. Thus, for the further experiments we use the delta TFIDF weighting scheme.

Table 2. Accuracy comparison for different discounting schemes schemes

Discounting Scheme	Ave. accuracy (%)	Δ
No discounting	**85.1**	-
Uniform	84.4	-0.6
Sigmoid	84.45	-0.55
Cosine square	82.55	-2.55
Sine square	82.5	-2.6
3 quantiles	83.75	-1.35
4 quantiles	82.65	-2.45
5 quantiles	81.5	-3.6

Discounting scheme. Next, we have examined whether capturing a subgraph position within the text brings improves the classification results. We have compared uniform, sigmoid, cosine and sine discounting functions as well as the quantile strategy using 3, 4, and 5 quantiles as described in Section refsec:subgraphs. The evaluation results are presented in Table 2

We have obtained similar results as in Pang et al.[11] and did not confirm the importance of the last part of the review [3]. Capturing position using quantiles and discount functions does not improve the classification accuracy. Using sigmoid function yields the best performance (84.45%) as compare to other schemes and it is better than the baseline accuracy, while the accuracy is slightly lower than using no discounting scheme. Perhaps, a more refined way of notion the subgraphs positions is needed to obtain better results. Our explanation is that authors use their own style of writing and therefore express their opinions in different parts of reviews.

We have also tried to index the dataset using subgraphs of size 2. However, the obtained results were much lower. The best achieved accuracy was 76.5%.

4 Related Work

An early research by Pang et al. [11] using bag-of-words representation of text with binary features and SVM classifier became a baseline for many other works in the domain of sentiment classification. The authors improved their system in [10] using subjectivity detector and minimum cuts technique. Using subjectivity detector allowed to decrease the noise and focus only on sentences expressing sentiments. This raised the accuracy from 82.7% to 86.4%. Many further researchers utilized advanced techniques and additional lexicons to augment the feature space or refine the selected features and thus increasing the classification accuracy. Whitelaw et al. [12] used appraisal group features in combination with bag-of-words model and obtained a higher accuracy of 90.2% on the movie-review dataset. Aue et al. [2] used SVM with log likelihood feature selection and obtained an accuracy of 90.2% on the same dataset.

Sentence dependency tree has been widely used in the sentiment analysis domain. A recent research by Arora et al. [1] noted the problems of the standard bag-of-words text representation. The authors suggested their algorithm to extract subgraph features using genetic programming. However, the obtained features were not used to replace the standard n-gram model, but rather as a complementary set of features. Another recent research by Nakagawa et al. [8] utilized dependency tree to obtain features that were used to train a CRF classifier for sentiment polarity detection. In [13], authors use dependency tree to extract feature-opinion pairs, where the first member of the pair is a feature term (such as movie") and the second is an opinionated term (such as "masterpiece"). The dependency tree is used to establish relations between feature words and opinion keywords. In [4], dependency tree is used to normalize headlines to grammatically correct form for further sentiment tagging. In [7], authors use dependency tree to analyze the sentence construct along with WordNet[3] to perform sentence level sentiment classification.

5 Conclusion

With the population of blogs and social networks, opinion mining and sentiment analysis became a field of interest for many researches. An early work on supervised sentiment classification using n-gram model gave promising results and motivated many researches to use this model. However, bag-of-words text representation cannot capture complicated sentiment expressions and is hard to scale for bigger problems, such as identifying the polarity intensity or opinion holder/target tagging. A new text model is needed to improve the performance of sentiment analysis and opinion mining systems.

In our research, we develop a new text representation based on sentence parsing dependency tree. We represent a text as a collection of subgraphs, where nodes are words (or a wildcard) and edges represent relations between them. Such a representation allows to fill the information loss occurred when representing a text as a collection of n-grams without relation information.

We have tested our model on the movie-review dataset which was often used in sentiment analysis community. An SVM classifier with features based on the extracted subgraphs yields a better performance than traditional system based on the unigram model. The highest accuracy we obtained is 85.1%. However, we think it can be improved by using advanced techniques such as feature selection algorithms or utilizing additional sentiment lexicons.

We have examined different weighting for the feature construction and concluded that the best performance is achieved using subgraphs of size 1 and delta TFIDF scheme. We have also tried to note the position of the subgraphs within the given text by utilizing different discount functions and dividing texts into quantiles. However it did not provide us with any significant results.

In the future work, we plan to further develop our model, to apply it to more advanced sentiment analysis and opinion mining applications.

---

[3] A large lexical database of English: http://wordnet.princeton.edu/

References

1. Arora, S., Mayfield, E., Penstein-Rosé, C., Nyberg, E.: Sentiment classification using automatically extracted subgraph features. In: Proceedings of the NAACL HLT 2010 Workshop on Computational Approaches to Analysis and Generation of Emotion in Text, CAAGET 2010, Morristown, NJ, USA, pp. 131–139. Association for Computational Linguistics (2010)
2. Aue, A., Gamon, M.: Customizing Sentiment Classifiers to New Domains: a Case Study. In: Proc. International Conference on Recent Advances in NLP (2005)
3. Becker, I., Aharonson, V.: Last but definitely not least: on the role of the last sentence in automatic polarity-classification. In: Proceedings of the ACL 2010 Conference Short Papers, ACLShort 2010, Morristown, NJ, USA, pp. 331–335. Association for Computational Linguistics (2010)
4. Chaumartin, F.-R.: Upar7: a knowledge-based system for headline sentiment tagging. In: Proceedings of the 4th International Workshop on Semantic Evaluations, SemEval 2007, Morristown, NJ, USA, pp. 422–425. Association for Computational Linguistics (2007)
5. de Marneffe, M.-C., Maccartney, B., Manning, C.D.: Generating Typed Dependency Parses from Phrase Structure Parses. In: LREC (2006)
6. Fan, R.-E., Chang, K.-W., Hsieh, C.-J., Wang, X.-R., Lin, C.-J.: Liblinear: A library for large linear classification. J. Mach. Learn. Res. 9, 1871–1874 (2008)
7. Meena, A., Prabhakar, T.V.: Sentence level sentiment analysis in the presence of conjuncts using linguistic analysis. In: Amati, G., Carpineto, C., Romano, G. (eds.) ECIR 2007. LNCS, vol. 4425, pp. 573–580. Springer, Heidelberg (2007)
8. Nakagawa, T., Inui, K., Kurohashi, S.: Dependency tree-based sentiment classification using crfs with hidden variables. In: Human Language Technologies: The 2010 Annual Conference of the North American Chapter of the Association for Computational Linguistics, HLT 2010, Morristown, NJ, USA, pp. 786–794. Association for Computational Linguistics (2010)
9. Paltoglou, G., Thelwall, M.: A study of information retrieval weighting schemes for sentiment analysis. In: Proceedings of the 48th Annual Meeting of the Association for Computational Linguistics, ACL 2010, Morristown, NJ, USA, pp. 1386–1395. Association for Computational Linguistics (2010)
10. Pang, B., Lee, L.: A sentimental education: Sentiment analysis using subjectivity summarization based on minimum cuts. In: Proceedings of the ACL, pp. 271–278 (2004)
11. Pang, B., Lee, L., Vaithyanathan, S.: Thumbs up?: sentiment classification using machine learning techniques. In: Proceedings of the ACL 2002 Conference on Empirical Methods in Natural Language Processing, EMNLP 2002, Morristown, NJ, USA, vol. 10, pp. 79–86. Association for Computational Linguistics (2002)
12. Whitelaw, C., Garg, N., Argamon, S.: Using appraisal groups for sentiment analysis. In: Proceedings of the 14th ACM International Conference on Information and Knowledge Management, CIKM 2005, pp. 625–631. ACM, New York (2005)
13. Zhuang, L., Jing, F., Zhu, X.-Y.: Movie review mining and summarization. In: Proceedings of the 15th ACM International Conference on Information and Knowledge Management, CIKM 2006, pp. 43–50. ACM, New York (2006)

A Local Information Passing Clustering Algorithm for Tagging Systems

Yu Zong[1,2], Guandong Xu[3], Ping Jin[1,*], Peter Dolog[3], and Shan Jiang[1]

[1] Department of Information and Engineering, West Anhui University, Luan, China
[2] Department of Computer Science and Technology,
University of Science and Technology of China, Hefei, China
[3] IWIS-Intelligent Web and Information Systems, Aalborg University,
Computer Science Department Selma Lagerlofs Vej 300 DK-9220 Aalborg, Denmark

Abstract. Under social tagging systems, a typical Web2.0 application, users label digital data sources by using tags which are freely chosen textual descriptions. Tags are used to index, annotate and retrieve resource as an additional metadata of resource. Poor retrieval performance remains a major problem of most social tagging systems resulting from the severe difficulty of ambiguity, redundancy and less semantic nature of tags. Clustering method is a useful tool to increase the ability of information retrieval in the aforementioned systems. In this paper, we propose a novel clustering algorithm named LIPC (Local Information Passing Clustering algorithm). The main steps of LIPC are: (1) we estimate a KNN neighbor directed graph G of tags and calculate the kernel density of each tag in its neighborhood; (2) we generate local information, local coverage and local kernel of each tag; (3) we pass the local information on G by I and O operators until they are converged and tag priory are generated; (4) we use tag priory to find out the clusters of tags. Experimental results on two real world datasets namely MedWorm and MovieLens demonstrate the efficiency and the superiority of the proposed method.

1 Introduction

With the development of Web2.0 application services, tag-based services, e.g., Del.icio.us (http:// del.icio.us), Last.fm (http://www.last.fm), and Flickr (http:// flick.com), have undergone tremendous growth in the past years. Tags are simple, uncontrolled and ad-hoc labels that are assigned by users to describe or annotate any kind of resource. The low technical barrier of tag based recommender system and easy usage of tagging have attracted a large amount of research interest. The user-contributed tags are not only an effective way to facilitate personal organization but also provide a possibility for users to search for information or discover new things.

However, the ambiguity, redundancy and less semantic nature are the major problems suffering all tagging systems. For example, for one same resource, different users will use their own textual description to annotate, resulting in the tagging behavior

---

* Corresponding author.

J. Xu et al. (Eds.): DASFAA Workshops 2011, LNCS 6637, pp. 333–343, 2011.

much sparse and less semantic. In order to deal with these difficulties, recently clustering method has been introduced into tag based recommender system to find meaningful information conveyed by tags. As the user tagging behaviors can be modeled as data record with triple attributes, i.e. user, resource and tag, clustering on tagging data could be conducted on these three attributes respectively. The efficiency of clustering of tagging data is the ability of aggregating tags into topic domains. In past years, many studies have been carried out on tagging clustering. [1,2] demonstrated how tag clusters serving as coherent topics can aid in the social recommendation of search and navigation. In [3] topic relevant partitions are created by clustering resources rather than tags. By clustering resources, it improves recommendations by distinguishing between alternative meanings of query. While in [4], clusters of resources are shown to improve recommendation by categorizing the resources into topic domains. A framework named Semantic Tag Clustering Search which is able to cope with the syntactic and semantic tag variations is proposed in [5]. P. Lehwark et al. use Emergent-Self-Organizing-Maps (ESOM) and U-Map techniques to visualize and cluster tagged data and discover emergent structures in collections of music [6]. State-of-the-art methods suffice for simple search, but they often fail to handle more complicated or noisy Web page structures due to a key limitation. Miao et al. propose a new method for record extraction that captures a list of objects in a more robust way based on a holistic analysis of a Web page [7]. In [8], a co-clustering approach is employed, that exploits joint groups of related tags and social data resources, in which both social and semantic aspects of tags are considered simultaneously.

In the context of conventional tag clustering, the first step is to find out the clustering structure from tagging data, and then make use this structure for further applications such as forming recommended information. In this case, the quality of the clustering result is critical for the recommender system based tag. Most of the researches on tagging clustering are directly use the traditional clustering algorithms on tag data. These clustering algorithms often focus on local aspect of tagging data and cannot capture the global information of tagging. However, various tags used in tagging data apparently possess different significance in tag groups due to the semantic or domain topic tendency of tags. The basic idea of this paper is originated from the latent significance of each tag in tagging activities to creating tag clusters. Particularly in this paper, in contrast, we propose a clustering algorithm named Local Information Passing Clustering algorithm (LIPC). In LIPC, We first estimate a KNN neighbor directed graph G of tags, the kernel density of each tag in its neighborhood is calculated in the same time; We then use Local coverage and Local kernel to capture the local information of each tag; thirdly, we define two operators, that is, I and O, to pass the local information on G; then a tag priory is generated when I and O are converged; at last, we use the tag priory and their coverage values to find out the clusters of tags. Experimental results demonstrate the efficiency and the improved outcome of tag clusters by using the proposed method.

The contributions of our paper are as follows:

— We address the problem of improving the quality of the group information abstracted from the tags.

- We propose a new tag clustering algorithm named LIPC, and we define two indexes to capture the local information, and two passing operator named I and O are defined to transit these local information on the graph G and the tag priority is generated.
- We empirically investigate the effect of K on the generation of tag's priority and clusters.
- We conduct comprehensive experiments on two real world datasets. The evaluation results demonstrate the effectiveness of the proposed solutions against the traditional k-means clustering.

The remainder of this paper is organized as follows. We introduce the preliminaries in Section 2. The details of Local Information Passing Clustering Algorithm are discussed in Section 3. Experimental evaluation results are reported in Section 4. Section 5 concludes this paper and outlines the future work.

2   Preliminaries

2.1   Social Tagging System Model

In this paper, our work is to deal with tagging data. A typical social tagging system has three types of objects, users, tags and resources which are interrelated with one another. Social tagging data can be viewed as a set of triples [9,10]. Each triple (u,t,r) represents a user u annotates a tag t to a resource r. A social tagging system can described as a four-tuple, where exists a set of users, U ; a set of tags, T ; a set of resources, R ; and a set of annotations, A^N . We denote the data in the social tagging system as D and define it as $D = <U,T,R,A^N>$. The annotations are represented as a set of triples contains a user, tag and resource defined as: $A^N \subseteq <u,t,r>: u \in U, t \in T, r \in R$. Therefore a social tagging system can be viewed as a tripartite hyper-graph [11] with users, tags and resources represented as nodes and the annotations represented as hyper-edges connecting one user, tag and resource.

2.2   Tag Vector and Tag Similarity

X. Li, et al. analyze the bookmark data set and find a phenomenon that the distribution of URLs, Users and Tags follows power law distribution. This phenomenon indicates that most URLs are only bookmarked once and most users only bookmark one URL, in the same way, most tags are only annotated on one URL [12]. Recently, an experiment on detecting the pair-wise relationship between tags and resources and between tags and users has shown that only a small part of resources are annotated frequently by many tags, whereas as a large number of resources are annotated once, and that the same observation of power law distribution also exists in the relationship between tags and users. Most of applications on tags are using tags to describe resources or users, that is, a resource or user is defined as a tags vector. In this model, thus the tag vector is in a very high dimension due to the free style of tag texts. And most of tags are redundant and ambiguous, in turn; bring in a difficulty of similarity calculation. Therefore, tag clustering is able to capture the topic domains of tags, which is expected to partially handle the above problems. In a tagging system, resources are mostly fixed and unique. Tag can be described by a set of resources which this tag has been assigned to it by users. In this way, a tag vector was constructed by using resources as dimensions, e.g., $t_i = (r_1,...,r_m)$. The similarity between any two tags is defined as Definition 1.

Definition 1. Given two tags $t_i = (r_{i1},...,r_{im})$ and $t_j = (r_{j1},...,r_{jm})$, the similarity $Sim(t_i,t_j)$ is defined as the ratio of co-occurred resource.

$$Sim(t_i,t_j) = \frac{|t_i \cap t_j|}{|t_i \cup t_j|} \tag{1}$$

3   Local Information Passing Clustering Algorithm

In real world, how should we know other peoples whom we didn't know before? The recommendations from our friends are commonly used method. In web world, users are always using tags to appraise a resource and other users can accept the resource according to the annotated tags. This behavior of web could be regarded as the copy of real world, that is, the social network. Similarly, the tags could be regarded as the recommendation information. If we assume that the most similar K tags are the K friends of one tag, we can use the behavior of social network recommender system in the real world to simulate the tag's recommendation. But these recommendation information are always locally, so we need to define operators to pass these information to all the tags. In this section, we first use the KNN neighbor method to find out the K nearest neighbors of one tag and then construct a KNN directed graph G . Local information are defined by using the kernel density estimator method. In order to pass local information, we define two operators I and O to transit the local information to all the tags and the priority of each tag is generated. The purpose of this paper is to find the groups with similar tags, so we devise a clustering algorithm based on tag priority to generate tag clusters.

3.1   KNN Directed Graph and Local Information

According to Definition 1, a similarity matrix S could be constructed from the tagging data. From S, we can find KNN neighbors of each tag and then a KNN directed graph G could be created. Graph $G = <V,\{E\}>$, where V is the tag set and $\{E\}$ is the directed edge set between tags, $< p,q > \in \{E\}$ denotes that tag q is a KNN neighbor of tag p .

Fig.1 shows an example of a part of graph G . On one hand, the black node is tag p , and five heavy dark nodes with black line are the KNN-neighbors of p and there have arches between p and them. On the other had, p is the KNN-neighbor of each of light line node and there have arches between these nodes and p . In this way a KNN directed graph G could be constructed and the adjacency matrix A of G is defined as Definition 2.

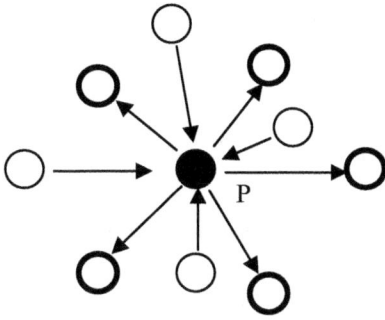

Fig. 1. An example of a part of Graph G

Defintion 2. Given a KNN directed graph G, the adjacency matrix is defined as A, where $A(p,q)=1$, if the directed arch $<p,q>$ exists, and $A(p,q)=0$, otherwise.

The KNN kernel estimate method [13] has mainly been used in capturing local character and density distribution. In this paper, we first use it to calculate the KNN kernel density of each node, and then define two important indexes named Local Coverage (LC) and Local Kernel (LK) to capture the local information of each node.

Defintion 3. Given a node $p \in G$ and its KNN neighborhood $N(p)$, the Local Coverage of p is defined by the KNN kernel density of its neighbors:

$$LC(p) = \sum_{q \in N(p)} f(q),$$

(2)

where $f(q)$ denotes the KNN kernel density value of node q.

Because $LC(p)$ is defined as the sum of KNN kernel density of p's KNN neighbors, the higher value of $LC(p)$ indicates more nodes with higher KNN kernel density in its neighborhood, and the probability of it locating in a high density area is higher.

Defintion 4. Given a node $p \in G$ and a node set Q which contains p, the Local Kernel of p is defined as the KNN kernel density of Q:

$$LK(p) = \sum_{q \in Q} f(q).$$

(3)

According to Definition 4, we can find that $LK(p)$ is the sum of KNN kernel density of nodes which directed to p. The higher value of $LK(p)$ indicates more nodes with higher KNN kernel density connected to p and p has a higher dependability to represent a local center.

3.2 Local Information Passing Clustering Algorithm

$LC(p)$ and $LK(p)$ capture the local information of each node in G. In this section we define a passing method to pass the local information to all the nodes and generate the authority and coverage of all the nodes based on $LC(p)$ and $LK(p)$. We define I and

O operators to transit the local information $LC(p)$ and $LK(p)$ respectively, as shown in Definition 5.

Defintion 5. Assume vector LC_i indicates the local coverage value of all nodes in ith iteration and vector LK_i indicates the local kernel value of all nodes in ith iteration respectively. The I and O operators are defined as:

$$I : LC_i = A^T \times LK_{i-1}, \quad O : LK_i = A \times LC_{i-1}. \tag{4}$$

The function of operator I is to pass the LK information of nodes which are directed to p, while, the function of operator O is passing the LC information of node p. After the convergence of operation I and O, we use LC_i to define the priority of each node in G.

Algorithm 1 gives the main steps of LIPC. In step1, we first find the KNN neighbor of each tag based on the tag similarity defined as Definition 1, and then, the local kernel density of each tag has been generated by using kernel density estimate operator. According to Definition 2, a KNN directed graph G could be constructed (step 2). For each node $v \in G$, we initialize the value of lc and lk according to Definition 3 and 4. In this way, the local information of each node has been captured and then we execute I and O operators to transit this local information on the graph G until these two operators are converged. We sort the vector LK_i to generate the priority of each tag. The node priority of tag shows the importance of the tag to its cluster. To generate the clusters: (1) we select a tag, $v \in G$, with highest priority as the centre of a cluster, (2) we use Depth First Search (DFS) method to find the corresponding cluster members which lc values are smaller than that of v. Steps (1) and (2) are iteratively executed until all the nodes in G are assigned to its corresponding cluster.

```
Algorithm 1
Input Tags, K
Output: cluster result, C.
1 generates KNN neighbors of a tag based on definition 1, and calculate
the KNN kernel density of the tag by using kernel density estimate
operator;
2 construct a KNN directed graph G according to definition 2;
3 generate the local information of each node by calculate lc and lk
based on definition 3 and 4 respectively;
4 Iteratively calculate LCᵢ and LKᵢ by using  I  and O operators
until they are converged;
5 Sorts LKᵢ ;
6 for each no visited node v∈G
6.1  v∈max(LKᵢ) ;
6.2  Cᵥ = DFS(G,v) , where Cᵥ denotes a cluster with v as centre;
6.3  C←C∪Cᵥ ;
7 Return C .
```

Table 1. The example of tag's priority

Tag	Priority
Favorite	2.39
Favorites	2.20
Like	2.02
Preference	1.98
JAVA	2.89
JavaScript	2.84

We run LIPC on Dmoz dataset (`http://www.michael-noll.com/dmoz100k06/`) and the priority of some tags are shown in Table 1.

From table1, we can find that the priority of Favorite is higher than Favorites, Like and Preference. This indicates that Favorite more likely selected as the centre than others, that is, Favorite could by generally describing other tags. And the same phenomenon shows for JAVA and JavaScript tags.

4 Experimental Evaluations

To evaluate our approach, we conducted extensive experiments. We performed the experiments using Intel Core 2 Duo CPU (2.4GHz) workstation with a 4G memory, running windows XP. All the algorithms were written in Matlab 7.0. We conducted experiments on two real datasets, MedWorm (http://www.medworm.com/) and MovieLens (http://www.movielens.org/).

4.1 Experimental Datasets

In order to evaluate our approach, we crawled the article repository in MedWorm system during April 2010 and downloaded the contents into our local experimental environment. After stemming out the entity attributes from the data, four data files, namely user, resource, tags and quads, are obtained as the source datasets. The first three files are recorded the user, tag and document information, whereas the fourth presents the social annotation activities where for each row, it denotes a user u annotates a resource r by a tag t.

The second dataset is MovieLens which is provided by GroupLens (http://www. grouplens.org/). It is a movie rating dataset. Users were selected at random for inclusion. All users selected had rated at least 20 movies. Unlike previous MovieLens datasets, no demographic information is included. Each user is represented by an id, and no other information is provided. The data are contained in three files, movies.data, rating.dat and tags.dat. Also included are scripts for generating subsets of the data to support five-old cross validation of rating predictions. The statistical results of these two datasets are listed in Table 2. These two datasets are pre-processed to filter out some noisy and extremely sparse data subjects to increase the data quality.

Table 2. Statistics of Experimental Datasets

Property	MedWorm	MovieLens
Number of users	949	4,009
Number of resources	261,501	7,601
Number of tags	13,507	16,529
Total entries	1,571,080	95,580
Average tags per user	132	11
Average tags per resource	5	9

In this paper, we use resources to describe tags, that is, each tag described by a set of resource which assigned to it by users. In order to reduce the length of tag vector, we first omit the resources which tags are less than the average tags per resource in table 2.

4.2 Evaluation Measurements

Our aim of the proposed method is to find out the similar tags which assign to different resources. Here we assume a better tag cluster composed by lots of similar tags and these tags are dissimilar to tags which belong to other different tag clusters. In particular, we use Similarity and Dissimilarity to validate our method.

Definition 6. Given tag cluster C, the Similarity is defined as

$$Similarity = \frac{1}{|C|}\sum_{k=1}^{|C|} Sim(t_i,t_j), \quad t_i,t_j \in C_k \tag{5}$$

Definition 7. Given tag cluster C, the Dissimilarity is defined as

$$Dissimilarity = \frac{1}{|C|}\sum_{k=1}^{|C|} Dism(k) \tag{6}$$

where $Dism(k) = \sum_{k'=1}^{|C|} Sim(t_i,t_j), \quad t_i \in C_k, t_j \in C_{k'}, k \neq k'$.

According to the requirement of clustering, we know that higher Similarity value and smaller Dissimilarity value indicate better clustering results.

4.3 Experiments and Discussion

Due to the priority of each tag has a close relationship with the kernel density which depends on the KNN neighbor, there have a relation between the number of K and the tag's priority. In order to present this relationship, we manually extract thirty tags which form two clusters from MedWorm data set. The priority of each tag has shown in Fig. 2 (the value of priority is multi by 100) with K equal to 4, 8, and 12 respectively.

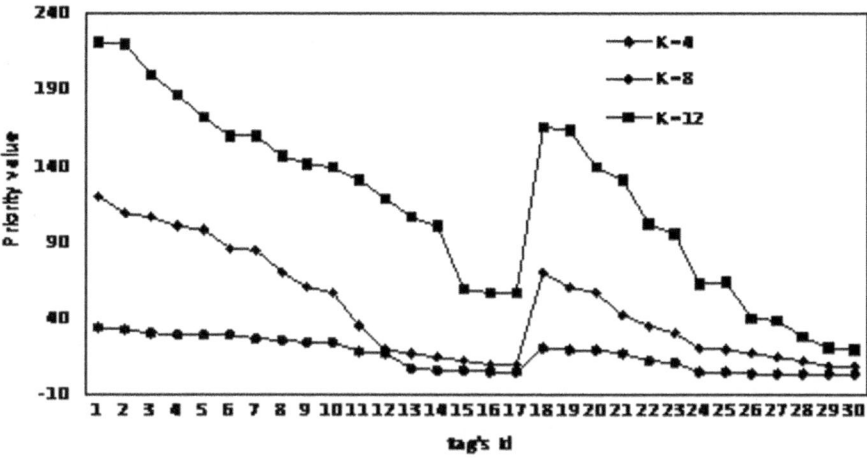

Fig. 2. The relationship between tag's priority and K

From fig. 2, we can find that the change of K has no influence on the tag's priority, as well as the structure of cluster.

Table 3. The comparison of Similarity and Dissimilarity on two datasets

	Similarity		Dissimilarity	
	MedWorm	MovieLens	MedWorm	MovieLens
K=4	0.973	0.867	0.0245	0.450
K=8	0.974	0.870	0.0240	0.448
K=12	0.970	0.869	0.0246	0.452

As we discussed in the previous sections, tag clustering can make tags be organized into groups over clusters. That is to say, by clustering, tags can be centralized into groups. In the following we will conduct the experiments to evaluate the effect on the tag cluster's quality. Table 3 gives the comparison result of Similarity and Dissimilarity on Medworm and MovieLens datasets. From Table 2, we can first find that the cluster results on two datasets are nearly coincident under different K settings. This phenomenon validates our previous experiment of the relationship between tag's priority and K's setting. As we defined in Definition 6 and 7, the higher Similarity value and the smaller Dissimilarity value indicates high quality of the tag cluster. On these two datasets, the Dissimilarity values are all small, and at the same time, the Similarity values are all high. This phenomenon shows that the quality of tag clusters obtained from Medworm and MovieLens by using our method are high. Interestingly, the clustering results derived from Medowrm look better than that of Movielens, which might be due to the tags used in Medowrm dataset is focused on a more specialized medical domain, while the domain topics span more diversely on movielens dataset. This finding is also verified by the measures of Dissimilarity.

In order to show the effectiveness of the proposed method, we execute traditional clustering algorithm K-means and our proposed algorithm LIPC on these two real world datasets as well. The experimental results are shown in Table 4.

Table 4. The comparison of LIPC and K-means on two datasets

	Similarity		Dissimilarity	
	MedWorm	MovieLens	MedWorm	MovieLens
K-means	0.873	0.856	0.0645	0.735
LIPC	0.966	0.885	0.0239	0.409

From Table 4, we can find that the quality of clustering results obtained by LIPC is better than that of K-means. In particularly, in one hand, the Dissimilarity values of LIPC on two datasets are smaller than that of K-means, and on the other hand, the Similarity values of LIPC on two datasets are, on the contrary, larger than that of K-means. This phenomenon indicates that LIPC algorithm has ability of finding better clustering results than that of K-means.

5 Conclusion

Tag clustering is a useful method to find out interest tag cluster embedded in tag datasets and it has a potential in improving the effectiveness and accuracy of tag based recommender system. In this paper, we propose a local information passing clustering algorithm for tags which is based on calculating the priority of tag. We first use the KNN neighbor and Kernel density estimate method to find out the local information of each tag, and then, and define I and O operator to transit the local information over all the tags and further generate tag priority, at last, we use the tag priority to find the representative centre of various clusters. Experimental results conducted on two real world datasets have demonstrated the effectiveness and advantage of the proposed method in comparison to other traditional clustering approaches.

The future work can be continuing on along the directions of the visualization of tag clusters and the improvement on recommendation.

References:

[1] Gemmell, J., Shepitsen, A., Mobasher, M., Burke, R.: Personalization in folksonomies based on tag clustering. In: Proceedings of the 6th Workshop on Intelligent Techniques for Web Personalization and Recommender Systems (July 2008)
[2] Shepitsen, A., Gemmell, J., Mobasher, B., Burke, R.: Personalized recommendation in social tagging systems using hierarchical clustering. In: RecSys 2008: Proceedings of the 2008 ACM Conference on Recommender Systems, pp. 259–266. ACM, New York (2008)
[3] Hayes, C., Avesani, P.: Using tags and clustering to identify topic-relevant blogs. In: International Conference on Weblogs and Social Media (March 2007)

[4] Chen, H., Dumais, S.: Bringing order to the web: automatically categorizing search results. In: CHI 2000: Proceedings of the SIGCHI Conference on Human Factors in Computing Systems, pp. 145–152. ACM, New York (2000)

[5] van Dam, J.W., Vandic, D., Hogenboom, F., et al.: Searching and browsing tag spaces using the semantic tag clustering search framework. In: IEEE Fourth International Conference on Semantic Computing (2010)

[6] Lehwark, P., Risi, S., Ultsch, A.: Visualization and Clustering of Tagged Music Data, pp. 673–680. GfKl, Berlin (2007)

[7] Miao, G.X., Tatemura, J.C., Hsiung, W.P., et al.: Extracting data records from the web using tag path clustering. In: Proceedings of the 18th International Conference on World Wide Web, Spain (April 2009)

[8] Giannakidou, E., Koutsonikola, V., Vakali, A., et al.: Co-clustering tags and social data sources. In: 9th International Conference on Web-age Information Managemnet, pp. 317–324 (July 2008)

[9] Guan, Z., Bu, J., Mei, Q., et al.: Personalized tag recommendation using graph-based ranking on multi-type interrelated objects. In: Allan et al. [1], pp. 540–547

[10] Guan, Z., Wang, C., Bu, J., et al.: Document recommendation in social tagging services. In: Rappa, M., Jones, P., Freire, J., Charkrabarti, S. (eds.) WWW, pp. 391–400. ACM, New York (2010)

[11] Mika, P.: Ontologies are us: A unified model of social networks and semantics. In: Gil, Y., Motta, E., Benjamins, V.R., Musen, M.A. (eds.) ISWC 2005. LNCS, vol. 3729, pp. 522–536. Springer, Heidelberg (2005)

[12] Lin, X., Guo, L., Zhao, Y.E.: Tag-based social interest discovery. In: Proceeding of the 17th International World Wide Web Conference (2008)

[13] Liu, H., Lafferty, J., Wasserman, L.: Sparse nonparametric density estimation in high dimensions using the rodeo. In: 11th International Conference on Artificial Intelligence and Statistics, AISTATS (2007)

What's in a Name: A Study of Names, Gender Inference, and Gender Behavior in Facebook

Cong Tang[1], Keith Ross[2], Nitesh Saxena[2], and Ruichuan Chen[3]

[1] Institute of Software, EECS, Peking University, China
MoE Key Lab of High Confidence Software Technologies (PKU), China
tangcong@infosec.pku.edu.cn
[2] CSE, Polytechnic Institute of NYU, Brooklyn, USA
{ross,nsaxena}@poly.edu
[3] MPI-SWS, Kaiserslautern, Germany
rchen@mpi-sws.org

Abstract. In this paper, by crawling Facebook public profile pages of a large and diverse user population in New York City, we create a comprehensive and contemporary first name list, in which each name is annotated with a popularity estimate and a gender probability.

First, we use the name list as part of a novel and powerful technique for inferring Facebook users' gender. Our name-centric approach to gender prediction partitions the users into two groups, A and B, and is able to accurately predict genders for users belonging to A. Applying our methodology to NYC users in Facebook, we are able to achieve an accuracy of 95.2% for group A consisting of 95.1% of the NYC users. This is a significant improvement over recent results of gender prediction [14], which achieved a maximum accuracy of 77.2% based on users' group affiliations.

Second, having inferred the gender of most users in our Facebook dataset, we learn several interesting gender characteristics and analyze how males and females behave in Facebook. We find, for example, that females and males exhibit contrasting behaviors while hiding their attributes, such as gender, age, and sexual preference, and that females are more conscious about their online privacy on Facebook.

1 Introduction

The current Online Social Networks (OSNs) allow users to control and customize what personal information is available to other users. For example, a Facebook user (Alice) can configure her account in such a way that her friends can see her photos and interests, but the general public can see only her name.

However, Alice probably assumes that if she makes available only her name to the general public, third parties have access only to her name and nothing more. Unfortunately for Alice, third parties, by crawling OSNs and applying statistical and machine learning techniques, can potentially infer information – such as gender, age, relationship status, and political affiliation – that Alice has not explicitly made public[14]. To the extent this is possible, third parties not only could use the resulting information for

J. Xu et al. (Eds.): DASFAA Workshops 2011, LNCS 6637, pp. 344–356, 2011.

online stalking and targeted advertising, but could also sell it to others with unknown nefarious intentions. This information may also be useful to Facebook itself, e.g., to provide a personalized service to its users, and to understand user characteristics and behaviors.

Prior work has considered this problem in the context of Facebook and other OSNs [14]. Their approach is based on a general observation that it may be possible to infer private information about Alice by exploiting information provided by Alice's friends or based on Alice's affiliations with various Facebook groups (public information). For example, if the majority of Alice's friends reveal that they are in their twenties and are Republicans, then it is highly probable that Alice is also in her twenties and is a Republican. Similarly, if Alice is a member of a girls' high school, then she is likely a female. For predicting gender, different inference models based on machine learning techniques were studied in [14]. However, this work only had limited success at gender prediction, with a maximum accuracy of 77.2% based on users' group affiliations. Moreover, and perhaps more importantly, this method of predicting gender can be circumvented by hiding group affiliations from public profiles, as also mentioned in [14].

Our approach to gender inference is based on users' first names. Our observation is that since name is a fundamental attribute of a Facebook user, which can not possibly be hidden from general public (and users also do not intent to use fake names, otherwise it will be hard to locate the user), a name-centric approach to gender inference would be difficult to evade. To develop such an approach, it is necessary to analyze users' names.

Our Contributions: We make three-fold contributions:

• *Facebook-Generated Name List:* By crawling Facebook public profile pages for 1.67 million users in New York City, we create a comprehensive and contemporary name list, in which each name is annotated with a popularity estimate and a gender probability. Note that traditionally it has been laborious, via census or otherwise, to obtain a contemporary list of people's names. We study the properties of this annotated name list. After combining nicknames with their "canonical names," we find that the resulting name popularity has a Zipf distribution, and that more than 94% of the names can be assigned a specific gender with high probability.

• *Name-Centric Gender Inference:* Our name-centric approach to gender prediction partitions the users into two groups, A and B, and is able to accurately predict gender for users belonging to A. Applying our methodology to NYC users in Facebook, we are able to achieve an accuracy of 95.2% for group A consisting of 95.1% of the NYC users. This is a significant improvement over recent results of gender prediction in [14], which achieved a maximum accuracy of 77.2% based on users' group affiliations.

• *Gender Behavior and Characteristics:* Having inferred the gender of most users in our Facebook dataset, we learn several interesting gender characteristics and analyze how males and females behave in Facebook. We find, for example, that females and males exhibit contrasting behaviors while hiding their attributes, such as gender, age, and sexual preference, and that females are more conscious about their online privacy on Facebook.

2   Related Work

We review prior work most closely related to the theme of our paper. Most of the prior work is concerned with the problem of inference of one or more private attributes, which is related to our second contribution in this paper. We are not aware of any prior research that analyzes and builds on users' names over OSNs (our first contribution).

Zheleva and Getoor [14] proposed techniques to predict the private attributes of users in four real-world datasets (including Facebook) using general relational classification and group-based classification. In addition to gender inference (which is the focus of our work), they also looked at prediction of political views. Their accuracy for gender inference with their Facebook dataset, however, is only 77.2% based on users' group affiliations, and the sample dataset used in their study is quite small (1,598 users in Facebook). Moreover, their inference methods can be prevented by hiding group affiliations from public profiles, as mentioned in [14]. In contrast, our inference methodology – based on users' names – would be difficult to circumvent, and we demonstrate its validity on a much larger dataset and achieve much better accuracies.

Other papers [8,13,9,7] have also attempted to infer private information inside social networks. Methods they used are mainly based on link-based traditional Naive Bayes classifiers. However, none of them used name-list to infer users' genders, and we achieve much better accuracies compared to these methods for gender inference.

Jernigan and Mistree [4] demonstrated a method for accurately predicting the sexual orientation of Facebook users by analyzing friendship associations. In particular, they have been successful at predicting whether a Facebook user might be homosexual by correlating similar information provided by user's friends.

Most recently, Mislove et al. [11] proposed a method of inferring user attributes by detecting communities in social networks, based on the finding that users with common attributes form dense communities. However, people with same attributes, such as gender and birthday, may not form communities, and thus these attributes may not be accurately predicted using this approach.

3   Crawling and Data Gathering

We develop a multi-threaded crawler that visits Facebook user profile pages and stores these pages in a file system. In July, 2009 we obtained a list of Facebook IDs of users in NYC ("New York, NY" network). We were able to do that because at that time, users, by default, were assigned to geographical networks. For each ID, we visit each of its friends, then each of its friends' friends, and so on, until we obtain all NYC users reachable. Because of size of Facebook's social network, the crawler was restricted to profiles only inside NYC. The crawler obtained the profile of pages for 1.67 million users. At the time of the crawl, there were approximately 2 million NYC users. We suspect that most of the non-crawled users are bogus users (see below). Therefore, we crawled nearly all the Facebook users in NYC.

Eliminating Bogus Users: Although many Facebook users have hundreds of friends and 50% of users visit the site daily (as discussed in [1]), many of the users may be *bogus or dormant*: users who signed up, created a few friends, and disappeared quickly. It may be difficult to predict anything about such users. In order to prevent these bogus

Table 1. Properties of the dataset from NYC before and after elimination of bogus users

Property name	Before	After
# users in NYC	1, 668, 602	1, 282, 563
# users who specified gender	864, 543	679, 351
% users who specified gender	51.81	52.97
# users who identified as males	456, 591	349, 730
# users who identified as females	407, 952	329, 621

users from skewing the results of our study, we remove, from our dataset, the users with less than 5 friends across Facebook.

The size of our compressed dataset is 1, 282, 563. Out of the 679, 351 users who specified their genders, the percentage of males is 52.97%. Table 1 shows the properties of the dataset before and after the elimination of bogus users. In this paper, we do all processing on the reduced data set after elimination of bogus users.

4   Using Facebook to Generate an Annotated Name List

We demonstrate that the Facebook network can be used to generate an up-to-date list of first names of the users. In our name list, each first name is annotated with the number of users having this name, the number of male users who have identified themselves with this name, and the number of female users who have identified themselves with this name. To guide the design of our gender inference scheme (as we will discuss in Section 5), we have carefully studied the properties of this list. Our name list and its properties are also of independent interest for other applications, including naming newly born babies and studying naming trends.

We first extract the first names for each of the 1.28 million users and create a crude annotated name list. Note that a Facebook user can choose to "Display Full Name" either as "First Last" or "Last First". We carefully handle this issue. We then process the crude list to remove entries that are not really names. We remove all one-letter names, all names without a vowel, and names that have been referenced only once. Notice that, for the gender inference analysis in Section 7, we still infer the gender of users whose names have been removed from the list.

After this pre-processing, we obtain a list having 23, 363 names. For each name in the list, we determine the number of users having this name, the number of times it is labeled as male, and the number of times it is labeled as female. We provide this name list online, publicly available at: http://sites.google.com/site/facebooknamelist/.

4.1   Combining Names with Their Nicknames

As one would expect, we found that many Facebook users identify themselves by using nicknames as their first names. The nicknames, however, might behave as noisy data in our analysis. To avoid this, we design a method that combines nicknames with their "canonical names".

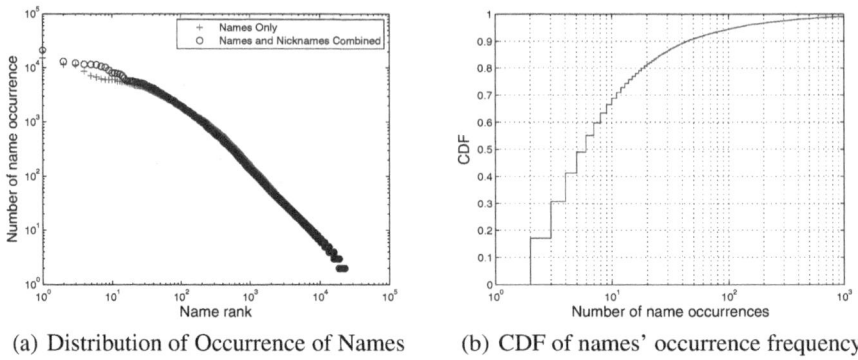

(a) Distribution of Occurrence of Names   (b) CDF of names' occurrence frequency

Fig. 1. Properties of names

We first create a nickname list which contains 535 nicknames based on resources available on the Internet (e.g., http://www.yeahbaby.com/, http://www.moonzstuff.com/articles/nicknames.html). For each nickname, we list its canonical names. For example, *Dave*'s canonical name is *David*, and *Stan*'s canonical names are *Stanford* and *Stanley*. Next, we combine the frequency of occurrence of each nickname with frequency of occurrence of its respective "canonical names". Specifically, if a nickname only has one "canonical name", we simply add its frequency of occurrence with the frequency of occurrence of its "canonical name"; if a nickname has multiple canonical names, we calculate its frequency of occurrence based on the frequency of occurrence of each of its "canonical names". For example, let x, y and z be the frequency of occurrence of *Stanford*, *Stanley* and *Stan*, respectively. When combining *Stan* with *Stanford* and *Stanley*, we redefine $x = x + z \cdot \frac{x}{x+y}$, and $y = y + z \cdot \frac{y}{x+y}$. After combining nicknames with names, we obtain a name list with $22,878$ entries.

4.2 Analysis of Annotated Name List

Our annotated name list is large and comprehensive (reflecting the broad and diverse demographics of NYC); moreover, this name list is annotated with the number of declared males and females corresponding to each name.

Note that there is a government online service [3] that provides a list of the most popular names for a particular year of birth in the US. However, our annotated name list contains information about NYC Facebook users born both in and outside the US. Moreover, from the public online service, one can only get at most top $1,000$ names for each year, from which we can obtain a total of $1,736$ male names and $2,023$ female names. Since our name list consists of $22,878$ entries, it is much larger and more diverse than the name list we get publicly from [3]. We now study several interesting properties of this name list.

Popularity of names. Figure 1(a) shows the distribution of names' occurrence frequency, which roughly follows the power-law distribution with a Zipf parameter $\alpha = 1.3$. We get a more flat Zipf curve after the name/nickname combination. Interestingly, after the combination, there is a single most popular name – *Michael* – which occurs

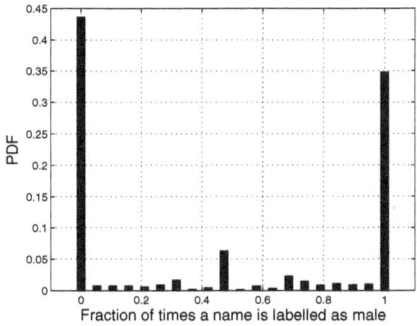

Fig. 2. PDF of fraction of times a name is labeled as male

more than $20,000$ times in our name list; then, the next 7 most popular names – *David*, *Elizabeth*, *Jennifer*, *Robert*, *John*, *Joseph* and *Daniel* – occur more than $10,000$ times each. Indeed, these popular names are classic and common American names. From Figure 1(b), we investigate the distribution of names from another perspective. We find that around 18% of names occur only twice. (Note that, when generating the name list, we have removed names that are referenced only once.) Moreover, 80% and 90% of names occur no more than 20 and 50 times, respectively, in our name list.

Gender consistency of names. Due to various reasons, e.g., the cross-gender names and possible mislabeling, some names may have been labeled as both male and female. Specifically, for each name in our name list, let N_m be the number of users who indicate they are male, and N_f be the number of users who indicate they are female. The fraction of times that a specific name is labeled as male is $f_m = N_m/(N_m+N_f)$. From Figure 2, it is clear that most names are associated with a specific gender; only about 6% of names are ambiguously labeled (i.e., $f_m = 0.5$). This observation will play a central role in our gender inference methodology (as we will discuss in Section 5).

The above analysis of our annotated name list provides some useful insights for gender inference. But a methodology solely based on the name list will clearly have some difficulties in predicting two types of names: names that have never been labeled and names that are used for both genders. For these two types of names, we have no choice but to resort to other inference methods. In particular, we adopt machine learning techniques (as we will discuss later) to predict these unlabeled and ambiguous names.

5   Design of Gender Predictors

In this section, we propose seven predictors for gender inference. These predictors use different features and algorithms, and use different methods of gender inference. We first investigate gender inference using the offline name list and our Facebook generated name list (as discussed in Section 4).

Besides name list, we take into account additional information, such as users' local and friends information, to improve our prediction. We adopt machine learning algorithms to classify users based on gender. Finally, we combine our annotated name list predictor with these classification algorithms.

5.1   Offline Name List Predictor (OFL)

We created a first-name list using USA baby name list [3], which consists of $1,736$ male names and $2,023$ female names. Some names in the list, such as "Chris", can be both a male's as well a female's name (e.g., Christopher and Christine, respectively). Such ambiguous names may decrease the gender prediction accuracy, and thus we remove names that are labeled as both male and female from the list. After that, we obtain $1,520$ male names and $1,807$ female names. We then compare each NYC Facebook user to the name list: if user's name can be found in the list, we predict its gender accordingly; otherwise, we only make a random guess to predict the gender.

5.2   Facebook Generated Name List Predictor (FB)

Our annotated name list (discussed in Section 4) is much larger and more comprehensive than the aforementioned offline name list. We compare the two lists and find many unpopular first names in our annotated name list that have not been listed in the offline name list. We use Facebook generated name list to predict user's gender.

We assign probability to each name in the list based on the fraction of times a specific name is labeled as male, i.e., $f_m = N_m/(N_m + N_f)$. For example, if a name "Tom" has been labeled 95 times as male and 5 times as female, "Tom" is predicted to be a male with probability 95%. We randomly guess for names do not appear in the list.

5.3   Local Information Predictor (LCL)

Generally, additional information available from user's Facebook profiles, such as relationship status and sexual preferences, can be helpful to our prediction methodology. We select 12 features of a user from his/her profile page, which are six relationship status (single, in a relationship, engaged, married, it's complicated, and in a open relationship), two sexual preference settings (interested in men/women), and four "Looking for" attributes (looking for friendship, dating, relationship and networking). Each (binary) feature has a value of 1 if the user corresponds to this feature, and 0 otherwise. For example if the feature "Relationship status: single" is 1, the user has indicated he/she is single. We then build our feature vector for a classifier using these twelve features. We choose training data from the profiles of users who have identified their genders, and feed the feature vectors to traditional classifiers.

5.4   Friend Information Predictor (FRND)

In this predictor, we take each user's friends' information into account. We introduce a new feature which is the fraction of a user's male friends. For the friends who have not specified their gender, we pre-assign genders to them using FB predictor.

5.5   Hybrid Predictors

Name List and Local Information Predictor (FB-LCL). We combine the FB predictor and the LCL predictor to obtain the FB-LCL predictor. This predictor uses a feature vector for the classifier using the 12 features from the LCL predictor and 2 extra features obtained from the Facebook generated name list: number of times the name is labeled as male, and number of times it is labeled as female.

Name List and Friend Information Predictor (FB-FRND). We combine the two aforementioned features obtained from the Facebook generated name list and the feature used in FRND predictor into the feature vector for FB-FRND predictor.

Name List, Local and Friend Information Predictor (FB-LCL-FRND). We combine the FB-LCL predictor and the FRND predictor into a single predictor: FB-LCL-FRND. This predictor extends the FB-LCL predictor's feature vector with features used in the FRND predictor.

6   Evaluation of Gender Predictors

6.1   Experimental Setup

We ran experiments for each of the seven predictors (discussed in Section 5). For the LCL, FB-LCL, FB-FRND and FB-LCL-FRND predictors, we choose users who have specified their genders from our data set, generate corresponding feature vectors for each predictor, then split the feature vectors into test set and training set by randomly marking each user's gender as unknown with a given probability. In the following experiments, we use a probability of 50%. We use the Weka toolkit [6] to build classifiers for all of the above training sets. We explored a variety of classifier types and selected Multinomial Naive Bayes (MNB) [10] which yielded the best overall performance in preliminary tests using the training set. In FRND predictor, instead of using MNB classifier, we use a decision tree based classifier J48 [12].

6.2   Effectiveness of Gender Predictors

We outline our inference results as follows.

- The results show that the OFL predictor achieves an accuracy of 75.5% by using the offline name list, in which 55.2% of users' names can be found.
- Our Facebook generated name list significantly improves the inference accuracy to 92.6%, in which 91.7% of users' names can be found.
- LCL predictor provides a higher accuracy (66.9%) than FRND predictor (60.0%) in classification based gender inference.
- Introducing users' local information by using the FB-LCL predictor provides a small gain, increasing the accuracy to 94.8%.
- Introducing friends' information by using the FB-FRND predictor also provides a small gain, increasing the accuracy to 94.1%.
- Friends' information does not provide any additional gain when using the FB-LCL-FRND predictor (94.6% accuracy), because there is some noise along with the friends' information that decreases the prediction accuracy.

Impact of Features in the LCL predictor. We run experiments to determine the local features which are most important and useful for gender inference. Specifically, we test four different feature vectors outlined as follows:

1. *Feature Vector 1* is composed of 6 relationship status features of the user whose gender is to be predicted.

2. *Feature Vector 2* is composed of 2 sexual preference features.
3. *Feature Vector 3* is composed of 4 'looking for' features.
4. *Feature Vector 4* is composed of all the 12 features.

From the results we can see that *Feature Vector 2* can lead to the highest accuracy (66.9%) among the first 3 feature vectors (which are 52.8%, 65.2% and 54.2% sequentially). This result is perhaps not surprising because sexual preference is generally more correlated to gender than relationship status and what people are looking for. This observation will help us improve our following inferences.

Impact of friends number in the FRND predictor. We try to determine the performance of the FRND predictor on users with different number of friends. We generate four training set containing the users who have no less than 1, 5, 10 and 20 NYC friends, and then apply the FRND predictor to them. The accuracies are 60.0%, 60.8%, 61.4%, 61.8% sequentially. We can see that increasing the NYC friends number threshold from 1 to 20 provides small gains.

Validating the FB predictor Using Boston Network. We validate our FB predictor using another network – "Boston, MA" network (Boston). We obtain $156,940$ users in Boston Facebook, in which there are 53.7% males and 46.3% females. Since in the Boston database, we only crawled users' names (At the time of the crawl, Facebook has removed the feature that publicly accessing profile pages of users in the same network.), so for each user, we apply the FB predictor to predict the gender. The prediction accuracy is 92.7%. We find that $144,946$ users' names in Boston can be found in our Facebook generated name list. These results show that the FB predictor performs well on and extends to other networks beyond NYC.

7   Inferring Gender for NYC Facebook Users

We first provide the approach to partition the users into two groups, A and B. The users belonging to Group A are further divided into various subsets, and we are able to apply different gender predictors to each of these subsets to get better results than using a single predictor. For users in Group B we have to randomly guess. Finally, we provide our inference results and ideas to further improve the inference accuracy.

7.1   User Partitioning

Inspired by the analysis of our Facebook generated annotated name list presented in Section 4.2, we first partition the users into two groups. For the first group A, we are mostly certain about users' genders, and for the second group B, we are randomly guessing. Users belonging to Group B should satisfy all the following conditions:

– Names never appeared in our annotated name list;
– Do not specify their local information;
– Have very few friends in NYC (we will set a friend number threshold later).

Our detailed partitioning method is described as follows. Let U be the set of all users. Let V be the set of users who have a name in our name list and are not in the ambiguous gender group, i.e., with an $f_m > T_1$ or $f_m < 1 - T_1$, where f_m is the fraction of times that a specific name is labeled as male, and the ambiguous threshold T_1 is in $(0.5, 1]$. Let W be the set of users in U who specified their local information. Let X be the users who have no less than T_2 friends in NYC, where T_2 is a threshold for number of friends. So, we divide the users into two groups: Group A consists of the set $V \cup W \cup X$, and Group B consists of the rest, i.e., $U - A$.

7.2 Applying Gender Predictors to Group A

We adopt different gender predictors (discussed in Section 5) to various subsets of users belonging to Group A.

1. For users in $V \cap W$, since their names can be found in the non-ambiguous group of our name list, and have specified their local information, we can adopt the FB-LCL predictor to achieve a high prediction accuracy.
2. For users in $V - W$, whose names can be found in our name list but have not specified local information, by using the FB predictor, we will achieve a high prediction accuracy, if we set an appropriate value for the threshold T_1.
3. For users in $W - V$, it is not effective to use only the name-list based predictors, since their names either have never been labeled or exist in the ambiguous name group. We instead employ a local information based predictor – LCL – for users belonging to the set $W - V$.
4. For users in set $X - V - W$, it is not effective to use the name-list based or local information based predictors. We can, however, predict users' genders using the FRND predictor.

7.3 Gender Inference Results

Parameter Selection. In our experiments, we consider two different thresholds: $T_1 = 0.65$ and $T_1 = 0.8$. We place the users from U, who specified their sexual preference information, in the set W, based on the result in Section 6.2. Then, we choose $T_2 = 5$, based on the results from Section 6.2. We eventually get a Group A which consists of 96.3% of the users, when $T_1 = 0.65$ and 95.1% of the users when $T_1 = 0.8$.

Table 2. Accuracies of Gender Inference

Group	Fraction of Users with $T_1 = 0.65$	Training and test dataset size with $T_1 = 0.65$	Accuracy with $T_1 = 0.65$	Fraction of Users with $T_1 = 0.8$	Training and test dataset size with $T_1 = 0.8$	Accuracy with $T_1 = 0.8$
$V \cap W$	21.1%	$244, 438$	97.3%	20.2%	$234, 562$	98.6%
$V - W$	68.1%	$365, 006$	96.8%	65.4%	$350, 023$	98.5%
$W - V$	2.69%	$30, 195$	89.7%	3.54%	$40, 073$	89.6%
$X - V - W$	4.4%	$39, 712$	61.7%	5.94%	$54, 693$	63.0%
A	**96.3%**	**679,351**	**94.6%**	**95.1%**	**679,351**	**95.2%**

We then adopt our gender predictors to those users in Group A. We choose inference dataset from users who have identified their genders, and split the dataset into training set and test set by randomly marking each user's gender as unknown with a probability 50%. We list the size of each inference dataset in Table 2.

Results. Table 2 provides a summary of our inference results. In addition to accuracies, we also indicate the fractions of the users belonging to various sets, for the two threshold values $T_1 = 0.65$ and $T_1 = 0.8$. We find that for $T_1 = 0.65$, Group A consists of 96.3% of users and has an accuracy of 95.5%. Also, for $T_1 = 0.8$, Group A consists of 95.1% of users and has an accuracy of 95.2%. These results represent a significant improvement over recent results of gender prediction of [14], which achieved a maximum accuracy of 77.2% based on users' group affiliations. After final inference, the male and female composition of the NYC Facebook network turns out to be 49.8% and 50.2%, respectively. This composition is different from the composition prior to our inference, which is 51.5% males and 48.5% females.

We note that recently Facebook has updated its privacy settings [2]. Under the new default settings, most personal attributes, such as relationship status, "interested in", and "looking for", are only visible to users' friends. Though there is now less default information in Facebook, Our inference method continues to work well. This is because we can still visit users' profile pages, and obtain their names and friend lists. Once we obtain the name and friend list, we can predict users' genders.

8   Gender Characteristics and Behavior

8.1   Privacy of Attributes

We apply our inference method to each user in Group A (as discussed in previous section) with parameters $T_1 = 0.8$ and $T_2 = 5$. We compute the percentage of male and female users who hide several of their attributes. The results are shown in Figure 3. Based on the two-proportions z-tests, we confirm that there is a highly significant ($p < 0.0001$) effect of gender on the privacy of each attribute (females showing a higher tendency to hide their attributes). In other words, a larger fraction of females hide their attributes such as gender, age, birthday and relationship status, compared to the male users. Thus, we can conclude that females are more conscious (and intuitively so) in terms of their online privacy on Facebook than their male counterparts. We also

Table 3. Pairwise correlation coefficients for private attributes

Attribute Pair	Males	Females	Attribute Pair	Males	Females
Gender, age	**0.539**	**0.523**	Age, looking for	0.244	0.252
Gender, birthday	**0.731**	**0.77**	Birthday, relationship status	0.486	**0.51**
Gender, relationship status	**0.51**	0.5	Birthday, sexual preference	0.407	0.383
Gender, sexual preference	0.433	0.376	Birthday, looking for	0.392	0.432
Gender, looking for	0.437	0.444	Relationship, sexual preference	**0.582**	**0.561**
Age, birthday	**0.738**	**0.669**	Relationship status, looking for	**0.625**	**0.681**
Age, relationship status	0.325	0.311	Sexual preference, looking for	**0.579**	**0.558**
Age, sexual preference	0.284	0.265			

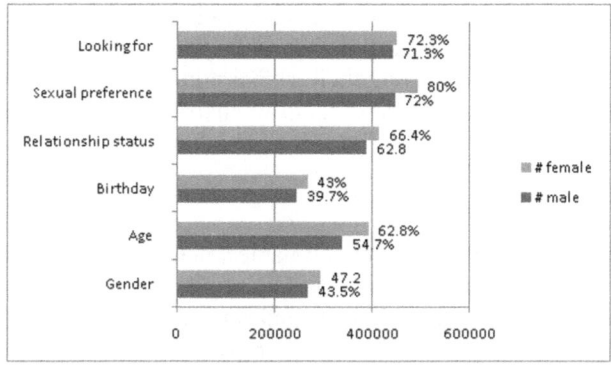

Fig. 3. Number and Percentage of male/female users who hide attributes

examine possible correlations between the hiding of different attributes. For both males and females, we calculate the pairwise Pearson's correlation coefficients, as shown in Table 3. In Social Sciences, correlation coefficients ranging from -0.3 to -0.1 and 0.1 to 0.3 are generally regarded as small, -0.5 to -0.3 and 0.3 to 0.5 as medium, and co-efficients larger than 0.5 and smaller than -0.5 as high [5]. From the results in Table 3, we find that the strongest correlations are between "hiding of gender" and "hiding of birthday", and "hiding age" and "hiding birthday", both for males and females. This is followed by correlations between "hiding relationship status" and "hiding looking for", and "hiding relationship status" and "hiding sexual preference". "Hiding of relation-ship status" and "hiding of sexual preference", and "hiding of gender" and "hiding of age" are also strongly correlated. These correlations are more or less consistent for both males and females, and imply that users who hide one attribute is also likely to hide several of other attributes. Looking further into these correlations, we find two indepen-dent clusters consisting of private attributes for both genders: (gender, age, birthday) and (relationship status, sexual preference, looking for).

8.2 Targeted Advertising and Privacy Implications

We provide examples of how third parties could use our results for gender-specific online stalking and targeted advertising. These third parties can use the resulting gender information from our gender inference methods combined with users' attributes, to help improve the accuracy of targeted advertising.

For example, an online dating company might be very interested in marketing their services and websites to single males and females who are looking for dating. In NYC Facebook, we find that there are $35,076$ males and $14,865$ females matching this cri-teria. A cosmetic company might be interested in marketing there products to young females, while we find that there are $106,007$ females that are in their 20s in NYC Facebook. A "gifts for lovers" company might want to know the information of people that are in a relationship. Our statistics show that there are $46,522$ males and $48,328$ females who specified that they are in a relationship.

There are several other interesting and concerning implications of our results. For example, there are 752 males and 463 females in NYC Facebook who are married but looking for dating; there are $9,077$ males indicating their sexual preference as men, however, 18.2% of them are in a relationship, and 5.88% of them are married; similarly, there are $18,259$ females specifying their sexual preference as women, but 18.3% of them are in a relationship, and 17.1% of them are married. All these statistics and others can potentially be used by malicious parties with unknown nefarious intentions.

9   Conclusions

The focus of this paper was on Facebook names, name-centric gender inference and gender behavior. By crawling Facebook public profile pages for 1.67 million users in New York City, we create a comprehensive and contemporary name list. We studied the properties of this annotated name list, and compared it with a popular name list that has been obtained via offline mechanisms. Based on our name list, we developed a new and powerful technique for inferring gender for users who do not explicitly specify their gender. Applying our methodology to NYC users in Facebook, we are able to achieve an accuracy of 95.2% for group A consisting of 95.1% of the NYC users. Having inferred the gender of most users in our Facebook dataset, we learn gender characteristics and analyze how males and females behave in Facebook.

References

1. Facebook statistics,
   http://www.facebook.com/press/info.php?statistics
2. Facebook updates privacy settings,
   http://blog.facebook.com/blog.php?post=197943902130
3. Popular baby names, http://www.ssa.gov/OACT/babynames/
4. Carter Jernigan, B.F.M.: Gaydar: Facebook friendships expose sexual orientation. First Monday 14(10) (2009)
5. Cohen, J., Cohen, P., West, S., Aiken, L.: Applied multiple regression/correlation analysis for the behavioral sciences. Erlbaum Hillsdale, NJ (1983)
6. Hall, M., Frank, E., Holmes, G., Pfahringer, B., Reutemann, P., Witten, I.H.: The weka data mining software: an update. SIGKDD Explor. Newsl. (2009)
7. He, J., Chu, W.W., Liu, Z(V.): Inferring privacy information from social networks. In: Mehrotra, S., Zeng, D.D., Chen, H., Thuraisingham, B., Wang, F.-Y. (eds.) ISI 2006. LNCS, vol. 3975, pp. 154–165. Springer, Heidelberg (2006)
8. Heatherly, R., Kantarcioglu, M., Thuraisingham, B., Lindamood, J.: Preventing Private Information Inference Attacks on Social Networks. Tech. Rep. UTDCS-03-09, University of Texas at Dallas (2009)
9. Lindamood, J., Kantarcioglu, M.: Inferring Private Information Using Social Network Data. Tech. Rep. UTDCS-21-08 (2008)
10. McCallum, A., Nigam, K.: A comparison of event models for naive bayes text classification. In: AAAI 1998 Workshop on Learning for Text Categorization (1998)
11. Mislove, A., Viswanath, B., Gummadi, K.P., Druschel, P.: You are who you know: Inferring user profiles in online social networks. In: WSDM (2010)
12. Quinlan, J.R.: Improved use of continuous attributes in c4.5. Journal of Artificial Intelligence Research 4, 77–90 (1996)
13. Xu, W., Zhou, X., Li, L.: Inferring Privacy Information via Social Relations. In: 24th ICDE Workshop, pp. 154–165 (2008)
14. Zheleva, E., Getoor, L.: To join or not to join: the illusion of privacy in social networks with mixed public and private user profiles. In: WWW (2009)

Realtime Social Sensing of Support Rate for Microblogging

Jun Huang and Mizuho Iwaihara

Graduate School of Information, Production and Systems,
Waseda University, Japan
junhuang@akane.waseda.jp, iwaihara@waseda.jp

Abstract. This paper proposes realtime estimation of support rate based on social sensors. Nowadays, micro blogs like Twitter have gained wide popularity, especially among the youth for its capability of updating personal opinions in a realtime manner. Academically, they have received tremendous attention as well. We argue that realtime events that have great influence on the attitudes of Twitter users can be detected by strategically monitoring tweets on certain topics. Building on the collected data, sentiment analysis enables us to calculate percentage of positive tweets, namely, support rate. In particular, given Twitter's realtime nature, the support rate calculation shall also be done in realtime. Drawing on World Cup 2010, we collect a large amount of tweets and carry out analysis so as to extract sentiment information of the audience and go further to show the realtime support rate of the participators.

Keywords: Realtime, event detection, support rate, sentiment classification.

1 Introduction

As a popular micro-blogging service provider, Twitter has grasped tremendous attention from the public. People can update their status anywhere in the form of tweet which is a message within 140 characters with the aid of computer or mobile phones [10]. Therefore, millions of people are benefited from this social network service that enables them to make new friends and keep connection with their friends, classmates or colleagues.

Topic of a tweet can be variable. Mostly, tweets help people to express their opinions of a certain topic, like a new product, a game or presidential campaign. By extracting the sentiment information of such tweets, support rate for each aspect (percentage of positive tweets to a certain aspect) of the topic can be calculated. These sentiment information and support rate are rather important since feedbacks are aggregated without manual intervention.

Moreover, given Twitter's realtime characteristics, any change of the support rate affected by the sudden presence of a big event, even the most trivial one, could be caught immediately, like a goal during a soccer game or a speech given by a president candidate during election debate.

J. Xu et al. (Eds.): DASFAA Workshops 2011, LNCS 6637, pp. 357–368, 2011.

Enormous efforts have been made in the area of sentiment analysis [2, 4]. Traditionally, methods mainly focus on static corpus, like reviews [3], yet not without flaws. Due to the realtime characteristic of Twitter, the number of tweets centering on one specific topic could grow at an astonishing pace. So far, the largest number of tweets per second was 3283 during the game between Japan and Denmark in World Cup 2010. Lacking the ability of tracking continuous realtime change of the sentiment information (support rate) for a great deal of tweets is the major disadvantage of conventional methods [7].

Social sensor is a new concept proposed in recent years, which treats Twitter users as virtual sensors to detect occurrences of events such as earthquake and typhoon [1]. Although studies on Twitter and Twitter users have been carried out widely [12], such kind of event detection is under the assumption that the sensors are not independent to each other [1]. Another limitation of the previous researches is that they can utilize only tweets having geo information, causing a huge waste of tweets lacking geo information.

We aim at social sensors for detecting non-geographical events and tracking continuous realtime change of support rate, which have not been studied extensively. Contributions of this paper are summarized as follows:

1. Our support rate calculation is based on realtime data collected from Twitter. It inherits the distinctive feature of realtimeness, which means the support rate will be updated immediately when a new tweet taking about specific topic arrives;

2. A realtime event detection mechanism based on social sensors is deployed in our approach, thus the change of support rate caused by a significant event can be caught promptly;

3. With the realtime support rate and realtime event detection, prediction for future support rate will be easier. In our estimation, we achieved an acceptable accuracy in sentiment classification and the realtime support curve shows that the support rate clearly reflects of events like a goal in a football game.

A wide variety of applications can be built based on our realtime support-rate estimation, for instance, realtime census, better advertizing effects and more precise marketing with the knowledge on the attitudes of the users and low-cost social feedbacks without costly surveys.

The rest of the paper is organized as follows: in the next section we give a formal definition of the problem we address in this paper. Data preparation is described in Section 3. Section 4 illustrates our approach of extracting sentiment information and calculating the realtime support rate using machine learning method Support Vector Machine (SVM) [6, 11]. Experiments and evaluations will be presented in Section 5. The final section is devoted to discussion on prospective improvements of the present work.

2   Problem Setting

Before giving a formal definition of the problem we address in this paper, we first present several definitions.

Definition 1 (Topics and Aspects). Topics in our research should be sentiment-oriented, which involves several *aspects* $a_1, a_2, ... a_i$ in the real world or *sentiment concepts*.

For example, political candidates in an election, such as the recent election of the party leader of the ruling party of Japan can be regarded as a proper topic, in which two candidates Kan Naoto and Ozawa Ichiro are regarded as two opposite aspects. Take the area of electronic products as another example. For the topic of electronic book readers, iPad and Kindle can be regarded as two opposite aspects.

Definition 2 (Sentiment Value). Given an aspect a of a specific topic and a tweet t_i talking about a, the *polarity* s_i of t_i with respect to a is an integer -1 or 1. Tweet t_i is said to be *positive* if the overall sentiment expressed in t_i is positive, with $s_i = 1$ to denote positive emotions; while t_i is nonpositive if the overall sentiment expressed in t_i is neutral or negative, with $s_i = -1$ to denote neutral and negative emotions. Table 1 shows examples of tweets during the game England vs. Germany and the game England vs. Slovenia in World Cup 2010.

However, our research excludes the usage of irony which in most cases, positives words are deployed to express negative feelings. Also tweets with mixed polarities are excluded from our discussion. Only tweets with apparent polarities are used in our training data, we manually pick them up from the tweets we collected after applying several simple filters which we will mention in Section 4.1.

Definition 3 (Labeled Tweet/Unlabeled Tweet). Given an aspect a of a specific topic p, suppose that t_i is a tweet talking about p and s_i is the polarity of t_i with respect to a. The pair (t_i, s_i) of t_i and s_i is called a *labeled tweet*. If s_i is not assigned to t_i, it is called an *unlabeled tweet*

Definition 4 (Support Rate). Given an aspect a_i of a topic p, and all the related labeled tweets (t_i, s_i) collected in the time period T, we define the *support rate* of the aspect a_i during T as the percentage of positive tweets $p_T(a_i)$, where

$$p_T(a_i) = \frac{\#of(t_i, 1)}{\#of(t_i, 1) + \#of(t_i, -1)} * 100\% \tag{1}$$

In this formula, $\#of(t_i, 1)$ represents the number of positive tweets with respect to aspect a_i during T, while $\#of(t_i, -1)$ denotes the number of nonpositive tweets, respectively.

Table 1. Example tweets during World Cup 2010

Sentiment	Sentiment Value	Query word	Tweet
Positive	1	England	Cummm onnn #ENGLAND do us proud! http://tweetphoto.com/28693928
Neutral	-1	England	Watching the US vs Algeria and England vs Slovenia games with the team
Negative	-1	England	London bridge has fallln down .. So does england!!!

Based on the definitions above, we now define the problem we try to address in this paper as follows:

Problem Definition (Realtime Support Rate Calculation). Given tweets collected continuously with specific topics, the task is to assign a proper sentiment value to each of the tweets and calculate the support rate of current period. Meanwhile a realtime event detection mechanism should be built to show the influence to the support rate of events which have great impacts on people's attitudes towards the topics.

In order to solve this problem, we propose an approach which aims to achieve two subtasks: (1) to learn an accurate classifier to do classification and (2) to design an algorithm for detecting events promptly. In the first subtask, we need to prepare tweets which already have been manually assigned corresponding sentiment polarities. The function of extracting features is used as a bridge to make features transferrable between tweets written in natural language and vectors used in SVM. In the second subtask, we aim to use social sensors to confirm presence of the predefined features of the target events and detect occurrences of such events without any duplication of a single event.

3   Data Preparation

We developed a web Twitter mining tool by using Twitter API [14]. After setting query words, requests with one query word each will be sent to the server of Twitter, and then responses with 15 entries in an atom file will be supplied as input to the analysis program. With the limitation of connections to Twitter, about 2250 latest tweets talking about a given topic can be collected every hour.

Each entry of a tweet contains 12 attributes among which 3 will be examined: *id*, *published time*, and *title*.

- **Id:** This attribute is used for removing duplicates of the tweets.
- **Published time:** We sorted the tweets according to this attribute to calculate the support rate with the unit of minute.
- **Title:** It is not the real title for a tweet, just an attribute which covers all the textual information posted by Twitter users.

Topics that are chosen should be sentiment-oriented, to which people may have strong positive or negative feelings. A soccer match would be a good candidate because there exist multiple possibilities of happenings that may instantly influence the event and furthermore affect the support rate. For example, a goal, a beautiful shot or an unexpected winner could largely influence the attitude of the audience. In this paper, we take soccer games as the target of support rate estimation. The two participators of a soccer game, namely, two teams will be regarded as two opposite aspects.

There are two important soccer events in 2010, UEFA Champion League Final 2010 in May and World Cup 2010 in June and July in South Africa. We gathered tweets focusing on these matches. During each match, the names of the two teams are set as keywords for querying and over 5000 tweets are collected for one match.

4 Approach

4.1 Preprocess

Twitter users post their messages via variety of devices, and the language used in a tweet is usually informal. This results in a high frequency of misspelling and full of repeated letters to express their emotions. In Table 1, the tweet "Cummm onnn #ENGLAND do us proud! http://tweetphoto.com/28693928" is a typical example. "Cummm onnn" should actually be "Come on". In the tweets we collected, words like "Goooooooal" or "Yeahhhhhhh" are very common, yet such repeated characters should be removed. Also, letters "h", "w" and "l" occurred repeatedly in the tail of a word need to be removed. After preprocessing there will be no more than two consecutive occurrence of a letter. Table 2 shows some examples of preprocess.

Table 2. Examples of preprocess

Before	After
Cummm onnn	Cum onn
Goooooooal	Gooal
Yeahhhhhhh	Yeahh

Note that the above preprocessing is language-dependent; we choose English as our target language, and collect tweets in English. We need to modify the features depending on the target language, but required linguistic characterization is shallow, so that the modification should be minor.

Notice that, even after preprocess, there still exists misspellings, and this will cause a matching problem when we transform a tweet into vectors indicating the presence of features. In the case that follows, a non-strict matching function is needed, which we will discuss in the end of Section 4.2. Also, users often include links in their tweets; such kinds of URLs are often little correlation with sentiments, so we remove these links to reduce the calculation of feature matching in our approach.

4.2 Classification via Support Vector Machine

Support Vector Machine is a popular classification method. By preparing positive and nonpositive tweets as training data, it automatically produces a classifier that classifies tweets into two categories (positive and nonpositive). A particular type of Support Vector Machine, LibSVM2 [13] is used in our experiments.

Before defining features we designed for classification, we need to explain several elements which are related with the sentiment values.

Aspects: As defined in Section 3.2, aspects in the topic of a football game are the names of the two teams. We send the team names to Twitter as query words. The number of occurrences and the position of the query words have high weight for deciding sentiment values. It is easy to understand the number of occurrences of a certain aspect has high weight, because the repeated words can be treated as a symbol that the author is getting excited. The position of the aspect in a tweet is taken into consideration under the assumption that during an exciting game, people would not express their opinion in the usual manner, like "I like X's performance today."(X means the query word) Instead of that, users may prefer to use a shorter and more straightforward sentence to deliver their feelings and in most cases, these sentences begin with the aspect or have the aspect in the tail. For instance, "Come on! X!", "GoGoGo, X" and "X is awesome!"

Supporting Entities: Each aspect may be accompanied by particular types of entities. By mentioning these entities, people show their sentiment tendency, such as new coming events and people involved in these events. In the case of a football game, a goal can be treated as the most obvious event which will arouse people's desire to express their support to one of the two teams, and the player who just scores will also be mentioned in the tweets. This results from the behaviors of the Twitter users when they get excited for the game, especially, when there is a goal or a beautiful shot. They will post their feelings to such an event and usually a player related with the event will be mentioned. There are a large amount of such examples in the tweets we have collected, like "GOAL! Milito breaks the deadlock! 1-0 to Inter!", "Blimey, Inter being outplayed by Bayern, now 0-1 up with a lovely goal from Milito", where Diego Milito is a famous forward in Inter. By mentioning his name, the audience show their support to Inter. Therefore we can assume that occurrences of supporting entities such as player names have a positive contribution to the support rate of their team.

Table 3. 12 Features used in classification

Feature	Value Type
Tweet Length	Integer
Number of occurrence of first aspect	Integer
Position of first occurrence of first aspect	Integer
Position of last occurrence of first aspect	Integer
Number of occurrence of second aspect	Integer
Position of first occurrence of second aspect	Integer
Position of last occurrence of second aspect	Integer
Number of occurrence of players of first team	Integer
Number of occurrence of players of second team	Integer
Occurrence of event symbol	Integer
Current scores of first team, if mentioned	Integer
Current scores of second team, if mentioned	Integer

Utilizing these aspects and supporting entities, we can design the features for a football game. Table 3 shows twelve features used in our estimation.

Algorithm. 1 describes the process of extracting feature values.

Algorithm. 1. Feature Value Extraction Algorithm

1. Obtain a new tweet t discussing the topic;
2. Use regular expression to catch the current scores if mentioned. Else, mark the corresponding feature values as -1;
3. Split this tweet into a series of tokens $w_1, w_2, ... w_n$, denote n as the length of the tweet;
4. For each w_i do:
    (a) For each aspect a do:
        1) compare w_i with a using edit distance to check whether there is a match;
        2) check whether w_i contains a;
        3) if the results of 1) or 2) is true add 1 to corresponding features and update the feature stands for the position of the aspect;
    (b) For each entity e do:
        1) compare w_i with e using edit distance to check whether there is a match;
        2) check whether w_i contains e;
        3) if the results of 1) or 2) is true add 1 to corresponding features or update the feature stands for event symbol;
End

Aspects here stand for two team names and supporting entities include the name list of the players and event symbols like "Goal". Tokens here can be a single word or a combination of several words; this depends on how to split the tweets. In our case, we use a blank character as the separator.

When a comparison is carried out between the current token and the aspect or supporting entity, because of casual language used in Twitter and frequent misspelling mentioned before, edit distance between two given phrases should be calculated. If it is less than one, presence of the current target word can be confirmed. In the example of Table 2, after preprocess we obtain a token "Gooal", and a comparison is carried out between the event symbol "Goal" and "Gooal". Since the chance of misspelling occurring in the first and last letter in a word is tiny, a higher weight is set to these letters while calculating the edit distance. In this case, after we set higher weight to "G" and "l", the distance turns out to be zero and the event symbol is confirmed.

After executing the feature value extraction process, the vector for the input data is obtained. By applying the classifier to the vector, a sentiment value of 1 or -1 will be sent to the support rate calculation program as the output.

Notice that the above feature designing is topic-dependent. For other domains, modification is needed according to the different aspects and supporting entities. The process of feature designing should obey the principle that features should be easy to extract to satisfy the need of realtimeness.

4.3 Event Detection

Another target of our approach is to show the reflection of significant events which may to a large extent change the attitude of audience as mentioned at the beginning of this section. A proper event detection mechanism can help to confirm the occurrence of the event without reading original tweets.

A target event to be detected can be characterized by three key features: *top event*, *supporting entity* and *numerical indicator*. Top events are expressions that directly refer to the event needed to be detected. Supporting entities are the same as those discussed in Section 4.2. Numerical indicators are such that a change of their numerical value indicates that a top event has happened.

In our estimation, we choose goals as the target event. We also try to find which team scores, when and who. In the following, corresponding to the features we defined above, the top event could be a word like "Goal", the player who just scores is a supporting entity and current score report is a perfect candidate of numerical indicator.

From our observation, a surge of tweets talking about the goal occurs when someone scores, and active Twitter users will post tweets immediately after they detect a goal, like the behavior of sensors. To confirm a goal, we must consider the reliability of multiple sensor values to avoid duplication of the same goal. It is also possible that we may miss the information of a new goal if it comes with a short interval from the last one.

To solve these problems, we design the three features as follows:

- **Top event:** Presence of "Goal" is the most obvious feature of a goal;
- **Numerical indicator:** A score report, in the forms of "x-x", "x vs. x" or "x to x", caught by regular expression;
- **Supporting entity (optional):** One or more occurrences of the name of a player in a tweet co-occurring with the top event (a "Goal" like word occurs). We can surmise that repeated occurrence of a player name after a goal indicates that the player did the goal.

Note that the above features are language-dependent; we choose English as our target language, and collect tweets in English. We need to modify the features depending on the target language, but required linguistic characterization is shallow, so that the modification should be minor.

Although by detecting the presence of these features in a tweet a goal can be confirmed, checking whether the next several tweets also describe the same goal is necessary to avoid false positives. Dealing with these three feature values obtained from the tweets, we can maintain a score report of our own and check whether the current tweet is talking about a new goal. Algorithm. 2 explains details of the goal detection mechanism:

Algorithm. 2. Goal Detection Algorithm

1. Obtain a new tweet t containing the current query word;
2. If regular expression returns the current scores,
    Compare it with our own score report;
    if one of the two numbers is larger than ours;
        do step 3 and update our score report;
   Else, go back to step 1;
3. Split this tweet into a series of tokens $w_1, w_2, ... w_n$;
4. For each w_i do:
   (a) Compare w_i with all of player names using edit distance;
   (b) Check whether w_i contains any of the player names;
   (c) If there is a match, add 1 to the occurrence of the current player name;
5. Save all the occurrences of player names find the player with highest occurrence;
End

This algorithm works well with the real data. The regular expression is treated as a filter that tweets do not contain a score report can skip steps 3 to 5, which speeds up the overall processing.

5   Experiments and Evaluations

In this section we use tweets collected during World Cup 2010 to conduct our experiments. Our evaluations mainly focus on the following aspects:

- **Accuracy of support rate:** Percentage of correctly classified tweets;
- **Accuracy of event detection:** Percentage of correctly detected goals;
- **Feasibility of realtime calculation:** Due to the time limitation of World Cup, we collect tweets first, and carry out off-line analysis. Then we evaluate whether the computation time is fast enough to deliver results in realtime.

5.1   Training Data

Tweets collected during UEFA Champion League Final are used for training; we use different games for training and test. We collected 11,250 tweets during the game between Inter and Bayern[1]. Two simple filters are applied to reduce tweets to essential ones so that manual assignment of sentiment values becomes effective:

1. Retweets are removed. Retweeting is the process of making comments on other tweets. These comments may contain their own sentiment information which may have conflicts with the original ones. Any tweets with "RT" will be removed from the training data;

2. Duplicates are removed. As mentioned in Section 3, there can be some duplicates in the collected tweets. To avoid giving extra weight to a particular tweet, we need to remove these duplicates. It is time consuming if we compare all id attributes to remove these repeated tweets. Fortunately, since we send the queries simultaneously and Twitter only returns the latest tweets, the comparison could narrow to 2 atom files, 30 tweets.

After applying these filters, we randomly pick up 200 tweets from the rest and manually label them certain sentiment values. 93 out of 200 tweets are assigned to be positive.

5.2   Support Rate Results

In the experiments we conduct, we choose to show the results of two dramatic matches: England (1) vs. Germany (4)[2] and Brazil (1) vs. Netherlands (2)[3]. Surge of the support rate of events like a goal is clearly recognized in these results. Fig.1 and Fig. 2 show the support rate of England and Brazil respectively:

---

[1] F.C. Internazionale Milano vs Fussball-Club Bayern München, May 23, 2010.

[2] http://www.fifa.com/worldcup/matches/round=249717/match=300061501/index.html

[3] http://www.fifa.com/worldcup/matches/round=249718/match=300061507/index.html

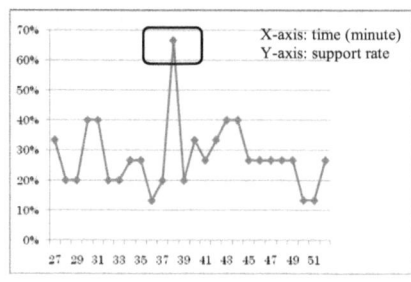

Fig. 1. Realtime support rate of England   Fig. 2. Realtime support rate of Brazil

   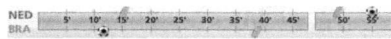

Fig. 3. The play by play record provided by FIFA

The results generated by the classifier trained with SVM are reasonable. In Fig.1, the support rate of England was relatively lower than Germany when Germany took one goal lead at the beginning. It experienced fluctuation for a few minutes which however, was followed by a sudden drop when Podolski made Germany's second goal at 32'. Yet, a significant rise of England's support rate occurred when they dramatically scored two goals in just one minute. But the high support rate of almost 70% did not last long as the second goal was disallowed and the following tweets could not be classified into a positive category since most of them were discussing the mistake made by the referees.

In Fig.2, as the five championships owner with an early goal in the game, Brazil received a higher support rate form beginning. After the half time break (45'-60'), the average support rate was near 65%, and it kept at a high level with little fluctuation. However, the turning point of this game came at 70' (55', if excludes the half time break), because of a mistake made by the goalkeeper of Brazil, Melo headed the goal in the direction of their own, Own goal! All the following tweets concentrated on this dramatic own goal, therefore, the support rate dropped to the bottom.

5.3 Analysis

There are two reasons that we did not show the support rate through the whole game: (1) We can explain the reflection in a clear way if we pick the period near the interesting goals, and (2) manual classification of the tweets for the entire game is costly.

Table 4. Accuracy of the classification

Team	#of tweets	#of tweets properly classified	Accuracy
England	390	267	68.4%
Brazil	330	232	70.3%

Table 4 shows the accuracy of the experiments that we manually checked, by calculating the percentage of the tweets which classified into the correct category.

Although our method is still in an early stage, we obtained an acceptable result of accuracy around 70%.

5.4 Verifying Realtimeness

In this part, we show a breakdown of execution time to check realtime requirements. Our cost model consists of two parts: time t_1 for feature extraction and time t_2 for classification. With the prepared classifier, time cost for classification is so tiny that can be ignored. The major time cost is in the process of feature extraction.

Table 5 shows the average, maximum and minimum time t_1 in our experiments.

Table 5. Average/max/min classification time

Team	#of tweets	time cost (ms)	Average(ms)	Max	Min
England	390	3264	8	11	4
Brazil	2730	14753	5	7	3

In the case of Brazil, we use all the tweets collected through the game to check the performance in the batch job with a large amount of tweets. In both of these two cases, the time cost for each tweet is less tan 0.01s, a satisfactory result as a realtime system.

5.5 Results of Event Detection

The results for goal detection is impressive, all 4 goals are correctly detected, which means the accuracy is 100%, and we find several interesting points:

1. Though the second goal of England scored by Lampard is canceled by the referee, and most of the tweets were discussing this mistake which means features defined for a goal were not working on these tweets. We still detect this goal successfully, because there are several tweets mentioned this goal and provided the score report of "2-2" which it should be "2-1". For instance, "@nmones should of been 2-2 but England had a goal not even!"

2. Brazil's own goal was scored by Wesley Sneijder (Netherlands). However this goal was counted to Melo (Brazil) because the number of the occurrence of Melo was 21, which was much larger than the number of the occurrence of Sneijder , which was only 5.

6 Conclusion and Future Work

In this paper, we investigated a new social mining method which captures realtime nature of Twitter. Considering each Twitter user as a sensor, we proposed a new concept of support rate which was calculated based on sensory observation. Drawing on a soccer game, we collected a large amount of tweets and classified these tweets into two categories to compute the support rate for the two competitors using SVM. Also we proposed an event detection mechanism for detecting occurrences of a target event such as a goal and showed the reflection of the target event to the support rate. In our experiments, we drew curves of support rate according to the percentage of positive tweets, and compared it with the timeline of real games. The results were reasonable; especially a goal was clearly recognized.

Machine learning can help to extract the polarities of tweets with an acceptable accuracy. Nevertheless since only events like a goal are detected, others such as a beautiful shot are yet to be explored. In future work, we consider detecting more varieties of events using social sensors. Also, we plan to expand our approach to more domains by utilizing different feature settings and supporting entities.

References

1. Sakaki, T., Okazaki, M., Matsuo, Y.: Earthquake Shakes Twitter Users: Realtime Event Detection by Social Sensors. In: WWW2010, Raleigh, North Carolina, April 26-30 (2010)
2. Pan, S.J., Ni, X., Sun, J.-T., Yang, Q., Chen, Z.: Cross-Domain Sentiment Classification via Spectral Feature Alignment. In: WWW 2010, Raleigh, North Carolina, April 26-30 (2010)
3. Dave, K., Lawrence, S., Pennock, D.: Mining the peanut gallery: opinion extraction and semantic classification of product reviews. In: WWW 2003 (2003)
4. Pang, B., Lee, L., Vaithyanathan, S.: Thumbs up? Sentiment classification using machine learning techniques. In: EMNLP (2002)
5. Tsochantaridis, I., Hofmann, T., Joachims, T., Altun, Y.: Support Vector Learning for Interdependent and Structured Output Spaces. In: ICML (2004)
6. Klinkenberg, R., Joachims, T.: Detecting Concept Drift with Support Vector Machines. In: Proceedings of the Seventeenth International Conference on Machine Learning (ICML). Morgan Kaufmann, San Francisco (2000)
7. Sahami, M., Dumais, S., Heckerman, D., Horvitz, E.: A Bayesian Approach to Filtering Junk {E}-Mail. AAAI Technical Report WS-98-05 (1998)
8. Popescu, A.-M., Etzioni, O.: Extracting Product Features and Opinions from Reviews. In: EMNLP (2005)
9. Campbell, A.T., et al.: Transforming the Social Networking Experience with Sensing Presence from Mobile Phones. In: Proc. of ACM SenSys 2008, Raleigh, North Carolina, USA (2008) (Demo abstract)
10. Milstein, S., Chowdhury, A., Hochmuth, G., Lorica, B., Magoulas, R.: Twitter and the micro-messaging revolution: Communication,connections and immediacy.140 characters at a time. O'Reilly Media, Sebastopol (2008)
11. Joachims, T.: Text categorization with support vector machines. In: Proc. ECML1998, pp. 137–142 (1998)
12. Jansen, B., Zhang, M., Sobel, K., Chowdury, A.: Twitter power:tweets as electronic word of mouth. Journal of the American Society for Information Science and Technology (2009)
13. http://www.csie.ntu.edu.tw/~cjlin/libsvm/
14. http://apiwiki.twitter.com/Twitter-API-Documentation

Searching Consultants in Web Forum*

Zhao Zhang, Weining Qian, and Aoying Zhou

Institute of Massive Computing,
East China Normal University, Shanghai 200062
{zhzhang,wnqian,ayzhou}@sei.ecnu.edu.cn

Abstract. Web forums attract many users contribute their rich experiences and professional skills in various filed by initial post or reply post. So, the web forum has been one of the main platforms to exchange information. Actually, different forum user have different professional background, a forum user has expertise in some specific areas, which is possible another forum user lacked. The aim of this paper is to find knowledgeable forum users for seeker provided query in web forum.

Keywords: Web Forum, Searching Consultants, Ranking Consultants.

1 Introduction

Online forums are one of main platforms to exchange information. More and more people contribute their experiences in various fields and learn from other's experiences. That is there are many people who have expertise in some specific areas. These experts tend to fall into two groups. One is they are real professional, for example, their occupation could be doctor or lawyer. Another is they have rich experiences in a certain field or area, for example, they just finished decorating their home or they are parents of five children. However, there exist some freshman in a certain area, for example, young mother or people who are prepared for decorating their new house. They have no idea about how to take care a baby or how to decorate their home. They come to online forum to learn some useful experience. In this paper, we hope to help forum users to find consultants in some specific fields.

In web forum, web users show their expertise knowledge by two way, answering other's question or posting their experience in web forum. The statistics on consumption web forum showed 71 percents topics is question topics in web forum. So, we can find the candidate consultants given query based on question topics, because the candidate consultants could be authors answering the question topic with relevance to the query. It's key challenge that efficiently find and rank the candidate consultants.

---

* This work is partially supported by National Science Foundation of China under grant number 61070051, and National Basic Research (973 program) under grant number 2010CB731402.

J. Xu et al. (Eds.): DASFAA Workshops 2011, LNCS 6637, pp. 369–377, 2011.

Some researches have been done on finding and ranking experts in some co-pora. SmallBlue [5][7] is to locate knowledgeable colleagues, communities, and knowledge networks in companies. SmallBule find consultants depending mainly on strong social networks among employees in enterprise. However, there are only weak social networks in web forum builded by post-reply among authors. Actu-ally, it's newbies that need to find consultants of specific area. These newbies did not post enough messages to build social netwok in web forum. So, we mainly depend on web forum contents to find consultants in this paper. In reference [11], authors construct expertise networks and rank the experts by PageRank or hits algorithm in a java forum. However, this kind approach of expertise networks is not suitable for web forum, because the expertise networks is weak and not associative among consultants.

In this paper, we present our work on searching consultant given query q in consumption web forums. Our achievements are listed as the following: (1)giving formal definitions for search consultants task in web forum. (2) presenting an effi-cient rank approach for consultant candidates. (3) The algorithm is implemented and its effectiveness is tested on a real forum data set.

The rest of this paper is organized as follows. Section 2 presents problem statement. Section 3 models the approach of finding consultant. Section 5 gives experiments on a real forum data set. Section 6 introduces related work. Section 7 gives conclusion and on-going work.

2   Problem Statement

There are some knowledgeable authors in some specified areas in web forum, we call these authors as consultants. In this paper, we want to find these consultants in certain fields defined by a query.

Table 1 lists the main symbols we use throughout the paper.

Table 1. Main Notations Used in this Paper

Symbols	Definition and description
T	the set of topics in web forum
A	the set of authors in web forum
q	the query containing several keywords
c_i	the forum author
$p(c_i|q)$	the probability of c_i being a consultant given query q
$p(c_i, q)$	the preference of author c_i for query q
$p(c_i|t_i)$	the probability of c_i being a consultant given topic t_i
$s(t_i, c_i)$	association score for between topic t_i and author c_i
F_i	the ith being consultant indicator
α_i	the weight for F_i
T_q	topics with relevance to query q
L_a	candidate consultants list given q and topic t_i

2.1   Objects in Web Forum

A web forum can be viewed as the combination of three entities. They are author, post and topic. Each web forum contains many topics, and each topic contains several posts published by authors. The meanings of the terms are listed as the following. **Author** A registered user who left message. We use set A to denote all authors on web forum. **Post** A text which published by a user. **Topic** A set of posts under the same topic. We use set T to denote all topics in web forum.

The three objects in web forum are presented in Fig. 1.

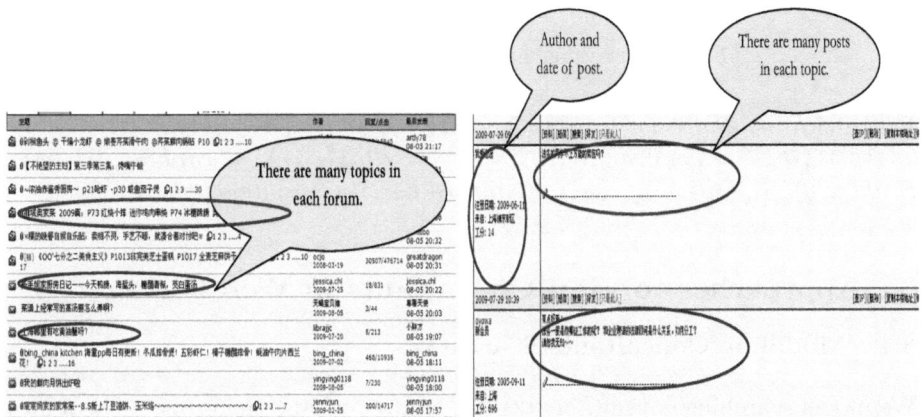

Fig. 1. An Example of Web Forum

The relationships between authors and topics are shown in Fig. 2, where the dashed edge denotes initial post, the real edge denotes reply post.

2.2   Definitions

The definitions finding consultants given query in web forum are listed as the following.

Definition 1. *The query q is comprised by several keywords.*
$\{q|k_i \bigcup ... \bigcup k_j\}$, *where k_i is keywords.*

Definition 2. *Consultants are knowledgeable forum users in specific area identified by query q in the set of authors A.*
$\{c_i|i = 1, ...n, c_i \in A \bigwedge p(c_i|q) \geq \varepsilon\}$, *where where $p(c_i|q)$ = probability of author c_i being an Consultants given query q.*

Definition 3. *Ranking author c_i and c_j given the query q.*
$p(c_i|q) \geq p(c_j|q) \Rightarrow rank(c_i, q) \leq rank(c_j, q)$, *where $p(c_i|q)$ − probability of author c_i being an Consultants given query q, and $rank(c_i, q)$ = position of c_i on the ranked list of consultants given topic q.*

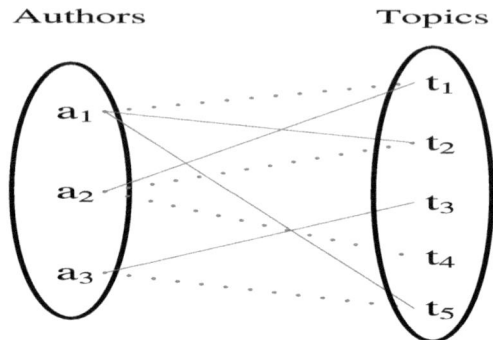

Fig. 2. Relationships between Authors and Topics

Definition 4. *Rank list R of consultants.*
$R = \{(c_i, ...c_j...c_k)|rank(c_i, q) \leq ... \leq rank(c_j, q) \leq rank(c_k, q)\}$, *where*
$rank(c_i, q) = position\ of\ c_i\ on\ the\ ranked\ list\ of\ Consultants\ given\ topic\ q.$

3 Approaches to Find Consultants in Web Forum

3.1 Modeling Consultants Search

We model searching consultants task as follows: what's the probability of forum author c_i being a consultant given the query q. In other words, we need compute the probability $p(c_i|q)$. We can use equation 1 represents $p(c_i|q)$ based on Bayes' Theorem.

$$p(c_i|q) = \frac{p(q|c_i)p(c_i)}{p(q)} \qquad (1)$$

In equation 1, $p(q)$ is the probability of q. Given the q, $p(q)$ is a constant. $p(c_i)$ is the priori probability of c_i being a consultant. We assume $p(c_i)$ is uniform. So, we only need compute $p(q|c_i)$ for ranking forum user c_i being consultant of specific area.

We first find topics which is relevant to the query, then find out who is most strongly associated with the relevant topics. That is represented by equation 2.

$$p(q|c_i) \propto p(c_i, q) \sum_{t_i \in T} p(c_i|t_i) \qquad (2)$$

In equation 2, $p(c_i, q)$ is preference probability given the query. $p(c_i, q) = \frac{n}{m}$, where n denotes the posts of author c_i with relevant query q, m denotes posts of consultant c_i. t_i is a topic with relevance query q. And $p(c_i|t_i)$ represents the probability of c_i being a consultant in topic t_i.

$$p(c_i|t_i) = \frac{\sum_{t_i \in T} s(t_i, c_i)}{|T|} \qquad (3)$$

Algorithm 1. Finding Rank list of consultants in web forum

Input: the query q, $\alpha_1, \alpha_2, \alpha_3$,the set of topics T on web forum
Output: $R = \{(c_i, ...c_j...c_k)$

1 Initialize topic list with relevance to query q, T_q;
2 **for** t_i in T_q **do**
3 | Append author name to list L_a except for original post author;
4 **end**
5 **for** c_i in L_a **do**
6 | n=the posts of author c_i in T_q;
7 | m=the posts of author c_i in T;
8 | $p(c_i, q) = \frac{n}{m}$;
9 | **for** t_i in T_q **do**
10 | | **if** c_i in T_q **then**
11 | | | $n =$ the number of question topics containing c_i in T_q;
12 | | | $m=$ the number of all topics containing c_i in T_q;
13 | | | $F_1 = \frac{n}{m}$; n=the posts of the author c_i in t_i ;
14 | | | m=all posts in t_i;
15 | | | $F_2 = \frac{n}{m}$;
16 | | | n=the posts of the author c_i containing keywords of query q;
17 | | | m=the posts of the author c_i in t_i;
18 | | | $F_3 = \frac{n}{m}$;
19 | | | $s(t_i, c_i) = \alpha_1 \times F_1 + \alpha_2 \times F_2 + \alpha_1 \times F_3$;
20 | | **end**
21 | | $p(c_i|t_i) = p(c_i|t_i) + s(t_i, c_i)$;
22 | **end**
23 | $p(q|c_i) = p(c_i, q) \times p(c_i|t_i)$;
24 **end**
25 Generate rank list $R = \{(c_i, ...c_j...c_k)$ for all authors in topic T_q, where $p(q|c_i) >= ... >= p(q|c_i) >= ... >= p(q|c_k)$;
26 return $R = \{(c_i, ...c_j...c_k)\}$

Where $|T|$ is the number of topics, and $s(t_i, c_i)$ is a association score between a topic t_i and a author c_i. we use equation 4 to compute this association probability. We consider $s(t_i, c_i)$ can be captured based on heuristic rule. To turn these associations into probabilities, we divide $\sum_{t_i \in T} s(t_i, c_i)$ by $|T|$.

$$s(t_i, c_i) = \frac{\sum_{i=1}^{3} \alpha_i F_i(t_i, c_i)}{\sum_{i=1}^{3} \alpha_i} \tag{4}$$

Where F_i is a heuristics indicator which a author is consultant. α_i is weight for each indicator F_i. In this paper, we use 3 indicators. F_1 measure how much the author c_i subjective know about the query. F_2 measure query driven an author's passion for the question topic. F_3 measure query driven an author's knowledge about the query. To turn these associations into probabilities, we divide $\sum_{i=1}^{3} \alpha_i F_i(t_i, c_i)$ by $\sum_{i=1}^{3} \alpha_i$.

F_1: In all topics, ((the number of all topics of c_i)-(the number of question topics of c_i with relevance query q)) /(the number of all topics of c_i with relevance query q).

F_2: (the posts of the author c_i)/(all posts) in each topic with relevance the query q.

F_3: (the posts of the author c_i containing keywords of query q)/(the posts of the author c_i) in each topic with relevance the query q.

3.2   Algorithms

After modeling the search consultants, we only need to compute $p(q|c_i)$ for ranking author c_i given query q. In this section, we give an algorithm to describe this model in details. We first find topics T_q relevant to the query, that is all authors in topic T_q are the candidate consultants. We append these candidate consultants to L_a except for original post author in this topic. For each author c_i in L_a, we compute $p(q|c_i)$ according to equation 3.

In equation 3, $s(t_i, c_i)$ represents association score between author c_i and t_i. It's be estimated by consultant indicator F_i and weight α_i of F_i. In this paper, we use three indicators, F_1 measure how much the author c_i subjective know about the query. F_2 measure query driven an author's passion for the question topic. F_3 measure query driven an author's knowledge about the query. The probability $p(c_i, q)$ captures the preference of author c_i for query q. Afterward, we sum up to obtain probability $p(q|c_i)$ for every c_i in L_a. Finally, we rank all c_i in L_a based on probability $p(q|c_i)$. The details are algorithm 1.

4   Experiments

In this section, we verify the advantages of the proposed model for ranking consultants given query on a real-world web forum data set.

4.1   Data Collection

Crawling a Decorating board from a popular web forum (bbs.libaclub.com) in China from April 6 to June 24 2010, we got 9327 topics including 6057 question topics, and extracted names of the author who involved in these topics. Details about the data set is in table 2.

4.2   Experiment Results

In this section, we show our experiment results from two aspects.

(1)Capture Consultants Rank List

we pick six typical queries for testing our algorithm 1 based on decorating home phases. In algorithm 1, we set $\alpha_1 = 1, \alpha_2 = 1, \alpha_3 = 2$ based on heuristic method. We get probability of every candidate consultant given each query. The results are shown in Fig. 3. We know some authors have high probability of being consultant from Fig. 3, and we consider these authors are senior consultants we search for. We think top 15 authors are senior consultants for each query in Fig. 3.

Table 2. Dataset

Board	Decorating
Topics	9327
Question topics	6057
Authors	10670
Average posts of each author	5120
Time span	April 6 to June 24 2010

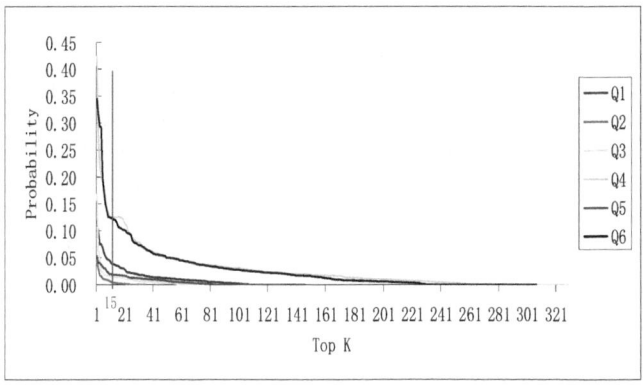

Fig. 3. Probability an Author being Consultant

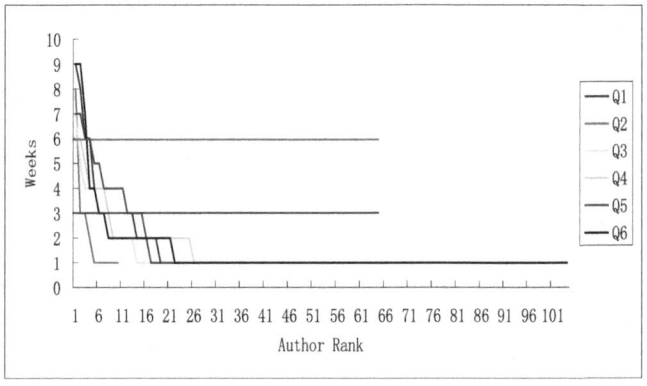

Fig. 4. Consultants evolution with time

(2)Find Stable Consultants

The consultants is evolving with time. However, some consultants are stable. We track the evolution of top 15 consultants given six query for 10 weeks. Fig. 4 shows the stable consultants. In Fig. 4, horizontal axis denotes the author rank based on the times of this author appear in 10 weeks, vertical axis denotes the times of this author appear in 10 weeks. We think the authors which the times appear in 10 weeks is more than 6 are core consultants we seek.

5   Related Work

In the past few years, some research was devoted to finding experts or consultants. SmallBlue[7][8] finds experts in large organizations through data mining, information retrieval, and artificial intelligence. SmallBlue[7][5] main depend on social network among company, it focus on "who knows what? ","who knows whom? "and "who knows what about whom? ".

EOS[6]is a researcher social network system. It has gathered information about computer science researchers from the Web and constructed a social network among the researchers through their co-authorship. Neil Rubens et. al.[10] try to find groups of experts aimed at Collaborative Work.

Jun Zhang et. al.[11] propose an expertise network, and present Hits and PageRank algorithms to find experts in this expertise network for java forum.

Krisztian Balog[2][1] discuss people search in the enterprise by a generative probabilistic modeling framework for capturing the expert finding and profiling tasks in a uniform way.

In References [3][4][9], authors discuss how to identify experts in E-mail Corpora.

6   Conclusion and Future Work

There are some people who have expertise knowledge of specific area on web forum. There exists some freshman in some certain fields in web forum, they hope to get help from those experts. In this paper, we study how to find and rank consultants given the query q on web forum. The major contributions of the paper include:

(1)**Propose the formal definitions of the consultants given query.** How to judge an author whether is consultant

(2)**Give a Model to search and rank consultants of specific area.** We model the procedure of search and rank consultants by Bayes' theorem, and present a method to compute association between a topic and an author based on three heuristic rules.

(3)**We implement the algorithm of finding and ranking consultants in a real forum data.** We implement the searching and ranking consultants algorithm, and its effectiveness is tested on a real forum data set.

Our further research direction is to find more candidate consultants indicators on web forum, and obtain the weight of each indicator by machine learning.

References

1. Balog, K.: People search in the enterprise. In: Kraaij, W., de Vries, A.P., Clarke, C.L.A., Fuhr, N., Kando, N. (eds.) SIGIR, p. 916. ACM, New York (2007)
2. Balog, K.: People search in the enterprise. SIGIR Forum 42(2), 103 (2008)
3. Balog, K., de Rijke, M.: Finding experts and their details in e-mail corpora. In: Carr, L., Roure, D.D., Iyengar, A., Goble, C.A., Dahlin, M. (eds.) WWW, pp. 1035–1036. ACM, New York (2006)

4. Campbell, C.S., Maglio, P.P., Cozzi, A., Dom, B.: Expertise identification using email communications. In: CIKM, pp. 528–531. ACM, New York (2003)
5. Ehrlich, K., Lin, C.-Y., Griffiths-Fisher, V.: Searching for experts in the enterprise: combining text and social network analysis. In: GROUP, pp. 117–126 (2007)
6. Li, J.-Z., Tang, J., Zhang, J., Luo, Q., Liu, Y., Hong, M.: Eos: expertise oriented search using social networks. In: WWW, pp. 1271–1272 (2007)
7. Lin, C.-Y., Cao, N., Liu, S., Papadimitriou, S., Sun, J., Yan, X.: Smallblue: Social network analysis for expertise search and collective intelligence. In: ICDE, pp. 1483–1486. IEEE, Los Alamitos (2009)
8. Lin, C.-Y., Ehrlich, K., Griffiths-Fisher, V., Desforges, C.: Smallblue: People mining for expertise search. IEEE MultiMedia 15(1), 78–84 (2008)
9. Mock, K.J.: An experimental framework for email categorization and management. In: SIGIR, pp. 392–393 (2001)
10. Rubens, N., Vilenius, M., Okamoto, T., Kaplan, D.: Cafe: Collaboration aimed at finding experts. I. J. Knowledge and Web Intelligence 1(3/4), 169–186 (2010)
11. Zhang, J., Ackerman, M.S., Adamic, L.A.: Expertise networks in online communities: structure and algorithms. In: Williamson, C.L., Zurko, M.E., Patel-Schneider, P.F., Shenoy, P.J. (eds.) WWW, pp. 221–230. ACM, New York (2007)

Comparing Similarity of HTML Structures and Affiliate IDs in Splog Analysis

Taichi Katayama[1], Akihito Morijiri[1], Soichi Ishii[2],
Takehito Utsuro[1], Yasuhide Kawada[3], and Tomohiro Fukuhara[4]

[1] University of Tsukuba, Tsukuba, 305-8573, Japan
[2] Tokyo Denki University, Tokyo, 101-8457, Japan
[3] Navix Co., Ltd., Tokyo, 141-0031, Japan
[4] National Institute of Advanced Industrial Science and Technology,
Tokyo 135-0064, Japan

Abstract. Spam blogs or splogs are blogs hosting spam posts, created using machine generated or hijacked content for the sole purpose of hosting advertisements or raising the number of in-links of target sites. Among those splogs, this paper focuses on detecting a group of splogs which are estimated to be created by an identical spammer. In this paper, we compare two clues: namely, similarity of HTML structures of splogs and affiliate IDs automatically extracted from splogs. We first show that the similarity of HTML structures of splogs is quite effective in splog detection, as well as in identifying spammers. We then show that the identity of affiliate IDs extracted from splogs can identify spammers much more directly than similarity of HTML structures, although it is not easy to achieve high coverage in extracting affiliate IDs. Finally, we show that the coverage of the intersection of the two clues, similarity of HTML structures and affiliate IDs, is relatively low, and it is necessary to apply them in a complementary strategy.

Keywords: spam blog detection, HTML structures, affiliate IDs.

1 Introduction

Weblogs or blogs are considered to be one of personal journals, market or product commentaries. While traditional search engines continue to discover and index blogs, the blogosphere has produced custom blog search and analysis engines, systems that employ specialized information retrieval techniques. With respect to blog analysis services on the Internet, there are several commercial and non-commercial services such as *Technorati*[1], *BlogPulse* [1][2], and *kizasi.jp*[3]. With respect to multilingual blog services, *Globe of Blogs*[4] provides a retrieval function

---

[1] http://technorati.com/
[2] http://www.blogpulse.com/
[3] http://kizasi.jp/ (in Japanese).
[4] http://www.globeofblogs.com/

J. Xu et al. (Eds.): DASFAA Workshops 2011, LNCS 6637, pp. 378–389, 2011.
© Springer-Verlag Berlin Heidelberg 2011

of blog articles across languages. *Best Blogs in Asia Directory*[5] also provides a retrieval function for Asian language blogs.

As with most Internet-enabled applications, the ease of content creation and distribution makes the blogosphere spam prone [2,3,4]. Spam blogs or splogs are blogs hosting spam posts, created using machine generated or hijacked content for the sole purpose of hosting advertisements or boosting the ranking of target sites. [3] reported that for English blogs, around 88% of all pinging URLs (i.e., blog homepages) are splogs, which account for about 75% of all pings. Based on this estimation, as stated in [5], splogs can cause problems including the degradation of information retrieval quality and the significant waste of network and storage resources. Several previous works [3,4] reported important charac-teristics of splogs. [4] reported characteristics of ping time series, in-degree/out-degree distributions, and typical words in splogs found in TREC[6] Blog06 data collection. [3] reported the results of analyzing splogs in the *BlogPulse* data set. In the context of semi-automatically collecting Web spam pages/hosts includ-ing splogs, [6] discussed how to collect spammer-targeted keywords to be used when collecting a large number of Web spam pages/hosts efficiently. [7] ana-lyzed (Japanese) splogs based on various characteristics of keywords contained in them.

Along with those analysis on splogs reported in previous works, several splog detection techniques (e.g., [8,9,5]) have been proposed. [9] studied features for splog detection such as words, URLs, anchor texts, links, and HTML meta tags in supervised learning by SVMs. As features of SVMs, [5] studied temporal self similarities of splogs such as posting times, post contents, and affiliated links. [8] also studied detecting link spam in splogs by comparing the language models among the blog post, the comment, and pages linked by the comments.

Unlike those studies in the previous works on splog detection, [10] proposed to employ the similarity of HTML structures of splogs in splog detection by ma-chine learning. In measuring the similarity of HTML structures, a list of DOM (Document Object Model) elements (minimum unit of content) is extracted from the DOM tree of an HTML document. [10] reported that they manually exam-ined the HTML documents of splogs estimated to be created by an identical spammer and found that they tend to have similar DOM trees.

In this paper, following [10], we evaluate the similarity of HTML structures of splogs proposed in [10] with much larger splog / authentic blog data set. More specifically, out of 10 major Japanese blog hosts, we focus on the three blog hosts with the highest splog rates. Then, we apply the similarity of HTML structures of splogs to randomly selected splogs / authentic blogs, and show that the measure is quite effective in splog detection as well as in identifying spammers. Unlike [10], this paper shows that, given a small set of seed splog sites, we can easily collect splogs which have HTML structures similar to a seed splog with quite high precision. We next compare the similarity of HTML structures

---

[5] http://directory.bestblogs.asia/
[6] http://trec.nist.gov/

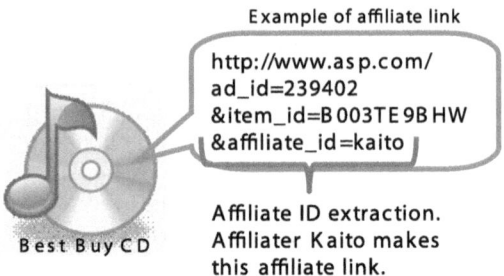

Fig. 1. Example of an Affiliate Link Containing an Affiliate ID

of splogs with another clue: affiliate IDs[7] automatically extracted from splogs. We show that the identity of affiliate IDs extracted from splogs can identify spammers much more directly than similarity of HTML structures, although it is not easy to achieve high coverage in extracting affiliate IDs. Finally, we show that the coverage of the intersection of the two clues, similarity of HTML structures and affiliate IDs, is relatively low, and it is necessary to apply them in a complementary strategy.

2   Similarity of HTML Structures

Following [10], this section briefly introduces how to measure the similarity of HTML structures of splogs.

2.1   Extracting DOM Sequences of an HTML Document

First, from an HTML document of a splog / an authentic blog, a sequence of DOM (Document Object Model) elements is extracted[8]. In this procedure, as shown in Figure 2, given an HTML document s, it is first converted into a tree structure of HTML tags. This tree structure is referred to as the DOM tree of the HTML document s. Then, following the definitions of *HTML elements*[9], we extract only a small portion of them including certain *block elements*. More specifically, first, we divide the DOM tree into subtrees with *BODY* tags as well

---

[7] Figure 1 shows an example of an affiliate link. In this link, ad_id, item_id, and affiliate_id are contained. We extract a value associated with affiliate_id.

[8] The underlying motivation of measuring differences in HTML layouts for splog detection is somehow similar to that of [11]. However, the evaluation procedure of [11] is different from ours, in that they artificially created a mixture of Web spam pages and authentic Web pages, and then apply their similarity measure within the task of clustering Web spam pages / authentic Web pages. Compared to the similarity measure of [10], that of [11] is more coarse-grained when incorporating HTML tags in the similarity.

[9] W3C HTML 4.01 Specification (http://www.w3.org/TR/HTML401/).

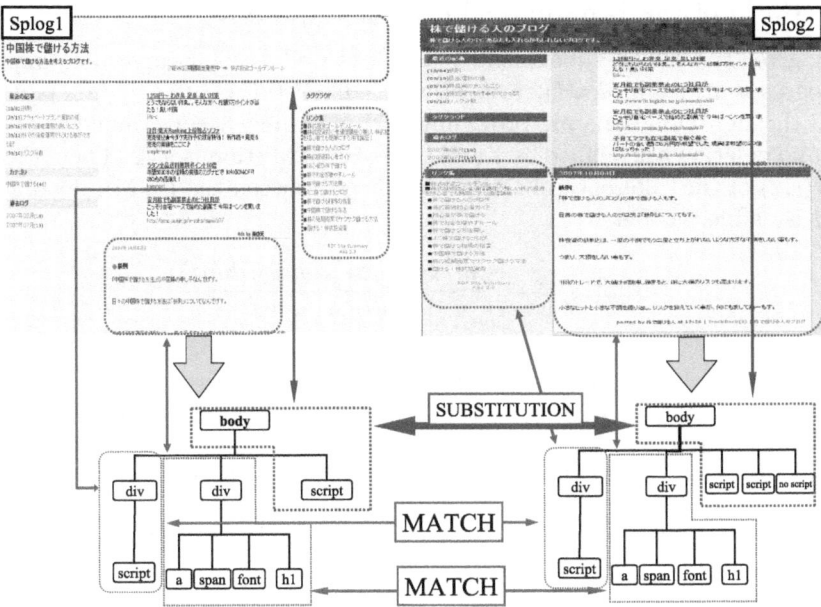

edit distance=2

Rdiff=2/(3+3)=0.33

Fig. 2. Extracting DOM Sequences of HTML Documents and Measuring the Ratio of their Differences: An Example

as P tags and DIV tags which indicate block elements[10]. We remove subtrees with other tags from the DOM tree. Each of the extracted subtrees can be considered as minimum unit of content and we refer to it as a DOM element. Then, the DOM tree is traversed in the breadth-first strategy, and DOM elements are collected into a sequence $dm(s)$.

Figure 2 shows an example of extracting sequences of DOM elements from two splog HTML documents "splog 1" and "splog 2". In this case, from each of the two HTML documents, two DOM elements with a DIV tag at their root nodes are extracted. Each of the two DOM elements with a DIV tag is identical between "splog 1" and "splog 2". Also, from each of the two HTML documents, a DOM element with a $BODY$ tag at its root node is extracted. Those two DOM elements with a $BODY$ tag at their root nodes have their internal nodes different from each other.

---

[10] Even though HTML elements with $SCRIPT$ tags and $STYLE$ tags are not usually rendered by visual browsers, we keep subtrees with those tags within block elements, since their differences should be accounted for in our task of splog detection.

2.2   Ratio of the Differences in DOM Sequences

Next, given two HTML documents s and t, as well as their corresponding DOM sequences $dm(s)$ and $dm(t)$, respectively, edit distance $(dm(s), dm(t))$ is measured through DP (dynamic programming) matching, where the costs of insertion and deletion are defined as 1 and that of substitution is as 2. Then, the ratio $\mathrm{Rdiff}(s, t)$ of the difference of DOM sequences between the HTML documents s and t is defined as below:

$$\mathrm{Rdiff}(s, t) = \frac{\text{edit distance } (dm(s), dm(t))}{|dm(s)| + |dm(t)|}$$

In the case of the two splog HTML documents in Figure 2, the two DOM sequences differ in one DOM element out of the three and the ratio of their difference is calculated as 0.33.

Next, we measure the similarity of DOM sequences between an HTML document s and a set T of HTML documents. Here, we introduce the notion of the smallest ratio $\mathrm{Rdiff}(s, t \in T)$ between s and any member t of T ($s \neq t$) and refer to it as $\mathrm{MinDF}(s, T)$.

$$\mathrm{MinDF}(s, T) = \min_{t \neq s} \mathrm{Rdiff}(s, t \in T)$$

3   Automatic Collection of Splogs with High Similarities of HTML Structures

3.1   Seed Splog Data Set

We first collected 6.2 million blog sites through a system called KANSHIN [12] for the years $2004 \sim 2009$. Next, we focused on major 10 blog hosts, and 5.2 million blog sites out of the whole 6.2 million. For each of the 10 blog hosts, we then randomly selected about 400 blog sites and manually annotated splog / authentic blog distinction to each blog site. For each blog host, Figure 3 shows the number of the collected blog sites as well as splog rate. In the rest of the paper, we focus on the three blog hosts with the highest splog rates, i.e., "S", "F", and "C".

Let the meta-variable H range over the three blog hosts "S", "F", and "C". For each host H, we denote the set of seed splog sites as $SP_{seed}(H)$, where those seed splog sites are used for collecting splogs which have high similarities of HTML structures with at least one of the seed splog sites. The seed splog sites used in this paper are those random samples collected above, where additional splogs are further added if the number of splogs is less than 100. The numbers of seed splog sites are 208 for the host "S", and 100 for each of "F" and "C".

3.2   The Procedure

For the three blog hosts, the total number of blog sites is about 1.7 million. We randomly selected 10% of them, i.e., 170,000 blog sites, out of which about

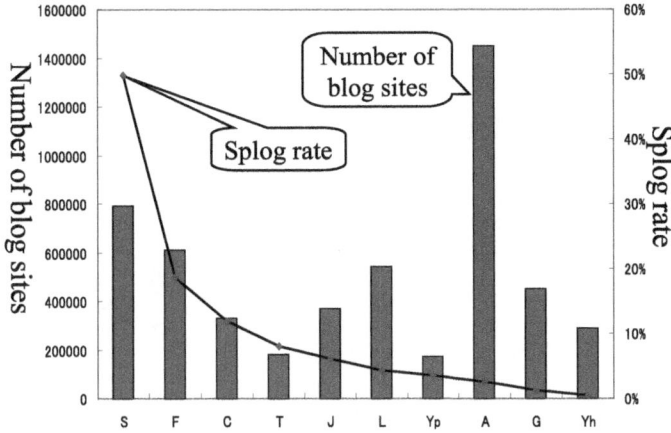

Fig. 3. Number of Blog Sites and Splog Rate per Blog Host

130,000 blog sites were accessible on the Web in June 2009. Next, for each blog host H, we collect all the blog sites b of H from the members of the 130,000 blog sites, and calculate the value $\mathrm{MinDF}(b, SP_{seed}(H))$. Then, we collect any blog site b from the 130,000 blog sites, which have the values of $\mathrm{MinDF}(b, SP_{seed}(H))$ as $k - 0.05 \leq \mathrm{MinDF}(b, SP_{seed}(H)) \leq k$ into the set $B(H, k)$. Similarly, we can define the splog subset of $B(H, k)$ as $SP(H, k)$ below:

$$B(H, k) = \Big\{ \text{blog (i.e., splog or authentic blog) site } b \ \Big| \ \text{blog host of } b \text{ is } H,$$

$$b \notin SP_{seed}(H), \quad k - 0.05 \leq \mathrm{MinDF}(b, SP_{seed}(H)) \leq k \Big\}$$

$$SP(H, k) = \Big\{ \text{ splog site } s \ \Big| \ s \in B(H, k) \Big\}$$

In the case of the much smaller data set examined in [10], splogs created by an identical spammer tend to have the similarity of HTML structures less than 0.2. Based on this result, in the following, we examine the blog sites b which have the values of $\mathrm{MinDF}(b, SP_{seed}(H))$ as lower than 0.3. The numbers of such blog sites are 7,787 for the blog host "S", 628 for the blog host "F", and 7,352 for the blog host "C", respectively. For the host "F", we used all of the 628 blog sites in further analysis. From those blog sites for the hosts "S" and "C", we randomly sampled 1,013 sites and 558 sites respectively for further analysis. We then manually annotated splog / authentic blog distinction to each blog site. Finally, for each blog host H and for each of the values of $k = 0.05, 0.1, 0.15, 0.2, 0.25, 0.3$, we measure the splog rate and plot them as "Splog rate (similar to seed)" in Figure 4. In the figure, we also show random splog rates taken from Figure 3 as "Splog rate (random)", which are measured over the randomly sampled blog sites, but not over those which have HTML structures similar to each other.

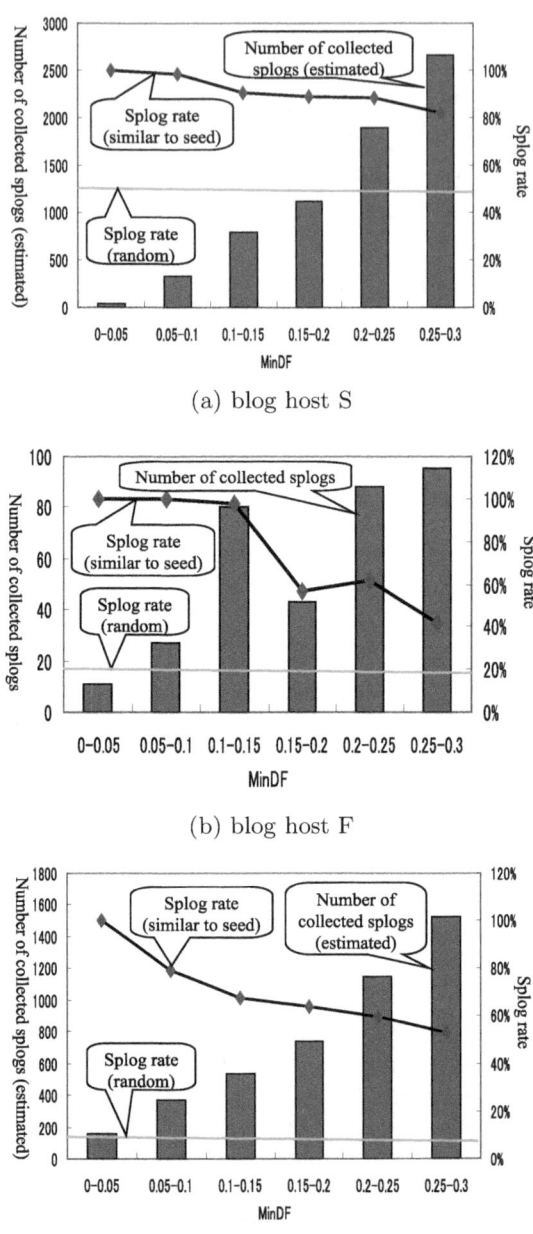

(a) blog host S

(b) blog host F

(c) blog host C

Fig. 4. Splogs Rates for Blog Sites Similar to Seeds and Number of Collected Splogs

Figure 4 also shows the numbers of the collected splog sites for the blog host "F", or the estimated numbers of the collected splog sites for the blog hosts "S" and "C".

3.3 Analysis on Splog Rate

The results of analyzing splog rates in Figure 4 can be summarized as below:

1. For all of the three hosts, "Splog rate (similar to seed)" are much higher than "Splog rate (random)". This result clearly shows that the proposed method of collecting blog sites with high HTML similarity successfully collects splogs which are similar to seed splogs.

2. It is clearly shown that the numbers of splogs automatically collected are quite large compared to the numbers of seed splogs. After detailed analysis, we found the followings: For the blog host "S", we can collect 1,160 splog sites which are similar to at least one of 50 seed splog sites in the range of $\mathrm{MinDF}(s, SP_{seed}(H))$ less than 0.15, with splog rate as high as 95%. In the range of $\mathrm{MinDF}(s, SP_{seed}(H))$ less than 0.3, we can collect 6,824 splog sites which are similar to at least one of 136 seed splog sites with splog rate as high as 87%. For the blog host "F", we can collect 118 splog sites which are similar to at least one of 21 seed splog sites in the range of $\mathrm{MinDF}(s, SP_{seed}(H))$ less than 0.15, with splog rate as high as 98%. For the blog host "C", we can collect 155 splog sites which are similar to at least one of 17 seed splog sites in the range of $\mathrm{MinDF}(s, SP_{seed}(H))$ less than 0.05, with splog rate as high as 100%.

4 Splogs and Affiliate IDs

In an HTML page that contains affiliate links or banners, one or more affiliate codes are embedded. Affiliate codes are described as JavaScript code or a URL. Various information is embedded in the affiliate code such as advertiser's ID, commodity's ID, affiliator's ID, and so on. For the former case, some affiliate service providers (ASPs) provide an affiliate ID in the code. An example is Google's AdSense[11]. In the JavaScript code of the AdSense, there is a unique ID. We extract this ID as an affiliate ID. For the latter case, some ASPs embed several IDs in a URL. Figure 1 shows an example of an affiliate link, where we extract a value associated with affiliate_id.

Spammers often create and maintain multiple blog sites in order to obtain a large quantity of commissions from ASPs. In addition to this, some ASPs allow a user to create multiple affiliate IDs. Spammers create multiple IDs using this framework. If we can find an ID as a spam ID, we can exclude all of blog sites that contain this ID automatically.

Actually, in our results of preliminary examination on affiliate IDs, 95.1% of affiliate IDs that appear more than 10 blog sites are spam IDs. Thus, in section 5.2, we extract affiliate IDs which appear in more than 10 blog sites, and manually classify these IDs into spam and authentic IDs.

---

[11] http://www.google.com/adsense/

5   Analysis on Identifying Spammers

Based on the results of section 3, this section studies how to identify spammers in the set of given splogs. We especially compare the similarity of HTML structures and affiliate IDs. We show that the coverage of their intersection is relatively low, and it is necessary to apply them in a complementary strategy.

5.1   Identifying Spammers Based on the Similarity of HTML Structures

First, in the results of collecting splog sites with high similarity of HTML structures in section 3, we manually examine whether the spammers who create those splog sites are identical or not. Rough idea of this analysis is shown in Figure 5. In this analysis, based on the discussion in section 3.3, we decided that we focus on splog sites which are similar to at least one of the seed splog sites in the range of $\mathrm{MinDF}(s, SP_{seed}(H))$ less than 0.15. We regard those splog sites as the subset L in Figure 5. When judging whether two splog sites are actually created by an identical spammer, we follow the criteria below:

1. There exists on the Web a document which is estimated to be pasted onto the two splog sites. When pasting a document onto a splog post, usually, proper nouns are randomly replaced with other proper nouns, or nouns are randomly replaced with their synonyms using certain dictionaries.

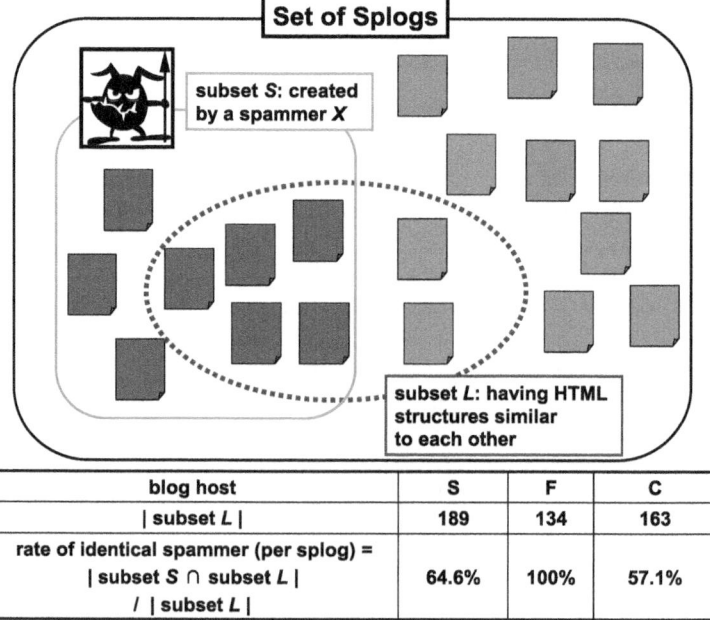

blog host	S	F	C
| subset L |	189	134	163
rate of identical spammer (per splog) = | subset S ∩ subset L | / | subset L |	64.6%	100%	57.1%

Fig. 5. Rate of Identical Spammer

2. The two splog sites have similar frame layouts with a Web browser, or have similar layouts of HTML objects in a frame.
3. The two splog sites share an outlink to an identical URL.

The results are shown in Figure 5, which can be summarized as below:

1. For the blog host "F", all the collected splog sites can be judged as being created by spammers, each of who creates one of the seed splog sites.
2. For the blog hosts "S" and "C", the rate of identifying identical spammer is relatively low compared to that of the blog host "F". This is mainly because certain number of splog sites which contain just a few posts are included in the set of seed splog sites. Although it is easy to manually judge that such splog sites with just a few posts were not created by an identical spammer, their HTML structures tend to be highly similar to each other.

5.2 Comparison of the Similarity of HTML Structures and Affiliate IDs

Next, we compare the similarity of HTML structures and affiliate IDs with respect to the coverage when identifying spammers. Rough idea of this analysis is shown in Figure 6, where we focus on a set of splogs created by an identical spammer.

First, in the results of collecting splog sites with high similarity of HTML structures in section 3, we automatically extract affiliate IDs and measure the rate of including affiliate IDs. In this analysis, we focus on splog sites which are similar to at least one of the seed splog sites in the range of $\mathrm{MinDF}(s, SP_{seed}(H))$ less than 0.15. We regard those splog sites as the subset L in Figure 6. As shown in the left-bottom in Figure 6, the rates of including affiliate IDs are less than or around 25%. This is mainly because affiliate IDs of two major ASPs are encrypted and are hard to automatically extract. Another reason is that certain splog sites sometimes have only outlinks for SEO (search engine optimization) purposes, but no affiliate links, and affiliate links are included only in pages outlinked from splog sites.

Next, in the data set of 130,000 blog sites described in section 3, we extract affiliate IDs which appear in more than 10 blog sites, and manually classify these IDs into spam and authentic IDs. For each of the blog hosts "S" and "F"[12], we show the numbers of affiliate IDs and splog sites as shown in the right-bottom in Figure 6. We regard those splog sites as the subset A in Figure 6. Out of those splog sites, we then measure the rate of pairs of splog sites which have the similar HTML structure to each other. In this analysis, we focus on pairs of splog sites which have the similarity in the range of Rdiff less than 0.15. As shown in Figure 6, these rates are around 15%.

Based on these results, we can conclude that the coverage of the intersection of the two clues, similarity of HTML structures and affiliate IDs, is relatively low, and it is necessary to apply them in a complementary strategy.

---

[12] For the blog host "C", the number of affiliate IDs which appear in more than 10 blog sites is too small, and we omit it.

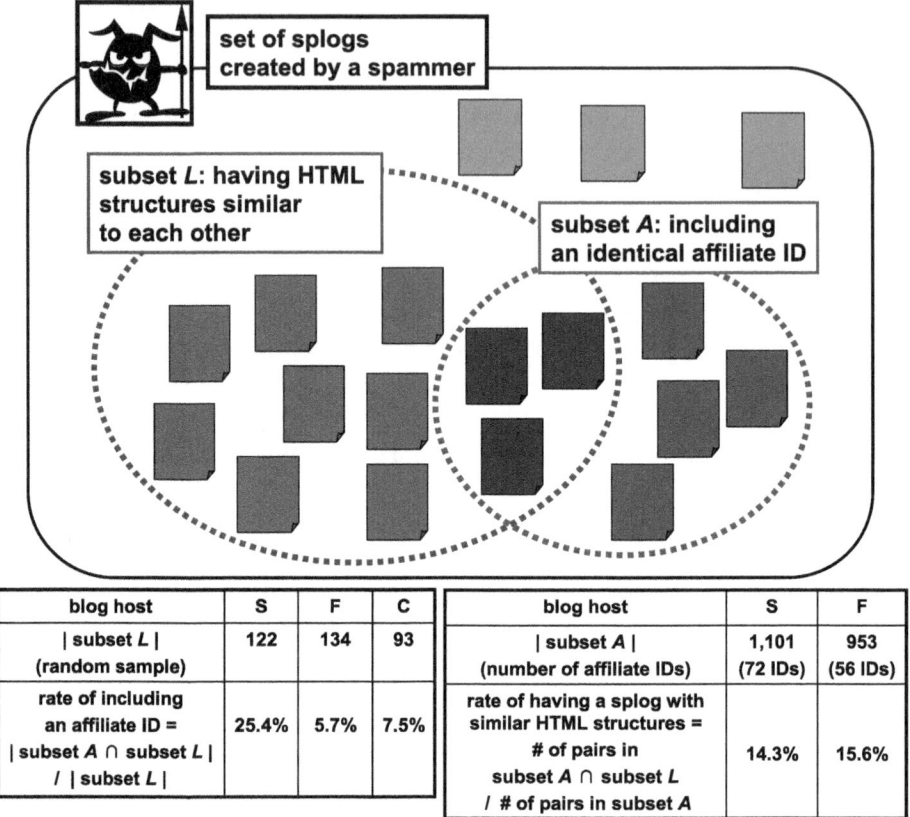

blog host	S	F	C
| subset L | (random sample)	122	134	93
rate of including an affiliate ID = | subset A ∩ subset L | / | subset L |	25.4%	5.7%	7.5%

blog host	S	F
| subset A | (number of affiliate IDs)	1,101 (72 IDs)	953 (56 IDs)
rate of having a splog with similar HTML structures = # of pairs in subset A ∩ subset L / # of pairs in subset A	14.3%	15.6%

Fig. 6. Comparison of the Similarity of HTML Structures and Affiliate IDs

6   Concluding Remarks

This paper focused on a group of splogs which are estimated to be created by an identical spammer. We compared two clues: namely, similarity of HTML structures of splogs and affiliate IDs automatically extracted from splogs. We first showed that the similarity of HTML structures of splogs is quite effective in splog detection, as well as in identifying spammers. We then showed that the identity of affiliate IDs extracted from splogs can identify spammers much more directly than similarity of HTML structures, although it is not easy to achieve high coverage in extracting affiliate IDs. Finally, we showed that the coverage of the intersection of the two clues, similarity of HTML structures and affiliate IDs, is relatively low, and it is necessary to apply them in a complementary strategy. Although the principle of the techniques employed in this paper is totally language independent and they are definitely applicable to splogs in languages other than Japanese, the applicability of the proposed approach is somehow dependent on the spammers' behavior and the way affiliate IDs are embedded in HTML pages in each language. Future works include inventing a technique of detecting splogs which can not be detected either of the two clues, where at present we are working on applying a machine learning technique [10] with training samples automatically collected using spam affiliate IDs studied in this paper.

References

1. Glance, N., Hurst, M., Tomokiyo, T.: Blogpulse: Automated trend discovery for Weblogs. In: Proc. Workshop on the Weblogging Ecosystem: Aggregation, Analysis and Dynamics (2004)
2. Gyöngyi, Z., Garcia-Molina, H.: Web spam taxonomy. In: Proc. 1st AIRWeb, pp. 39–47 (2005)
3. Kolari, P., Joshi, A., Finin, T.: Characterizing the splogosphere. In: Proc. 3rd Workshop on the Weblogging Ecosystem: Aggregation, Analysis and Dynamics (2006)
4. Macdonald, C., Ounis, I.: The TREC Blogs06 collection: Creating and analysing a blog test collection. Technical Report TR-2006-224, University of Glasgow, Department of Computing Science (2006)
5. Lin, Y.R., Sundaram, H., Chi, Y., Tatemura, J., Tseng, B.L.: Splog detection using self-similarity analysis on blog temporal dynamics. In: Proc. 3rd AIRWeb, pp. 1–8 (2007)
6. Wang, Y., Ma, M., Niu, Y., Chen, H.: Spam double-funnel: Connecting web spammers with advertisers. In: Proc. 16th WWW, pp. 291–300 (2007)
7. Sato, Y., Utsuro, T., Fukuhara, T., Kawada, Y., Murakami, Y., Nakagawa, H., Kando, N.: Analyzing features of Japanese splogs and characteristics of keywords. In: Proc. 4th AIRWeb, pp. 33–40 (2008)
8. Mishne, G., Carmel, D., Lempel, R.: Blocking blog spam with language model disagreement. In: Proc. 1st AIRWeb (2005)
9. Kolari, P., Finin, T., Joshi, A.: SVMs for the Blogosphere: Blog identification and Splog detection. In: Proc. 2006 AAAI Spring Symp. Computational Approaches to Analyzing Weblogs, pp. 92–99 (2006)
10. Katayama, T., Yoshinaka, T., Utsuro, T., Kawada, Y., Fukuhara, T.: Detecting splogs using similarities of splog HTML structures. In: Proc. 4th ICUIMC, pp. 256–263 (2010)
11. Urvoy, T., Lavergne, T., Filoche, P.: Tracking Web spam with hidden style similarity. In: Proc. 2nd AIRWeb, pp. 25–30 (2006)
12. Fukuhara, T., Kimura, A., Arai, Y., Yoshinaka, T., Masuda, H., Utsuro, T., Nakagawa, H.: KANSHIN: A cross-lingual concern analysis system using multilingual blog articles. In: Proc. 1st INGS 2008, pp. 83–90 (2008)

Crowd-Powered TV Viewing Rates: Measuring Relevancy between Tweets and TV Programs

Shoko Wakamiya[1], Ryong Lee[2], and Kazutoshi Sumiya[2]

[1] Graduate School of Human Science and Environment, University of Hyogo, Japan
[2] School of Human Science and Environment, University of Hyogo, Japan
nd09a025@stshse.u-hyogo.ac.jp,
{leeryong,sumiya}@shse.u-hyogo.ac.jp

Abstract. Due to the advance of many social networking sites, social analytics by aggregating and analyzing crowds' life logs are attracting a great deal of attention. In the meantime, there is an interesting trend that people watching TVs are also writing Twitter messages pertaining to their opinions. With the utilization of bigger and broader crowds over Twitter, surveying massive audiences' lifestyles will be an important aspect of exploitation of crowd-sourced data. In this paper, for better TV viewing rates in the light of the evolving TV lifestyles beyond home environments, we propose a TV rating method by means of Twitter where we can easily find crowd voices relative to TV watching. In the experiment, we describe our exploratory survey to exploit a large amount of Twitter messages to populate TV programs and on-line video sites.

Keywords: TV Viewing Rates, Micro-blogging, Social Network.

1 Introduction

The recent advances in social networking sites such as Facebook[1] and Twitter[2] encourage crowds to share their updates in almost real time across the open space. At the moment, a new kind of interaction between the TV stations and general audiences increasingly appears stimulating beneficial interactions between both sides. Generally, in the side of TV stations, they want to listen to their audiences' opinions on their contents. Conversely, audiences would like to often participate in the TV program expressing their thoughts or feelings directly to the content providers. Accordingly, in terms of conventional TV viewing surveys, social media must be a valuable source to gather much bigger and wider audiences rating, with less additional costs to selected participants who worked for the conventional TV ratings.

In fact, current TV ratings in the USA and Japan are measured mostly based on Nielsen ratings, which were developed by Nielsen Media Research[3] many years ago. This method measures TV ratings in three different ways: First, "Set Meter," which

---

[1] Facebook: http://www.facebook.com/
[2] Twitter: http://twitter.com/
[3] Nielsen Media Research: http://www.nielsen.com/

J. Xu et al. (Eds.): DASFAA Workshops 2011, LNCS 6637, pp. 390–401, 2011.

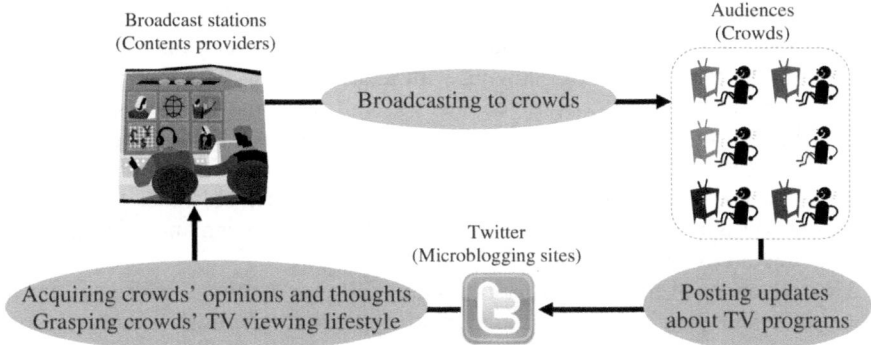

Fig. 1. Twitter-based interactive TV

means an electronic device to monitor what TV programs the selected homes view. The collected viewing logs are transmitted in the night or in real time to the Nielsen center or other media research companies to derive a statistical summary. Next, "People Meter," which is a specially designed remote controller, to recognize the members of a household who watch the TV programs by selecting one of the identification buttons on the remote controller, eventually enabling analysts to survey various demographic groups such as younger vs. older generations. Lastly, "Viewer Diary," the oldest way, is based on audiences' self-recording on paper-based questionnaires about what they have watched individually. The first two methods, which are most often employed methods, need to have the specific devices set up on television sets.

Apparently, unlike a few decades ago, we are no more bound to watch TVs at our homes. We can carry the TVs to any outdoor place through smart phones. Additionally, watching TVs is also not limited to the broadcasting time—rather, video recorders or the recently introduced time shift functions in TVs can help us make up for the missed programs consequently. Furthermore, the concept of TV is now extending its realm with many online video sites such as YouTube[4] and Japan's NicoNicoDouga[5]. The increasing scope of watching programs on the TV makes it much difficult to assess the ratings through home-centric measurements.

In order to overcome the above problems of conventional TV rating methods, we focus on a new source by crowds. Obviously, among the numerous postings on today's social networking sites, there are many useful crowd life logs related to media consuming. In practice, recent TV programs are adopting Twitter as a backward-channel to directly obtain audiences' opinions about on-aired programs. In a sense, this movement can be seen as a form of interactive TV as shown in Fig. 1. Based on this conception, we may trace personal media life patterns from these logs and probably rank TV programs or songs that audiences mostly prefer. For example, in the case of YouTube, the most popular video sharing website, their pages have a tweet-writing button to each video viewing page to let people remain or share their viewing experience.

---

[4] YouTube: http://www.youtube.com/
[5] NicoNicoDouga: http://www.nicovideo.jp/

In this paper, we present a novel TV rating method by utilizing the audiences' media-life logs over microblogging sites. In particular, we focus on Twitter where we can find many posts about TV viewing with additional tags of where and when audiences enter their logs. However, the site was not designed for this specific goal to collect the TV-related Twitter messages—so-called tweets—identifying those that are relevant to TV programs, which are the target of this work. Therefore, we need to confirm the tweets to a particular region in order to filter out other tweets from regions outside our interest. In particular, we will present a semantic linking between tweets relevant to TV programs.

The remainder of this paper is organized as follows: Section 2 addresses our initial motivation and provides a TV rating platform utilizing crowd power via Twitter. Section 3 describes the detailed methodology for semantic linking from tweets to relevant programs. Section 4 illustrates our experimental results conducted with a real dataset of tweets in Japan and actual electronic program list for a month. Section 5 concludes this paper with further work.

2   A Twitter-Based TV Rating Platform

In this section, we first describe our motivation to utilize Twitter as a back-channel from audiences to broadcast stations realizing the so-called interactive TV. Then, we introduce a TV rating platform on Twitter and highlight the most critical issue of constructing semantic links from tweets to relevant TV programs.

2.1 Looking for Audiences on Twitter

Generally, Twitter is a microblogging site targeting for various real life applications. Thus, in order to use the Twitter platform as a back-channel for TV rating, we should find out tweets related TV programs. In order to look for those TV-related tweets, we may use some specialized hashtags which are popularly used on the site as an index to enable retrieval by other services or users. Generally, Twitter users can simply create a hashtag by prefixing a word with a hash symbol "#hashtag." For instance, various hashtags, such as "#tvasahi," "#2010wc," "#worldcup," "#jfa2010," and "#wcj," "#wc2010" had been used during the 2010 FIFA World Cup. Actually, twtv.jp[6] has already used these tags to collect the messages sent intentionally to the stations.

However, the hashtag-based TV rating method is not sufficient to aggregate large amounts of public opinions, since hashtags are not always given for all existing TV programs and it needs an effort to intentionally add specific tags relative to TV stations or programs in the current situation where users must write hashtags in a tweet with different devices such as smart phones or PCs during their watching. Furthermore, it is unlikely that all existing broadcast stations have their own Twitter accounts and hashtags. Therefore, it is important to develop a method to capture much more TV-related messages. In fact, this requires a kind of semantic linking between the freely written messages under the length limitation of 140 bytes and the TV programs. We will present the details of our method to link tweets to the corresponding programs, if there are any relevancy between them.

---

[6] twtv.jp: http://twtv.jp/

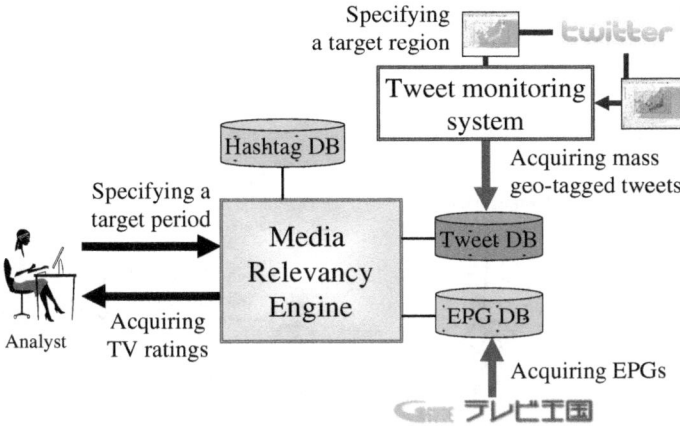

Fig. 2. Tweet-based TV rating platform

2.2 Twitter-Based TV Rating Platform

In order to develop a TV rating system utilizing crowd life logs about TV viewing on Twitter, we propose a tweet-based TV rating platform as illustrated in Fig. 2. With this platform, we support analysts enabling them to easily investigate crowds' media consumption. Especially, since we are looking into the TV viewing logs on Twitter, we need information on on-air programs and on-line video sites to identify what crowds are looking.

Furthermore, to realize monitoring to local TV ratings, we are dealing with geo-tagged tweets which have information on when and where a tweet is written. For this, we developed a tweet aggregation system in our previous work [4] intentionally to collect such specific type of tweets effectively. For the simplicity, we will not describe the detail of the system and methodology here. Instead, with a geo-tagged tweet database collected by the system, we will investigate the TV-relevant ones and utilize them to populate TV programs.

Specifically, when we identify the most relevant program with a tweet, we approach two different levels of identification processes. First, as a primitive and essential step, we look up a prepared hashtag list which includes hashtags and the source information. However, as aforementioned, hashtag-based linking from tweets to relevant programs will eventually suffer the lack of relevance enough to measure final TV ratings, because unlike on-line TVs or video sharing sites, on-air TVs required a user effort to manually write such hashtags into the writing tweets as shown in Fig. 3. In general, people prefer to write just a title of program or a few keywords representing program. In other words, we cannot ignore such freely written texts which are connected to much more hidden TV audiences. Therefore, we further have to semantically examine the relevance between tweets and possibly relevant programs from those raw texts. As for on air broadcasting TV programs, we use an Electronic Program Guide (EPG), which typically provides people with scheduling information for current and upcoming programs.

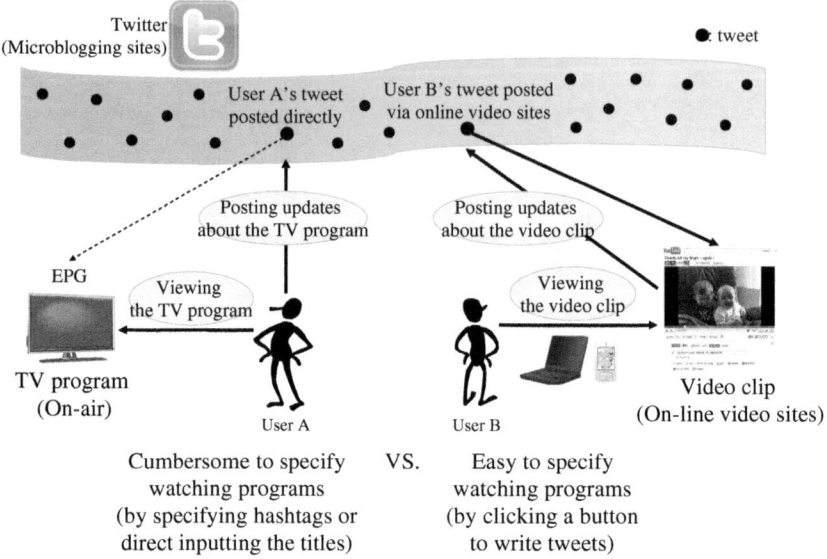

Fig. 3. Difference of crowds' behavior for viewing on-air TV programs and on-line video site

2.3 Related Work

Recently, microblogging services represented by Twitter grow more popular and have been aggregating lots of researchers' attentions as a critical research topic in various fields. As initial work, the usage and role of Twitter in creating a social community on the basis of its basic functions were examined by Java et al. [6], Zhao et al. [10], Krishnamurthy et al. [7], and Cha et al. [1]. In these studies, Twitter was investigated for its social networking role, that is, how it would be used to send massive amounts of short messages about social activities. Obviously, even major global news channels refer to Twitter as an important social channel, and many people are aware of its role as an uncontrolled and uncensored communication channel.

Furthermore, several research studies focused on the role of tweets as a novel media to represent crowd opinions. O'Connor et al. [8] compared the measures of public opinion from polls with ones from the analysis of tweets. Diakopoulos et al. [2] demonstrated an analytical methodology including visual representations and metrics that aid in making sense of the sentiment of social media messages around a televised political debate. In this paper, by finding tweets relative to TV watching, we estimate the public TV viewing rates. In addition, as a study focusing on the integration of Twitter and TV programs as same our approach, Sawai et al. [9] have proposed a method to recommend TV programs based on relations among users over social networking.

3 Semantic Linking from Tweets to Relevant TV Programs

In order to construct semantic linking from user-written tweets to possibly relevant TV programs, we analyze tweets in two different levels of processing. For this, we developed a Media Relevance Engine as depicted in Fig. 4. First, for each tweet, it goes the first relevance assessing by checking included hashtags. In this stage, we use a list of hashtags, where each one consists of triple attributes of <"hashtag," "station," "program"> as shown in Table 1 (a). However, all these tuples do not need to be filled, since a hashtag can only refer to a TV station or a program. Additionally, in case of on-line videos, the "station" attribute will specify the site name. In the next stage, "term-based identification" step will identify TV-relevant tweets by examining where each one has some specific strings which can determine a station or a program uniquely. For instance, a partial string of "http://www.youtube.com/..." will give a hint that it is a tweet relative to a YouTube video as described in Table 1 (b).

However, after this step, there are still lots of unidentified tweets which are obviously failed to determine any relevance as a TV-relevant tweet, but they can be if we look into the content in detail. For this, we present a Semantic Media Linkage Engine to examine the tweet's media relevance by content analysis as illustrated in Fig. 4. In this part, we will find semantic relevance of tweets to on-air programs appeared on the EPG lists. As for EPGs, the items that were broadcasted during the specified period are obtained from the local EPG database, which has also been storing EPG items from TV Kingdom[7] as shown in Table 2. We use these tweets and EPG items as the datasets of the specified period. Then, we need to compute the relevance between tweets and EPGs. Basically, for a tweet relative to a TV program, we have to find the best matching case in the EPGs. Since tweets in this stage, we have to exploit other information on textual, spatial, or location information of tweets.

Table 1. Example of hashtags and key-terms

(a) Example of hashtags used for on-air TV stations or program

hashtag	station	program
#nhk	nhk	-
#tbs	tbs	-
#tvtokyo	tvtokyo	-
#etv	nhk_edu	-
#keion	-	Keion!
#precure	-	Precure
#gegege	nhk	Wife of gegege
#ryomaden	nhk	Ryomaden

(b) Example of hashtags and key-term lists for on-line video sites

On-line video sites	Hashtags, terms, and URLs for Linkage
nicovideo	#nicovideo
	niconico
	http://www.nicovideo.jp/watch
YouTube	#youtube
	YouTube
	@YouTube
	http://www.youtube.com/watch
DailyMotion	#dailymotion
	dailymotion
	http://www.dailymotion.com/video
Gyao	#GyaO
	@Yahoo_Gyao
	http://gyao.yahoo.co.jp/player
Veoh	#veoh
	veoh
	http://www.veoh.com/browse/videos

---

[7] TV Kingdom: Japan TV Program Guide, http://tv.so-net.ne.jp/

Table 2. Example of local EPG database

region	station	date	time		title	genre
			start	end		
CATV Tokyo area J:COM Tokyo Suginami)	NHKGeneral Tokyo	Sep. 1, 2010	0:00	0:15	News and weather information	news
CATV Tokyo area J:COM Tokyo Suginami)	NHKGeneral Tokyo	Sep. 1, 2010	1:05	1:50	Chase! A to Z	documentary
CATV Tokyo area J:COM Tokyo Suginami)	NHKGeneral Tokyo	Sep. 1, 2010	1:50	2:00	Scoop! Contributed video clips (Tokudane! Toukou DO-ga)	talk show
CATV Tokyo area J:COM Tokyo Suginami)	NHKGeneral Tokyo	Sep. 1, 2010	2:00	2:45	Try and convince (Tameshite Gatten)	talk show / lifestyle

Actually, the other information is all required to assess the relevance in a comprehensive way. For instance, will be in hashtags identification, as drawn in Fig. 5, a user is location in the middle of a city and these are four different broadcast stations around there. But only three stations tv_a, tv_b, and tv_c are accessible from the location of the user. If a tweet written by this user is matched with some program information broadcasted from the surrounding four stations, we can think that the user's message can be to these programs. However, the station tv_d cannot support this assumption, since it is out of the period. Furthermore, in terms of broadcasting time, it is likely that the programs broadcasted in the nearly same time range with the written time would be desirable. Therefore, we need to compute the relevance of tweets to find out relevant on-air programs in the respects of textual, spatial, and temporal relevance as follows.

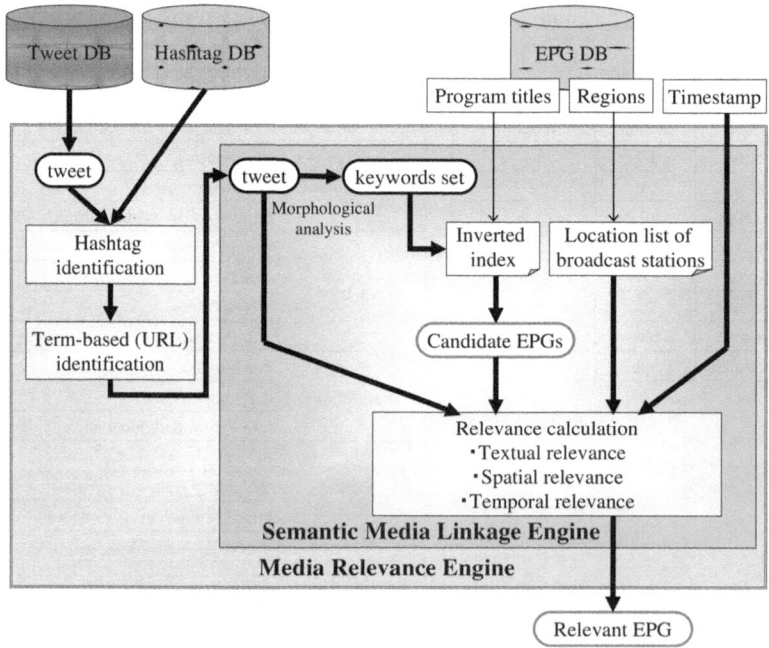

Fig. 4. Detecting relevance between a tweet and EPG

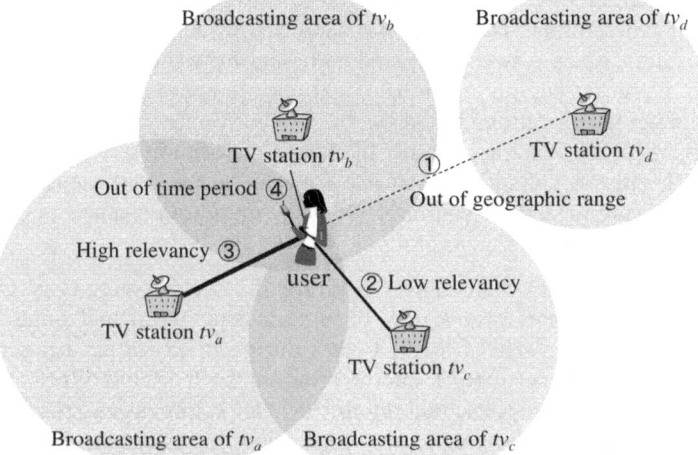

Fig. 5. Computation of semantic linkage by textual, spatial, and temporal relevance

- **Textual Relevance**

In order to find a corresponding EPG item relative to a tweet about a TV program, we first applied a words-based similarity computation: both sides are textual message. In the estimation of the correspondence, we compute it with the following formula based on the Jaccard similarity coefficient [3],

$$
textual\_relevance = \frac{\left|mp(tw_i) \cap mp(e_j.title)\right|}{\left|mp(tw_i)\right| + \left|mp(e_j.title)\right| - \left|mp(tw_i) \cap mp(e_j.title)\right|} \times \sum_{k}^{\left|mp(e_j.title)\right|} \frac{1}{df(k)} \quad (1)
$$

where tw_i is a tweet, e_j is an EPG item, $e_j.title$ is the title in the EPG item, and mp is a morphological analysis function where the output consists of nouns found in the given message. df is a function that calculates document frequency. Each tweet should be compared with all the program titles in the local EPG database. For the rapid searching for seemingly relevant EPGs we use an inverted index [11] to reduce the number of calculations required for determining relevance between a tweet and program titles in comparison to directly using the EPG database wherein the total of possible combinations would be enormous. Then, in order to detect relevant EPGs related to titles of TV programs, we applied the formula (1) in the computation. In the formula, with the df, we also considered the frequency of keywords of EPGs' titles. For example, keywords that are frequently used in EPGs such as "news," "drama," and "sports" should have less weight since these generic terms would retrieve many unrelated EPGs.

- **Spatial Relevance**

According to EPG items in the local EPG database, the same titles of EPGs are often found, because some TV programs can be broadcasted repetitively by multiple stations. In this case, we should identify the station that broadcasted the program at the time of tweet occurrence. The number of TV programs extracted by the inverted index usually corresponds with many different local stations. However, a user can exist at a place in a given moment so that a TV-relevant tweet should be matched to one of the possible local stations. Therefore, we should consider the physical distance between the location where a tweet is posted and that of the station that broadcasted the TV program. Because specific locations of stations are not included in EPG items, we roughly estimate their locations based on "region" attributes of the EPG items using Google maps API [5]. For this, we use the stations' location list that was generated beforehand. Then, we calculate distances between a location where a tweet was posted and each station, and the station that has the minimum distance is selected.

- **Temporal Relevance**

There is also an important consideration regarding the tweet posting time. Usually, we can think that TV-relevant tweets may be written near the actual on-air time. For instance, audiences may write a lot of tweets during or just after a popular drama. Sometimes, before a very popular sports program such as the World Cup, many tweets may occur far before the actual on-air time. Therefore, as regards the relevance between tweets and TV programs, the time elapsing between them is also an important factor.

$$temporal\_relevance\,(tw_i, e_j) = \begin{cases} 0 \\ \quad when \;\; e_j.start\_time \le tw_i.timestamp \; \le e_j.end\_time \\ \log(e_j.start\_time - tw_i.timestamp \; + 0.1) \\ \quad when \;\; tw_i.timestamp \; < e_j.start\_time \\ \log(tw_i.timestamp \; - e_j.end\_time + 0.1) \\ \quad when \;\; e_j.end\_time < tw_i.timestamp \end{cases} \qquad (2)$$

- **Final Rating by the Triple Relevance Measures**

Based on the above criteria, we computed the final relevance using the following formula:

$$relevance\_score(tw_i, e_j) = \frac{textual\_relevance(tw_i, e_j)}{(spatial\_relevance(tw_i, e_j) + 1) \times (temporal\_relevance(tw_i, e_j) + 1)} \qquad (3)$$

After computing the relevance scores, we obtain a list of tweet-EPG mapping and populate TV programs based on the following popularity score. In this formula, *#tweets* denotes the number of tweets for a program e_j, while *#users* means the distinct number of users. In fact, we consider the biases occurring by aggressive users to write many tweets for a program should be normalized.

$$popularity(e_j) = \sqrt{\#tweets \times \#users} \qquad (4)$$

4 Experiment

4.1 Experimental Dataset

In order to achieve our purpose to rank TV programs by means of Twitter users, we prepared a dataset for a period between Sept. 1–30, 2010: (1) tweets that occurred in that period in Japan, and (2) EPGs of all TV stations (except CS satellite broadcast) in

Japan. In that period, we collected 6,276,769 geo-tagged tweets, which were all mapped onto location points on a map. However, it was still burdensome to use this tweet dataset in our preliminary test. For the practical findings of TV-relevant tweets, we empirically chose tweets whose relevance to TV watching was seemingly higher using the prepared hashtag lists (for on-air TVs and on-line videos) and a set of filtering terms such as "テレビ," "TV," "てれび"—Japanese expressions for "television," and "視聴," "番組," "見てる," "見ている"—expressions for "watching" or "viewing." By filtering using these terms, we could obtain a reduced tweet dataset (119,575 tweets, about 1.9% of the collected dataset). These potential tweets were written by 33.392 distinct users (on average, 3.58 tweets were made per a user.)

In our experiment, we identified TV-relevant tweets and successively ranked the TV programs. In addition, we also prepared 838,636 EPGs for the same period. The TV program list we compiled covers 110 geographic regions in Japan with 188 different TV stations. (Here, for nationwide stations such as NHK, which may have many local stations appearing in the EPGs, we dealt with them all as different channels for convenience.) On the list, 24,841 distinct TV programs were identified (actually, 188 unique TV stations exist, but with the combinations with different regions—by region, TV stations—we could virtually determine 875 different channels.) Hence, each station has 31.9 programs a day (during on air-time) on average.

4.2 Experimental Results

In the first place, we extracted 60,318 tweets by means of hashtag identification. Among them, we analyzed what TV stations and programs were popularly referred to by hashtags. For the situations, we could obtain an expected result of shown in Fig. 6 (a) where NHK station (tagged by #nhk) than a half of the total hashtags about stations. Likewise, we could find other major TV stations such as Nippon Network Television Corp. (#ntv), Nippon Network Television Corp. (#tbs), and tv asahi (#tvasahi). In addition, for on-air programs detected by hashtags, we could also obtain the result as shown in Fig. 6 (c), where the most popular programs are ranked; Keion! (#keion), Wife of Gegege (#gegege), and life history of Ryoma Sakamoto (#ryomaden). Furthermore, for online video sites referred to by hasetags and specific URLs, we found an interesting result; the most popular one was NicoNicoDouga and YouTube was ranked in the next as shown in Fig. 6 (b).

Lastly, we made a comprehensive ranking for on-air programs and on-line videos as shown in Table 3. For on-air programs, we investigated them into two types of "on-air hashtag" identified by hashtags and "on-air". For online videos, we focused on the very detailed URL's which are usually directing a unique video page corresponding to a program in on-air TVs. In the table, we listed the result in a decreasing order of popularity scores. As a result, 7 of the top 15 popular programs came from on-line videos and the others were related to on-air programs. Especially, 3 of the results were found by the aid of extended searches through our proposed semantic linkage process. Consequently, we could say the proposed method could support the TV ratings finding out hidden audiences who were not using hashtags.

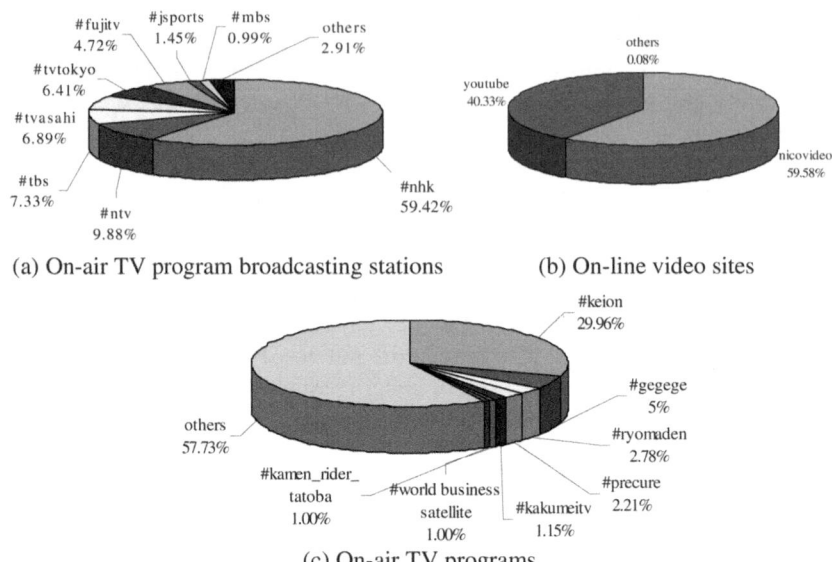

(a) On-air TV program broadcasting stations   (b) On-line video sites

(c) On-air TV programs

Fig. 6. Ranking result of (a) broadcasting stations (b) on-line video sites (c) on-air TV programs based on popularity

Table 3. Ranking results of Twitter-based TV ratings for on-air TV programs and on-line videos

programs	type	#tweets	#users	popularity
#keion (anime)	on-air hashtag	978	141	4409.49
http://nico.ms/lv25848987	on-line	147	70	848.70
#gegege (drama)	on-air hashtag	138	59	693.09
http://nico.ms/sm11982230	on-line	57	56	422.79
#ryomaden (drama)	on-air hashtag	76	47	409.74
precure (Anime)	on-air hashtag	145	27	325.12
If I become a prime minister (talk show)	on-air	45	44	295.16
Keion! "Examination" (Anime)	on-air	45	43	288.45
http://nico.ms/sm12005146	on-line	39	37	231.06
http://nico.ms/lv25538558	on-line	61	26	203.07
http://nico.ms/sm12052293	on-line	33	32	183.83
Leading actor —setting beaty salon— (comedy)	on-air	32	30	169.71
#kakumeitv	on-air hashtag	59	22	168.99
http://nico.ms/sm12098837	on-line	31	28	155.90
http://nico.ms/sm12017331	on-line	30	28	153.36

5   Conclusions

In this paper, we introduced a novel approach to improve TV viewing rates borrowing crowd-powered media consuming logs via Twitter. Especially, we also provide a platform looking for audiences of on-air TVs and on-line video sites together. We furthermore presented the very detailed methods and experimental results based on a real dataset of tweets and electronic program guide. In the future work, we will explore much deeper crowds' media lifestyles and their opinions to media contents to activate fruitful interactions between TV content providers and audiences by opinion mining and sentiment analysis for tweets.

Acknowledgments. This research was supported in part by a Grant-in-Aid for Scientific Research (B)(2) 20300039 from the Ministry of Education, Culture, Sports, Science, and Technology of Japan.

References

1. Cha, M., Haddadi, H., Benevenuto, F., Gummadi, K.P.: Measuring User Influence on Twitter: The Million Follower Fallacy. In: Proc. of the 4th International AAAI Conference on Weblogs and Social Media, ICWSM 2010 (2010)
2. Diakopoulos, N.A., Shamma, D.A.: Characterizing debate performance via aggregated twitter sentiment. In: Proc. of the 28th International Conference on Human Factors in Computing Systems (CHI 2010), pp. 1195–1198 (2010)
3. French, J.C., Powell, A.L., Schulman, E.: Applications of approximate word matching in information retrieval. In: Proc. of the 6th International Conference on Information and Knowledge Management, CIKM 1997 (1997)
4. Fujisaka, T., Lee, R., Sumiya, K.: Monitoring Geo-Social Activities through Micro-Blogging Sites. In: Proc. of the 1st International Workshop on Social Networks and Social Media Mining on the Web, SNSMW 2010 (2010)
5. Google maps API, http://code.google.com/intl/ja/apis/maps/
6. Java, A., Song, X., Finin, T., Tseng, B.: Why we twitter: understanding micro-blogging usage and communities. In: Zhang, H., Spiliopoulou, M., Mobasher, B., Giles, C.L., McCallum, A., Nasraoui, O., Srivastava, J., Yen, J. (eds.) WebKDD 2007. LNCS, vol. 5439, pp. 56–65. Springer, Heidelberg (2009)
7. Krishnamurthy, B., Gill, P., Arlitt, M.: A few chirps about twitter. In: Proc. of the 1st Workshop on Online Social Networks (WOSN 2008), pp. 19–24 (2008)
8. O'Connor, B., Balasubramanyan, R., Routedge, B., Smith, N.: From Tweets to Polls: Linking Text Sentiment to Public Opinion Time Series. In: Proc. of the 4th International AAAI Conference on Weblogs and Social Media, ICWSM 2010 (2010)
9. Sawai, R., Ariyasu, K., Fujisawa, H., Kanatsugu, Y.: TV Program Recommendation Method Using SNS Based on Collaborative Filtering, IPSJ SIG Technical Reports, Vol. 2010-DBS-151, No. 43 (2010) (in Japanese)
10. Zhao, D., Rosson, M.B.: How and why people Twitter: the role that micro-blogging plays in informal communication at work. In: Proc. of the ACM 2009 International Conference on Supporting Group Work (GROUP 2009), pp. 243–252 (2009)
11. Zobel, J., Moffat, A.: Inverted files for text search engines. ACM Computing Surveys (CSUR 2006) 38(2), Article 6 (2006)

GreenOrbs: Lessons Learned from Extremely Large Scale Sensor Network Deployment

Yunhao Liu

Hong Kong University of Science and Technology
liu@cse.ust.hk

"The world has just ten years to bring greenhouse gas emissions under control before the damage they cause becomes irreversible." This is a famous prediction raised by climate scientists and environmentalists recently. It reflects the increasing attention in the past decade from human beings on global climate change and environmental pollution. On the other hand, forest, which is regarded as the earths lung, is a critical component in global carbon cycle. It is able to absorb 10% – 30% of CO_2 from industrial emissions. Moreover, it has large capacity of water conservation, preventing water and soil loss, and hence reducing the chance of nature disasters like mud-rock flows and floods. Forestry applications usually require long-term, large-scale, continuous, and synchronized surveillance of huge measurement areas with diverse creatures and complex terrains. The state-of-arts forestry techniques, however, support only small-scale, discontinuous, asynchronous, and coarse-grained measurements, which at the same time incur large amount of cost with respect to human resource and equipments. WSNs have great potential in resolving the challenges in forestry. Under such circumstances, GreenOrbs is launched. The information GreenOrbs offers can be used as evidences, references, and scientific tools for human beings in the battle against global climate changes and environmental pollution.

The prototype system is deployed in the campus woodland of Zhejiang Forestry University. The deployment area is around 40,000 square meters. The deployment started in May 2009 and included 50 nodes. In November 2009 it was expanded to include 330 nodes. The system scale reaches 400 in April 2010. The duty cycle of nodes is set at 8%. The network diameter is 12 hops. The sensor data are published online via the official GreenOrbs website.The Tianmu Mountain deployment includes 200 nodes and has been in continuous operation since August 2009. The deployment area is around 200,000 square meters. The duty cycle of nodes is set at 5%. The network diameter is 20 hops.

We learned a lot of lessons during the deployment of GreenOrbs. This experiment results in several publications, including ACM Sensys 2009, 2010, ACM Sigmetrics 2010, ICNP 2010, INFOCOM 2010, etc. In this keynote, we will focus on several open issues for extremely large scale deployment of sensor networks including routing, diagnosis, localization, link quality, and etc.

J. Xu et al. (Eds.): DASFAA Workshops 2011, LNCS 6637, p. 402, 2011.
© Springer-Verlag Berlin Heidelberg 2011

Adapting Skyline Computation to the MapReduce Framework: Algorithms and Experiments*

Boliang Zhang[1], Shuigeng Zhou[1], and Jihong Guan[2]

[1] School of Computer Science, and Shanghai Key Lab of Intelligent Information Processing, Fudan University, Shanghai, China
[2] Dept. of Computer Science & Technology, Tongji University, Shanghai, China
{boliangzhang,sgzhou}@fudan.edu.cn, jhguan@tongji.edu.cn

Abstract. This paper addresses the problem of skyline computation under the MapReduce framework. As a parallel programming model for data-intensive computing applications, MapReduce runs on a cluster of commercial PCs with the main idea of task decomposition and result reduction. Based on different data partitioning strategies, three MapReduce style skyline computation algorithms are developed: MapReduce based BNL (MR–BNL), MapReduce based SFS (MR–SFS) and MapReduce based Bitmap (MR–Bitmap). Extensive experiments are conducted to evaluate and compare the three algorithms under different settings of data distribution, dimensionality, buffer size and cluster size.

Keywords: Cloud computing; MapReduce; Skyline computation.

1 Introduction

Skyline computation, aiming at multi-objective decision originally, has a variety of applications in database area nowadays. Suppose you go to some seaside city for a holiday [1], and you need to find a hotel that is both cheap and close to the beach. Apparently, you can not have it in both ways. In fact, those hotels that are not worse than others in both ways are acceptable. We call the set of interesting hotels above the *skyline*, each of which is a *skyline point*. Data visualization is another important application of skyline. For instance, we can show the landscape outline of Manhattan by computing the set of buildings that are higher or closer to the river, which constitutes the skyline of Manhattan. This is just where the name "skyline" comes.

The skyline operator was first proposed by Börzsönyi et al. [1] in 2001. In the past years, approaches to skyline computation have been published extensively in major database conferences, including SIGMOD, VLDB and ICDE. Although

* This work was supported by National Natural Science Foundation of China under grants No. 60873040 and No. 60873070. Jihong Guan was also supported by the Shuguang Scholar Program of Shanghai Education Development Foundation under grant No. 09SG23.

J. Xu et al. (Eds.): DASFAA Workshops 2011, LNCS 6637, pp. 403–414, 2011.
© Springer-Verlag Berlin Heidelberg 2011

skyline computation has been studied well under traditional definition, there are still some issues worthy of further investigation. To name a few, 1) *quality*. Usually, the number of skyline points grows exponentially with data dimensionality, and it also depends on data distribution. So how to retrieve a small number of high quality (or most representative) skyline points is absolutely not a trivial issue. 2) *Progressiveness and user preferences.* Many skyline applications require progressive response and to consider users' specific preferences while delivering the results. 3) *Efficiency.* Due to the dimension curse, skyline computation is very expensive for large-scale and high-dimensional datasets. Thus, how to speed up skyline computation is still a meaningful research direction.

This paper addresses the efficiency issue of skyline computation from parallel computation perspective. Concretely, we employ the MapReduce framework [2] to support skyline computation. MapReduce, as a parallel programming model developed by Google, runs on a large cluster of commercial PCs to process large datasets. The MapReduce program will carefully partition the input file, automatically schedule tasks over the cluster, deal with machine failures and manage communications between nodes. With its scalability, along with the reliability provided by GFS (Google File System) [3], we are able to process large scale of data well.

The combination of skyline computation and the MapReduce framework in this paper is mainly motivated by the two facts: 1) with the development of cloud computing, supporting data management and query processing in cloud platforms is emerging as a new trend of database research and commercialization, and 2) As a typical parallel programming model, MapReduce is receiving increasingly recognition in both academia and industry, and more and more applications are being developed under this framework. As most existing skyline computation algorithms are devised for centralized or distributed applications, and the MapReduce framework is a new parallel programming model with unique features, the combination can not be done well in a simple and straightforward way. Contributions of this paper include three aspects:

- Adapting skyline computation to the MapReduce framework for improving efficiency, to the best of our knowledge, is possibly the first of such work.
- Three MapReduce style algorithms for skyline computation are developed based on different data partitioning strategies.
- Extensive experiments are conducted to evaluate and compare the three algorithms under different settings of data distribution, dimensionality, buffer size and cluster size.

The rest of this paper is organized as follows. Section 2 describes the basic information of skyline computation and the MapReduce framework. Section 3 introduces three MapReduce style algorithms for skyline computation in cloud platforms. Section 4 presents the results of experimental evaluation. Section 5 reviews the related work. Section 6 draws conclusion and highlights future work.

2   Preliminaries

In this section, we present the definition and some basic properties of skyline, and a brief introduction to the MapReduce framework.

2.1   Skyline: Definition and Properties

Given a d-dimensional dataset $S = (S_1, S_2, \ldots, S_d)$, suppose that there exists an ordering relation on each dimension, denoted by $\preceq$ and $\prec$. For arbitrary values x and y in S_i, $x \preceq y$ means that x is not worse than y, and $x \prec y$ means that x is better than y.

Definition 1 (Dominate). *Given two data points in a d-dimensional dataset, say p and q, we say p dominates q iff for every $i \in [1, d]$ we have $p_i \preceq q_i$ and there exists at least one j such that $p_j \prec q_j$.*

Definition 2 (Skyline point). *Given a data set S, a point $p \in S$, if there exists no other point q such that q dominates p, we say p is a skyline point in S. All skyline points in S constitute the skyline of dataset S.*

From the definition above, we can easily infer some properties of skyline. If p dominates q in S, then for any scoring function $\varphi : S \to R$ that is monotonically increasing in all dimensions, we have $\varphi(p) > \varphi(q)$. Furthermore, for every φ, if $\varphi(p)$ reaches the maximum score, then p must be a skyline point. The opposite also holds.

2.2   The MapReduce Framework

In a MapReduce program, users specify a *map* function to generate a large amount of key/value pairs, and a *reduce* function to collect all intermediate values associated with the same intermediate key. Fig. 1 shows the workflow of a MapReduce program. Before the map phase starts, the input file is divided into logical splits for parallelization. Before the reduce phase, a shuffle procedure is carried out to release network pressure.

For instance, suppose that we want to count the occurrences of each word in a document. The *map* function simply outputs the word associated with 1, and the *reduce* function merges the same word and counts its occurrences.

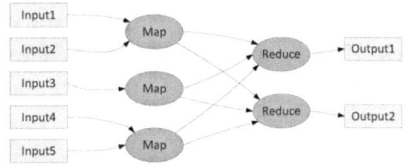

Fig. 1. The workflow of a MapReduce program

3   MapReduce-Based Skyline Computation Algorithms

Since MapReduce is a parallel programming model, the first consideration is how we partition the computation task. A straightforward way is to divide the original dataset into several subsets. Both NN [4] and D & C [1] algorithms use this idea. We call it horizontal partitioning. Another way is to partition data in a vertical fashion, i.e., each subset representing one single dimension of the dataset. Balke et al. [5] exploited sorted lists of all dimensions. Bitmap [6] utilizes this scheme to establish its bitmap structure.

Followed the two data partitioning strategies mentioned above, we design three algorithms based on three existing centralized approaches. They are MapReduce based BNL (MR-BNL), MapReduce based SFS (MR-SFS) and MapReduce based Bitmap (MR-Bitmap). All of them are comprised of two cascading MapReduce jobs. For convenience, we assume that in a d-dimensional dataset S, the size of S is N, and there are R distinct values per dimension, denoted as $[1, 2, \ldots, R]$.

3.1   MR-BNL

BNL [1] is an iterative algorithm that repeatedly reads a set of data records for processing. It keeps a window containing incomparable records in main memory to collect candidate skyline records. When a record p is read, it is compared with records in the window. According to the value of p, either of the following situations may happen. Record p is discarded or added into the window, or the points in the window dominated by p are deleted. With fixed data size N and dimension d, the runtime of BNL depends on the window size and the ordering degree of the original data. The window size influences the number of iterations, which is also a dominant factor since I/O cost far exceeds CPU cost.

Our MapReduce based BNL algorithm (MR-BNL) is a two-stage method. The first stage is to divide the whole data into small disjoint subsets. For each subset, we run a BNL procedure to compute the skyline. In the second stage, the local skylines are merged and filtered, thus the global skyline is obtained.

The NN algorithm [4] divides the global space into 2^d subspaces according to the first nearest neighbor. The D & C algorithm [1] recursively partitions the data space according to the median of a certain dimension. In MR-BNL, we combine these ideas together such that the input dataset is divided into 2^d subspaces based on a carefully chosen point. Considering load balance, the median of each dimension is chosen as the point of division. Thus, each dimension is divided into two parts: the higher part with larger values in the dimension and the lower part with smaller values. If we mark the higher as 1 and the lower as 0, then each subspace can be identified by a d-bit *flag*. In the merging stage, we exploit the flags to reduce unnecessary comparisons.

Fig. 2 shows a two-dimensional dataset. The lines $x = 0.5$ and $y = 0.5$ divide the space into four parts, whose flags are 00, 01, 10 and 11, respectively. The data records in subspace 00 dominate that in subspace 11, while subspace 01 is incomparable with subspace 10. There are total $C_4^2 = 6$ pairs of subspaces, with 16.67% incomparable pairs. This ratio grows with dimension d, denoted by

$\gamma(d)$. $\gamma(d)$ rises to 88.92% when d reaches 10, which is calculated by a simple program. Since bitwise operation costs only a little, the cost of comparisons in the merging stage can be approximately reduced by a factor of $\gamma(d)$.

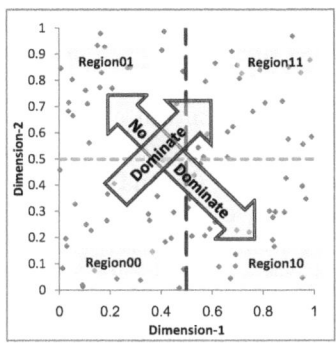

Fig. 2. Subspace and dominating relationship

Algorithm 1. MR-BNL

    INPUT: the original data set S
    OUTPUT: the skyline of data set S
1: *Division Job*
2:    *Map Task*
3:    **for each** point P_i **in** dataset S
4:      **compute** P_i's subspace flag F_i
5:      **output** (F_i, P_i)
6:    *Reduce Task*
7:    **for each** subspace flag F_i
8:      **compute** local Skyline SP_k using **BNL**
9:      **output** (F_i, SP_k) in file t
10:
11: *Merging Job*
12:    *Map Task*
13:    **for each** point P_i **in** file t
14:      **output** (**null**, (F_i, P_i))
15:    *Reduce Task*
16:    **compute** global Skyline SP_k using **BNL** with **pre-comparison**
17:    **output** $(SP_k, \textbf{null})$

The procedure of MR-BNL is outlined in Algorithm 1. In line 14, the key of the map output is null so that all records are gathered in one reduce call. The *pre-comparison* procedure does judgements in advance by using subspace flags.

The bottleneck lies in the second stage where the merging is done by one single node. The performance may be significantly degraded in the case that the size of local skyline produced by the first stage is huge.

3.2 MR-SFS

One shortcoming of BNL is that a record, once inserted into the window, might be discarded later. Therefore, this non-skyline record occupies window space, which may increase the iteration times. Sort-Filter-Skyline (SFS) [7] was devised to overcome the drawback. The input data is sorted at first in a topological order compatible with the skyline criterion. Thus, once a record r is inserted into the window, no record following it will dominate r and r is a skyline point. Therefore, at the end of each iteration, we can output all records inside. The iteration times of SFS is optimal, i.e., $\lceil \frac{N}{W} \rceil$ where N represents the number of data records and W is the window size in number of records.

We implement the MapReduce based SFS algorithm by doing some modification over MR-BNL: presorting. However, MapReduce can only sort by the keys of intermediate key/value pairs. In order to achieve secondary sort for the values set in the reduce call, we exploit the grouping and comparing features of MapReduce and the APIs provided by the Hadoop platform [8]. First, we assign key/value pairs of the same key to one reducer. Second, we put together the values of the same key. Third, user defined *comparators* are used to determine the order of records in the reduce phase.

To implement the three steps above, for any record r, we embed its key and value together, forming a new *Composite*. Then, we create two *comparators* and one *partitioner* for the *Composite*. In step 1, the *partitioner* partitions *Composite* in accordance with its key. In step 2, we use a *KeyComparator* to compare only the key of *Composite*. In step 3, we use a *KeyValueComparator* to compare not only the key but also the value of *Composite*, and yet, the key has the priority. Algorithm MR-SFS can be easily transformed from MR-BNL according to the above methods. Due to space limit, here we omit the pseudo-codes of MR-SFS.

3.3 MR-Bitmap

In Bitmap [6], every point is mapped into a bit vector, and the whole dataset forms the bitmap structure. The bitmap structure reflects the order of data records in all dimensions, ignoring their magnitudes. Every comparison in Bitmap is a lightweight bitwise operation. However, in order to examine a record, we need to compare it with all the other records. Bitmap supports progressive skyline computation since a point can be returned to the user immediately once it is identified as a member of the skyline. But the spatial cost of Bitmap is quite large. In particular, let M_i be the number of distinct values in dimension i, the size of Bitmap structure $S_b = N \times \sum_{i=1}^{d} M_i$ (bit). In the extreme case, for each dimension i, if $M_i = N$ (N is size of data set), then the space cost will be dN^2. If the bitmap structure can not be kept in main memory, the performance will be terribly degraded due to frequent I/O accesses.

The basic idea of MR-Bitmap keeps in line with that of Bitmap. In the first job, we build the bitmap structure. In the second job, we examine each point

---

Algorithm 2. MR-Bitmap

    INPUT: the original data set S
    OUTPUT: the skyline of data set S
1: **Building Job**
2:    *Map Task*
3:    **for each** point P_i **in** data set S
4:      **for each** attribute A_j **in** P_i
5:      **output** $(j, (A_j, P_i\text{'s byte offsets}))$
6:    *Reduce Task*
7:    **for each** dimension k
8:      **generate** sorted attribute list L_k in descending order
9:      **for each** distinct value v **in** L_k
10:        **generate** bit-slice and **write** to **HDFS**
11:
12: **Examining Job**
13:    *Map Task*
14:    **for each** point P_i in data set S
15:      **assign** P_i to a reducer R_i
16:    *Reduce Task*
17:    **for each** reducer R_i
18:      **load** bitmap b into memory as much as possible
19:      **for each** point P_i **in** R_i
20:        **check** P_i using bitmap b

---

based on the bitmap structure. Algorithm 2 outlines the procedure of the MR-Bitmap algorithm.

Line $8 \sim 10$ is the procedure of generating bitmap, and each reducer task is responsible for one dimension. Each distinct value in a dimension will produce a bit-slice. Line 15 determines how many reducers will be launched for the examining process. Let R_n be this number. Each reducer needs to read bitmap into memory. If R_n is too large, loading cost becomes dominant, which may degrade efficiency. On the contrary, if R_n is too small, the degree of parallelism is too low. In practice, R_n is set to the number of cluster nodes. In line 20, we use the same procedure as in [6] to judge whether point P_i is a skyline point.

If the bitmap structure cannot fit into main memory, we adopt cache-like strategy. When accessing some bit-slices, we check whether it is hit in memory, if not, we read it from disk. So it is natural to store the bitmap in column manner.

4 Performance Evaluation

In this section, we will evaluate and compare the proposed algorithms by a series of experiments with different data distributions, dimensionalities, buffer sizes and cluster sizes.

4.1   Experimental Setting

We set up a cluster of 8 commodity PCs, each of which has an Intel Duo Core 3.00GHz CPU, 4GB memory and Windows XP OS. We use Hadoop 0.20.2, and compile the source codes under JDK 1.6 in Eclipse 3.3.2.

We use similar datasets as in [1], each of which contains 100000 data records. The size of the record is fixed to 100 bytes. For each dimension, we generate integers from 1 to 100. Three data distributions are considered as follows:

- **Independent:** all dimensions follow uniform distribution, independently and identically distributed.
- **Correlated:** points perform well in one dimension are also good in the other dimensions.
- **Anti-related:** points perform well in one dimension are bad in one or all of the other dimensions.

Evaluation Metrics. We evaluate the algorithms by skyline ratio, runtime and I/O cost in different situations, including data distributions, dimensionality, distinction, buffer size and cluster size. Followings are the details:

- Skyline ratio: the number of skyline points over the total number of points.
- Distinction: the number of distinct values per dimension.
- Buffer size: a logical concept extended from the window in BNL, which means how many records can be kept in main memory. For MR-Bitmap, the bitmap structure can be kept in main memory in general.
- Cluster size: the number of nodes in a cluster, including one master node.

4.2   Experimental Results

Skyline Ratio. First, we vary the dimensionality d from 2 to 10 for three data distributions, and the results are presented in Fig. 3. We can see that the skyline ratio of Correlated grows negligible with d. On the contrary, the skyline ratio of Anti-related grows exponentially with d, reaching 58% when $d = 10$. As for Independent, its curve lies between the other two distributions. We then consider the relation between skyline ratio and data distinction. We vary the number of distinct values per dimension R from 20 to 20000 in a 5-dimension independent dataset. The result is shown in Fig. 4. As R increases, skyline ratio also increases.

Runtime and I/O cost. Here we measure the runtimes of three algorithms for three distributions with d=2, 5, 8 and 10, and the I/O cost of anti-related. Here I/O cost includes read/write bytes upon both local disk and HDFS. The buffer size is set to 5 MB and the cluster size is 4. Fig. 5 shows I/O cost of Anti, while Fig. 6, 7 and 8 show the runtimes of different distributions.

When the bitmap structure can fit into memory, the runtime of MR-Bitmap is linear with dimensionality and irrelevant to data distribution. The runtime of MR-Bitmap only depends on the size of bitmap and dimensionality. With respect to I/O cost, MR-Bitmap outnumbers the others a lot since it needs to write and

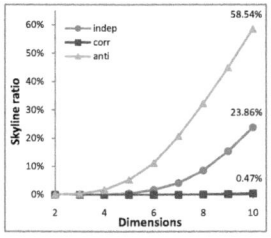
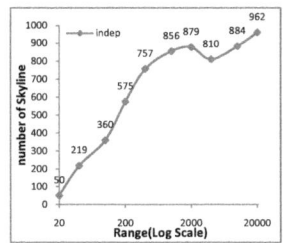
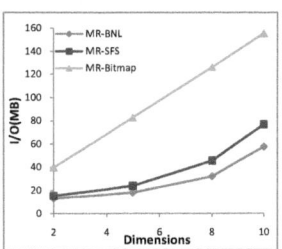

Fig. 3. Skyline ratio vs. dimensionality

Fig. 4. Skyline ratio vs. data distinction

Fig. 5. I/O cost of Anti

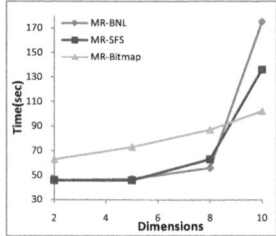
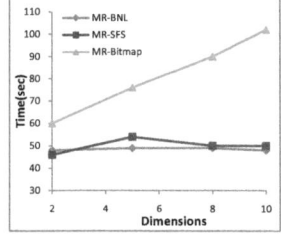
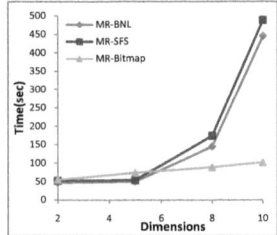

Fig. 6. Runtime of Indep   **Fig. 7.** Runtime of Corr   **Fig. 8.** Runtime of Anti

read bitmap to and from HDFS. The curves of MR-BNL and MR-SFS are very close. Unfortunately, we do not see the superiority of SFS in performance. On one hand, the sorting strategy of MR-SFS causes extra I/Os. As we can see in Fig. 5, I/O cost of MR-SFS outnumbers that of MR-BNL. On the other hand, the buffer size is half of the total number of records, so the number of iterations is small. Furthermore, Both MR-BNL and MR-SFS are sensitive to data distribution. MR-Bitmap performs better when dimensionality is large, because MR-BNL and MR-SFS concentrate on one node in the second phase, degenerating to centralized ones.

Effect of Buffer Size. We change the buffer size from 0.1 MB to 10 MB in a 8-dimensional anti-related dataset under a cluster of 4 nodes. The result is illustrated in Fig. 9. The runtimes of MR-BNL and MR-SFS drop drastically when the available buffer size increases from 0.1 MB (1000 records at a time). When the buffer size reaches 3 MB and larger, the runtime keeps almost constant since the number of iterations changes little. For MR-Bitmap, we conduct only two experiments, corresponding to 5% and 10% of bitmap not being loaded into main memory respectively. The results are disappointedly 958s and 1567s. The key point lies in that examining records may access HDFS several times randomly when the bitmap structure can not be totally loaded into main memory.

Effect of Distinction. For an 8-dimension anti-related dataset with a cluster of 4 nodes, we vary the number of distinct values per dimension R from 10 to

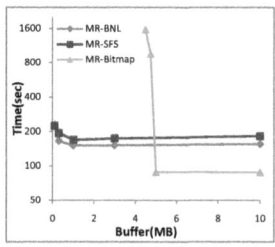
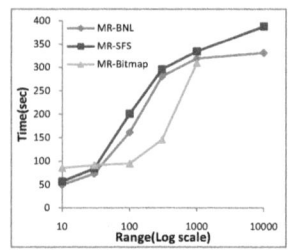
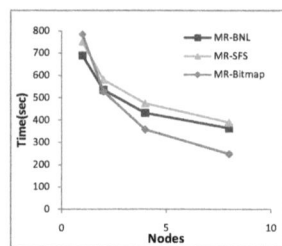

Fig. 9. Runtime vs. buffer size   **Fig. 10.** Runtime vs. distinction   **Fig. 11.** Runtime vs. cluster size

10000. We do not test MR-Bitmap when $R=10000$ since the bitmap is too large. The results are presented in Fig. 10, MR-Bitmap outperforms MR-BNL and MR-SFS when R increases from 30 to 1000. When R exceeds 1000, the bitmap structure is large and MR-Bitmap has the largest I/O cost.

Effect of Cluster Size. At last, we change the nodes in the cluster. Theoretically, the runtime of a purely parallel algorithm is in inverse proportion to the cluster size. In practice, the performance will be negatively impacted by the cost of scheduling among nodes. In this experiment, we use a 10-dimension anti-related dataset, with values changing from 1 to 1000 in each dimension. We expand the size of the cluster from 1 to 8, Fig. 11 shows the results. As expected, 1) the increase of cluster size improves stably but not linearly the computation efficiency, 2) MR-Bitmap outperforms MR-BNL and MR-SFS in efficiency because the former is concurrent in both stages.

5   Related Work

Here we review the related work from two aspects: skyline computation and MapReduce-based data management and query processing.

5.1   Skyline Computation

Up to now, a number of algorithms for skyline computation have been developed, such as block-nested-loops (BNL) [1], sort-filter-skyline (SFS) [7], Bitmap [6], nearest neighbor (NN) [4] and branch and bound skyline (BBS) [9]. It was admitted that BBS performs best regarding to runtime and space cost when dimensionality is not too large. In addition, there are some works that either expand the notion of skyline or solve the problem in different application scenarios.

Chan et al. [10] proposed k-dominant skyline and developed efficient ways to compute it in high-dimensional space. Lin et al. [11] proposed n-of-N skyline query to support online query on data streams, i.e., to find the skyline of the set composed of the most recent n elements. In the cases where the datasets are very large and stored distributedly, it is impossible to handle them in a centralized fashion. Balke et al. [5] first mined skyline in a distributed environment by partitioning the data vertically. Wang et al. [12] and Deng et al. [13] proposed skyline computation under P2P and road networks, respectively. Recently, Zhu et al. [14] presented an efficient skyline computation algorithm with low network bandwidth consumption in general distributed environments.

5.2 Data Management and Query Processing under the MapReduce Framework

Existing MapReduce-like systems, as efficient implementations towards specific problems, mainly focus on mass data analysis and processing. The data of such applications often has weak relationships and regular structures, which is beneficial to being processed by highly parallel subtasks. Furthermore, the processing or analysis tasks are not complicated in general. In this aspect, Google's Sawzall [15] and Yahoo!'s Pig [16] are two typical systems. And yet, for complex query processing under MapReduce framework, such as skyline and kNN, few works have been reported.

Some recent works are also worthy of being mentioned. For sharing input among a batch of queries, Nykiel et al. [17] proposed the MRshare framework, which merges several MapReduce jobs into a single one through dynamic programming to optimize overall efficiency. Unsatisfied with HadoopDB's modification over Hadoop, Dittrich et al. [18] proposed the Hadoop++, which merely overrides a few user defined functions (UDFs) to obtain advantages brought about by index and join techniques in traditional DBMSs. Considering that MapReduce paradigm does not provide adequate support for iterative programs, Bu et al. [19] put forward HaLoop, a modified version of Hadoop, which serves applications with iterative approaches, such as k-means, pagerank and recursive queries.

6 Conclusion

This paper addresses skyline computation in cloud computing environments. Based on the MapReduce framework, we have presented three MapReduce style algorithms for skyline computation: MR-BNL, MR-SFS and MR-Bitmap. To evaluate these algorithms, we have conducted a series of experiments under different experimental settings. Experimental results show that MR-BNL and MR-SFS are good in most cases but still suffer from dimensional curse in parallel environments. MR-Bitmap performs well regardless of data distributions, especially when the bitmap can fit into the memory of a single node.

It is worthy of being mentioned that this work can be expanded in a number of directions. First, from the perspective of parallel computing, how to extract as many independent subtasks as possible from the problem is crucial and needs to be further investigated. This is also true in the problem of this paper. Second, the progressiveness of Skyline computation should be thought highly of, since users are very likely to make a decision after receiving only a few recommendations rather than waiting for the whole results. To shorten response time is not an easy task because the fixed start-up time of a Hadoop job is around 15 seconds. Last but not the least, to design efficient algorithms in cloud computing environments by using indices and exploiting features of splitting and sorting, are all promising research topics.

References

1. Börzsönyi, S., Kossmann, D., Stocker, K.: The Skyline operator. In: Proceedings of ICDE, pp. 421–430 (2001)
2. Dean, J., Ghemawat, S.: MapReduce: Simplified data processing on large cluster. In: Proceedings of OSDI, pp. 137–150 (2004)
3. Ghemawat, S., Gobioff, H., Leung, S.T.: The Google file system. In: Proceedings of SOSP, pp. 29–43 (2003)
4. Kossmann, D., Ramsak, F., Rost, S.: Shooting stars in the sky: An online algorithm for Skyline queries. In: Proceedings of VLDB, pp. 275–286 (2002)
5. Balke, W. T., Güntzer, U., Zheng, J.: Efficient Distributed Skylining for Web Information Systems. In: Hwang, J., Christodoulakis, S., Plexousakis, D., Christophides, V., Koubarakis, M., Böhm, K. (eds.) EDBT 2004. LNCS, vol. 2992, pp. 256–273. Springer, Heidelberg (2004)
6. Tan, K.L., Eng, P.K., Ooi, B.C.: Efficient progressive Skyline computation. In: Proceedings of VLDB, pp. 301–310 (2001)
7. Chomicki, J., Godfrey, P., Gryz, J., Liang, D.: Skyline with presorting. In: Proceedings of ICDE, pp. 717–719 (2003)
8. White, T.: Hadoop: The Definitive Guild. O'Reilly, Sebastopol (2009)
9. Papadias, D., Tao, Y., Fu, G., et al.: Progressive Skyline Computation in Database Systems. ACM TODS 30(1), 41–82 (2005)
10. Chan, C., Jagadish, H.V., Tan, K.L., et al.: Finding k-dominant Skylines in high dimensional space. In: Proceedings of SIGMOD, pp. 503–514 (2006)
11. Lin, X., Yuan, Y., Wang, W., et al.: Stabbing the sky: Efficient Skyline computation over sliding windows. In: Proceedings of ICDE, pp. 502–513 (2005)
12. Wang, S., Ooi, B.C., Tung, A., et al.: Efficient Skyline query processing on peer-to-peer networks. In: Proceedings of ICDE, pp. 1126–1135 (2007)
13. Deng, K., Zhou, X., Shen, H.: Multi-source Skyline query processing in road networks. In: Proceedings of ICDE, pp. 796–805 (2007)
14. Zhu, L., Tao, Y., Zhou, S.: Distributed Skyline Retrieval with Low Bandwidth Consumption. IEEE Transactions on Data and Knowledge Engineering 21(3), 321–334 (2009)
15. Pike, R., Dorward, S., Griesemer, R., et al.: Interpreting the data: Parallel analysis with Sawzall. Journal of Scientific Programming 13(4), 277–298 (2005)
16. Olston, C., Reed, B., Srivastava, U., et al.: Pig latin: a not-so-foreign language for data processing. In: Proceedings of SIGMOD, pp. 1099–1110 (2008)
17. Nykiel, T., Potamias, M., Mishra, C., et al.: MRShare: Sharing Across Multiple Queries in MapReduce. In: Proceedings of VLDB, vol. 3(1), pp. 494–505 (2010)
18. Dittrich, J., Quiane-Ruiz, J.-A., Jindal, A., et al.: Hadoop++: Making a Yellow Elephant Run Like a Cheetah (Without It Even Noticing). In: Proceedings of VLDB, vol. 3(1), pp. 518–529 (2010)
19. Bu, Y., Howe, B., Balazinska, M.: HaLoop: Efficient Iterative Data Processing on Large Clusters. In: Proceedings of VLDB, pp. 285–296 (2010)

Efficient Event Stream Processing: Handling Ambiguous Events and Patterns with Negation

Murali Mani

University of Michigan, Flint
`mmani@umflint.edu`

Abstract. Event stream processing, where we detect patterns on incoming event streams, has tremendous value in early determination of critical conditions, enabling on-time response for several important applications. Event stream processing has two significant differences from prior work on XML/relational stream processing: *ambiguous events*, where an event can match multiple query symbols/conditions in the pattern; and *negation*, used in event stream processing patterns to specify the non-occurrence of a pattern. In this paper, we develop a formal framework to define the semantics of event patterns, including negation, and describe how to construct a deterministic finite state automaton even in the presence of ambiguous events. Using our framework, we can construct an efficient deterministic finite state automaton for detecting patterns with any complex nesting of negations over an event stream which may have ambiguous events. Our preliminary experimental studies illustrate the significant benefits of our approach to existing approaches.

Keywords: Event Stream Processing, Finite State Automaton, Regular Expression, Query Optimization, Relational Algebra.

1 Introduction

Event stream processing, where we match complex patterns against an input event stream, is critical for several modern-day applications, such as patient health monitoring, fraud detection etc. Database researchers have studied processing of streaming data in the past, especially in the context of relational stream processing [2] and XML stream processing [6]. Relational stream processing typically deals with joining multiple input streams; XML query languages include XPath or XQuery. In event stream processing, the query is typically specified as a regular expression and we are looking for events in the input stream that match the specified query [12].

Let us consider a simplified example of event stream processing from stock market analysis. Consider every event in the event stream to specify a transaction, and has two attributes: price and volume. An example event stream is given below:

```
event timestamp=1 price=75 volume=90
event timestamp=2 price=110 volume=130
event timestamp=3 price=120 volume=80
event timestamp=4 price=80 volume=140
event timestamp=5 price=125 volume=120
```

J. Xu et al. (Eds.): DASFAA Workshops 2011, LNCS 6637, pp. 415–426, 2011.
© Springer-Verlag Berlin Heidelberg 2011

A pattern that the user is interested in could be to determine events with volume > 100 that occur after an event with price > 100. The result of the above query includes events with timestamps 4 and 5. Event with timestamp 4 has vol > 100, and a prior event (with timestamp 3) has price > 100. Event with timestamp 5 matches because it has vol > 100, and a prior event (with timestamp 3) has price > 100.

One aspect of event stream processing that makes it particularly challenging is what we call as *ambiguous events*. Consider the above example itself; the pattern specifies two types of events: events with price > 100 (let us call them events of type a), and events with volume > 100 (let us call them events of type b). The pattern the user is looking for is a followed by b (there can be zero or more events between a and b events). In finite state automata, any input "symbol" will match exactly one of the symbols in the given alphabet: for instance, if the alphabet is $\{a, b\}$, any input symbol will be either a or b. However, that is not true for event stream processing. See that in the above example the events with timestamps 2 and 5 match both a and b; we call such input symbols that match multiple symbols in the pattern as *ambiguous events*.

Another challenging aspect of event stream processing is the use of *negation* in specifying a pattern query. In [12] an example pattern query that is discussed is (shelf, !counter, exit) that signals potential shoplifting. The above pattern is looking for items found at a shelf and later at the exit without going through a counter in between. Notice the negation specifies that the negated sub-pattern does not occur in between the two other sub-patterns. Processing such negated patterns efficiently is a significant challenge in event stream processing. Also existing works such as [12] do not define the semantics of patterns such as (a, !(!b, c), d) with nested negation.

There is some preliminary work in handling ambiguous events and patterns with negation in existing works such as SASE [12]. Here, the authors construct non-deterministic finite state automata for a pattern query; non-deterministic automata can handle ambiguous events. On the other hand, construction of deterministic finite state automaton to handle ambiguous events is non-trivial, yet can yield *significant* performance improvement. In SASE, simple negations are handled using joins outside the automaton: a pattern query is split into multiple positive and negative sub-patterns and later joined to determine the non-occurrence of the negative sub-patterns. For example, the pattern (shelf, !counter, exit) is split into a (shelf, exit) pattern, and a counter pattern. Events that match these patterns are joined to ensure counter did not occur between shelf and exit patterns. However, it might be more efficient if we could combine negated patterns and positive patterns into a single automaton.

In our work, we first develop a formal framework that defines the semantics of patterns, including negation. Using our framework we construct a non-deterministic finite state automaton for patterns, even with complex nested negation. We then describe how the non-deterministic automaton can be converted to a deterministic automaton, even in the presence of ambiguous events.

The contributions of this work include the following

(a) We develop a formal framework that defines the semantics of pattern queries (regular expressions extended with negation); we construct non-deterministic automaton for a pattern query, even if it has complex nested negation.

(b) We construct a deterministic automaton from the non-deterministic automaton, even when we have ambiguous events in the input stream.

(c)   We describe how an event processing system can utilize our results and search a richer space of execution plans for determining an efficient execution plan to be used for pattern matching.

(d)   We perform experimental studies where we consider different factors such as length of the pattern, number of events in the event stream. We note that our approach has significant performance benefits over alternative approaches.

Outline: The outline for the rest of the paper is as follows. In Section 2, define our problem and our assumptions. In Section 3, we define the semantics of pattern queries, and describe how to construct a non-deterministic automaton for a query, detailing how negation is handled. In Section 4, we describe how to construct a deterministic automaton, examining in detail how to handle ambiguous events. Our experimental results are described in Section 5. Related work is discussed in Section 6, and conclusions and future work are outlined in Section 7.

2 Background

In this section, we define the event stream processing problem, and our assumptions in this paper. We assume that the reader is familiar with deterministic finite state automaton (DFA) and non-deterministic finite state automaton (NFA) [7].

Event Stream: The event stream consists of a time ordered sequence of events. We assume that the events in the input event stream are time ordered. There has been some work on handling out-of-order events [8].

Pattern Query: Different languages have been studied for specifying queries over event streams. In [5], the authors consider event specification in active databases; query expressions are composed using operators such as and, not, relative. In [12], the query consists of an EVENT clause specifying a pattern, a WHERE clause specifying additional predicates, and a WITHIN clause specifying window constraints. In [1], the authors extend the WHERE clause in [12] to include specification of event selection strategy such as contiguity, skip till next match, skip till any match. In [4], the authors use three constructs to specify patterns: FILTER selects events in the input stream that satisfy specified predicates; NEXT joins two streams to select pairs of events, the event in the second stream comes after the event in the first stream; FOLD iterates multiple times over events till a terminating condition is satisfied, and then an aggregation is performed on the set of events involved in the iteration. In [9], the authors use a SQL-like language with a CLUSTER clause to specify grouping (called equivalence test in [12]), and a SEQUENCE clause to specify a pattern. In [3], the authors use operators such as SEQUENCE, UNLESS, NOT, CANCEL-WHEN etc.

  The pattern language that we use in this paper is similar to one used in SASE [12]. Our pattern query consists of events with predicates combined using regular expression operators such as , denoting sequence; + denoting choice or union; * denoting Kleene-*. We also have a negation operator ! denoting the non-occurrence of a pattern. However, equivalence tests as in SASE and window constraints are not considered in this work, as that is not our focus in this paper. It is possible to extend our work to include these features as well. For ease of understanding, while specifying a pattern, we will often specify only the expression in the EVENT clause in SASE; the predicates corresponding to each symbol are specified only when needed (these predicates are specified in the WHERE clause in SASE).

Let us look at some example pattern queries. A pattern query that we examined in Section 1 is *(a, b)*. The symbol *a* corresponds to events with price > 100; *b* corresponds to events with vol > 100. In SASE, this query would be specified as

```
EVENT SEQ(event a, event b)
WHERE a.price > 100 AND b.vol > 100
```

The set of symbols used in the pattern query is said to be the *alphabet* of the pattern query. In our above example pattern query, the alphabet = {*a, b*}. Note that a symbol in the alphabet is associated with a predicate, which is different from what is typically studied for regular expression patterns [7]. We can therefore have events in the input stream that are not in the alphabet of the query; in other words, we can have an input event that is neither *a* nor *b* (an event with price <= 100 and vol <= 100). Similarly we can have input events that are both *a* and *b* (an event with price > 100 and vol > 100). We call the latter as *ambiguous events*.

To summarize, a pattern query is specified as an expression over symbols using the regular expression operators: , denoting sequence; + denoting union or choice; * denoting Kleene-*; we also have ! denoting negation. The semantics of pattern queries are discussed in detail in Section 3.

3 Constructing NFA for Pattern Queries with Negation

In this section, we will define semantics of pattern queries, and explain how to construct a non-deterministic finite state automaton (NFA) for any pattern query, even if it includes negation. Recall from [7] that a NFA is a 5 tuple: $(Q, \Sigma \cup \{\varepsilon\}, \delta, q_0, F)$, where Q denotes set of states, Σ denotes the alphabet, $q_0 \in Q$ is the start state, $F \subseteq Q$ is the set of final states, and $\delta: Q \times \Sigma \cup \{\varepsilon\} \rightarrow \mathcal{P}(Q)$ denotes the set of transitions, and $\mathcal{P}(Q)$ denotes the power set of Q. As the automaton is non-deterministic, we can transition to multiple states from a state on seeing a symbol.

Consider the pattern: *a*, that matches all events of type *a* in the input stream. When constructing an automaton for any pattern, we need to consider the fact that events that can appear in the input stream (the input alphabet) are not known during automaton construction; therefore our automaton should handle events that are not of type *a* as well. The automaton for the pattern *a* is shown in Fig. 1 (a); this happens to be a DFA. An event matches the symbol *a* if it satisfies the predicate associated with *a*; an event matches the symbol *!a* if it does not satisfy the predicate associated with *a*.

Now, the patterns can be composed using the same operators as used in regular expressions [7] : , denoting sequence; + denoting choice or union; and * denoting Kleene-*. The construction of NFA for these patterns can be done as it is done for regular expressions [7]. For instance, one NFA for the pattern *(a, b*, c + d)* is shown in Fig. 1 (b). Note that this NFA includes ε-transitions [7]. Converting this NFA into a DFA, however, is quite different from how it is typically done [7], because of ambiguous events. For instance, suppose the automaton is in state 2, and we get an input event that satisfies both *b* and *c*, (i.e., the event satisfies the predicate associated

Fig. 1. (a) Automaton for the pattern *a*. The transitions marked *a* match input events of type *a*; the transitions marked *!a* match all other input events. 1 is the start state; 2 is the final state. (b) NFA for the pattern *(a, b*, c + d)*. Note that this is just one of the many possible NFAs.

with *b* and the predicate associated with *c*), then our DFA should still transition to only a single state. We will discuss construction of a DFA in Section 4.

In addition to the regular expression operators, event patterns may also include negation (denoted by !). Consider the pattern *(!b, a)*: this pattern matches events of type *a*, such that no *b* event has occurred before the *a*. There are two points to note: *!b* matches events with no prior *b* event in the event stream; in the pattern *(!b, a)*, the , is not really a sequence operator; rather we are looking for *a* events that also satisfy *!b* (i.e., events of type *a* such that no *b* has occurred prior in the event stream).

First, let us see how to construct automaton for *!p*, where *p* can be any pattern. According to our semantics, *!p* matches all events till *p* has been satisfied; once any input event satisfies *p*, all future events in the input do not satisfy *!p*. Therefore the semantics for *!p* is $(\Sigma^* - (p, \_, \_^*))$, where _ represents wild card and matches any event. See that $(p, \_, \_^*)$ matches all events that occur after pattern *p* has been matched, and $(\Sigma^* - (p, \_, \_^*))$ is its complement. Below, we give an example event stream, illustrating the events that match the pattern *!b*, and the events that match the pattern *!(b, a)*. Note that the first occurrence of *b* still matches the pattern *!b*, because according to our semantics, the pattern *!b* matches events such that *b* has not been matched prior to the event. Similarly, *!(b, a)* matches events such that *(b, a)* has not been matched prior to the event.

Input Event Stream: *a a c d a c b a c b d a c*
Events that match *!b*:   ✓ ✓ ✓ ✓ ✓ ✓ ✓ ✗ ✗ ✗ ✗ ✗ ✗
Events that match *!(b,a)*: ✓ ✓ ✓ ✓ ✓ ✓ ✓ ✓ ✗ ✗ ✗ ✗ ✗
(✓ (resp. ✗) below an event means the event matches (does not match) the pattern)

The automaton construction for *!p* requires construction of the complement automaton, as in [7]. As *!p* is equivalent to $(\Sigma^* - (p, \_, \_^*))$, first we construct a DFA (construction of DFA is detailed in Section 4) for $(p, \_, \_^*)$ and then we interchange the final and non-final states. The automaton for *!b* is shown in Fig. 2 (a).

Now let us examine a pattern such as *(!p, a)*, where *p* can be any pattern. As per our semantics, this pattern matches events that match *a*, such that the pattern *p* has not been matched prior to the event. The example below illustrates events that match the pattern *(!b, a)*. See that an event matches the pattern *(!b, a)* if the event matches *!b* and *a*. Therefore the semantics for *(!p, a)* is $!p \cap a$ (that is the intersection of the events that match *!p* and the events that match *a*).

Fig. 2. (a) Automaton (deterministic) for the pattern !b. All events in the input stream match the pattern, till an event of type b appears in the input stream, after which no event in the input stream matches the pattern. (b) Automaton (deterministic) for the pattern (!b, a). Note that this is the intersection of the automata for the patterns !b and a. During intersection, we combine the alphabet; for example, the symbol !ab matches events that do not match a and also match b.

Input Event Stream:	a a c d a c b a c b d a c
Events that match !b:	✓ ✓ ✓ ✓ ✓ ✓ ✓ ✗ ✗ ✗ ✗ ✗ ✗
Events that match a:	✓ ✓ ✗ ✗ ✓ ✗ ✗ ✓ ✗ ✗ ✗ ✓ ✗
Events that match !b, a:	✓ ✓ ✗ ✗ ✓ ✗ ✗ ✗ ✗ ✗ ✗ ✗ ✗

The automaton for the pattern $(!p, a)$ is constructed as follows. We first construct the automaton for $!p$ (as above), and the automaton for a. We then construct the intersection automaton of $!p$ and a as given below.

NFA intersect (NFA A1, NFA A2)
/*   intersect NFA A1 and A2, and return the intersection automaton.
    Let $A_1 = (Q_1, \Sigma_1 \cup \{\varepsilon\}, \delta_1, q_{01}, F_1)$, $A_2 = (Q_2, \Sigma_2 \cup \{\varepsilon\}, \delta_2, q_{02}, F_2)$. */

1. Construct alphabet $\Sigma_1.\Sigma_2$ consisting of all "compatible" alphabet symbols $(x_1.x_2)$, where $x_1 \in \Sigma_1$, and $x_2 \in \Sigma_2$. x_1 and x_2 are compatible if an input symbol can satisfy both x_1 and x_2 (e.g., $x_1 = ab$, and $x_2 = !bc$ are not compatible; $x_1 = ab$, and $x_2 = b!c$ are compatible and $x_1.x_2 = ab!c$).

2. Construct set of states $Q_1 \times Q_2$ as in [7]: $(q_{i1}, q_{i2}) \in Q_1 \times Q_2$, if $q_{i1} \in Q_1$ and $q_{i2} \in Q_2$. Similarly construct set of states $F_1 \times F_2$, and state $q_{01} \times q_{02}$.

3. Define set of transitions δ as follows: there is a transition $(q_{i1}, q_{i2}) \in Q_1 \times Q_2$ ($q_{i1} \in Q_1$ and $q_{i2} \in Q_2$) on seeing input symbol $x \in \Sigma_1.\Sigma_2$ (where x is $(x_1.x_2)$, $x_1 \in \Sigma_1$, $x_2 \in \Sigma_2$) to $(q_{j1}, q_{j2}) \in Q_1 \times Q_2$ (where $q_{j1} \in Q_1$ and $q_{j2} \in Q_2$), if we have transitions from q_{i1} on seeing x_1 to q_{j1} in A_1, and from q_{i2} on seeing x_2 to q_{j2} in A_2.

4. Return the automaton: $(Q_1 \times Q_2, \Sigma_1.\Sigma_2 \cup \{\varepsilon\}, \delta, q_{01} \times q_{02}, F_1 \times F_2)$

Let us illustrate the above algorithm with the example of the pattern $(!b, a)$, where b is events with price > 100; a is events with vol > 100. The automaton for $!b$ is shown in Fig. 2 (a); the automaton for a is as shown in Fig. 1 (a). The intersection automaton (automaton for the pattern $!b, a$) is shown in Fig. 2 (b). Step 1 constructs the alphabet for the intersection automaton; the alphabet for the automaton for $!b$ is $\{b, !b\}$, the alphabet for the automaton for a is $\{a, !a\}$. The alphabet for the intersection automaton is $\{ab, a!b, !ab, !a!b\}$; $a!b$ represents events with !(price > 100) and (vol > 100). There is a transition from state 1 (corresponding to state 1 in the automaton for $!b$ and state 1 in the automaton for a) to state 2 (corresponding to state 1 in the automaton for *!b* and state 2 in the automaton for *a*) on seeing the symbol *a!b*, because there is a transition in the automaton for *!b* from state 1 to state 1 on seeing *!b*, and there is a transition in the automaton for *a* from state 1 to state 2 on seeing *a*. Also state 2 in the intersection automaton is a final state because state 1 is a final state in the automaton for *!b*, and state 2 is a final state in the automaton for *a*.

The complete algorithm for construction of NFA for a pattern query is given below

---

NFA constructNFA (pattern *p*)
/*   Return NFA corresponding to the pattern *p* */

1.   Scan *p* and modify all *(!p, x)* with *(!p AND x)*
2.   Use the template given in Fig. 1 (a) to construct NFA for any symbol *x*
3.   Use the templates as in [7] to construct NFA for p_1, p_2 (denoting sequence), p_1+ p_2 (denoting union), and *p** (denoting Kleene-*)
4.   For pattern *!p*, construct the automaton for $(\Sigma^* - (p, \_, \_^*))$
5.   For pattern *(!p AND x)*, if A_1 is the automaton for *!p* and A_2 is the automaton for *x*, then construct intersect(*A1, A2*).
6.   Return the final NFA obtained for the entire pattern *p*.

---

4   Constructing DFA for Pattern Queries and Ambiguous Events

In Section 3, we examined how to construct a NFA for a given pattern query. Now, we will consider how to construct a deterministic finite state automaton (DFA) from the NFA, and also examine how the optimizer can choose a good execution plan.

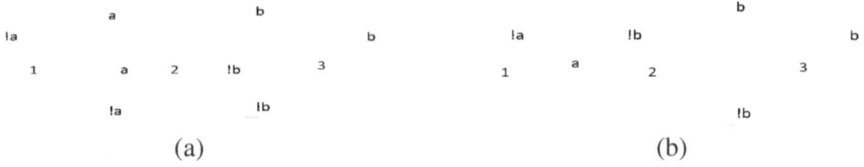

(a)   (b)

Fig. 3. (a) NFA for the pattern *(a, b)*. See that from state 2, on seeing an event that matches *!ab*, we can transition to state 1 and state 3. (b) DFA for the pattern *(a, b)* after minimization. From any state, on seeing an event, there is exactly one transition defined.

Consider the example pattern of *(a, b)* from Section 1 where *a* is events with price > 100, and b is events with volume > 100. An NFA obtained as described in Section 3 is shown in Fig. 3 (a). Consider the input event stream described in Section 1. The events that match the pattern are events with timestamp 4 and 5. Let us see how pattern matching can be performed using the NFA shown in Fig. 3 (a). We start from state 1; event with timestamp 1 matches *!a* (!(price > 100)); therefore we continue to be in state 1. The event with timestamp 2 matches *a* (price > 100), therefore we transition to state 2. The event with timestamp 3 matches *a* (price > 100), and *!b* (!(vol > 100)); we therefore continue to be in state 2. The event with timestamp 4 matches *!a* (!(price > 100)), and *b* (vol > 100). We can therefore transition to the set of states {1, 3}; 3 is a final state, hence this event satisfies the pattern. The event with timestamp 5 matches a (price > 100), and b (vol > 100). From state 1, we transition to state 2, and from state 3, we transition to state 3. Therefore the set of states after this event is {2, 3}; 3 is a final state, hence this event satisfies the pattern.

While we can execute the NFA, it would be more efficient to execute a DFA, which ensures that after any event, the automaton will transition to only one state (rather than a set of states as in an NFA). We confirm the significant performance benefits in our experimental study. However, the construction of DFA is especially hard because of ambiguous events; in the above example, we have an input event that matches the symbols, $!a$ and b (event with timestamp 4).

For construction of DFA, we first modify the alphabet of the NFA to consist of all combinations of input symbols; thus we handle ambiguous events. For the NFA in Fig. 3 (a), the alphabet is modified to have four symbols: ab (events with price > 100 and vol > 100), $a!b$ (price > 100 and !(vol > 100)), $!ab$ (!(price > 100) and vol > 100), and $!a!b$ (!(price > 100) and !(vol > 100)). The transitions are rewritten based on these symbols; for instance, the transition from state 1 to state 2 on symbol a is rewritten as two transitions: from state 1 to state 2 on symbol ab; from state 1 to state 2 on symbol $a!b$. After this modification of the alphabet, the construction of DFA is as described in [7]. Note that the number of states in the DFA is at most $(2^n - 1)$, where n is the number of states in the NFA (just as in [7]); however, the number of alphabet symbols in the DFA is at most 2^m, where m is the number of alphabet symbols in the NFA.

The DFA constructed from the NFA of Fig. 3 (a) is shown in Fig. 3 (b). We have performed minimization of the automaton (as described in [7]) and also combined symbols – for instance, if there are two transitions from state 1 to state 2 on symbols ab and $a!b$, we replace the two transitions with one transition from state 1 to state 2 on symbol a. Let us examine the execution of this DFA on the input event stream given above. We start from state 1; the event with timestamp 1 matches $!a$, hence we continue to be in state 1; the event with timestamp 2 matches a, hence we transition to state 2; the event with timestamp 3 matches $!b$, hence we continue to be in state 2; the event with timestamp 4 matches b, hence we transition to state 3 (3 is a final state and this event matches the pattern); the event with timestamp 5 matches b, hence we continue to be in state 3 (3 is a final state and this event matches the pattern). Note that at any point of time, the DFA transitions to at most one state.

Choosing a Good Execution Plan: For any query, there are multiple execution plans, and the optimizer chooses a "good" plan. This decision could be based on heuristics, on cost-based optimization or any alternative optimization strategy. For XML stream processing, the choice of plans and cost-based optimization is studied in the Raindrop project [10]. In SASE [12], the authors consider different optimization strategies. A pattern query can specify groupings on one or more attribute values, such groupings are called equivalence tests in [12]. The authors consider partitioning the event stream into groups based on equivalence tests, consider performing equivalence tests before or after pattern match, and consider pushing window constraints into pattern match.

In this work, we consider only pattern queries, and ignore other operators, such as selection, aggregation, join, window operations, transformations etc. Even for simple pattern queries there are multiple execution plans. Consider the pattern *(a, b)*; in Section 3, we constructed NFA for the pattern *(a, b)*, and in Section 4, we constructed DFA for this pattern. Both of these are valid execution plans. A third execution plan is that we divide it into two sub-patterns: *a* sub-pattern returns events that match *a*, and *b* sub-pattern returns events that match *b*. The events returned from these two sub-patterns (using automata) can then be joined based on ordering (or timestamps) to find *b* events that occur after *a* events. In [12], pattern *(a, !b, c)*, is executed by splitting it into two sub-patterns: *(a, c)* pattern, and *b* pattern. The results from these two sub-patterns are joined to check that there are no *b* events between *a* and *c* events. In Section 3 and in Section 4, we described how to construct a single automaton even for patterns with negation; this forms another execution plan.

5   Performance Evaluation

We compare different execution plans: using a single DFA/NFA for a pattern; splitting a pattern into multiple sub-patterns, finding sub-pattern matches using DFA/NFA and joining results of different sub-patterns (as in SASE [12]). We study various query patterns, and various input stream sizes. All experiments are conducted on a Dell Intel Dual-core 2.2GHZ computer with 4GB RAM running Windows 7.

We implemented the following operators: an input operator for reading events in the input stream (our input stream is a file) and writing them to a queue; a DFA (and a NFA) operator for reading events from an input queue, performing pattern match using a DFA (or a NFA) and writing the output to an output queue; a join operator which combines events from multiple input queues based on specified predicates and writes the output to an output queue; and an output operator that reads events from a queue, and writes them to an output stream (also a file). Our code is written in Java 1.6, where each operator in the query plan forms a separate thread. If the output of an operator *op1* serves as the input of another operator *op2*, then the output queue of *op1* is shared with *op2*; java monitors ensure correct implementation of queue sharing.

We implemented our own DFA and NFA construction. At run-time the state/(s) to transition to on seeing an input event is determined by a hash-lookup for both DFA and NFA. We implemented our own input stream generator; our test input streams are ordered and consist of only three event types, *a, b, c*, which are randomly generated using a uniform distribution. We consider different query patterns, for which we measure the performance of the different plans for different input sizes, ranging from 200,000 events to 1,000,000 events. In our experiments, we consider only the plan execution time and we do not measure the time to compile the query into the execution plan (we do not consider automaton construction time also).

Our first set of experiments considers the effect of the length of the query pattern without negation, and we compare two different plans: DFA and NFA. SASE [12] does not split a query pattern if it has no negation; therefore the NFA mimics SASE. We consider five different query patterns: *(a,b)*, *(a, b, c)*, *(a, b, c, a)*, *(a, b, c, a, b)*, *(a, b, c, a, b, c)*. The comparison is shown in Fig. 4. See that the performance of DFA does not depend on the size of the pattern; this is because irrespective of the pattern, the DFA performs a hash look up to determine the next state for every event, and checks whether the resulting state is a final state. The performance of the NFA can vary, because the number of states after every event could be different for different patterns; therefore the number of hash lookups can vary.

424   M. Mani

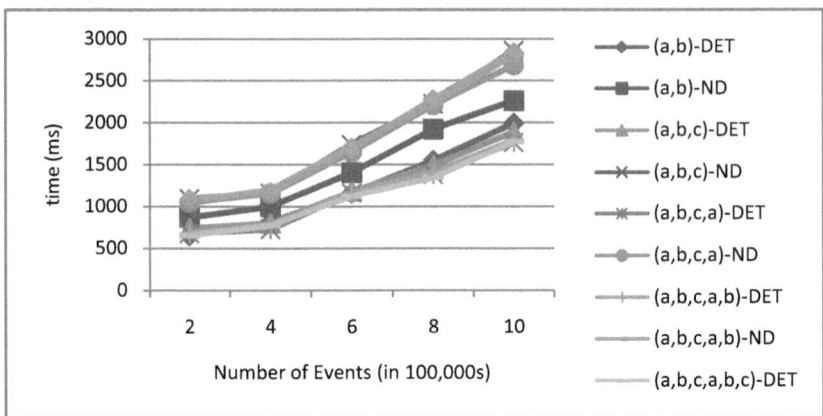

Fig. 4. DFA performs significantly better than NFA for different patterns. Also the performance of the DFA does not depend on the pattern, and depends only on the number of events.

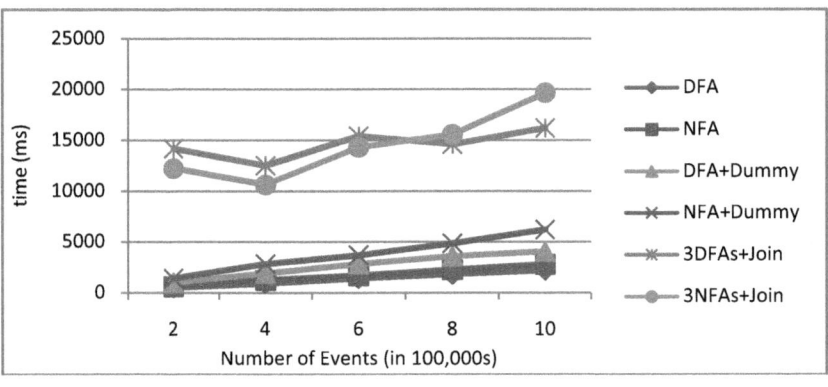

Fig. 5. Plans without joins perform much better than plans with joins. The best performance is for the plan with a single DFA.

Our next set of experiments compares the performance of using a single automaton for pattern match against using multiple automata for various sub-patterns and then joining the results from the sub-patterns (as in SASE [12]). We consider a negated pattern *(a, !b, c)*, and we compare six execution plans: (1) 3 NFAs corresponding to *a*, *b* and *c*, and then performing one join (2) 3 DFAs corresponding to *a*, *b* and *c*, and then performing one join (3) 1 NFA but with 2 dummy operators and a dummy join operator (4) 1 DFA but with 2 dummy operators and a dummy join operator (5) 1 NFA (6) 1 DFA. Note that the plans for options (3) and (4) have the same number of threads (six threads) as for options (1) and (2). However, plans (5) and (6) have only three threads (1 input operator, 1 automaton operator and 1 output operator). Options (3) and (4) are considered to negate the overhead of number of threads inherent in splitting a pattern into multiple sub-patterns. The performance of the different plans is shown in Fig. 5. The plans with joins perform significantly worse than plans without joins (even if the plans have the same number of threads using dummy operators).

6 Related Work

There has been quite a lot of interest in event stream processing in the last several years. One of the first to study pattern specification for events is [5]; the authors investigate the context of active databases, and come up with a rich array of operators for event expressions. Using their operators, shop-lifting query can be specified as *(relative (shelf, !before(counter)) AND exit)*; this is specified in our language and in SASE [12] as *(shelf, !counter, exit)*. In [5], the authors also consider construction of an automaton for a pattern query; however ambiguous events are not considered. Predicates are considered to a limited extent as join predicates (similar to equivalence tests in [12]), and are handled by partitioning the input stream into different groups.

In Cayuga [4], the authors identify that the transitions in the automaton should be defined based on predicates, and construct a NFA that can handle ambiguous events; however, ambiguous events are not explicitly identified. Negation is also not considered in Cayuga. Because the transitions in the automaton are based on predicates, Cayuga also investigates indexing schemes for efficient transition lookup.

In SASE [12], just like in [4], the authors define transitions in the automaton based on predicates, and construct a NFA. Here also, the authors do not identify the problem of ambiguous events, which are handled automatically by the NFA. The authors do define simple negation, but do not consider complex nested negation. A pattern query is split into multiple negated parts and non-negated parts, and the results of the different sub-patterns are joined. In [1], the authors extend [12] with different event selection strategies, such as contiguity, skip till next match, skip till any match etc.

In [9], the authors consider how to perform efficient pattern match over event streams by adapting KMP algorithm used for string matching, so that the symbols are based on predicates; this reduces the amount of backtracking required during pattern match. The techniques in [9] are orthogonal to our work, and our work can be extended to incorporate these pattern matching techniques as well. In [11], the authors study an orthogonal problem of event summarization, using hidden Markov models.

7 Conclusions and Future Work

In this paper, we studied efficient event stream processing. We identified two key challenges for event stream processing: ambiguous events, where an input event can match multiple symbols in the pattern, and negation used in pattern queries to specify non-occurrence of a pattern. We defined the semantics for pattern queries, including negation. Using our framework, we construct an NFA for pattern queries even with complex nested negation. This is different from existing work, where only simple negation is handled by splitting a pattern into multiple sub-patterns for the negated and the non-negated parts, and then joins are performed to combine the results of the different sub-patterns. Our approach enables us to construct a single automaton, and we do not need joins to obtain the results for a pattern query. Also existing work handle ambiguous events by using NFA; we investigate how to build a DFA that can handle ambiguous events. Experimental studies demonstrate the significant performance gains from our approach.

There are still several research directions that we would like to investigate. First is increasing the expressiveness of our pattern queries by including window constraints, arbitrary predicates, and aggregation operators. Handling arbitrary predicates will require us to extend our automaton with buffers, as investigated in SASE and Cayuga; however, we would like to still use a DFA. Processing multiple pattern queries simultaneously over an input stream also appears to be a useful problem to tackle.

Acknowledgements. We would like to acknowledge the significant contributions of Ms. Vinisha Lokesh, in the formulation of the research problem, and in conducting experiments. We would also like to acknowledge the members of the database research group at University of Michigan, Flint, for providing feedback on this work. This work is partially supported by UMFlint RCAC grant.

References

1. Agrawal, J., Diao, Y., Gyllstrom, D., Immerman, N.: Efficient Pattern Matching over Event Streams. In: ACM SIGMOD International Conference on Management of Data, Vancouver, Canada, pp. 147–160 (2008)
2. Babu, S., Widom, J.: Continuous Queries over Data Streams. SIGMOD Record 30(3), 109–120 (2001)
3. Barga, R.S., Goldstein, J., Ali, M., Hong, M.: Consistent Streaming Through Time: A Vision for Event Stream Processing. In: 3rd Biennial Conference on Innovative Data Systems Research (CIDR), Asilomar, CA (2007)
4. Demers, A., Gehrke, J., Panda, B., Riedewald, M., Sharma, V., White, W.: Cayuga: A General Purpose Event Monitoring System. In: 3rd Biennial Conference on Innovative Data Systems Research (CIDR), Asilomar, CA (2007)
5. Gehani, N.H., Jagadish, H.V., Shmueli, O.: Composite Event Specification in Active Databases: Model & Implementation. In: 18th International Conference on Very Large Data Bases (VLDB), Vancouver, Canada, pp. 327–338 (1992)
6. Green, T.J., Gupta, A., Miklau, G., Onizuka, M., Suciu, D.: Processing XML Streams with Deterministic Automata and Stream Indexes. ACM Transactions on Database Systems 29(4), 752–788 (2004)
7. Hopcroft, J., Motwani, R., Ullman, J.D.: Introduction to Automata Theory, Languages and Computation. Prentice-Hall, Englewood Cliffs (2006)
8. Li, M., Liu, M., Ding, L., Rundensteiner, E., Mani, M.: Event Stream Processing with Out-of-Order Data Arrival. In: 1st International Workshop on Distributed Event Processing, Systems and Applications (DEPSA), Toronto, Canada (2007)
9. Sadri, R., Zaniolo, C., Zarkesh, A., Adibi, J.: Expressing and Optimizing Sequence Queries in Database Systems. ACM Transactions on Database Systems 29(2), 282–318 (2004)
10. Su, H., Rundensteiner, E.A., Mani, M.: Automaton In or Out: Run-time Plan Optimization for XML Stream Processing. In: International Workshop on Scalable Stream Processing Systems, Nantes, France, pp. 38–47 (2008)
11. Wang, P., Wang, H., Liu, M., Wang, W.: An Algorithmic Approach to Event Summarization. In: ACM SIGMOD International Conference on Management of Data, Indianapolis, IN, pp. 183–194 (2010)
12. Wu, E., Diao, Y., Rizvi, S.: High-Performance Complex Event Processing over Streams. In: ACM SIGMOD International Conference on Management of Data, Chicago, IL, pp. 407–418 (2006)

Effective Keyword Search for Candidate Fragments of XML Documents

Yanlong Wen, Haiwei Zhang, Ying Zhang, Lu Zhang, Lei Xu, and Xiaojie Yuan

College of Information Technical Science, Nankai University, China
{wenyanlong,zhangying,zhanglu}@dbis.nankai.edu.cn,
{zhhaiwei,lxu,yuanxj}@nankai.edu.cn

Abstract. In this paper, we focus on the problem of effectively and efficiently answering XML keyword search. We first show the weakness of existing SLCA (*Smallest Lowest Common Ancestor*) based solutions, and then we propose the concept of *Candidate Fragment*. A *Candidate Fragment* is a meaningful sub tree in the XML document tree, which has the appropriate granularity. To efficiently compute *Candidate Fragments* as the answers of XML keyword search, we design *Node Match Algorithm* and *Path Match algorithm*. Finally, we conduct extensive experiments to show that our approach is both effective and efficient.

1 Introduction

XML has become the *de facto* standard for data representation and data exchange on the internet. With the rapid spread use of XML, it is one of the focus research problems among database research community and IR research community that how to effectively and efficiently get meaningful information from the XML data.

Keyword search is the most popular method to query information on the web search engines. Users can get their desired information by submitting simple keywords, and they need not to learn complex query syntax, such as XPath and XQuery, and understand the behind data schema. Keyword search on XML documents is different from traditional web information retrieval. The basic idea of XML keyword search system is to locate the most relevant XML fragments to answer each keyword query, while search engines usually return the whole web pages as the results to the user. It is still an open problem that which is the best semantics unit to answer XML keyword query.

When users submit keywords to search XML documents, they often hope to get compact and informative fragments which contain enough relevant information. So if we can properly partition the XML document tree into some user concerned, fine-grained, meaningful fragments, we can answer keyword query effectively and efficiently.

In this paper, we propose a novel method, which is different from SLCA-based proposals, to answer XML keyword query. We first identify *Candidate Nodes* (CANs) according to statistical information of XML documents, then construct each *Candidate Fragment* (CAF) from its center *Candidate Node*, which is the

J. Xu et al. (Eds.): DASFAA Workshops 2011, LNCS 6637, pp. 427–439, 2011.

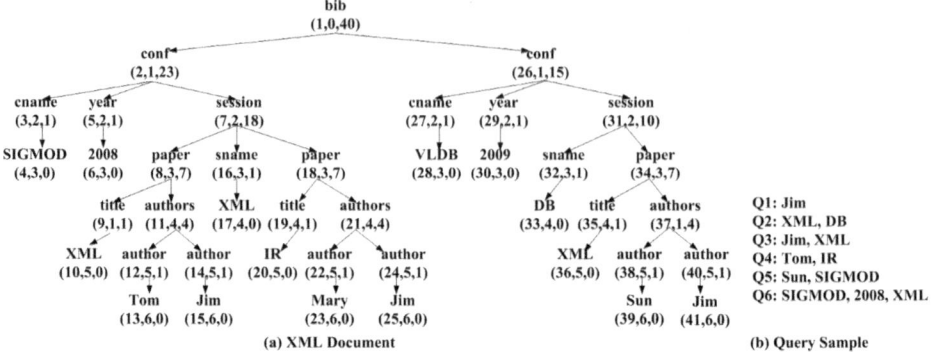

Fig. 1. Sample XML document with region codes and query examples

minimal semantic unit to answer keyword query in our approach. We create effective inverted index of all Candidate Fragments. When user submits a keyword query, we use *Node Match Algorithm* or *Path Match Algorithm* to return fragments to user. CAF contains more meaningful and fine-grained information than SLCA, and can solve the problems that may be caused by SLCA semantics.

Our contributions in this paper can be summarized as follows:

- We formally define *Candidate Node* (CAN) and *Candidate Fragment* (CAF), and propose a simple method to partition the XML document into CAFs without schema supporting.
- We propose *Node Match Algorithm* and *Path Match Algorithm* to answer XML keyword query efficiently.
- We conduct an extensive performance study in real datasets. The results show that our solution achieves high efficiency and effectiveness, and outperforms the SLCA proposal in elapsed time as well as query intention prediction.

The rest of this paper is organized as follows: We review the related work about XML keyword search in Section 2. Section 3 first presents the XML data model, then formally defines the concepts of CAN and CAF. We design Node Match algorithm and Path Match algorithm to efficiently answer keyword query in Section 4. In Section 5, we present our experimental evaluation to show the effectiveness and efficiency of our method. Finally in section 6, we conclude this paper and show our future work.

2   Related Work

Currently, most XML keyword search methods are based on the concept of LCA. [3] employs the ELCA semantics, and designs DIL (Dewey Inverted List) and RDIL (Ranked Dewey Inverted List) algorithms to efficiently answer XML keyword queries. [3] ranks results by generalizing the PageRank algorithm of Google. [1] employs the SLCA semantics, and designs Scan Eager (SE) and

Indexed Lookup Eager (ILE) algorithms, which are two state-of-art algorithms to compute SLCA. ILE algorithm is an effective algorithm when the keyword search involves at least one low frequency keyword, while SE algorithm performs better when the frequencies of keywords in the query do not vary significantly. It is obvious that LCA is the super set of ELCA, and ELCA is the super set of SLCA. [4] use the homogenous concept to define Valuable SLCA (VLCA) semantics which aims to improve the search effectiveness. However, in some situations, the homogenous concept is not appropriate to filter the LCA nodes, and can not guarantee the meaningfulness. XSeek[5] uses the inference rules to identify node type in the XML data tree and generates two types of nodes: return nodes that can be inferred explicitly by analyzing keyword match patterns, and return nodes that can be inferred implicitly by considering both keyword match patterns and the XML data structure. [6-7] designed multi-way based and hash based algorithms to fasten SLCA computation.

While SLCA semantics or its variants may cause the following problems:

1. SLCA semantics may return non-relevant result.
   Example 1. Consider that user submits query Q4 in Fig. 1(b) to search papers about IR written by Tom. The returned result set by SLCA is $\{session(7, 2, 18)\}$. But as we can see in Fig 1(a), Tom is not an author of $paper(18, 3, 7)$, so $session(7, 2, 18)$ is a non-relevant result. The same problem remains when using SLCA to answer query Q5 in Fig. 1(b).

2. SLCA semantics may miss relevant results.
   Example 2. User wants to query Jim's papers about XML and submit query Q3 in Fig. 1(b). SLCA will return set $\{paper(8, 3, 7), paper(34, 3, 7)\}$. But $paper(18, 3, 7)$ is one of the papers in the session about XML, and it also meets the query intention. SLCA semantics fails to locate $paper(18, 3, 7)$, and misses a relevant result.

3. SLCA semantics may not get answers with appropriate granularity.
   Example 3. When user submits query Q6 in Fig. 1(b) to search papers about XML published in $SIGMOD$ proceeding in 2008, SLCA semantics will return $\{conf(2, 1, 23)\}$, whose size is larger than preferred answers and contains much irrelevant information. When user submits query Q2, SLCA semantics will return $\{session(31, 2, 10)\}$, meanwhile lose relevant information about paper's publication VLDB 2009.

4. SLCA semantics may not answer single keyword query effectively.
   Example 4. When user wants to get Jim's paper by submitting query Q1 in Fig. 1(b), SLCA will return set $\{Jim(15, 6, 0), Jim(25, 6, 0), Jim(41, 6, 0)\}$. Each result in the result set only has one name of author, which is apparently not a complete answer. Thus SLCA semantics fails to deal with simple keyword query. The same problem remains when the search keywords occur in the same node of XML document.

In addition, there are some non-SLCA solutions. Query semantics that similar to our proposal are Meaningful Information Unit(MIU)[8] and Meaningful Information Segmentation(MIS)[9].

[8] first partitions XML schema graph into schema Minimal Information Units (sMIUs) through removing all the edges of high frequency and then removing all the isolated nodes. Then [8] partition XML data graph into data Minimal Information Units (dMIUs) according to the acquired sMIUs. MIU semantics requires the existence of a schema file, causes information loss while removing isolated nodes, and its size may be too large or too small when the tunable parameter is not well chosen.

[9] does a width-first traversal of the XML document tree to identify nodes with the same label, then partition an XML document into a series of meaningful and self-containing segments, called Minimal Information Segments(MISs), and return MIS-subtrees which consist of MISs that are logically connected by the keywords. As the authors have mentioned in [9], in practice the MIS may be very large and contain non-relevant information.

In this paper, we propose a novel semantics called *Candidate Fragment* (CAF) to answer XML keyword queries. CAF semantics uses statistics information of XML elements, and keeps all information of original document. Thus CAF does not need XML schema support.

3   Query Semantics

In this section, we first introduce the XML data model, then formally define the concepts of *Candidate Node* (CAN) and *Candidate Fragment* (CAF), and illustrate how to effectively answer keyword queries with CAFs.

3.1   XML Data Model

In this paper, an XML document is modeled as a labeled, directed tree, a keyword query is a set of keywords.

Definition 1 (Keyword Query). *A keyword query is a set of different keywords, denoted as* $Q = \{w_i | i = 1, \ldots, k\}$, w_i *is a keyword.*

Definition 2 (XML Document Tree). *An XML document can be denoted by* $T=(V, E, tag(v), pre(v), level(v), size(v), path(v), type(v), atn(v))$.

- V *is the set of nodes. We assign each node* v *with an unique region code* $(prev, level, size)$ *according to [10]. prev is the id of* v *when pre-traversing the document tree, level is the level of node* v *and size is the number of* v's *descendants.*
- $E \subseteq V \times V$. E *is the set of containment edges in the document tree.*
- $tag(v)$ *is a function, returns the label of node* v.
- $pre(v)$ *is a function, returns the id of* v. *If* r *is the root node of the document tree, then* $prev(r) = 1$.
- $level(v)$ *is a function, returns the level of* v. *If* r *is the root node of the document tree, then* $level(r) = 0$.
- $size(v)$ *is a function, returns the number of* v's *descendants.*

- *path(v) is a function, returns the id path from root r to node v. If node r is the root node of the document tree, then path(r) = 1.*
- *type(v) returns the label path from root r to node v. If node r is the root node of the document tree, then type(r) = tag(r).*
- *atn(v) returns the number of different attribute types of node v.*

In this paper, we use the following inference rules, which are similar to [5] and [13], to identify value node, attribute node in XML document. A node represents a value if it is a leaf node. A node denotes an attribute if it has only one child, which is a value node. A node denotes a group attribute if all its children denote attributes with the same type.

Example 5. In Fig. 1(a), $XML(10, 5, 0)$ is a value node, $title(9, 1, 1)$ is an attribute node. For node $paper(8, 3, 7)$, its *tag* is *paper*, its *id* path is 1.2.7.8, its *type* is *bib.conf.session.paper*, and its *atn* is 2 because its child $title(9, 1, 1)$ is an attribute node and $authors(11, 4, 4)$ is a group attribute node.

3.2 CAF Semantics

When search XML documents, we prefer fragments with meaningful information and appropriate granularity. Such fragments could be entities or their complex descriptions in the real world, and are often described by multiple attributes. In this paper, we have verified the above observations by analyzing many XML data sets. We first formally give the concepts of Candidate Node and Candidate Fragment.

Definition 3 (Candidate Node, CAN). *let c be a XML tree node, $C = \{v|v \in V \land type(v) = type(c)\}$, c is a candidate node, iff, $avg(atn(v)|v \in C) >= \alpha$ and $avg(size(v)|v \in C) >= \beta$. Here, α is the threshold of attribute type number and β is the threshold of node size.*

As we know in the real world, an entity is usually described by two or more attributes, it means that $\alpha >= 2$ and $\beta >= 4$ in most cases (because 2 attributes of an entity will be denoted by 2 attribute nodes and 2 value nodes).

Table 1. Internal node types of XML document Tree

type	avg(atn)	avg(size)
1 bib	0	40
2 bib.conf	**2**	**19**
3 big.conf.cname	0	1
4 bib.conf.year	0	1
5 bib.conf.session	1	14
6 bib.conf.session.sname	0	1
7 bib.conf.session.paper	**2**	**7**
8 bib.conf.session.paper.title	0	1
9 bib.conf.session.paper.authors	1	4
10 bib.conf.session.paper.authors.author	0	1

Example 6. According to *Definition 3*, CAN should be chosen from non-leaf nodes. In Fig. 1(a), there are ten internal node types in the XML document tree. We show average attribute type number and average size of each type in Table 1. Then we can identify nodes in the XML document tree with type *bib.conf* or *bib.conf.session.paper* are CANs.

Definition 4 (Candidate Fragment, CAF). *A Candidate Fragment is a compact sub tree in XML tree, contains only one Candidate Node, and represents a meaningful entity or its complex attribute in the real world.*

Given a CAN c, we can get its CAF by the following two steps.

- Step 1: $ST(c)$ is the sub tree rooted with c. CAN $c_1, c_2, \ldots, c_m$ are descendant nodes of c. Node $a_1, a_2, \ldots, a_k$ are child nodes of c and ancestor nodes of $c_1, c_2, \ldots, c_m$. We can get sub tree $ST'(c)$ by removing the parent-child edges between c and $a_1, a_2, \ldots, a_k$.
- Step 2: Get CAN c's id-path $path(c) = t_1 \cdot t_2 \cdot \ldots \cdot t_{level(c)}$. If there exists CANs among c's ancestor nodes, let the lowest CAN's *prev* be t_p, else let $p = 0$. The nodes in the sub path $t_{p+1} \cdot \ldots \cdot t_{level(c)-1}$ construct fragment PF. Connect PF with $ST'(c)$ according to their ancestor-descendant relationship, then we get CAF $ST''(c)$.

Example 7. In Fig. 1(a), $conf(2, 1, 23)$ has descendants $paper(8, 3, 7)$ and $paper(18, 3, 7)$, which are CANs. Node $session(7, 2, 18)$ is the child of $conf(2, 1, 23)$, and is the parent of $paper(8, 3, 7)$ and $paper(18, 3, 7)$. Remove the edge between $conf(2, 1, 23)$ and $session(7, 2, 18)$. $Path(conf(2, 1, 23))$ is 1.2 and there is no CAN among $conf(2, 1, 23)$'s ancestors. Keep the fragment from node with *prev* 1 to $conf(2, 1, 23)$. And then connect it with $conf(2, 1, 23)$'s remaining part to form a CAF, which is shown in Fig. 2(a).

Example 8. In Fig. 1(a), the descendants of the $paper(8, 3, 7)$ are not CANs. So we can ignore step 1 according to *Definition 4*. $Path(paper(8, 3, 7))$ is 1.2.7.8, and its ancestor $conf(2, 1, 23)$ is a CAN. Keep the fragment from node with *prev* 7, and then connect it with $paper(8, 3, 7)$ to form a CAF, which is shown in Fig. 2(b).

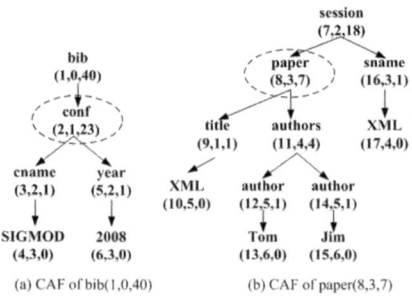

Fig. 2. CAF sample

Definition 5 (Fragment ID Path, FIP). *Given node* c, a_1, a_2, ..., a_k *are CANs among ancestor nodes of* c. *Then the Fragment ID Path of* c *is* $prev(a_1) \cdot prev(a_2) \cdot ... \cdot prev(a_k) \cdot prev(c)$.

Example 9. In Fig. 1(a), the FIP of $paper(8, 3, 7)$ is 2.8, and the FIP of $conf(2, 1, 23)$ is 2.

Definition 6 (Keyword Match Result). *Given CAN list* CN, *Fragment ID Path list* FP *and Query* Q. *When there is only one type among CAN list or user only decides to search CAFs with all keywords, the keyword match result is* $M = \{v | v \in CN, (\forall w_i \in Q) contains(v, w_i)\}$. *Otherwise, the keyword match result is* $M = \{fp | fp \in FP, (\forall w_i \in Q \exists v \in fp) contains(v, w_i)\}$. *Here,* $contains(v, w_i)$ *denotes that CAF, which center CAN is* v, *contains word* w_i.

Example 10. For Query $Q = \{2008, SIGMOD, Jim, XML\}$, the keyword match result is $M = \{2.8, 2.18\}$, when user allows keyword matches in different fragments.

Algorithm 1. Node Match Algorithm

Input : Keyword Query $Q = \{k_1, k_2, ..., k_m\}$ and XML Document T
Output: CAN list $R = \{r_1, r_2, ..., r_t\}$, which matches all keywords

1 $R = \Phi$;
2 **for** $i=1$ **to** m **do** $S_i \leftarrow GetMatchNode(k_i)$;
3 $Sort(S_1, S_2, ..., S_m)$;
4 **for** $i=1$ **to** $S_1.Length$ **do**
5 | $found = 1, finish = 0$;
6 | **for** $j=2$ **to** m **do**
7 | | **while** $S_j \neq \Phi \wedge S_j[1] < S_1[i]$ **do** $RemoveHead(S_j)$;
8 | | **if** $S_j \neq \Phi$ **then**
9 | | | **if** $S_j[1] \neq S_1[i]$ **then** found=0,break;
10 | | **else**
11 | | |_ $found = 0, finish = 1$,break;
12 | **if** $found$ **then** $R = R \cup \{S_1[i]\}$;
13 |_ **if** $finish$ **then** break;

---

4 Query Algorithms

In the pre-processing stage, when parsing an XML document, we compute average attribute numbers and average size of nodes for different node types. According to *Definition 3*, we can identify CANs and then form CAFs based on the two step algorithm. We construct an inverted index for CAFs to fasten keyword query. Each index item is a pair of *keyword* and *prev*. Here, *keyword* is a query term and *prev* is the *id* of CAN. Match list of each keyword is ordered by *id* of CAN.

Algorithm 2. Merge Match Algorithm

    Input   : CAN List S_1, S_2
    Output: Match result list $L = \{r_1, r_2, \ldots, r_t\}$

 1 $L = \Phi$;
 2 **while** $S_1 \neq \Phi \land S_2 \neq \Phi$ **do**
 3   **if** $isSamePath(S_1[1], S_2[1])$ **then**
 4         $RemoveDescendants(S_1, S_1[1])$;
 5         $RemoveDescendants(S_2, S_2[1])$;
 6   **if** $S_1[1] = S_2[1]$ **then**
 7             $L = L \cup \{S_1[1]\}$;
 8             $RemoveHead(S_1)$;
 9             $RemoveHead(S_2)$;
10   **else**
11             $L = L \cup \{Max(S_1[1], S_2[1])\}$;
12             $RemoveHead(S_{Max})$;

13   **else**
14   **if** $topCN(S_1[1]) > topCN(S_2[1])$ **then**
15   **while** $topCN(S_1[1]) > topCN(S_2[1])$ **do** $RemoveHead(S_2)$;
16   **else**
17   **while** $topCN(S_2[1]) > topCN(S_1[1])$ **do** $RemoveHead(S_1)$;

In this section, we will present *Node Match algorithm* and *Path Match algorithm*. When there is only one type of CAN or different keyword is prohibited by user at different fragments, *Node Match algorithm* can be used to answer keyword search. Otherwise, *Path Match Algorithm* could take place.

4.1 Node Match Algorithm

Algorithm 1 is used to compute all CAFs that contains all keywords. As shown in algorithm 1, we first initialize the result list to be empty in line 1. In line 2-3, we scan inverted index to get matched CAN list, and sort nodes in each CAN list by their ascendant prev order. For each node in S_1, which has the minimal length, we scan other list in line 6-11, remove nodes with smaller *id* than current node in S_1 in line 7. If each head node has the same *prev* with the current node in S_1, then add it into result list R in line 12. Then go to next loop.

Example 11. For Q3 = $\{Jim, XML\}$ in Fig. 1 (b), the match list to Jim is $S_{Jim} = \{8, 18, 34\}$, and the match list to XML is $S_{XML} = \{8, 18, 34\}$. We will return list $\{8, 18, 34\}$ as the result with Algorithm 1.

4.2 Path Match Algorithm

We give the solution to compute all the Fragment ID Path list in Algorithm 3, which matches all the keywords. Algorithm 2 deals with two keyword match list S_1 and S_2, and return list L as the result. Algorithm 3 uses Algorithm 2 to cover all the keywords, and return Fragment ID Path list as the answer.

Algorithm 3. Path Match Algorithm

Input : Keyword Query $Q = \{k_1, k_2, \ldots, k_m\}$ and XML Document T
Output: Fragment ID Path list $RP = \{fp_1, fp_2, \ldots, fp_t\}$, which matches all
        keywords

1 $R = \Phi$;
2 **for** $i=1$ **to** m **do** $S_i \leftarrow GetMatchNode(k_i)$;
3 $Sort(S_1, S_2, \ldots, S_m)$;
4 $R = S_1$;
5 **for** $i=2$ **to** m **do** $R = MergeMatch(R, S_i)$;
6 $RP = \Phi$;
7 **for** $i=1$ **to** $R.Length$ **do** $RP = RP \cup GetFP(R_i)$;

Algorithm 2 solves the two keyword query problem. We first initialize the prev list L empty in line 1. Then in line 2-17, we traverse the two lists. When the top elements of the two lists are on the same path of the XML document tree, we remove their descendant nodes in line 4-5. If the top elements of the two lists are the same element, we add it to the result list and remove it in line 8-9. Otherwise, we add the max one into the result list and remove it. If the top elements of the two lists are not on the same path, remove the left one, then go to next loop.

As shown in Algorithm 3, we first initialize the result list empty in line 1. In line 2-3, we scan inverted index to get matched CAN list, and sort nodes in each CAN list by their ascendant prev order. In line 4, We chose node list S_1 as the initial result list, which has the minimal length. Then we use Algorithm 2 to recursively merger result in line 5. In line 6, we initialize Fragment ID Path list RP empty. Then we get each Fragment ID Path fp in list R and add it into RP list in line 7.

Example 12. For Q6 $=\{SIGMOD, 2008, XML\}$ in Fig. 1(b), the match lists to Jim, 2008, XML are $S_{SIGMOD} = \{2\}$, $S_{2008} = \{2\}$, $S_{XML} = \{8, 18, 34\}$. We will return list $\{2.8, 2.18\}$ as the result with Algorithm 3.

5   Experimental Evaluation

5.1   Experimental Setup

We use a laptop with Intel dual-core 2.0GHz CPU, 2G memory, 300 GB SATA hard disk, and Windows XP Professional as the operating system. The compared algorithm is ILE, which is an effective algorithm to answer SLCA. We implement our Node Match Algorithm and Path Match Algorithm to answer CAF. All algorithms are implemented using C sharp programming language.

5.2   Datasets and Keyword Queries

We use DBLP (130MB)[11], SIGMOD Record (500KB)[11] and Product Review (14MB)[12] datasets in our experiments. The main characteristics of the datasets

Table 2. Statistics of datasets, ATN denotes attribute type number

Dataset	Size (KB)	Elem. #	Depth	Avg ATN	Avg Size	CAF #	CAF %
DBLP	130726	3332130	6	0.85	5.14	328858	9.87
SIGMOD Record	468	11526	6	0.73	11.97	1571	13.63
Product Review	15170	427278	9	0.42	13.59	26526	6.21

is shown in Table 2. The 5th column shows average attribute type number of Element nodes in XML document. The 6th column is the average size of Element nodes in XML document. Number of CAFs identified by our methods is listed in the 7th column. And the last column is the ratio of CAF number to Elements number. From the statistics of each dataset, we can know that entities or their complex descriptions (denoted by CAFs in our solution) in XML document are usually a small proportion of the elements.

As we have mentioned in Section 3.2, an entity in the real world is usually described by at least two different attributes. So in the following experiments, we assume that each CAN should have at least two kinds of different attributes and four descendants, i.e., $\alpha = 2$ and $\beta = 4$.

In the following sub sections, we conduct both query effectiveness and query efficiency experiments to compare CAF semantics with SLCA semantics. In the experiment, we choose 6 different types of keyword querys to each dataset which are named QDi to DBLP, QSi to SigmodRecord, and QPi to ProductReview, the keywords and their frequencies can be found in Table 3.

Table 3. Keyword Queries

Query	Keyword (Frequency)
QD1	Jim(872),Gray(680), author(716595)
QD2	Ullman(333), database(8584)
QD3	database(8584), information(20570)
QD4	book(1602), Ullman(333), database(8584)
QD5	Ricardo(417), information(20570), retrieval(4008)
QD6	Modern(192), information(20570), retrieval(4008)
QP1	digital(1377), camera(2175)
QP2	digital(1377), camera(2175), travel(11313)
QP3	sports(1554), outdoors(1660)
QP4	product(1409), name(1419), rating(1339), review(25136)
QP5	product(1409), review(25136)
QP6	product(1409), rating(1339), review(25136), user(28444)
QS1	data(180), base(22)
QS2	article(1505), database(348), title(1504)
QS3	issue(67), volume(67), article(1504), title(1504), author(3737)
QS4	Stephen(7), Database(348)
QS5	article(1505), Jim(28), transaction(30)
QS6	volume(67), 11(32), article(1505)

5.3 Query Effectiveness

The query intention of each keyword query in Table 3 can be found in Table 4, which is got by a user survey. Table 4 compares the user's query intention and suggestions of SLCA semantics and CAF semantics (*Path Match Algorithm*). As shown in the results, SLCA returns incomplete results when answering some keyword queries (e.g., QD1-QD3, QD5, QD6, QP1-QP5, QS1, etc), and returns irrelevant results when answering keyword queries like QS3-QS6, etc. It is because the weakness of SLCA semantics itself, which we have discussed in Section 2.

Table 4. Effectiveness of user intention prediction

Query	Intention	SLCA result	CAF result (Path Match)
QD1	article	author	article
QD2	book, inproceedings, article	title, book, inproceedings, article	book, inproceedings, article
QD3	phdthesis, book, article, incollection, inproceedings, proceedings	title, book, article, incollection, inproceedings, proceedings	phdthesis, book, article, incollection, inproceedings, proceedings
QD4	book	book	book
QD5	article, book, incollection	title, article, book, incollection	article, book, incollection
QD6	article, inproceedings, book	title	article, inproceedings, book
QP1	product	product, name, bestuses, pros, pro, user, weblink	product, <product,review>
QP2	product	product, reviews, review, bestuses	product
QP3	product	product, reviews, review, bestuses	<product, review>
QP4	product	product, name, bestuses, pros, pro, user, weblink	<product,review>
QP5	product	product, review	<product, review>
QP6	product	product	<product, review>
QS1	article	title	article
QS2	article	article	article
QS3	article	issue	<issue, article>
QS4	article	article, articles	article
QS5	article	article, articles	article
QS6	article	issue	<issue, article>

Note CAF semantics answers QP1, QP3-QP6 with <product, review>, and QS3, QS6 with <issue, article>. It means that CAFs have Ancestor-Descendant relationship, such answer should also be relevant to the user's keyword query. As shown in Table 4, CAF semantics can give the user preferred results in most of cases.

5.4   Query Efficiency

Fig. 3 shows the running time of ILE, *Path Match Algorithm* and *Node Match Algorithm* on different datasets and different queries with various term frequency. The result shows that our method achieves better performance than ILE algorithm, which is an efficient SLCA computation method.

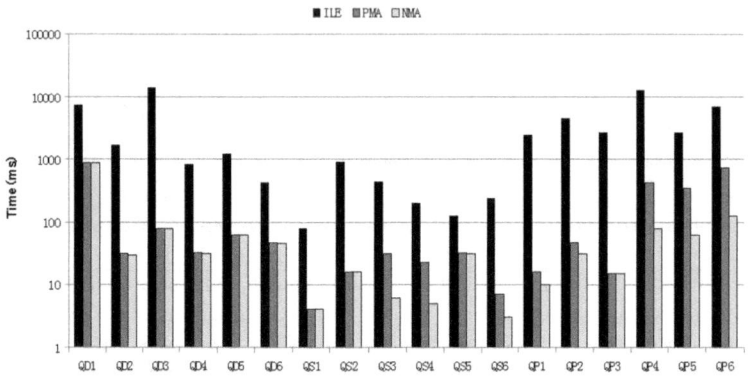

Fig. 3. Running Time, PMA/NMA denotes Path/Node Math Algorithm

There are two main reasons that our *Path/Node Match Algorithm* can gain better performance than SLCA's ILE solution:

- Our *Path/Node Match Algorithm* mainly adopts simple list merge strategy, while as mentioned in [1], ILE needs more computation for LCA solution, left match, right match, etc.
- SLCA solution needs to scan an inverted list index over all the match nodes in the XML document, while our CAF solution only scan a smaller proportion, because we only index keyword and *id* of CAF instead of the whole XML document's nodes.

6   Conclusion

In this paper, we focus on the semantics and search algorithm of XML keyword search. We analyze the main weakness of SLCA based semantics. Then we propose the CAF semantics to effectively answer a XML keyword query, and design node match algorithm and path match algorithm to efficiently return CAFs to users. Extensive experiments show that our method is significantly better than the SLCA semantics on both effectiveness and efficiency. In the future, we will consider IDREF in the XML document to deal with more CAF relation, design effective ranking model to solve the TOP-K problem, and study how to display the results to the users friendly.

Acknowledgments. This research is supported by the grant from the China 863 High-Tech Program(No: 2009AA01Z152).

References

1. Xu, Y., Papakonstantinou, Y.: Efficient keyword search for smallest LCAs in XML databases. In: SIGMOD, pp. 527–538 (2005)
2. Li, Y., Yu, C., Jagadish, H.V.: Enabling schema-free xquery with meaningful query focus. VLDB J. 17(3), 72–84 (2008)
3. Guo, L., Shao, F., Botev, C., Shanmugasundaram, J.: XRank: Ranked keyword search over xml documents. In: SIGMOD, pp. 16–27 (2003)
4. Li, G., Feng, J., Wang, J., Zhou, L.: Efficient keyword search for valuable LCAs over XML documents. In: CIKM, pp. 31–40 (2007)
5. Liu, Z., Chen, Y.: Identifying meaningful return information for XML keyword search. In: SIGMOD, pp. 329–340 (2007)
6. Sun, C., Chan, C.Y., Goenka, A.K.: Multiway SLCA-based keyword search in XML data. In: WWW, pp. 1043–1052 (2007)
7. Wang, W., Wang, X., Zhou, A.: Hash-search: An efficient SLCA-based keyword search algorithm on XML documents. In: Zhou, X., Yokota, H., Deng, K., Liu, Q. (eds.) DASFAA 2009. LNCS, vol. 5463, pp. 496–510. Springer, Heidelberg (2009)
8. Xu, J., Lu, J., Wang, W., Shi, B.: Effective keyword search in XML documents based on MIU. In: Li Lee, M., Tan, K.-L., Wuwongse, V. (eds.) DASFAA 2006. LNCS, vol. 3882, pp. 702–716. Springer, Heidelberg (2006)
9. Li, J., Wang, J., Huang, M.: XKMis: Effective and Efficient Keyword Search in XML Databases. In: IDEAS, pp. 121–130 (2009)
10. Boncz, P., Grust, T., Keulen, M.: Pathfinder: XQuery - the relational way. In: VLDB, pp. 1322–1325 (2005)
11. http://www.cs.washington.edu/research/xmldatasets/
12. http://wsdb.asu.edu/xsact/ProductReview.xml
13. Bao, Z., Ling, T.W., Chen, B., Lu, J.: Effective XML Keyword Search with Relevance Oriented Ranking. In: ICDE, pp. 517–528 (2009)

Optimized Data Placement for Column-Oriented Data Store in the Distributed Environment

Minqi Zhou and Chen Xu

Massive Computing Institute, East China Normal University
No. 3663 Zhongshan Rd.(North), Shanghai, China, 200062
mqzhou@sei.ecnu.edu.cn, chenxuecnu@gmail.com

Abstract. Column-oriented data storage becomes a buzzword nowadays for its high efficiency in massive data access, high compression ratio on individual columns and etc. However, the initial observations turn out to not be trivially true. The seek time and bandwidth of current hard disk drivers (HDD) become the bottleneck for massive data processing day by day, when comparing to other component enhancements of computers during the past four decades. In this paper, we provide a novel data placement strategy for massive data analysis (i.e., read-optimized) based on Gray Code, which enhances the ratio of sequential access to a great extent for diverse query evaluations (e.g., range query, partial match range query, aggregation query and etc). A centralized/distributed structured index is employed in the popularly deployed distributed file systems (e.g., GFS), which achieves the convenient management, efficient accessibility, high extendibility and etc. Detailed theoretical analysis on index extendibility, sequential access improvement and storage capacity usage in terms of proposed data placement strategies are provided as well as specific algorithms. Our extensive experimental studies confirm the efficiency and effectiveness of our proposed data placement methods.

1 Introduction

Column-oriented data storage becomes a buzzword nowadays, especially for read-optimized data management systems. More and more commercial database systems are deploying column-oriented data store for massive data analysis, such as MonetDB/X100 [1], Vertica [2], and etc. In terms of IDC's report in 2009, the analytical database market consists of $7.78 Billion of $20.4 Billion database market [3]. On first blush, it attracts much attention from both industry and academe, primary for its advantages in selecting a subset of relevant columns for efficient access, high compression ratio on individual columns because of the low entropy on a single column and etc.

However, the initial observations turn out to not be trivially true. Our overwhelming relational database management systems are designed nearly 40 years ago in terms of the hardware configurations at that time and mainly focus on centralized management. Nevertheless, the situation has been changed immensely during the past 40 years. Each hardware component in the computer literature developed in a imbalanced manner, especially for hard disk drivers. Fig. 1 briefly depicts the performance enhancement for each component during the past 4 decades. No doubt that both *seek time* and *bandwidth*

J. Xu et al. (Eds.): DASFAA Workshops 2011, LNCS 6637, pp. 440–452, 2011.

Components Year	Disk			CPU	Network
	Seek Time	Bandwidth	Capacity	Frequency	Bandwidth
1970s	50 ms	1 MB/s	20 MB	740 kHz	2.94 Mbit/s
2010s	5 ms	150 MB/s	2 TB	3.8 GHz*8(Cores)	100 Gbit/s
Enhancement	10 Times	150 Times	100000 Times	2500*8(Cores)	34013 Times

Fig. 1. Imbalanced Enhancement in Computer Components

of hard disk drivers development lack much behind (i.e., thousands of times) when comparing to other components. Obviously, the data access leaning upon current data storage media becomes the main bottleneck in terms of that of 40 years ago. Parallel processing is hopeful in alleviating this bottleneck, while not much since it is infeasible to attach thousands of hard disk drivers to a single process. Years ago, Jim Gray predicts the commonly used hard disk driver will become the tape (i.e., mainly used for sequential access) [4]. Therefore, reducing the ratio of random access in the massive data process is hopefully enhancing the performance dramatically.

For nowadays massive data analysis or processing, data are commonly distributed across hundreds or thousands of nodes (e.g., Google File System (GFS) [5]). In those systems, local data processing is mainly advocated (e.g., MapReduce [6]), which alleviates the burden of data transfer much. In other words, the cost for data transfer across nodes is rather high. Herein, the costs for data access, including both local I/O and network communication, outweigh other costs for massive data analysis. How to reduce the local I/O and network communication costs with clever data placement helps a lot for such kind of massive data analysis. In this paper, we address the problem of data placement in the scalable distributed parallel processing systems, which increases the ratio of local sequential access and reduces the volumes of data transmission across nodes. It can be formally defined as follows: *Given a scalable distributed system deployed with popularly used file systems (e.g., GFS, HDFS [7]), the task is to optimize the multi-dimension relational data placement across nodes in terms of data access cost.* Herein, the data access mainly focus on range selection on (a subset of) the multi-dimensionality.

To achieve the efficient data access for massive data analysis, we present a novel method towards data placement in the scalable distributed systems. Generally speaking, the reduction in data access costs mainly focus on the following two perspectives. Gray Code [8] is deployed to order the multi-dimensional data for local data placement on individual nodes, which improves the ratio of sequential access for diverse query evaluations (e.g., exact match query, range query, partial match range query, and etc). A content-aware bitmap index (encoded under Gray Code) is deployed to affiliate massive data analysis (e.g., aggregation query), which consequently results in communication reduction for information exchange. Our main contributions include while not limit to:

– We provide a novel data placement method leaning upon Gray Code encoding mechanism for massive data volumes across scalable large distributed systems, which achieves the scalable data storage, extendible index key regeneration, en-hanced ratio for sequential massive data access, and etc. Armed with the content-aware index (i.e., bitmap index) which has a centrilized/distributed structure, it achieves the capability for efficient diverse query processing (e.g., range query, aggregation query and etc).

− We provide the detailed analysis on the proposed data placement strategy, espe-cially for sequential massive data access improvement and storage usage efficiency, as well as instructions on diverse query evaluations.

− We conduct an extensive performance study which shows the effectiveness and efficiency of our methods for diverse query evaluation in the large scale distributed systems.

The rest of paper is organized as follows. In section 2, we review the related work. The formal problem statement is given in section 3. In section 4, we provide the detailed data placement algorithms, as well as query evaluation in section 5. We conduct extensive experimental studies in section 6 and conclude our paper in section 7.

2   Related Work

In the distributed system literature, many distributed file systems are designed, imple-mented, deployed during the past 3 decades, such as Andrew file system (AFS) [9], Coda file system [10], Network file system [11] and etc. Those distributed file systems are compact with local Unix file systems by providing a virtual layer which redirects the request for both local and distributed data access. Years ago, the overwhelming distributed file system (i.e., GFS [5]) is published by Google with extreme successful applications and wildly deployed across the world with its cloned open source version (i.e., HDFS [7]). This distributed file system has a master/slave structure, which simpli-fies the implementation to a great extend. The master (i.e., Namenode) which manages the meta data of the massive data stored in the system is the portal responding any clients issued data access requirements, while the slaves (i.e., Datanodes) physically store data in the file fashion consisting of blocks across nodes in the distributed systems.

Within a file, two physical models (i.e., row-store, column-store) can be selected to store relational data tuples which are presented in many logical models (e.g., n-ary model, binary model [12], hybrid model). N-ary is the most straightforward approach to express the relational data logically, which is widely used in most of the traditional relational database systems. It stores data tuples in a wide horizontal table, each column of which is an attribute of the data tuple. The binary model is called the decomposed storage model (DSM) and proposed in 1985, which projects the original table into a set of binary columnar projections. One column is a surrogate which identifies the tuples in the original table, and the other is one attribute from the original table. What should be noticed is that the binary model decouples the logical model and physical storage model. For physical storage of binary model, it is stored in a column-wise fashion (i.e., one column followed by another). C-store [13] is an extension to the binary model, which has a similar column-wise physical storage but may have more than two attributes. The hybrid model combines the n-ary model and the binary model, and widely deployed by Bigtable [14], HBase [15] and ect.

In this paper, we propose a novel data placement strategies based on the popularly deployed distributed file system (e.g., GFS, HDFS), which achieves the enhanced sequential access efficiency for massive data analysis.

3   Problem Statement

The data model supported in our system is the standard relational logical data model, which is the same as that used in the C-Store [13]. As for the self-containment of the paper, we give a brief introduction to this data model. The attributes in tables, denoted as logical tables (e.g., $T_i, 0 < i \leq n_t$), in our system can form a unique primary key or be a foreign key that references to the primary key in another table. The logical table may be massive in volume, which needs to be partitioned and stored across nodes in the system. Basically, two types of partitions are applied, i.e., vertical partition and horizontal partition. The vertical partition here means *projection*, denoted as $P_j^i, 0 < j < n_i$, which is physically stored on the disk. To form a projection, we project the attributes of the interest from a logical table (called anchored table), retaining any duplicate rows, and perform the appropriate sequence of value-based foreign-key joins to obtain the attributes from the non-anchored table(s). Hence, a projection has the same number of rows as its anchored table. As a projection may still too large in volume for convenient maintenance, we further horizontally partition the projection P_j^i into set of *segments*, denoted as $S_k^{i,j}, 0 < k < n_{i,j}$. Each segment $S_k^{i,j}$ is finally physically stored as a file in the distributed file system (e.g., GFS, HDFS) with a pre-configured specific number of sequential blocks. Data tuples within a segmentation are physically stored in a column-wise fashion, i.e., attributes within the segment are stored one after another sequentially. As the capacity limitation of the block in HDFS (e.g., 64MB, 128MB), the number of attributes in one projection is usually constrained to be blew 5 with the goal to enlarge the number of tuples that can be filled in one block, which may further result in the enhancement of sequential access on one column (or attribute).

The problem to be solved in this paper is to provide an elegant data placement to enhance the data access performance. Herein, the data placement includes both index placement and data tuples placement. Leaning on the proposed data placement, it can benefit diverse massive data analyses (e.g., aggregation query, approximate aggregation query) and queries processing (e.g., exact match query, range query, partial match range query) in a manner of efficient massive data access. As discussed in section 1, the seek time and bandwidth of current hard disk drivers become the bottleneck of massive data access. Herein, we mainly focus on the placement which is able to reduce the ratio of random access and achieve the bandwidth-efficient transmission during the massive data access.

4   Data Placement

In this section, we present the data placement algorithms in detail including both data tuples and index placement, which are hopefully in enhancing the performance (e.g., data access efficiency, especially for the ratio of sequential access) of diverse query evaluations (e.g., partial match query, range query, aggregation query and etc), which will be discussed further in section 5.

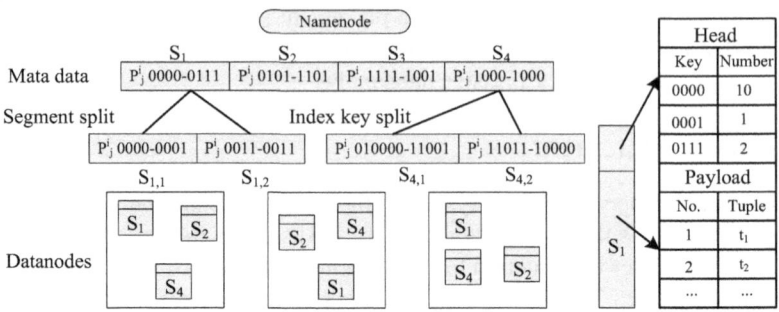

Fig. 2. The Architecture of the Storage

4.1   An Overview

As presented in section 3, each table T_i is projected into a set of projections, denoted as P_j^is, in terms of the data organization interests, each of which is further consists of a set of attributes $\{A_k^{i,j}\}$. Without loss of generality, we provide a detailed overview on data placement with respect to the projection P_j^i, assuming it consists a set of attributes $\{A_k^{i,j}\}$, where $0 < k < n_{i,j}$.

In consistent with the popularly deployed distributed file systems (e.g., HDFS), our index placement employs a master/slave structure, as shown Fig. 2. The master (e.g., Namenode) stores the content-aware index generated based on the Gray Code [8] encoding, and consequently redirects the requests for specific data segment access (i.e., a centralized search). The slaves (i.e., Datanodes) physically store the partitioned segments in a distributed manner in terms of the corresponding index sequence. Each segment has a head/payload structure. For the head, it stores the statistics information of stored tuples within the segment, while for the payload, it stores the data tuples ordered by the corresponding index sequence. Next, we give our detailed data placement strategies.

4.2   Content-Aware Bitmap Index Key Generation

The index key proposed in the paper has functionalities in two folds, one of which is to indicate the multi-dimensional content of the data tuple it refers to (i.e., property of bitmap index), and the other of which is to identify the location of the data tuples in the corresponding projection. Basically, the index key is constructed based on the index code of each attribute value in the data tuple. Next, we focus on the index code generation for each attribute value followed by the index key construction for the tuple.

For each attribute $A_k^{i,j}$, no matter numerical or categorical, we equally partition its normalized domain into a specific number of intervals (e.g., 8,16,32, and etc). After the partition, each interval is encoded under the Gray Code [8] sequentially. Fig. 3 shows an example, where the normalized domain (i.e., $[0,1]$) of attribute $A_k^{i,j}$ is partitioned into 8 equal intervals, each of which is encoded under the Gray Code sequence. Given a normalized value, $v_{l,k}^{i,j}$ on the attribute $A_k^{i,j}$, it can be indicated by its index code that is the Gray Code in which interval the value fills, denoted as $ic_{l,k}^{i,j}$. Take normalized value

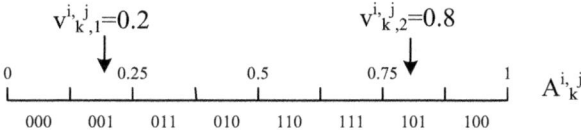

Fig. 3. Attribute Domain Partition

$v_{1,k}^{i,j} = 0.2$ and $v_{2,k}^{i,j} = 0.8$ in Fig. 3 as an example. As fill in the second and the seventh intervals irrespectively, they can be denoted by Gray Code 001 and 101 according.

With the same index code computation algorithm, each tuple $t_l^{i,j}$ in projection P_j^i is able to generate its index code on every attribute $A_k^{i,j}$ in terms of its individual attribute value, say $ic_{l,k}^{i,j}$, where $0 < k < n_{i,j}$, and $n_{i,j}$ is the number of attributes in projection P_j^i. After all these index codes have been generated, the index key, denoted as $ik_l^{i,j}$ for tuple $t_l^{i,j}$ is generated by concatenating them together in a shuffle-based manner. The so called shuffle-based concatenation manner is to concatenate one bit of index code of one attribute followed by one bit of another in a pre-defined concatenation sequence. Let's take tuple $t_l^{i,j}$ as an example, which has index codes "011", "010" and "101" on attributes $A_1^{i,j}$, $A_2^{i,j}$ and $A_3^{i,j}$ irrespectively. The index key for tuple $t_l^{i,j}$ is "001110101", which is concatenated by one bit of index code on attribute $A_1^{i,j}$ followed by one bit on $A_2^{i,j}$ as well as one bit on attribute $A_3^{i,j}$ after that and so forth.

Theorem 1. *The shuffled-based concatenation is a one-one map index key generation function.*

Given by the Theorem 1, the index key has the capability in indicating the content of the data tuples as its index codes do.

4.3   Index Construction

Armed with the content-aware index key discussed in the section 4.2, data tuples could be easily accessed. In this subsection, we focus on the index construction for all the tables stored in the system in a cost-effective manner.

The popular and widely deployed distributed file systems (e.g., HDFS [7], GFS [5]) have a master/slave structure, which hopefully simplifies the implementation of the file system. At the master side, it stores the meta data (i.e., file name, location, size, duplications and etc) and at the slave side, it stores the payload data (i.e., the actual data need to be stored). Based on the master/slave structure of the distributed file system, we proposed a two level index deployment, one of which is at the projection level and the other of which is at the segment level.

As discussed in section 3, each projection P_j^i is horizontally partitioned into a set of segments, each of which is stored as a file and allocated a predefined number of sequential blocks (e.g., 1 block, 2 blocks and etc) in the file system. Relying on the storage structure of the distributed file systems, the projection level index, which indicates the content of each segment within the projection, can be stored combining with the file name in the meta data without any additional storage cost. As proposed, data tuples within the segment (or projection) are stored sequentially in terms of their corresponding index keys in the Gray Code order (i.e., the index keys for data tuples in a segment are bounded within a sequential range). Therefore, using *start-end* key of the index key range, we are able to indicate the content of the data tuples stored within the segment. We use the segment (projection) name combining with the *start-end* key as a file name stored in the meta data region, which enables the index capability. As shown in Fig. 2, the file name of the segment (e.g., S_1) are stored as the meta data in the Namenode, including the projection id (i.e., P_j^i) and the *start-end* key (i.e., 0000-0111) which indicates the content of the data tuples within the segment.

To release the burden (i.e., storage, access and etc) of the Namenode, we store the segment level index across all the Datanodes which indicates the statistics information for data tuples within a segment. Two basic components construct the segment level index, which are stored in the head of each segment. One is the index key for which it has corresponding data tuple stored in the segment, and the other is the data tuple frequency for the corresponding index key in the segment. Basically, the segment level index has two functionalities, one of which is to indicate the number of the data tuples having the given index key, and the other of which is to indicate the location of the first data tuple for each index key in the segment. Take the head of S_1 in Fig. 2 as an example. The second element (i.e., $\{0001,1\}$) means there is 1 tuple that has index key 0001 and starts from the 11th tuple in the corresponding segment.

Being a centralized/distributed structure, the two level index structure has many advantages, 1) it is storage cost-efficient because using the existed meta data to store the content-aware index keys, 2) it has a concisely centralized structure which simplifies the location positioning in terms of the content of the index key, 3) it is much more scalable because the heavy loaded statics information is distributed all over the Datanodes.

4.4   Data Placement

Without loss of generality, the data placement can be presented as the progress of a tuple $t_l^{i,j}$ inserting into projection P_j^i, which consists one or more segments and physically stores in the distributed file systems (e.g., GFS, HDFS, etc). By generating its corresponding index key first, a data tuple is able to insert into a target segment consequently, which has been assigned and allocated with a specific number of blocks at the begin of segment creation. Function $TupleInsert()$ in Algorithm 1 shows this in detail, where issues three levels of tuple insertion (i.e., $ProjectInsert()$, $SegmentInsert()$ and $BlockInsert()$). During $SegmentInsert()$, it finds the target segment which has a index range covering the index key of the tuple to be inserted and invokes Function $SegmentCheck()$, which checks the segment whether needs to be split or not. The detailed segment split algorithm will be present in section 4.5. During $BlockInsert()$, it finds the target block to insert and adjusts the index in the segment level. Finally, the tuple finds the target position in the block to insert, and consequently result in sequentially stored in terms of tuple index keys in the segment.

The efficiency of the single tuple insertion algorithm proposed above could be enhanced, because it needs one random disk access for one tuple to be inserted. Bulk tuple insertion is also supported in our system, which enhances the disk access efficiency.

Algorithm 1. TUPLEINSERTION

ProjectionInsert(projection p, tuple t)

1: **for** k from 1 to n **do**
2:     ic_k=GenerateIndexCode($t.v_k$) //tuple t contains n attributes, generates each index code
3: **end for**
4: ik=GenerateIndexKey(ic,k)
5: SegmentInsert(t,ik)

SegmentInsert(tuple t, projection p, key ik)

1: $fileName$=FindSegment(ik) // $segment.start - segment.end$ covers ik
2: $newFileName$=SegmentCheck($fileName, ik$)
3: BolckInsert($newFileName, t, ik$)

BlockInsert(segment $fileName$, tuple t, key ik)

1: AdjustSegmentLevelIndex($fileName, ik$)
2: TupleInsert(t)

SegmentCheck(String $fileName$, key ik)

1: **if** Segment($fileName$) = full && $fileName.start <> fileName.end$ **then**
2:     BlockSplit($fileName$)
3: **else if** Segment($fileName$) = full && $fileName.start = fileName.end$ **then**
4:     IndexKeySplit($fileName$)
5: **end if**
6: **return** $newFileName$=FindSegment(ik)

BlockSplit(String $fileName$)

1: ik' = GetIndexKey($fileName, segmentLength/2$)
2: NewSegment($fileName, fileName.start, ik'$)
3: NewSegment($fielName, ik' + 1, fileName.end$)

IndexKeySplitSting $fileName$

1: $newFileName$=DoubleIndexRange($fileName$)
2: AdjustAllSegmentIndex($newFileName$)
3: AdjustIndexkey($fileName$)
4: BlockSplit($newFileName$)

For bulk insertion, data tuples to be inserted are pre-sorted in the memory on the Datenode which has a segment covering their corresponding index keys, before doing the block insertion. For example, supposing three tuples $t_1^{i,j}$, $t_2^{i,j}$ and $t_3^{i,j}$ with their corresponding index keys "0110", "1000" and "0001" irrespectively, the three tuples are pre-sorted in the memory on a Datenode under the order of $t_3^{i,j}$, $t_1^{i,j}$ and $t_2^{i,j}$. By exceeding the bulk insertion threshold, those pre-sorted tuples will merged into the specific segment. Merge-sort algorithm is deployed for bulk tuple insertion, where tuples pre-sorted in the memory and tuples in the segment are separated in two runs in terms of the index key range. When at the merge phase, the separated two runs are merged together to form a new segment.

4.5 Segment Split

In this subsection, we describe the segment split algorithm in detail. When the blocks pre-allocated for a segment are fully stored, the segment will split into two when new

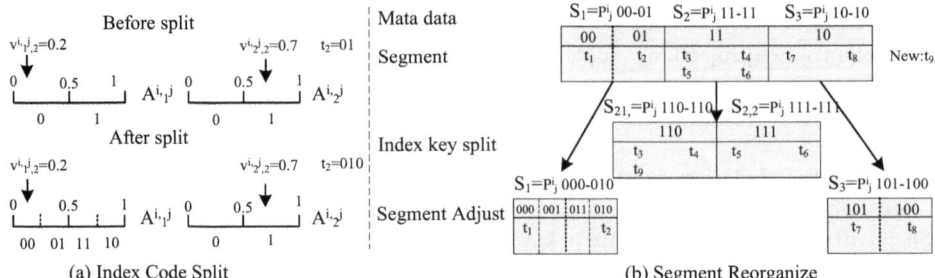

Fig. 4. Index Key Split

tuples need to insert. Generally speaking, there are two stages of segment split, which are denoted as *block split* and *index key split*.

If a segment into which a new data tuple should be inserted in terms of its index key is already full (blocks full), and its covered index key range has more than one index keys (i.e., the range can be split into two ranges), the block split issues (i.e., function $SegmentCheck()$ in Algorithm 1). For block split, it finds the mid index key (e.g., ik' of the data tuple in function $BlockSplit()$), which is located in the middle of the segment. Two new segments will be created in terms of the index key (e.g., one for range $[fileName.start, ik']$, and the other for $[ik' + 1, fileName.end]$). Data tuples will move to the corresponding segment sequentially.

If a segment into which a new data tuple should be inserted in terms of its index key is already full (blocks full), while its covered index key range has only one specific index key, the index key split takes place. As the range which covers only one index key can't be split further, it needs to be doubled first by extending one bit to its right end. In terms of the index key generation mechanism, we need to repartition the normalized domain of one selected attribute into a doubled intervals, and consequently result in the regenerated index keys having one more bit. Fig. 4(a) shows an example. When the index key split issues, the attribute $A_1^{i,j}$ is selected for repartition in terms of the index key concatenation sequence. The index code on $A_1^{i,j}$ for tuple t_2 is split from 0 to 00, and consequently result in its index key split from 01 to 010 as shown in the figure. Similarly, all the tuples within segment S_2 have to regenerate their corresponding index keys. After the regeneration, the fully stored segment will be split into two, which is the same as that do in block split. What should be noticed is that the Gray Code based index key split won't alter the tuples in which segment they store (e.g., t_1, t_2 are still stored in S_1 after index key split in Fig. 4(b)), even with similar sequence with the original segment, because of the existence of Theorem 2.

Theorem 2. (Extendibility) *The Gray Code based index key is extendable. Here, extendibility means the p-bit prefix of the new split $(p + 1)$-bit index key $((ik_l^{i,j})')$ equals to its p-bit index key $(ik_l^{i,j})$ before split (say $pre_p((ik_l^{i,j})') = ik_l^{i,j}$), where pre_p refers to the p-bit prefix.*

Armed with Theorem2, our data placement algorithms achieve the cost-efficient data insertion, say without any movement for tuples in the all segments which block split are not issued.

5 Query Processing

In this section, we present the query types supported in our system. Generally speaking, it supports efficient exact match, partial exact match, range query, partial range query, approximate aggregation query, aggregation query, projection join and etc.

For the purpose of clear presentation, all the descriptions for diverse query evaluations are based on the following projection, each tuple within which contains two numerical attributes (i.e., A_1 and A_2) both with continuous domain $[0, 10]$. The continuous domains (i.e., $[0, 10]$) are normalized into $[0, 1]$ for index key generation and segment storage.

5.1 Multi-dimensional Range Query and Multi-attribute Range Query

The multi-dimensional range query and multi-attribute range query have the similar definitions with respect to the exact match and partial exact match queries, expect for the query range. For example, we may have multi-dimensional range query $Q_1 : 1 \leq A_1 < 5, 6 < A_2 \leq 10$ and multi-attribute range query $Q_2 : 1 \leq A_1 < 5, A_2 = any$. Similarly, data tuple will be fetched in terms of the generated index keys. For example, we have $ik_{Q_1} = 01xx$ and $ik_{Q_2} = 0xxx$. Post filtering is also needed, but solely on the border of the query range.

5.2 Aggregation Query and Approximate Aggregation Query

We support all the commonly used aggregation operations (e.g., *max, min, average, sum, count*) in our system, while the approximate aggregation query means to get the approximate results but with high efficiency. Here, the approximate aggregation evaluation is mainly based on our proposed content-aware index. Let's take the operator *sum* as an example. For approximate result, it calculates a set of index keys that need access first, and consequently fetches the heads of the corresponding segments. In terms of the index key (i.e., content-aware) and the data tuple frequency, the approximate results can be derived by doing the multiply operation. Achieving the exact aggregation results, we have to fetch all the data tuples in the corresponding segments and make summation tuple by tuple.

Our novel data placement strategies achieve the efficient data access, especially for the massive data access by reducing the ratio of random access during one access requirement to a great extend.

6 Performance Evaluation

To evaluate the performance of our proposed data placement methods, we implement it on Hadoop with version 0.19.1, which an open source version from Yahoo!.

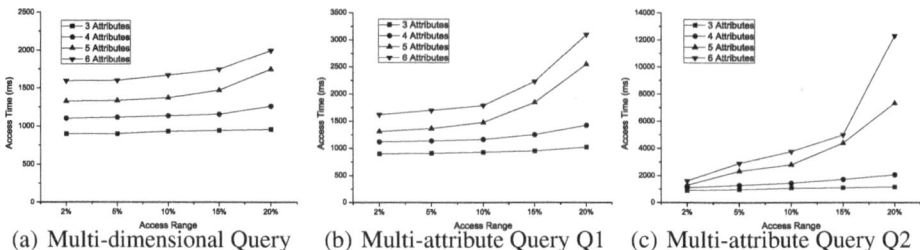

| (a) Multi-dimensional Query | (b) Multi-attribute Query Q1 | (c) Multi-attribute Query Q2 |

Fig. 5. Access Efficiency for Diverse Queries

The Hadoop is running on a cluster consisting of 22 blades, each of which has two quad-core CPUs, a 4GB memory, one 144GB hard disk driver, and runs RedHat Linux AS4. TPC-H is the benchmark for data analysis, which is used here for performance testing. 100 million data tuples are generated and stored in terms of our data placement strategy in the HDFS. Data access efficiency and approximate data aggregation are extensively tested in terms of the accessing time and aggregation accuracy irrespectively.

6.1   Evaluation on Access Efficiency

In this subset of experiments, we evaluate the access efficiency for multi-dimensional range query and multi-attribute range query. 4 bits of index code is generated here for each attribute. The range queries with range 2%, 5%, 10%,15% and 20% on all attributes (Fig. 5(a)), all attributes except one (Fig. 5(b)), one attribute (Fig. 5(c)) are issued on projections with 3,4,5,6 attributes irrespectively, where the ranges could arbitrary for not constrained attributes. As shown in Fig. 5, the data access time increases slightly when the access range increases, since many random accesses are reduced. Moreover, the access time also increases when the number of attributes increases, especially for the multi-attribute range query, since more tuples will be access if less constrains on attributes are issued.

6.2   Evaluation on Aggregation Accuracy

Variable numbers of bits for index codes affect the number of tuples having the corresponding index key. That's to say the more bits used for content indication, the more precious it is, and result in more accurate approximate aggregation evaluation, as shown in Fig. 6(a). The query range for aggregation also affects the approximate aggregation accuracy, since smaller ratio of tuples will be post filtered for larger query ranges. As for approximate aggregation evaluation, only indexes need to be fetched, and consequently result in low access time, as shown in Fig. 6(b).

6.3   A Comparison

A access efficiency comparison between Gray Code based data placement and binary code based is conducted here. For the binary encoding, it the use the same bits for index codes on each attribute, and concatenate each index code together sequentially.

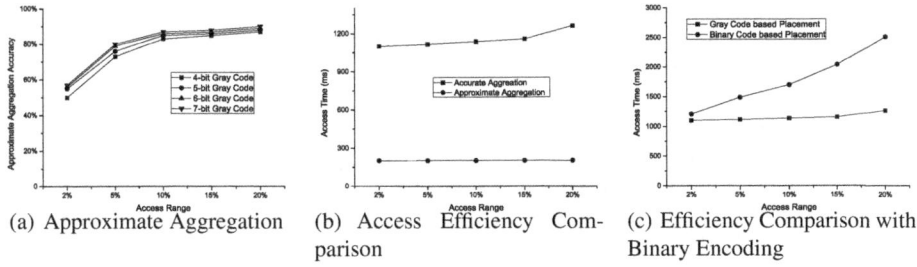

(a) Approximate Aggregation   (b) Access Efficiency Comparison   (c) Efficiency Comparison with Binary Encoding

Fig. 6. Comparisons

As shown in Fig. 6(c), nearly half of the access time can be reduce due to our data placement because of the enhanced ratio on sequential access, especially for lager volumes data access.

7   Conclusion

In this paper, we present a novel idea for data placement in the distributed environment. Our method achieves enhanced sequential access ratio and reduced data network transmission for diverse of query evaluations (e.g., exact match, partial exact match, range query, partial range query, approximate aggregation query, aggregation query, projection join and etc). Detailed theoretical analysis, together with extensive experiments prove that our methods are highly efficient and effective in the distributed systems.

Acknowledgement. This work is partially supported by Shanghai International Cooperation Fund Project under grant No.09530708400, Shanghai Leading Academic Discipline Project No. B412, Alibaba Young Scholars Support Program Fund under grant No. Ali-2010-A-12), National Science Foundation of China under grunt No.61003069, No.60833003 and No. 60925008, National Hi-Tech 863 program under grant No.2009AA01Z149.

References

1. Boncz, P., Zukowski, M., Nes, N.: MonetDB/X100: Hyper-pipelining query execution. In: Proceeding of CIDR 2005 (2005)
2. Vertica, "Vertica" (2008), http://www.vertica.com
3. Olofson, A.W.: IDC Excerpt Worldwide Database Management System 2009-2013 Forecast and, Vendor Shares. Technical Report 219232E (October 2008)
4. Gray, J.: A Conversation with Jim Gray. ACM Queue 1(4) (2003)
5. Ghemawat, S., Gobioff, H., Leung, S.: The Google file system. In: Proceedings of SIGOPS 2003, pp. 29–43 (2003)
6. Dean, J., Ghemawat, S.: MapReduce: Simplified Data Processing on Large Clusters. In: Proceedings of OSDI 2004 (2004)
7. Yahoo!, "Hadoop Distributed File System" (2008), http://hadoop.apache.org/hdfs/

8. Gray, F.: Pulse code communications. U.S. Patent 2632058 (1953)
9. Howard, J., et al.: An overview of the andrew file system. In: Proceedings of the USENIX 1988, pp. 23–26 (1988)
10. Kistler, J., Satyanarayanan, M.: Disconnected operation in the Coda file system. ACM Transactions on Computer Systems 10(1), 25 (1992)
11. Nelson, M., Welch, B., Ousterhout, J.: Caching in the Sprite network file system. ACM Transactions on Computer Systems 6(1), 134–154 (1988)
12. Copeland, G., Khoshafian, S.: A decomposition storage model. In: Proceedings SIGMOD 1985, pp. 268–279 (1985)
13. Stonebraker, M., Abadi, D., Batkin, A., Chen, X., Cherniack, M., Ferreira, M., Lau, E., Lin, A., Madden, S., O'Neil, E., et al.: C-store: a column-oriented DBMS. In: Proceedings VLDB 2005, pp. 564–275 (2005)
14. Chang, F., Dean, J., Ghemawat, S., Hsieh, W., Wallach, D., Burrows, M., Chandra, T., Fikes, A., Gruber, R.: Bigtable: A distributed storage system for structured data. ACM Transactions on Computer Systems 26(2), 4 (2008)
15. Yahoo!, "HBase" (2008), http://hbase.apache.org/

Two-Step Joint Scheduling Scheme for Road Side Units (RSUs)-Based Vehicular Ad Hoc Networks (VANETs)

G.G. Md. Nawaz Ali[1], Edward Chan[1], and Wenzhong Li[2]

[1] City University of Hong Kong, Kowloon, Hong Kong
[2] Nanjing University, Nanjing, China

Abstract. Recently, the use of Road Side Units (RSUs) has been proposed as a mechanism to handle the connectivity issues in VANETs for data dissemination. In this paper, we provide a model where an RSU deals with both download and upload queues. In VANETs, since vehicles are highly mobile, if as RSU fails to receive the updated information from a vehicle, all the subsequent vehicles receive the stale data from that RSU which substantially decreases the main objective of data dissemination. To find an efficient data dissemination procedure in this circumstances, we propose a second-step scheduling algorithm to form a two-step joint scheduling algorithm in where as the first-step scheduler we use existing on-demand real-time algorithm. We study the performance of a number of different joint scheduling algorithms by varying different on-demand scheduling algorithms as first-step scheduler using simulation experiments with various parameter settings and high workload. Finally, we recommend which two-step joint scheduling algorithm is suitable in this RSU-based VANETs environment.

Keywords: VANETs, Road Side Unit (RSU), on-demand scheduling algorithm, on-demand broadcast etc.

1 Introduction

Data dissemination in Vehicular Ad Hoc Networks (VANETs) received considerable attention by the researchers from the past decade. In VANETs, as many vehicles may request the same data item, so broadcasting is a popular approach for data dissemination. Recently, researchers have proposed the use of Road Side Units (RSUs) for supporting on-demand data broadcasts, particularly where strict time constraint is involved.

However, in such a scenario, when many vehicles need to upload and download data in the same RSU, an efficient scheduling strategy is required. Time constraint is an important issue here, because an RSU's transmission range is not large, and failure in reaching the client while the vehicle is in range will result in wasted transmission; similarly poor scheduling might prevent updates regarding time-sensitive data that are useful to other vehicles from being uploaded in time. For example, if a vehicle has observed a road accident while it

J. Xu et al. (Eds.): DASFAA Workshops 2011, LNCS 6637, pp. 453–464, 2011.
© Springer-Verlag Berlin Heidelberg 2011

approaches an RSU, it then can provide this information to that RSU. The RSU updates its database and provides this updated information to other vehicles, and upon getting this information these vehicles may change their routes or take appropriate actions.

A number of researchers have studied scheduling issues recently. Nadeem et al.[1] use periodic broadcast approach for data dissemination. Yi et al.[5] consider reliability and fairness of information distribution among Mesh Road Side Units (MRUs) but their approach does not deal with the strict time constraints of VANETs data dissemination from RSUs to vehicles. Zhang et al. [2] consider both upload and download services and try to balance the adaptivity of these two services according to the fluctuation of workload but they do not maintain the time constraint and update the database with most updated information.

We study the performance of different scheduling algorithms for both upload and download requests considering different constraints. In this paper our main contributions are:

- We propose a second step scheduling algorithm to form a two-step joint scheduling algorithm for selecting the most appropriate request from the upload and download queues.
- We apply different on-demand scheduling algorithms as first step scheduling algorithms with our proposed second-step algorithm to build different two-step joint scheduling algorithms aiming to find an efficient one.
- After analysis the simulation result and performance comparison, we recommend which two-step joint scheduling algorithm is best suited for getting the maximum performance in highly mobile and strict time constraint VANETs environment.

The rest of this paper is organized as follows. Section 2 surveys related work. Section 3 describes about our system model and preliminaries, section 4 shows the scheduling schemes with our proposed scheduling algorithm and section 5 describes the simulation model and experimental results. Finally, we conclude with the a discussion of our results and future work.

2 Related Work

Unlike unicasting, broadcasting maximize the channel bandwidth utilization because by a single broadcast many outstanding requests can be served. However, to get the maximum benefit from the broadcast, a suitable scheduling algorithm is needed. A number of push based model have been proposed by researchers. Wong and Ammar [6] investigate the First Come First Serve (FCFS) algorithm in videotex systems. Acharya et al. [7] introduce asymmetric communication environments where downstream link has greater capacity than downstream link. Other researchers propose pull based (also known as on-demand) model. Wong [8] uses the Longest Wait First (LWF) algorithm to find the next item for scheduling. Aksoy and Franklin [9] propose the $R \times W$ algorithm for large scale on-demand data broadcast, which incorporate popularity and request urgency for making scheduling decision. For heterogeneous workload Acharya and Muthukrishnan [10] introduce a new metric called stretch which is the ratio of response time to the service time and a corresponding algorithm called Longest Total Stretch First (LTSF). [3] proposes Slack Time Inverse number of Pending requests (SIN) for time critical on-demand broadcast. Chen et al. [4] introduce Preemptive Temperature Inverse Slack Time (PTIS) for handling multi-item data requests.

None of the above work considers the scheduling issue for both upload and download requests along with client mobility and strict time constraints.

3   Background and Preliminaries

3.1   System Model

In our model we assume that VANETs services are provided to the vehicles at the hot spot such as gas stations or intersection of the roads where the number of vehicles is higher. When a vehicle is in the transmission range of an RSU it can generate either upload or download requests. A download request means a vehicle wants the latest updated data item from the RSU server and upload means vehicle wants to upload the updated information of a data item to the RSU server. An RSU has two queues as shown in Fig. 1, one for upload requests and the other for handling download requests from the vehicles. A vehicle can generate request or receive response only until it is within the transmission range of an RSU.

3.2   Notation and Assumptions

Request: We denote each request i by 10 tuples as follows:
$R_i = (NO_i, ID_i, SIZE_i, TYPE_i, T_i^{in}, T_i^{out}, T_i^r, T_i^{stamp}, T_i^{deadline}, T_i^{serv})$.

NO_i: the number of the request;
ID_i: the ID of the data item it requests;
$SIZE_i$: the size of the data item;
$TYPE_i$: taking values in { *upload, download* }, indicating the type of uploading/downloading operations;
T_i^{in}: the time the vehicle enters the communication range of the RSU;
T_i^{out}: the time the vehicle leaves the communication range of the RSU;
T_i^r: the time the request is generated;
T_i^{stamp}: the time the updated information is generated by a vehicle;
$T_i^{deadline}$: the deadline assign by a request, beyond this time the request will be dropped;
T_i^{serv}: the time for uploading/downloading the data item, it can be evaluated by $SIZE_i$ divided by the available bandwidth;
Assume there are n requests. The set of requests is denoted by $REQ = \{R_1, R_2, \cdots, R_n\}$.

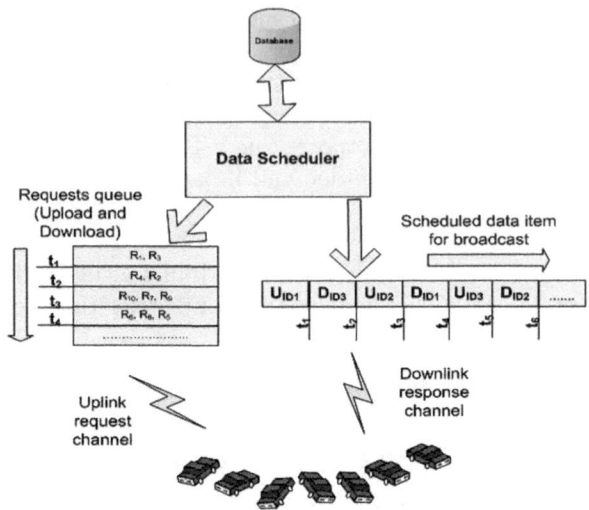

Fig. 1. Scheduling in an RSU

Schedule: When a vehicle submits a request, the request needs to be scheduled. Assume at time t, a set of requests R^t reside in the RSU received queue to be scheduled. The schedule to the requests should follows the following principles. First, since each request needs to occupy the communication channel for data transmission, it should makes sure that the uploading/downloading operation finishes before the vehicle moves out of the communication range. Second, since the uploading operation will update data content, if there exist both uploading and downloading requests to the same data, the uploading request should be served first to avoid the downloading of a stale data item.

If a request $R_i \in R^t$ is scheduled to be served at t, we call R_i is *satisfiable* if it meets the following conditions: (1) $t \geq T_i^{in}$ & $t + T_i^{serv} \leq T_i^{out}$; and (2) there is no uploading request to data ID_i in R^t. If either of them is violated, we called it *unsatisfiable*. We use $(R_i, t, R^t)_*$ to denote its satisfiability.

Due to the broadcast nature of wireless communication, for the downloading operation, when the data is broadcast, a set of requests waiting for the same data can be satisfied at the same time. We call such set of requests *shareable requests*. If a downloading request $R_i \in R^t$ is scheduled at time t, its shareable requests set is denoted as $SA(R_i, t, R^t)$, which can be defined as:

$$SA(R_i, t, R^t) = \{R_j | \forall j \neq i, R_j \in R \ \& \ ID_j = ID_i \ \& \ (R_j, t, R^t)_* = satisfiable\}$$

A schedule can be expressed as a sequence of the requests and their scheduling time. For example, assume the request set $R = \{R_1, R_2, R_3, R_4, R_5, R_6\}$. A possible schedule is $S = < R_2, t_1 > (R_3, R_5) \rightarrow < R_6, t_2 > () \rightarrow < R_1, t_3 > (R_4)$, which indicates R_2 is served at t_1, R_6 is served at t_2, R_1 is served at t_3, and the items in () are shareable requests being served at the same time.

Request's Life Time: A vehicle can generate request only within time range $[T_i^{in}, T_i^{out} - T_i^{serv}]$, where $T_i^{serv} = \frac{DataItemSize}{ChannelBandwidth}$.

Assume the radius of the transmission range of an RSU is R meter and average speed of vehicle within the transmission range of RSU is S meter/sec. So, if a vehicle reach at the transmission range of an RSU at time $t = 0$ and generate the first request at the same time, then the average deadline of the first request of a vehicle is, $AVERAGE\_DEADLINE = \frac{2R}{S}$. If we consider the speed variation factor of a vehicle from the average speed, the compensated deadline is: $DEADLINE = \frac{2R}{S} * uniform(\psi_{max}, \psi_{min})$; where ψ is the random number use for compensate the speed variation.

So, anytime T, the deadline of a request is

$$DEADLINE = \frac{2R}{S} * uniform(\psi_{max}, \psi_{min}) - T_i^r$$

where T_i^r is request generation time.

Suppose, the request i wants to update the server data item ID_i and the server's and vehicle's data item time stamp are $ID_{Server}^{T^{stamp}}$, $ID_i^{T^{stamp}}$ respectively. Then, the update request i will be received by the upload queue if and only if, $ID_i^{T^{stamp}} < ID_{Server}^{T^{stamp}}$, i.e. upload queue will accept those requests only which have recent information. However, any type request i will be discarded from the scheduler when $Current\_Clock > T_i^{deadline}$.

4 Scheduling Schemes

In a traditional single queue scheduling system, an on-demand scheduling algorithm is used to find the next suitable request for providing service. In systems with two queues, we have two suitable requests from the two queues, among which one is to be selected for providing service in the next service cycle. To find the most two appropriate requests from both queues we use an existing on-demand scheduling algorithm, we call this step **First-step scheduling**. We call the final selection of one request from these two queues **Second-step scheduling**.

4.1 First-Step Scheduling

For performing the first-step scheduling we use an on-demand scheduling algorithm to find the best candidate in a queue according to the selecting principle of the algorithm. For doing this selection procedure, our system calculates Upload_First_step_value and Download_First_step_value for each upload and download request respectively according to the principle of the used first-step scheduling algorithm. From this First_step_value, our system finds the best candidate from each queue.

To find the best performance of the two-step scheduling system, we need to find a good combination of first-step and second-step scheduling. We adopt the following on-demand scheduling algorithms as first-step schedulers in our system with our proposed second-step scheduling algorithm and then proceed to compare and analyse their performance.

1. **First Come First Served (FCFS):** This base line schedular selects the request according to the request arrival order. Using FCFS, First_step_value of a request i is: $First\_step\_value\_i = AT_i$; where AT_i is the arrival time of request i. Hence, the request $R_i \in R^t$ is selected as a best candidate from a queue at time t using FCFS principle can be defined as: $Req(R_i, t, R^t) = \{R_j | \forall j \neq i, R_j \in R \ \& \ max(AT_1, AT_2, ..., AT_j)\}$.

2. **Most Request First (MRF):** It selects requests according to the popularity of the requested data item. An data item in the database for which maximum number of requests waiting in the waiting queue will be selected for service. Once this data item is broadcast all shareable requests waiting for that data item will be satisfied. Here, First_step_value of request i is: $First\_step\_value\_i = Popu_{ID_i}$; where $Popu_{ID_i}$ is the current popularity of the data item requested by request i. So, the request $R_i \in R^t$ is selected as a best candidate from a queue at time t using MRF principle defined as: $Req(R_i, t, R^t) = \{R_j | \forall j \neq i, R_j \in R \ \& \ max(Popu_{ID_1}, Popu_{ID_2}, ..., Popu_{ID_j})\}$.

3. **Earliest Deadline First (EDF):** EDF selects requests according to the deadline of the requests. When a vehicle generates a request it calculates its deadline from the request generation time, vehicle speed and the transmission range of the RSU. As time passes the request's assigned deadline decreases. When the deadline reaches 0, the request will be discarded by the RSU. For EDF, First_step_value of request i is: $First\_step\_value_i = Deadline_i$; where $Deadline_i$ is the current deadline of request i. Here, the request $R_i \in R^t$ is selected as a best candidate from a queue at time t: $Req(R_i, t, R^t) = \{R_j | \forall j \neq i, R_j \in R \ \& \ min(Deadline_1, Deadline_2, ..., Deadline_j)\}$.

4. **Deadline Size Inverse Number of pending Requests(DSIN):** DSIN algorithm incorporates the deadline of the request, size and popularity of the requested data item. The request which has the tightest deadline and its requested data item has high popularity with small size will be selected from a queue as the best candidate in the first-step scheduling. Here, First_step_value of request i is determined by: $First\_step\_value_i = DSIN_{value\,i}$; where $DSIN_{value\,i}$ is the current $DSIN_{value}$ of request i and

$$DSIN\_Value = \frac{Deadline \times Size \ of \ the \ requested \ data \ item}{Number \ of \ pending \ requests \ of \ that \ data \ item}$$

The request $R_i \in R^t$ selected as the best candidate at time t can be defined as: $Req(R_i, t, R^t) = \{R_j | \forall j \neq i, R_j \in R \ \& \ min(DSIN_{value\,1}, DSIN_{value\,2}, ..., DSIN_{value\,j})\}$.

4.2 Our Proposed Scheduling Algorithm

After first-step scheduling, we get the requests in front of both the upload and download queues to be the best candidates in the respective queues for being served. The job of our proposed second-step scheduling algorithm is to find the final single request from the two queues to be served, as shown in Algorithm 1. The second-step scheduling algorithm compares the two non-empty

---

Algorithm 1. Second-step Scheduling Algorithm

Require: UQueue[] and DQueue[] are upload and download queues respectively having best selected candidate at the front after first-step scheduling;
Require: UIndex and DIndex are the current maximum queue size of upload and download queue respectively;
Require: i, j, and SelectedDataItem for keeping track of UQueue and DQueue and final selected data item respectively;
Require: i := 0, j :=0 and SelectedDataItem := -1
Require: Scheduled := **false**

  1.
  2. **if** $UQueue[UIndex] = \phi$ **and** $DQueue[DIndex] = \phi$ **then**
  3.   **return** /*both queues are empty*/
  4. **end if**
  5.
  6. **if** $UQueue[UIndex] \neq \phi$ **and** $DQueue[DIndex] = \phi$ **then**
  7. $SelectedDataItem := UQueue[i]$; /*Download queue is empty and select the first request from upload queue*/
  8. **end if**
  9.
10. **if** $UQueue[UIndex] = \phi$ **and** $DQueue[DIndex] \neq \phi$ **then**
11. $SelectedDataItem := DQueue[j]$; /*Upload queue is empty and select the first request from download queue*/
12. **end if**
13.
14. **if** $UQueue[UIndex] \neq \phi$ **and** $DQueue[DIndex] \neq \phi$ **then**
15.   **if** $First\_step\_value\_UQueue[i] \leq First\_step\_value\_DQueue[j]$ **then**
16. $SelectedDataItem := UQueue[i]$; /*Both queues have requests and select from upload queue*/
17.
18.   **else**
19.   **for** $k = i$ to UIndex **do**
20.   **if** $UQueue[k] = DQueue[j]$ **then**
21. $SelectedDataItem := UQueue[k]$; /*Both queues have requests and download requested data item also has upload request*/
22. $Scheduled := $ **true**;
23.   **end if**
24.   **end for**
25.
26.   **if** $Scheduled = $ **false then**
27. $SelectedDataItem := DQueue[j]$ /*Both queues have requests and select from download queue as there is no upload request for the same data item*/
28.   **end if**
29.   **end if**
30. **end if**
31.
32. Provide service to **SelectedDataItem**;

---

queues front requests' First_step_value by the line $First\_step\_value\_UQueue[i] \leq First\_step\_value\_DQueue[j]$ (Line 15, Algorithm 1). If upload queue front request's value is less than or equal to download queue front request's, it selects the upload queue front request for providing service, but if the upload queue front request's value is greater than the download queue, before selecting the download request to be served it checks the upload queue to see whether there is any request in the upload queue which requests the same data item that is requested by that download request to ensure the fresh data to the download request. If there is no such upload request in the upload queue, it provides the service to the selected upload request, otherwise download request will get chance to get service from the RSU. This is the case if we use the first-step scheduling algorithm like EDF, DSIN etc.(which use minimum First_step_value to find the best candidate in a queue). But for first-step scheduling algorithm like MRF, FCFS etc.(which use maximum First_step_value to find the best candidate in a queue), we need to change the condition for comparison between two non-empty queues front requests' First_step_value by $First\_step\_value\_UQueue[i] \geq First\_step\_value\_DQueue[j]$ (Line 15). This condition will choose the request which has the greater First_step_value.

However, for both types first-step scheduling algorithms, if either queue is empty, it selects the front request from the non-empty queue (Line 6–12).

4.3 Performance Metrics

To find the best combination of first-step and seconds-step scheduler among different first-step (DSIN, EDF, MRF, FCFS) second-step joint scheduling algorithms, we adopt the following performance metrics.

1. **Deadline Miss Rate:** It is defined as the percentage of requests missing the deadline to the total number of requests received by an RSU.
2. **Throughput:** The number of requests successfully served by a scheduler per unit time. If a scheduler serves a popular data item, many requests waiting for that item are satisfied, hence throughput will increase.
3. **Average Response Time:** This is the average time needed to get the response from an RSU after a request has been submitted by a vehicle.

Our target is to find an efficient two-step join scheduler which can achieve low deadline miss rate, low average response time and high throughput.

5   Performance Evaluation

5.1   Experimental Setup

Our simulation environment is similar to Fig. 1. We assume the inter-arrival time of request is exponentially distributed. Our default request generation interval (IGTM) is 0.25. In our model, a data item can be requested by more than one vehicle, and a vehicle can request services until it exceeds the transmission range

Table 1. Simulation parameters

Parameter	Default	Range	Description
NumVehicle	100	25 - 300	Number of vehicles
IGTM	0.25	0.1 - 1.0	Request generation interval of each vehicle
DBSize	500	–	Number of data items in the database
DownItemsize	–	10 - 512 K bytes	Size of each download data item
UploadItemsize	–	5 - 256 K bytes	Size of each upload data item
BrodcastBandwidth	100	–	Channel broadcast bandwidth K bytes/s
RSU Range	350 m [5]	350 - 400 m	RSU communication range
DataItemDistribution	RAND	INCRT, RAND, DECRT	Different kinds of data item size distribution
THETA	0.7	0.0 - 1.0	Zipf distribution parameter
GRate	–	0.5 - 1.3	Updated upload requests generation parameter

of an RSU. The request data access pattern is the commonly used Zipf distribution with θ ranging from 0.0 to 1.0. We perform our simulation experiment using CSIM19, other than the default parameters, we use parameters shown in the Table 1. We use 3 different kinds of data item distributions: RAND, INCRT and DECRT [11] for both upload and download data item to analyze their effect on the performance of our joint scheduling algorithms.

For experimental data generation, we let all the vehicles in the RSU transmission range repeatedly in similar fashion until we get the stable data from the same parameter settings. For collecting the mean generated data, we use 100 iterations for the same settings with a different seed value every time.

5.2   Effect of Deadline Miss Rate

Fig. 2 shows the deadline miss rate, throughput and average response time of different joint scheduling algorithms with increasing workload (by increasing number of vehicles with fixed Zipf distribution parameter θ and random data item size distribution). In Fig. 2(a), deadline miss rate of all the joint scheduling algorithm increases with increasing number of vehicles. However, popularity based algorithm DSIN_Second-step and MRF_Second-step keep their deadline miss rate lower than FCFS_Second-step and EDF_Second-step. This is because when the number of vehicles increases with default θ value 0.7, more vehicles requests sharable requests for the same popular data item, hence by serving one popular request many outstanding requests being satisfied, resulting in lower deadline miss rates than other non-popularity based algorithm. For a non-popularity based algorithm such as EDF_Second-step, when serving a non-popular urgent request, one or many popular requests may miss their deadline, which leads to a high deadline miss rate. The results are similar for FCFS_Second-step which always serves first received request irrespective of the request productivity and deadline.

Fig. 2(b) shows the throughput increases with increasing number of vehicles. As with skewed Zipf distribution by serving one popular requests a number of shareable requests satisfied with increasing number of vehicles in the RSU's transmission range, throughput increases. Nevertheless, DSIN_Second-step and MRF_Second-step can achieve greater throughput than EDF_Second-step and

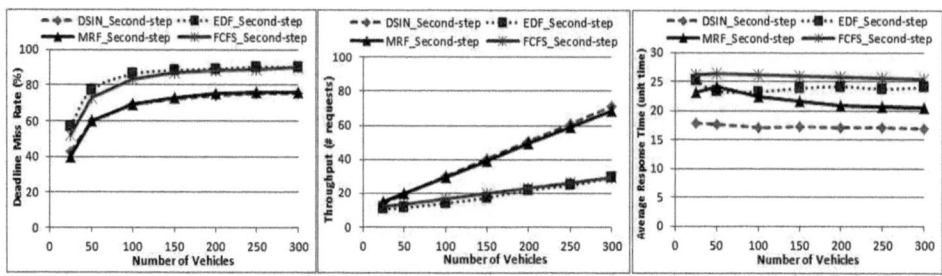

Fig. 2. Performance of different first-step and second-step joint scheduling algorithms for varying number of vehicles:(a)Deadline Miss Rate, (b)Throughput, (c)Average Response Time

FCFS_Second-step with increasing workload. However, for increasing number of vehicles, average response time does not change much (Fig. 2(c)) because of the Zipf skewness factor. But considering request deadline, data item size and popularity based algorithm DSIN_Second-step's average response time always lower than all others. In short, DSIN_Second-step algorithm performs better in high workload environment than others.

5.3   Effect of θ

Fig. 3 exhibits the effect of varying the Zipf parameter θ value from 0.0 to 1.0 in terms of deadline miss rate, throughput and average response time. When θ value is 0, vehicles requests are randomly distributed, hence popularity does not dominate for requests selection. However, with the increasing of θ values clients request pattern becomes more skewed, popular data items been requested by many vehicles' requests, then by a single broadcast many shareable requests are served. So, for increased θ value deadline miss rate decreases, throughput increases and average response time decreases. As both joint scheduler MRF_Second-step and DSIN_Second-step algorithm use popularity as requests selection criteria they get the high benefit for increasing θ. However, DSIN_Second-step outperforms all other joint schedulers considering all performance criteria.

5.4   Effect of Data Item Size Distribution

Fig. 4 depicts the effects of different data item size distribution in the joint scheduling algorithm in terms of our used performance metrics. Due to space constraints, we will only show the characteristics of DSIN_Second-step and EDF_Second-step schedulers as a representative of popularity based and non-popularity based algorithm respectively. With skewed θ value for increment size data item distribution, vehicles most likely request smaller sized popular data item and for decrement size distribution request bigger size popular data item. However, for random size distribution there is no such correlation between data item size and skewness of data access pattern. From Fig. 4(a) we

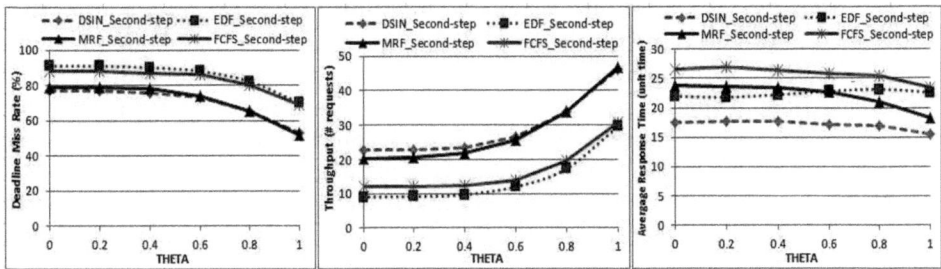

Fig. 3. Performance of different first-step and second-step joint scheduling algorithms for varying θ value: (a)Deadline Miss Rate, (b)Throughput, (c)Average Response Time

see that, DSIN_Second-step schedulers has significant influence with data distribution types. Here, for increment size distribution schedulers distinguishably outperforms other type distribution in terms of deadline miss rate, throughput and average response time. However, in Fig. 4(b) for EDF, although there is a small performance difference for different kind of distribution in terms of deadline miss rate and throughput, there is no corresponding difference in average response time. Hence we can conclude that if vehicles request smaller sized popular data item DSIN_Second-step scheduling algorithm can provide service with small deadline miss rate and response time with high throughput.

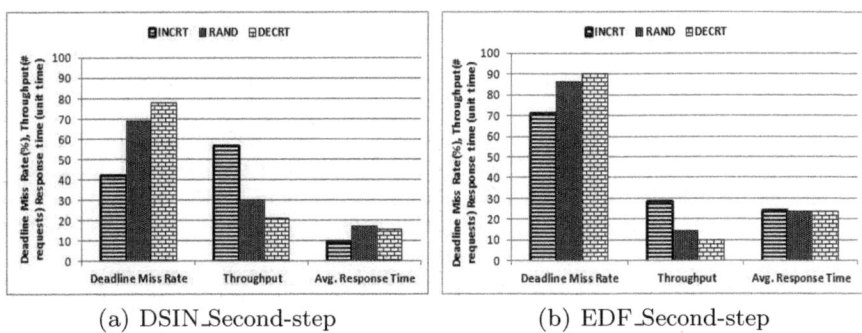

(a) DSIN_Second-step   (b) EDF_Second-step

Fig. 4. Effect of different data item size distributions in joint scheduling algorithms

6   Conclusion and Future Work

To update the RSU's database with the most updated information is an important issue for effective data dissemination in VANETs. In this paper, we propose a model where a vehicle can issue upload requests for updating the server information with its updated data as well as download request for downloading the server information. To do so, we implement a two-step joint scheduling procedure. In the first-step, we use existing on-demand algorithms to find the most appropriate requests form the each queue, and then in the second-step from these two queues we find one final request using our proposed second-step scheduling algorithm. Based on our simulation experiments we conclude that DSIN_Second-step scheduler outperforms other joint schedulers in terms of overall system performance in random data distribution; in addition, DSIN_Second-step scheduler has the best performance for increment size data distribution.

In our future work, we plan to apply our two-step joint scheduling algorithm to multiple RSUs with co-operative data access for load balancing among them.

References

1. Nadeem, T., Shanakr, P., Iftode, L.: A Comparative Study of Data Dissemination Models for VANETs. In: Vehicular Ad Hoc Network 2006 (2006)
2. Zhang, Y., Zhao, J., Cao, G.: On Scheduling Vehicle-Roadside Data Access. In: Vehicular Ad Hoc Network 2007 (2007)
3. Xu, J., Tang, X., Lee, W.: Time-Critical On-Demand Data Broadcast Algorithms, Analysis and Performance Evaluation. IEEE Transactions on Parallel and Distributed Systems 17(1), 3–14 (2006)
4. Chen, J., Lee, V.C.S., Chan, E.: Scheduling Real-time Multi-item Requests in Wireless On-demand Broadcast Networks. In: 4th International Conference of Mobile Technology, Applications and Systems (2007)
5. Yi, L.Z., Bin, L., Tong, Z., Wei, Y.: On Scheduling of Data Dissemination in Vehicular Networks with Mesh Backhaul. In: IEEE ICC 2008 (2008)
6. Wong, J.W., Ammar, M.H.: Analysis of Broadcast delivery in Videotex System. Journal of IEEE Transactions on Computers C-34(9) (1985)
7. Acharya, S., Alonso, R., Franklin, M., Zdonik, S.: Broadcast Disks: Data Management for Asymmetric Communication Environments. In: ACM SIGMOD Conference, CA (1995)
8. Wong, J.W.: Broadcast Delivery. Journal of IEEE 76(12) (1988)
9. Aksoy, D., Franklin, M.: $R \times W$: A Scheduling Approach for Large-Scale On-demand Data Broadcast. IEEE/ACM Transactions on Networking 7 (1999)
10. Acharya, S., Muthukrishnan, S.: Scheduling On-demand Broadcasts: New Metrics and Algorithms. In: MOBICOM 1998 (1998)
11. Xu, J., Hu, Q., Lee, W.C., Lee, D.L.: Performance Evaluation of an Optimal Cache Replacement Policy for Wireless Data Dissemination. IEEE Transactions on Knowledge and Data Engineering 16(1), 125–139 (2004)

A Content-Aware Adaptive Storage Approach for XML in PXRDB[*]

Xue Wang, Xiao Zhang[**], Xiaoyong Du, Shan Wang, and Kuicheng Liu

Key Laboratory of Data Engineering and Knowledge Engineering, MOE, China
Renmin University of China
No.37 Zhongguancun Street, Haidian District, Beijing, China
{xuew,zhangxiao,duyong,swang,kcliu}@ruc.edu.cn

Abstract. In many cases, it is pretty difficult to choose an efficient storage method, such as native, xml-enabled or hybrid, for storing XML documents in a relational database. We provide multiple storage approaches for XML documents in our hybrid XML-relational database PXRDB(Pure XML-Relational DataBase). Further, another problem is how to automatically choose storage method for a given XML document and whether different documents in same column can be stored in different formats. In this paper, we provide a content-aware adaptive storage approach for XML in PXRDB. This novel storage approach automatically selects one better storage scheme for a specific XML document from three candidate schemata, i.e., native storage, flat stream and multi-relations after fast-checking its content. Our approach frees end-users or administrators from either having no choice or having to specify the specific storage scheme for large number of XML documents manually. It also allows different XML documents in same relational column to be stored in different formats while being accessed indistinctively. By providing unified access interfaces, new storage approaches can be easily registered in our system. The performance evaluation illustrates our approach is feasible and effective.

Keywords: Content-aware, Adaptive, Storage Approach, XML-Relational Database.

1 Introduction

XML documents are used widely in web context and the volume of XML data keeps increasing rapidly. It is necessary and of importance to store and access the required XML data efficiently. In recent years many researches have been carried out on methods storing XML documents as xml-typed data in relational database so that applications can fully exploit functionality of RDBMS, such as concurrency control, recovery, scalability, and benefit from the highly optimized relational query processors.

[*] Partly supported by National 863 High Tech. Project (No. 2009AA01Z149), the Important National Science & Technology Specific Projects of China ("HGJ" Projects, Grant No.2010ZX01042-002-002-03), and National Natural Science Foundation of China (Grant No.61070054).
[**] Corresponding author.

J. Xu et al. (Eds.): DASFAA Workshops 2011, LNCS 6637, pp. 465–476, 2011.

In literature and practice, there are three basic approaches to store XML documents in RDBMS: 1) by storing the primitive XML documents in LOB format, 2) by shredding XML documents into relations and 3) in a native storage format. As for these methods, the LOB approach is the simplest one but prone to suffering from worse query performance because of less optimization. The relational way enables that the RDBMS manipulates the XML data in the same way as the relation. While the reconstruction of the original XML document is time-consuming and more space may be needed to store the data because of redundancy. Native storage method easily supports fragment update and document reconstruction but it has to be built from scratch and is very low efficient for managing relational data at the same time.

There is no 'one-size-fits-all' solution to determine how to store XML in a RDBMS so far[11]. The feasible way to address the above problems is to provide multiple storage models in one system. Accordingly, some factors must be taken into account. First, the system must be able to determine the storage format automatically and transparently for users. No matter which formats one document will be stored as, documents will be queried and updated indistinctively. Second, a storage-independent XQuery engine should be built and able to choose the best physical optimization strategies when working with the underlying XML physical model. In hybrid use-cases, both XQuery and SQL/XML can query XML. Last, the XQuery engine must be able to provide cross-language optimizations between XQuery and SQL.

Considering the above requirements, in PXRDB, an adaptive and unified storage approach CASF, standing for Choose Appropriate Storage Format, is designed and implemented to automatically choose a suitable storage for a given XML document. If the document is well-structured, it is mapped into relations directly. On the other hand, if the structure of document is irregular or there are many large texts existed in the document, either a native storage or LOB method is applied to provide faster query performance. We also build a storage-independent Query Engine in PXRDB to support both XQuery and SQL. Additionally, to accelerate query execution, various kinds of indexes are provided for XML data, too.

The main contributions of this paper are as follows:

- We present an algorithm to score each document for selecting appropriate storage scheme. This algorithm evaluates the structure of document. Well-structured documents will get a higher score than the bad-structured.
- We provide a novel adaptive storage approach CASF without manual intervention. Superior to those storage approaches that require significant manual work to choose storage scheme, CASF automatically stores an XML document in an appropriate method based on its content. It allows documents in same collection, i.e. one relational column, to be stored in different models, while being queried and updated in the same way.

The rest of this paper is organized as follows. In section 2, we present our adaptive storage method. Next, in section 3, we denote our storage selector algorithm that stores XML documents adaptively in PXRDB. In section 4 we conduct several experiments to evaluate performance of our approach. Then in section 5, we discuss the related work. Finally, section 6 concludes this paper.

2   Presentation of Adaptive XML Storage Schema

PXRDB supports multiple storage models. In this section, we will present our adaptive XML storage approach.

In PXRDB, three basic XML storage methods are provided, they are BLOB, native format and relational storage approach. An abstract data type, *XML-type*, is designed to store XML documents. Its physical storage and index models are use-case driven. Users do not need to know as which formats their documents will be stored. When an XML document is inserted, a selector function CASF, which will be discussed in next subsection, is used to choose an appropriate storage scheme for the document based on its content. Fig 1 shows the basic infrastructure of storage approach in PXRDB.

In Fig.1, the table with one XML column stores three XML documents in different formats. One uniform structural index is created on this XML column (collection) to provide indistinguishable XML data access and storage transparency. Apparently different from those existing storage approaches that require documents in one collection be stored in same format, our approach allows documents in one collection to be stored in different models. This approach offers more flexibility and can take more advantages of different storage methods.

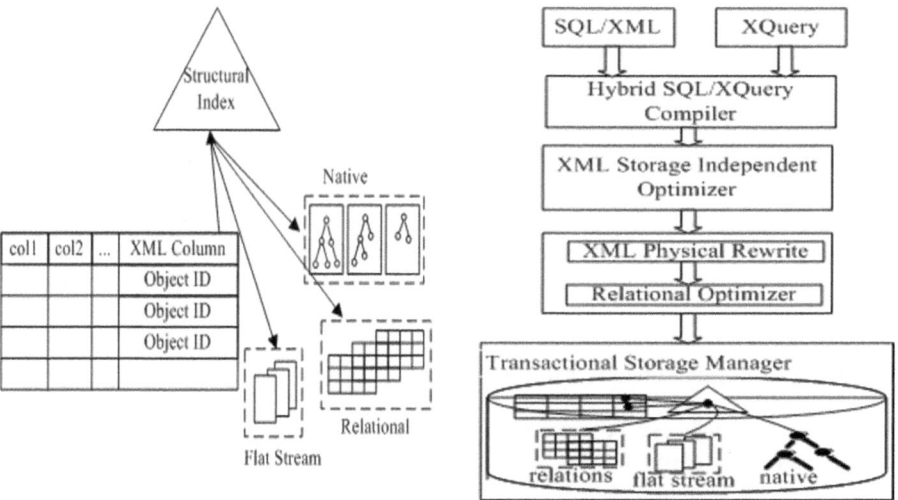

Fig. 1. Overview of Adaptive Storage Approach Schema

Fig. 2. System Architecture

The architecture of Unified XQuery/SQL Engine in PXRDB is shown in Fig. 2. First, both SQL and XQuery were compiled. Second, queries will be logically optimized. This logical evaluation is independent of physical storage. Third, for a given storage, the logical plan will be optimized if there are better physical rewritten strategies. Finally, all generated physical plan will be rewritten by relational optimizer. Before the stage of physical evaluation, query optimization is physical-independent as usual.

3 Storage Scheme Selector

For general-purpose reason, our storage approach CASF does not require that the input have schemas or DTDs, and the query workload for the certain XML data can be unknown to the execution engine.

Given one XML document without query workloads, one of the key techniques is to choose the appropriate storage approach. In order to introduce our approach more clearly, we give several definitions at first.

Definition 1[Unit]. *A **Unit** is a subtree rooted at one node called Unit's root: u(r,S,P,t) Where r is root of unit, S denotes the children sequence of the unit, P is the path from document root to r, and t presents tag name of r called tag name of u. NULL is a special unit. We use sample u to present a unit. Units with same tag name are of same type.*

Definition 2[Children Sequence of Unit]. *Given a unit u(r,S,P,t), we define the **Children Sequence of the Unit** u as a sequence, i.e., S(u)={$a_1,a_2.....a_n$}, where a_i satisfies the following condition: a_i.tagname≤a_{i+1}.tagname (according to the dictionary order) , a_i denotes one unit rooted at a_i and the parent of a_i is r.*

Definition 3[Singleton Node]. *A **Singleton Node** is an element appearing in the document only once.*

An XML document can be viewed as a collection that each element of which is a set of units with same tag name. From our observation, for a well-structured document, each occurrence of unit with same tag name will have similar structure and path. It will be easily shredded into a set of similar units that can be stored as records into tables, and each unit will be constructed as one row. Otherwise, if the document has overall very large depth or local very large depth frequently appears in the document or elements with same tag name have very different structure. This document will be stored natively. In addition, the document will be stored in BLOB format if 100% round trip is required.

3.1 Storage Scheme Selector Function

An XML document V can be viewed as a collection of which each element is a set of units with same tag name, that is, $V=\{U_1,U_2..U_m\}$ where $U_i=\{u_1,u_2....u_{ti}\}$ that satisfies the condition: u_1.tagname = u_2.tagname = ... = u_{ti}.tagname}. We use the similarity function below to evaluate similarity between two units with same tag name:

$$K(u_1,u_2) = \begin{cases} \sum_{i=1}^{\max(l(S(u_1)),l(S(u_2)))} K(G_i[S(u_1)],G_i[S(u_2)]) + s(u_1.tag,u_2.tag) \\ 0 \qquad\qquad\qquad\qquad \text{if } u_1 \text{ or } u_2 \text{ is NULL} \end{cases} \tag{1}$$

Where,

- ■ S(u) denotes the children sequence of u
- ■ l(S(u)) is the length of sequence S(u)
- ■ $G_i(S)$ is the *i*th nodes of S
- ■ s(u_1.tag, u_2.tag) is sign function defined as follows:

$$s(u_1.tagname, u_2.tagname) = \begin{cases} 1 \text{ if } u_1.tagname = u_2.tagname \\ 0 \qquad\qquad \text{else} \end{cases} \tag{2}$$

Since the size of the input unit is not constant, the similarity score is normalized as follows:

$$K'(u_1, u_2) = \frac{K(u_1, u_2)}{\sqrt{l(S(u_1)) * l(S(u_2))}} \tag{3}$$

The value of $K'(u_1, u_2)$ ranges from 0 to 1.

Example 1: According to the above equation (1) and (2), the similarity between the two person units in Fig.5 is computed as follows:

$$K(person_1, person_2) = K(name_1, name_2) + K(email_1, email_2) + K(link_1, link_2) + 1$$
$$= (1+1+1) + 1 + 1 + 1 = 6$$

Let $V = \{U_1, U_2, \ldots U_m\}$ be an XML document, let $U_i = \{C_1, C_2 \ldots C_{ni}\}$ $(t_1, t_2 \in C_i \wedge t_1.path = t_2.path)$ be a classified result of U_i using path of unit as the decision rule. The next equation (4) calculates the entropy of U_i as:

$$U_i.e = (\sum_{j=1}^{nj} -P_j * \ln(P_j)) / \ln(|U_i|)(P_j = |C_j| / |U_i|) \tag{4}$$

Let $V = \{U_1, U_2, \ldots U_m\}$ be an XML document, $R = \{R_1, R_2 \ldots R_n| R_i \cap R_j = NULL \wedge \bigcup_{i=1}^{n} R_i =$

$V \wedge R_i = \{Ut_1, Ut_2 \ldots Ut_i | u_p \in Ut_i, u_q \in Ut_j \wedge (u_p, u_q \text{ always happen together})\}\}$. R is a partition of set V. If two types of Units always appear in document simultaneously, it means one $u_p (\in U_i)$ always follows and only follows one $u_q (\in U_j)$, the relationship between u_p and u_q is 1:1. U_i and U_j belong to a same equivalence class. If document V is shredded into relational tables, each element R_i of R will be mapped into a separate table and if $Ut_1 \in R_i$ and $Ut_2 \in R_j (R_i \neq R_j)$, then Ut_1 and Ut_2 will not be mapped into one table. Otherwise there will exist a large number of redundancies because the relationship between two Units in different equivalence class of R is not 1:1. Here we do not take the Edge approach[3] into account because such general relational storage approaches have worse query performance than schema-aware relational approaches and have worse reconstruction performance than native approaches. If these general relational storage approaches will be adapted, the score function of the document should be adjusted. We do not discuss it in more details since it is beyond this paper.

Take all into account, the score of V is computed as follows:

$$score(V) = \frac{1}{3}\sum_{i=1}^{|V|}((\sum (U_i.s / |U_i|) / |V| + (\sum_{i=1}^{|V|}(1 - U_i.e)) / |V| + (|V| - |singletonNode|) / (|R| * V.avgdepth)) \tag{5}$$

The formula $(|V| - |singletonNode|) / (|R| * V.avgdepth)$ is employed to test in general whether all elements requested in one XPath query can be stored in a relational table. Here without query workload, we assume that the average depth of the document can denote the average length of XPath query. If one table can store more elements in average, XPath query over relational tables will require less join operations.

Let V.avgdepth be the average depth of document V, U_i.similarity is defined as follows:

$$U_i.similarity = (\sum K(u_p, u_e))/ |U_i|)(u_p \in U_i, u_e \text{ is standard Unit of } U_i) \qquad (6)$$

Where u_e is the standard unit of U_i, it will be discussed in next subsection.
The higher the score, the better structured the document.

Example 2: The score of the fraction shown in Fig.5 is computed as follows:

V={persons, names, familys, givens, emails, links}
R={R_1(person, person.id, name, family, given, email, link, link. subordinate)}
Average depth of V is 4/3.

$$score(V) = (\sum_{i=1}^{6}(U_i.\text{similarity})(U_i \in V)/6 + \sum_{i=1}^{6}(1-U_i.e)(U_i \in V)/6 + (6-0)/(1*V.avgdepth))/3$$

$$=(6/6+6/6+6/(4/3))/3=(6/6+6/6+4.5)/3 =2.1$$

3.2 Implementation of the Selector-CASF

We implement CASF for choosing appropriate storage format for a given XML document as in Fig. 3. It can identify whether the document is well-structured or not. Path and structural information of unit occurring in the document are recorded. For next occurrence of the unit with same tag name, it is compared with standard unit u_e, which is initialized as the first appearance of unit with this tag name, of this type recorded before to evaluate the similarity between them by using similarity function above. According to the similarity, procedure *extend*() (Line 6 in CASF algorithm) will extend structure of standard unit(u_e)so as to counteract effects of appearance orders of units.

Entropy function is used to measure chaos of the document's structure. The smaller the document's entropy is, the better-structured the document is.

More formally, for each unit in the document, we start with the procedure *create-Unit*(t)(Line 2) to create a unit with tag name of t, *findMatch*(D,t.tag)(Line 3) is used to look up weather there has been a unit set with tag name of t in D. If no such a unit set in D, a new set will be created and added into the collection D by *add*()(Line 7). Otherwise, if a unit set with t's tag name already exists, the similarity between t and u_e of this unit set is computed. At the end of the document, for each $U_i \in D=\{U_1,U_2,...U_n\}$, we groups units in U_i based on their paths. For each group, entropy value is computed using Formula 4.

The procedure *generateR*() (Line 12)shown in Fig. 3 partitions set D into equivalence class that different type of units in same equivalence class has the relationship 1:1 while different type of units in different equivalence class has the relationship 1:n or n:m. For more details, users can read shared-inlining algorithm presented in [3]. A threshold is set to limit the score of the document. Document with a score higher than threshold will be shredded into relations. Otherwise it will be stored as either native or BLOB formats. We will discuss how to set the threshold and the affection of threshold on document storage format in section 4.4. The *parseandStore*()(Line 15) function in Fig.3 is used to generate native or BLOB storage for the input document. For each native storage approach having its own implementation of parseandStore(), we will not provide detail algorithm.

Algorithm: CASF

Input: V XML Document

Output: S Native Storage or Relational schemas

D: collection of sets of units

U: set of units. // U.s is similarity of U, U.o presents |U|, U.e denotes entropy of U

1 **for** end of each element e in V

2   des=createUnit(e);   //create a new unit des

3   src= findMatch (D,e.tag);// search set of units with same tag as e.tag in D

4   **if** src **then**

5   src.s+=K'(src,des); src.o++;

6   extend(des,src); sequencialInsert(des,parent(des));

7   **else** add(D,des); // there are no units with same tag name as des, add a new
                             set(only including one element des) into D

8 **if** end of Document   **then**

9   **for** i=1 to |D|

10   compute U$_i$.s; D.s+= U$_i$.s;

11   compute U$_i$.e; D.e+= U$_i$.e;

12   S=generateR(D); singletonNode=count of singletonNode, avgDepth=
          average depth of document

13 score= 1/3*(D.s + D.e + (|D| - singletonNode / |S|) / avgDeph);

14 **if** score $< \lambda$ **then**   //λ is the threshold

15   S=parseandStore(V);

16 output(S);

Function: extend(des, src) //extend standard Unit of one Unit type

---

Input: src unit, des unit

1 **for** i=1 to |des.S|

2   **if** p$_i$ $\in$ des.S and p$_i$! $\in$ src.S **then**

3   sequencialInsert(pi,src);

4   **if** p$_i$ $\in$ des.S and p$_j$ $\in$ src.S and p$_i$.tag=p$_j$.tag and p$_i$.o>p$_j$.o **then**

5   p$_j$.o:=p$_i$.o;

Function: sequencialInsert(newChild, parent)//insert a new unit into its parent

---

Input: newChild unit, parent   unit

1 **for** each u$_i$ $\in$ parent.S

2   **if** u$_i$.tag > newChild.tag **then**

3   insert newChild before u$_i$; newChild.pos=i;

4   **else** i++;

5 **if** i = |parent.S| **then**

6   insert newChild at last position of parent.S;

7   newChild.pos=i;

Fig. 3. Algorithm CASF

It is easy to compute the complexity of the algorithm CASF. There are two nested loops, the outermost loop is processed at most |D| times(D presents collection of units), and the inner loops are processed at most max(|U$_i$|)(i=1..|D|) times. Hence, the algorithm has a worst-case runtime of O(max(max(|U$_i$|)(i=1..|D|)*|D|,n)). In most cases, n is far larger than |D| and |U$_i$|, so the upper bound on the temporal complexity is O(n).

4 Experiments

In this section, we evaluate the efficiency and accuracy of our adaptive storage approach. Specifically, we focus on classification error of our storage approach. All experiments are conducted on a PC running windows XP with 1.86 GHz Intel Core Duo CPU and 2 GB of main memory.

4.1 Datasets

Document-centric data is more possible to be well-structured. We do not manually add more tag for these XML documents instead we take data-centric document as well-structured and regard them suitable for being shredded into relations. Meanwhile we take text-centric document as irregular-structured and suitable for native or BLOB format. We collect three types of XML documents, data-centric, text-centric and benchmark, as our datasets for evaluation to see whether CASF can free user from manual work of choosing storage formats and automatically make expected choice. Size of XML documents in these datasets varies from 3KB to 124MB with different schema characteristics. They are collected from UW XML Repository, IMDB and Wikipedia respectively. We totally extract 81 text-centric and 109 data-centric documents. We choose XBench as our benchmark dataset because it can generate both text-centric(TC) and data-centric(DC) documents and these generated documents have a relative complex schema as well.

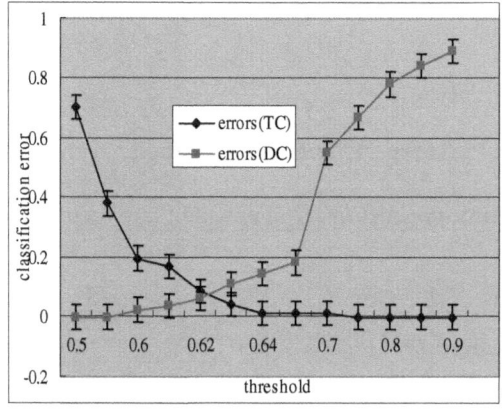

```
<person id="Big.Boss">
<name><family>Boss</family>
<given>Big</given>
</name>
<email>chief@foo.com</email>
<link
subordinates="five.worker"></link>
</person>
<person id="one.worker">
<name><family>Worker
    </family>
        <given>One</given>
</name>
<email>one@foo.com</email>
<link manager="Big.Boss"></link>
</person>
```

Fig. 4. Classification Error Percentage   **Fig. 5.** An XML Document Fraction

4.2 Accuracy of Choosing Suitable Storage

This subsection is to test the accuracy of our adaptive storage algorithm CASF. We run CASF first on a training dataset including 10 data-centric documents and 10 text-centric documents to verify the threshold. Then it is tested using the records in the test dataset comprising 99 data-centric documents and 71 text-centric data.

The statistics of the dataset are shown in Table 1. On average, text-centric documents have a larger depth value than data-centric documents. Column of different elements in Table 1 presents numbers of elements with different tag name appearing in the document. We can conclude from Table 1 that numbers of different elements in the document has no relationship with the document's type.

Table 1. Statistics of Datasets

	Count of documents	Avg(maxDeph)	AvgDepth	Different Elements
Data-centric dataset	109	12.24	6.32	270.59
Text-centric dataset	81	28.19718	14.84507	119.69014

Table 2 denotes summary statistics of documents score accepted on training dataset.

Table 2. Summary Statistics of Document Score

	Minimum	25% quantity	Median	75% quantity	Maximum
TC(10)	0.45132	0.48514	0.51789	0.56504	0.61350
DC(10)	0.56821	0.64620	0.67616	0.75196	2.66667

Seen from table 2, threshold is set to 0.62. Classification accuracy on data-centric and text-centric documents is 0.9 and 1 respectively. Table 3 illustrates the recognition rate under threshold 0.62to classify documents in test dataset.

Fig.4 shows classification error rate when document's score as classification rule. The Y-axis of Fig.4 stands for percentage of classification error while x-axis is the threshold. If score of one document is larger than threshold, it will be taken as data-centric and stored as relational format. Otherwise it will be taken as text-centric and stored as native or BLOB format. When threshold is set larger than 0.6 in Fig. 4,most text-centric document can be classified correctly that means scores of most text-centric document are smaller than 0.6. When threshold is set smaller than 0.65, classification error of data-centric document will decrease until approaching 100% which means scores of most data-centric documents are higher than 0.65.

Table 3. Confusion Matrix on Test Dataset

	Data-centric=yes	Text-centric=yes	Total	Recognition rate
Data-centric=yes	95	4	111	96.0
Text-centric=yes	4	67	69	94.5

As in Table 3, a classification accuracy of 92.4% is obtained over text-centric dataset. The reason why it is not very high lies in blurred boundary between data-centric and text-centric documents. In fact, many documents are not strictly data-centric or document-centric. We list the statistics information of four of these so-called "text-centric" documents in Table 4.

Table 4. Summary Statistics of 4 "Text-centric" Documents

ID	Max depth	Average depth	Different elements	Singleton node	Score
D1	21	13	29	11	0.620655
D2	28	13	60	29	0.626105
D3	17	11	39	15	0.630942
D4	12	6	23	11	0.706324

As shown in Table 4, there are few different elements appeared in them, and most of these elements only occur one time. These four documents are actually between data-centric and document-centric, vice versa, they are called data-centric while being classified into the group of text-centric documents.

5 Related Work

Two basic approaches are used to construct the XML storage manager in a DBMS. One is the extended relational approach and the other is the native approach.

The native approach is deployed in many XML data management systems[20, 15, 16, 21, 7]. Among them, TIMBER and Berkeley DB XML break the XML document into nodes and store the node information in a B^+tree. Natix partitions XML tree into subtrees each of which can be stored in a record fitting into one disk page. System RX[17,18] employs a similar technique to Natix and builds a structure Regions Index to connect each subtree located in different disk page.

Comparably, relational approaches are storing XML documents in the RDBMS by either shredding the XML documents into multiple relational tables or storing them in the LOB column(s)[1,2,8]. The DTD of the XML documents and query of applications can be used to improve the performance of the XML DBMS. [4, 5] indicate the shared-inling[6] algorithm outperforms other strategies in query efficiency when DTDs are available.

In practice, Oracle 11g[11,19] provides multiple XML storage approaches in one system. Oracle 11g provides relational storage, LOB and Binary-XML storage formats, but users have to specify storage method for their XML documents themselves and for one XML column only one storage approach can be deployed i.e. all documents in this column will be stored in same format. Also, Oracle does not allow schema-oblivious XML documents to be stored as relational tables. It also provides a physical independent XQuery/SQL/XML Engine, which defines and optimizes both XQuery and SQL/XML into the same logical algebra presentation and does physical optimizations based on the underlying XML storage, index and view models.

6 Conclusion

This paper presents a content-aware storage approach CASF implemented in our prototype DBMS PXRDB, which can automatically choose appropriate storage schemes for document. Our solution addresses the following key problems of storing XML data:

1). Multiple approaches can be applied automatically to the XML documents during insertion. 2). Different documents in the same column can be stored in different methods for the certain reasons. 3). Extensibility of the storage manager for XML data that is a new one can be easily integrated into PXRDB at run time. CASF has linear temporal complexity O(n) in the worst case, where n is the number of nodes in the input document.

We have conducted several experiments to evaluate our algorithm. The results show that CASF can accurately distinguish well-structured documents from those irregular or bad-structured ones while user's view is kept unified. And the experiments indicate that CASF outperforms its competitor in some circumstances.

References

1. Rys, M.: XML and Relational Database Management Systems: inside Microsoft SQL Server 2005. In: Proc. ACM SIGMOD Int. Conf. on Management of Data, pp. 958–962 (2005)
2. Microsoft. White Paper: What's New for XML in SQL Server, White Paper (2008)
3. Shanmugasundaram, J., Tufte, K., He, G., Zhang, C., DeWitt, D., Naughton, J.: Relational Databases for Querying XML Documents: Limitations and Opportunities. In: VLDB 1999 (1999)
4. Tian, F., DeWitt, D., Chen, J., Zhang, C.: The Design and Performance Evaluation of Alternative XML Storage Strategies. ACM Sigmod Record 31(1) (March 2002)
5. Florescu, D., Kossmann, D.: A Performance Evaluation of Alter native mapping Schemas for Storing XML Data in a Relational Database. In: Proc. of the VLDB 1999 (1999)
6. Shanmugasundaram, J., Tufte, K., He, G., et al.: Relational Databases for Querying XML Documents: Limitations and Opportunities. In: VLDB 1999 (1999)
7. Boncz, P.A., Grust, T., Keulen, M., Manegold, S., Rittinger, J., Teubner, J.: MonetDB/XQuery: a fast XQuery processor powered by a relational engine. In: SIGMOD Conference 2006, pp. 479–490 (2006)
8. Ennser, L., Delporte, C., Oba, M., Sunil, K.: Integrating XML with DB2 XML Extender and DB2 Text Extender. IBM Redbooks (2001)
9. Profressional XML, Wrox Press (2000)
10. DB2 goes hybrid Integrating native XML and XQuery with relational data and SQL 2006 (2006)
11. Liu, Z.H., Chandrasekar, S., Baby, T., Chang, H.J.: Towards a Physical XML independent XQuery/SQL/XML Engine. In: Proc. 34th Int. Conf. on Very Large Data Bases, pp. 1356–1367 (2008)
12. Chebotko, A.: Reconstructing XML subtrees from Relational Storage of XML documents. In: ICDE 2004 (2004)
13. Florescu, D., Kossmann, D.: A Performance Evaluation of Alternative Mapping Schemes for Storing XML in A Relational Database. Technical Report 3680, INRIA (1999)
14. Kanne, C.-C., Moerkotte, G.: A Linear Time Algorithm for Optimal Tree Sibling Partitioning and Approximation Algorithms in Natix. In: Proc. 32th Int. Conf. on Very Large Data Bases, (September 2006)
15. Fiebig, T., Helmer, S., Kanne, C.-C., Mildenberger, J., Moerkotte, G., Schiele, R., Westmann, T.: Anatomy of a Native XML Base Management System. The VLDB Journal 11(4), 292–314 (2002)
16. Kanne, C.-C., Moerkotte, G.: Efficient Storage of XML Data. In: Proc. 16th Int. Conf. on Data Engineering, pp. 198–209 (2000)

17. Beyer, K.S., Cochrane, R., Josifovski, V., Kleewein, J., Lapis, G., Lohman, G.M., Lyle, B., Ozcan, F., Pirahesh, H., Seemann, N., Truong, T.C., der Linden, B.V., Vickery, B., Zhang, C.: System RX: One Part Relational, One Part XML. In: Proc. ACM SIGMOD Int. Conf. on Management of Data, pp. 347–358 (2005)
18. Funderburk, J.E., Kiernan, G., Shanmugasundaram, J., Shekita, E., Wei, C.: XTABLES: Bridging relational technology and XML. IBM Systems Journal 41(4), 616–641 (2002)
19. Zhang, N., Agarwal, N., Chandrasekar, S.: Sum Idicula.: Binary XML Storage and Query Processing in Oracle 11g. In: Proc. 35th Int. Conf. on Very Large Data Bases (June 2009)
20. Jagadish, H.V., Al-Khalifa, S., Chapman, A., Lakshmanan, L.V.S., Nierman, A., Paparizos, S., Patel, J.M., Srivastava, D., Wiwatwattana, N., Wu, Y., Yu, C.: TIMBER: A Native XML Database. VLDB Journal 11(1), 274–291 (2002)
21. Meier, W.: eXist: An Open Source Native XML Database, http://exist-db.org

The Flamingo Software Package on Approximate String Queries*

Chen Li

Department of Computer Science
UC Irvine, CA 92697, USA
chenli@ics.uci.edu

Abstract. An important operation in data cleaning is similarity search on textual strings. A simple example is "finding actor names similar to **schwarzeneger**," given the fact that few people know the exact spelling of our former governor in California. It is challenging to support this operation efficiently on large amounts of data. Despite its importance, the problem did not receive enough attention in the research community a decade ago. In this talk, I will give an overview of recent results on this problem, and describe the development history of the Flamingo package, an open-source software that supports efficient approximate string queries. I will also describe my outreach activities to apply our research results of data cleaning in real applications, which led to a startup called Bimaple that specializes in powerful instant search on large data sets.

Keywords: Data Cleaning, Flamingo Package, Approximate String Search.

---

* This research is partially supported by the US NSF CAREER award IIS-0238586, the NSF award IIS-0742960, the NSF award IIS-0844574, the NSF award 1030002, the NSF award 0331707, the National Nature Science of China 60828004, a Google Research Award, and a gift fund from Microsoft.

J. Xu et al. (Eds.): DASFAA Workshops 2011, LNCS 6637, p. 477, 2011.

A Framework for Data Quality
Aware Query Systems

Naiem K. Yeganeh and Mohamed A. Sharaf

School of Information Technology and Electrical Engineering
The University of Queensland, St Lucia. QLD 4074, Australia,
naiem@itee.uq.edu.au, m.sharaf@uq.edu.au

Abstract. Data Quality (DQ) is increasingly gaining more importance as organizations as well as individuals are relying on data available from various data sources. User satisfaction from query result is directly related to the quality of data returned to user. In this paper we present a framework for DQ aware query systems focused on three key requirements of profiling DQ, capturing user preferences on DQ and processing data quality aware queries.

1 Introduction

User satisfaction from a query response is a complex problem encompassing various dimensions including both the efficiency as well as the quality of the response. Quality in turn includes several dimensions such as completeness, currency, accuracy, relevance and many more [26].

Consider for example a virtual store that is integrating a comparative price list for a given product (such as Google products, previously known as froogle) through a meta search (a search that queries results of other search engines and selects best possible results amongst them). The search engine obviously does not read all the millions of results for a search and does not return millions of records to the user. It normally selects top k results (where k is a constant value) from each search engine and finally returns top n results after the merge.

In the above scenario, when a user queries for a product, the virtual store searches through a variety of data sources for that item, ranks them and returns the results. For example the user may query for "Canon PowerShot". In turn the virtual store may query camera vendor sites and return the results. The value that the user associates with the query result is clearly subjective and related to the user's intended requirements which go beyond the entered query term, namely "Canon PowerShot" (currently returns 91,345 results from Google products). For example the user may be interested in comparing product prices, or the user may be interested in information on latest models.

More precisely, suppose that the various data sources can be accessed through a schema consisting of attributes ("Item Title", "Item Description", "Numbers Available", "Price", "Tax", "User Comments"). A user searching for "Canon PowerShot" may actually be interested in:

J. Xu et al. (Eds.): DASFAA Workshops 2011, LNCS 6637, pp. 478–489, 2011.
© Springer-Verlag Berlin Heidelberg 2011

1. Browsing products: such a user may not care about the "Numbers Available" and "Tax" columns. "Price" is somewhat important to the user although obsoleteness and inaccuracy in price values can be tolerated. However, consistency of "Item Title" and completeness within the populations of "User Comments" in the query results, is of highest importance.

2. Comparing prices: where the user is sure about the item to purchase but is searching for the best price. Obviously "Price" and "Tax" fields have the greatest importance in this case. They should be current and accurate. "Numbers Available" is also important although slight inaccuracies in this column are acceptable as any number more than 1 will be sufficient.

The above examples indicate that getting satisfactory query result is subjected to three questions: how good is each data source? what does the term "good" mean to the user? and how to rank the sources? To answer above questions, we face the following three challenges.

First challenge is to measure the quality of data. In order to estimate the quality of data we should collect descriptive statistical information about data. These statistics can in turn be used in query planning, and query optimization. Descriptiveness of the collected information contributes to the effectiveness of the system, making predictions on the quality of the source/result-set closer to reality but it comes with a trade-off with storage. However, in today's technology, data storage is rarely a problem, and user satisfaction with the query results can be deemed more important than storage.

Second challenge is to capture user preferences on DQ. Modelling user preferences is a challenging problem due to its inherent subjective nature. Additionally, DQ preferences have a hierarchical nature, since there can be a list of different DQ requirements for each attribute in the query. Several models have been developed to model user preferences by decision making theory and database communities. Models which have been based on partial orders are shown to be effective in many cases [21]. Different extensions to the standard SQL have also been proposed to define a preference language [13].

Third challenge is to develop data quality aware query planning methods that allow for efficient data source selection in the presence of pre-specified user preferences over multi-dimensional DQ measures. In particular, techniques for ranking data sources based on a multi-criteria decision making are needed and in this paper we discuss examples of such techniques.

In this paper we present a framework for DQ aware query planning in data integration systems in presence of multiple redundant data sources available for answering the same query where data sources are selected in the order that they satisfy user requirements on DQ. We define this framework around the three challenges mentioned earlier in this section which leads us to three key requirements: 1) Profiling DQ which is the process of extracting statistics about the quality of each data source for a given query. 2) Capturing user preferences on DQ and 3) Planning queries that are extended with DQ preferences.

The rest of this paper is organized as follows: In Section 2, existing literature related to the three key requirements of the framework is studied. In Section 3, first the architecture and framework elements are defined. Rest of the section is dedicated to discuss solutions and options for all three challenges. At the end of each section, relevant future works and open questions are discussed. Finally in Section 4 we conclude the paper.

2   Existing Literature

Consequences of poor quality of data have been experienced in almost all domains. From the research perspective, data quality has been addressed in different contexts, including statistics, management science, and computer science [22]. To understand the concept, various research works have defined a number of quality dimensions [22] [25].

Data quality dimensions characterize data properties e.g. accuracy, currency, completeness etc. Many dimensions are defined for assessment of quality of data that give us the means to measure the quality of data. Data Quality dimensions can be very subjective (e.g. ease of use, expandability, objectivity, etc.).

To address the problems that stem from the various data quality dimensions, the approaches can be broadly classified into investigative, preventative and corrective. Investigative approaches essentially provide the ability to assess the level of data quality and is generally provided through data profiling tools. Many sophisticated commercial profiling tools exist [6]. There are several interpretations of dimensions (which may vary for different use cases), e.g. completeness may represent missing tuples (open world assumption) [1]. Accuracy may represent the distance from truth in the real world. Such interpretations are difficult if not impossible to measure through computational means and hence in the subsequent discussion, the interpretation of these dimensions is assumed as a set of DQ rules.

A variety of solutions have also been proposed for preventative and corrective aspects of data quality management. These solutions can be categorized into the following broad groups : Semantic integrity constraints [3]. Record linkage solutions. Record linkage has been addressed through approximate matching [8], de-duplicating [9] and entity resolution techniques [2]. Data lineage or provenance solutions are classified as annotation and non-annotation based approaches where back tracing is suggested to address auditory or reliability problems [23]. Data uncertainty and probabilistic databases are another important consideration in data quality [15]. In [12], the issue of data imputation for incomplete datasets is studied, whereas maximizing data currency has been addressed in [19].

Nevertheless, data quality problems can not be completely corrected and in presence of errors and further consideration is required to maximise user satisfaction for the quality of data received.

Profiling Data Quality. Measurements made on a dataset for DQ dimensions are called DQ metrics and the act of generating DQ metrics for DQ dimensions is called DQ profiling. Profiling generally consists of collecting descriptive statistical information about data. These statistics can in turn be used in query planning, and query optimization. Data quality profile is a form of meta data which can be made available to the query planning engine to predict and optimize the quality of query results.

Literature reports on some works on DQ profiling. For example in [16] DQ metrics are assigned to each data source in the form of a vector of DQ metrics and their values (source level). In [25] a vector of DQ metrics and their values is attached to the table's meta data to store additional DQ profiling; e.g. $\{(Completeness, 0.80), (Accuracy, 0.75), \ldots\}$ (relation level). In [28] an additional DQ profile table is attached to the relation's meta data to store DQ metric measurements for the relation's attributes (attribute level). In this paper we categorize these approaches as source level, relation level, and attribute level DQ profiling.

We assume that a set of DQ metrics M is standardized between data sources, however data sources may have different approaches (i.e. different rules) to calculate their DQ metrics (e.g. a UK based data source has a different set of rules from an Australian based data source for checking accuracy of address). None of the above profiling techniques is sufficient to estimate the quality of a query results. In Section 3.1 we discuss this limitation in detail.

User Preferences on Data Quality. The issue of user preferences in database queries dates back to 1987 [14]. Preference queries in deductive databases are studied in [7]. In [13] and [4] a logical framework for formulating preferences and its embedding into relational query languages are proposed.

Several models have been developed to model user preferences by decision making theory and database communities. Models which have been based on partial orders are shown to be effective in many cases [21]. Typically models based on partial order let users define inconsistent preferences. Current studies on user preferences in database systems assume that existence of inconsistency is natural (and hard to avoid) for user preferences and a preference model should be designed to function even when user preferences are inconsistent, hence; they deliberately opt to ignore it. Nevertheless, all studies do not always agree with this assumption [10]. Human science and decision making studies show that people struggle with an internal consistency check and they will almost always avoid inconsistent preferences if those individuals are given enough information about their state in their decision (e.g. visually). In fact, existence of inconsistency in user preferences dims the information about user preferences captured by the query. In Section 3.2 we discuss the possibility of proposing suitable languages and user interfaces to capture user preferences specifically for DQ and avoid inconsistency when possible.

Query Planning and Data Integration. From a query planning perspective, a data source is abstracted by the source descriptions. These descriptions specify the properties of the sources that the system needs to know in order to use their data. In addition to source schema, the source descriptions might also include information on the source: response time, coverage, completeness, timeliness, accuracy, and reliability, etc.

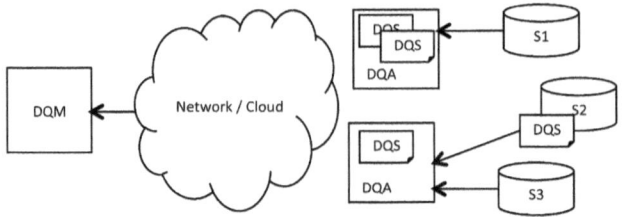

Fig. 1. Data Quality Aware Query Answering Architecture

When a user poses a data integration query, the system formulates that query into sub-queries over the schemas of the data sources whose combination will form the answer to the original query. For applications with a large number of sources, typically the number of sub-query plans is very large and plan evaluation is costly, so executing all sub-query plans is expensive and often infeasible. In practice, however, only a subset of those sub-queries is actually selected for execution [5][17][18].

Each executed sub-query is also optimized to produce a physical query execution plan with minimum cost. This plan specifies exactly the execution order of operations and the specific algorithm used for every operation (e.g., algorithms for joins). Existing techniques tend to separate the query selection process into two separate phases using two alternative approaches: 1) select-then-optimize, and 2) optimize-then-select. In the first approach, a subset of sub-queries is selected based on coverage, then each selected sub-query is optimized separately [5][17], whereas in the second approach, each sub-query is optimized first, then a subset of optimized sub-queries is selected to maximize coverage while minimizing cost [18]. However, in both approaches, the selection of query plans is primarily based on data coverage and/or query planning costs without further considerations for DQ. To the contrast, in this paper, we describe our approach for query plan ranking based on DQ that allows for efficient data source selection in the presence of pre-specified user preferences over multi-dimensional DQ measures.

3 Framework for DQ Aware Query Systems

Before starting the discussion about the framework for DQ aware query planning in data integration systems, we assume the following architecture which is an extension of general data integration architecture with Data quality components. Figure 1 represents the architecture which consists of: Data sources (S_1, S_2, ...), Data Quality Services (DQS), Data Quality Agents (DQA) and a Data Quality Aware Mediator (DQM) – meta search engines are a good example of query mediators.

Data sources should expose their schema to the mediators and also Data Quality Agents. Data Quality Services are generally containers for DQ metrics and their definitions. They may be either integrated with data source or separated as external service. They may also query some master data source for calculating DQ metrics. Notion of the Data Quality Services is widely used in industrial products such as IBM Data Quality Stage, recent MS SQL Server Data Quality Services, etc. Data Quality Agents are services that manage DQ profiles. They may query relevant data sources and communicate with DQSs regularly to manage or generate DQ profiles. Query mediators are the most complex part of the architecture which orchestrate query planning and data integration for the end user. Data Quality Aware Mediators host three necessary parts: A service for parsing the DQ aware query languages, a service for planning the query and integrating data considering DQ preferences and requirements, and a local DQ Profile Dictionary which contains DQ profiles for all data sources. This profile dictionary can help the mediator to optimize its query plan and only query data sources that serve best for user preferences.

3.1 Data Quality Profiling

Attribute level profile over the whole dataset does not provide enough information to predict DQ of the query result set. For example, a dealer of new cars may also have used cars in its database. Since they are more particular about data entry of their new cars, DQ of the used car subset of database is much less than DQ of the new cars subset of DB. The only situation where attribute level metric value of the whole dataset will be similar to the attribute level metric value of any subset of the dataset is when distribution of dirty data within data-set is evenly random. Data quality profile that stores attribute level DQ statistics for the whole dataset will require very little amount of storage for each data source, an attribute level DQ profile is shown in Figure 2 (b). The DQ profile in Figure 2 is generated from the data set of Figure 2 (a). It is the result of measuring completeness of each attribute in the dataset presented in Figure 2 (a). In this example completeness is considered as number of null values over the number of records in the data for each attribute. For example the completeness of the "Image" attribute of the given data set is %50 since there are 3 null values over 6 records. The three columns in the profile table of Figure 2 (b) identify object (an attribute from the relation) against which the metric Completeness is measured, and the result of this measurement.

Item Title	Item Desc	Num Available	Price	Image	User Comments
S2IS	Canon PShot	8	310	1.jpg	
	Camera		1000		
S3IS	Canon PShot		375		<4 Reviews>
XL-2		4	184	a.bmp	
DMC-TZ5K	Panasonic Lmx		340		
DSC-W55	S ony Cshot	2	260	9.jpg	

(a)

Object	Metric	Value
Shop.ItemTitle	Completeness	0.83
Shop.ItemDesc	Completeness	0.83
Shop.NumAvailable	Completeness	0.50
Shop.Price	Completeness	1.00
Shop.Image	Completeness	0.50
Shop.User Comments	Completeness	0.16

(b)

Fig. 2. A traditional data quality profile for a sample dataset

Given the DQ profile of Figure 2 (b) as the profile table for data source S_1, a query engine can estimate that the quality of the attribute "Image" resulting from data source S_1 will be about %50 and can use this information to rank data source S_1 based on the projected quality of the result from a given query. In [28] a service oriented architecture for DQ aware query systems (DQAQS) is proposed in which a DQ profiling service is deployed for generation of DQ profiles from data sets, and can utilize the profiling service for quality aware query results.

Traditional DQ profiles which are similar to the one in Figure 2 (b) are incapable of returning reliable estimates for the quality of the result set most of the times. For example; even though the completeness of the Image attribute for data source S_1 is %50, the completeness of the Image attribute for query results (with conditions) from this data source can be anything between %0 and %100: Completeness of the Image attribute for "Cannon" cameras is %50, this value is %100 for "Sony" cameras, and %0 for "Panasonic" cameras. In reality, for example a Cannon shop may have records of other brands in their database, but they are particularly careful about their own items, which means the quality of Cannon items in database will significantly differ from the overall quality of the database. In this example, the selection conditions (brand is "Cannon" or brand is "Sony") are fundamental parts of the query. However, they have not been considered in previous work on data quality profiling.

Development of DQ profiling methods to generate a minimal DQ profile such that quality of any projected subset of the original dataset can be estimated from it is a challenging problem which we call advanced DQ profiling.

A preliminary consideration towards advanced DQ profiling is introducing an extension of traditional DQ profiles, referred to as *Conditional DQ Profiles*, that are capable of correct estimation of the quality of conjunctive query results. Conditional DQ profile consists of a set of: *Conditions* → *DQ Measurements*, where Condition refers to a query's selection condition (like the WHERE clause in SQL) and DQ measurement infers the quality of the selection query result (e.g. Brand=Cannon → Completeness=%50). Attribute level DQ profile is a special case of conditional DQ profile where the condition is empty. In order to understand conditional DQ profile, we should first define metric function:

Definition 1. *Let* $\{a_1, \ldots, a_m\}$ *be all attributes of the relation R, and metric m be a set of rules. We define* **metric function** $m_a(t)$, $t \in \zeta, \zeta \subseteq R$ *as 1, if the value of attribute a from tuple t, does not violate any rule in m, and 0 otherwise, where ζ is an arbitrary subset of R*

We define the problem of conditional DQ profiling as follows: Given dataset D of relation R, and attribute a find the minimum set of conditional definition tuples Condition → Measurement called DQ profile Pr of R such that $m_a(\sigma_\Phi(D))$ for an arbitrary selection operation $\sigma_\Phi(D)$ where Φ is a selection condition consisting of $\wedge$ and $\vee$ operators can be predicted.

We assume that any attribute $a \in R$ has a limited domain of values. We define $dom(a)$ as limited domain of the values of the attribute a in addition to the special value "-" as don't-care (which can sit instead of any value).

For simplicity, we also assume that conditions within Φ are only consist of equality comparisons since in a finite domain, range comparisons can be defined as a set of equality comparisons.

Brute-force search can be used over the complete domain of possible selection conditions over dataset D to pre-compute the metric function for each possible condition. Assuming results of the brute-force search and the selection condition is stored in DQ profile Pr, we can query Pr to exactly predict DQ of the result set for any given query with any selection condition. For this reason, the search space not only includes all possible equality comparison for attribute a (i.e. $\{\Phi = a \; equals \; d | d \in dom(a)\}$), but also it includes all possible $\wedge$ combination of the equality conditions. We observe that $m_a(\sigma_{\Phi_1 \vee \Phi_2}(D))$ can be directly calculated from $m_a(\sigma_{\Phi_1})$ and $m_a(\sigma_{\Phi_2})$, hence, profile data is independent from the $\vee$ operator.

The conditional DQ profile can become extremely large. Indeed, the conditional DQ profile may become larger than the original database. Methods to drastically prune the search space considering trade-offs between the accuracy, performance, and storage are currently part of our future work.

3.2 Capture User Preference on Data Quality

Preference modelling is in general a difficult problem due to the inherent subjective nature. Typically, preferences are stated in relative terms, e.g. "I like A better than B", which can be mapped to strict partial orders [13]. In quality-aware queries, a mixture of preferences could be defined, e.g. the accuracy of the price and completeness of the user comments should be maximized.

The notion of *Hierarchy* in preferences is defined in the literature [4] as prioritized composition of preferences. For example; completeness of user comments may have priority over completeness of prices. We use the term Hierarchy to define prioritised composition of preferences which can form several levels of priority i.e. a hierarchy. The hierarchy over the preference relations is quantifiable such as: a is strongly more important than b, or a is moderately more important than b. [28] proposes an extension of SQL called DQ aware SQL that utilises the notion of hierarchical preferences.

Consider the Relation $ShopItem(Title, Price, UserComments)$ from source S. Using DQ aware SQL presented in [28] where a specialized HIERARCHY clause is defined to identify the hierarchy of preferences, following query can be user to describe user preferences on DQ.

```
SELECT Title AS t, Price AS p, [User Comments] AS u
FROM ShopItem WHERE ...
HIERARCHY(ShopItem) p OVER (t,u) 7, u OVER (t) 3
HIERARCHY(ShopItem.p) p.Currency OVER (p.Completeness) 3
```

In the HIERARCHY(a) a.x OVER (a.x',...) clause, a denotes the object that the hierarchy is applied for (e.g. ShopItem or column price). a.x denotes the preferred object, a.x',... is list of objects on which a.x is preferred and the number 1..9 denotes the intensity of the importance of the hierarchy. Number 1 denotes weak or slight importance and 9 denotes strong importance. In the above example two hierarchies are defined which indicate that i) for the ShopItem, price is strongly more important than title, and user comments attributes is slightly more important that title. ii) For the column price, Currency is slightly more important than completeness.

Inconsistency Detection in User Preferences. Due to the hierarchical nature of preferences, the uncertainty that happens as a result of inconsistency in user preferences is noticeable since uncertainties propagate to lower levels of preference hierarchy, thus eventually compromising query response. Hence, methods are needed to identify inconsistencies, as well as to notify user about it. Even though inconsistency in the user preferences could be accepted sometimes, informing the user of inconsistencies has no negative effects.

A preference query consists of a set of prioritized orders $\succ_{x,y} w$ where w is the weight of the priority and x and y are other preferences which can be recursively prioritized. Inconsistency detection problem can thus be defined as: Given a prioritized preference $\succ_{x,y} w$ within the preference query, any other recursively inferred prioritized preference should be same as $\succ_{x,y} w$. Searching for inconsistent set of pair prioritized preferences is not trivial.

Preferences can be modelled as directed weighted graphs which can then be efficiently searched for inconsistencies using a heuristic developed in [27]. However proposition of minimal changes to the query to fix the consistency problem still need to be addressed. In addition, an interactive graphical user interface for DQAQS that is able to effectively capture user DQ preferences is developed in [27]. In next section we leverage the quality aware SQL presented in this section to rank data sources based user preferences on DQ.

3.3   Query Planning

One major challenge for query planning in data integration systems is selecting data sources. One method to rank data sources considering the quality of data sources and user preferences on DQ can be based on a multi-criteria decision making technique. The general idea of having a hierarchy in decision making and definition of hierarchy as strict partial orders is widely studied. In [20] a decision making approach is proposed which is called Analytical Hierarchy Process (AHP). Processing the problem of source selection for Quality-aware queries can be delineated as a decision making problem in which a source of information should be selected based on a hierarchy of user preferences that defines what a good source is. In our approach, ranking consists of two phases; i) each DQ metric from the query is assigned a weight which is calculated using AHP technique and ii) sources are ranked against the metric weights and the quality aware metadata of the source.

In AHP, decision hierarchy is structured as a tree on objectives from the top with the goal (query in our case), then the objectives from a higher level perspective (attributes), and lower level objectives (DQ metrics) Fig. 3(a) shows a sample decision tree. This is then followed by a process of prioritization. Prioritization involves eliciting judgements in response to question about dominance

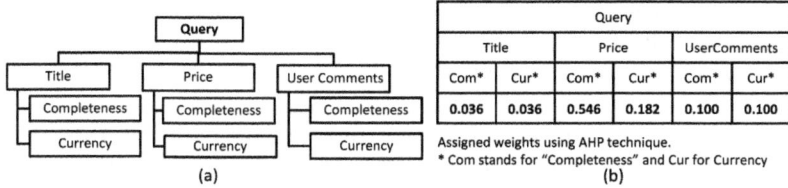

Fig. 3. (a)Decision hierarchy and (b)AHP weightings for querying a virtual shop

of one element over another when compared with respect to a property. Prioritizations form a set of pair-wise comparison matrices for each level of objective.

Within each objective level (query or attribute) elements of a pair-wise comparison matrix represent intensity of the priority of each item (e.g. DQ metrics) over another in a scale of 1 to 9. The Hierarchy clauses can be directly mapped to pair-wise comparison matrices, thus avoids the tedious task of eliciting judgements.

For the sample query in Section 3.2 when all required information for AHP technique (i.e. decision tree and pair-wise comparison matrices) are ready, AHP technique assigns a weight to each DQ metric for each attribute. Figure 3(a) shows the AHP decision hierarchy generated for the example query in Sec. 3.2 and Fig. 3(b) shows resulting weights. These weights are representation of user preferences on DQ and can be used against profiling data of each source for the metric of the same attribute to rank sources.

Having a fixed weight assigned to each decision criterion, which in our case is each column and metric, and a fixed respective profiling value for each source (from quality-aware meta data), sources can be ranked using a number of ranking methods [16]: Simple Additive Weighting (SAW), Technique for Order Preference by Similarity to Ideal Solution (TOPSIS), Elimination et Choice Translating Reality (ELECTRE) , and Data Envelopment Analysis (DEA). In [16] a comparative analysis of the mentioned ranking methods is provided which shows that the effectiveness of all above methods is not considerably different in regards to the source selection problem. The major difference of these methods is their computational complexity. Hence, we incorporate the SAW method which is easy and fast [11].

The SAW method involves three basic steps: Scale the scores to make them comparable, apply the weighting, and sum up the scores for each source. Data profiles of the sources, represent each column's quality metric in a value in [0,1].

$$dq(S_i) := \sum w_j v_{ij}$$

Where $dq(S_i)$ is the final weight of source S_i, w_j is the weight of (column.metric) j which has been calculated through the AHP process and $v_{ij} := \sigma_{(a.m)_j}$ is the profiling value for source i and (column.metric) j. Because $\sum_{j=1}^{m} w_j = 1$, the final scores are in [0,1].

Currently, we are working on DQ aware query planning techniques to handle more complex scenarios like multi-source queries (i.e. joins). Moreover, we believe there is a need to combine the previously mentioned different DQ metrics together with Quality of Service (QoS) metrics (e.g., query planning costs, response time, etc.) into a single overall metric. Such metric will enable comparing different execution scenarios (i.e., plans) in terms of both dimensions. Towards this, we are currently looking at alternative models for capturing user preferences, such as the economy-based model from the Mariposa DBMS [24] as well as Quality Contracts [19], together with ranking policies that work in conjunction with those models.

4   Conclusions

In this paper we proposed a framework for DQ aware query systems. In order to be able to achieve user satisfaction in relation to the quality of query results, the proposed framework utilises three elements of: 1) profiling DQ, 2) capturing user preferences, and 3) query planning. Each framework element holds a number of challenges. Our goal in this paper was to highlight the importance of DQ aware query systems and exhibit an overview of challenges that need to be addressed in order to realize such systems.

Acknowledgement. We would like to acknowledge Dr. Shazia Sadiq who has been highly involved in different stages of this work, in addition we would like to thank Prof. Xiaofang Zhou, and Dr. Ke Deng for helpful discussions that significantly contributed to the development of ideas in this research.

References

[1] Batini, C., Scannapieco, M.: Data Quality: Concepts, Methodologies and Techniques (Data-Centric Systems and Applications). Springer-Verlag New York, Inc., Secaucus (2006)

[2] Benjelloun, O., Garcia-Molina, H., Su, Q., Widom, J.: Swoosh: A generic approach to entity resolution. VLDB Journal (2008)

[3] Bohannon, P., Wenfei, F., Geerts, F., Xibei, J., Kementsietsidis, A.: Conditional functional dependencies for data cleaning. In: ICDE (2007)

[4] Chomicki, J.: Querying with Intrinsic Preferences. In: Jensen, C.S., Jeffery, K., Pokorný, J., Šaltenis, S., Hwang, J., Böhm, K., Jarke, M. (eds.) EDBT 2002. LNCS, vol. 2287, p. 34. Springer, Heidelberg (2002)

[5] Doan, A., Levy, A.Y.: Efficiently ordering query plans for data integration. In: ICDE (2002)

[6] Friedman, T., Bitterer, A.: Magic Quadrant for Data Quality Tools. Gartner Group (2006)

[7] Govindarajan, K., Jayaraman, B., Mantha, S.: Preference Queries in Deductive Databases. New Generation Computing (2000)

[8] Gravano, L., Ipeirotis, P.G., Jagadish, H.V., Koudas, N., Muthukrishnan, S., Srivastava, D.: Approximate String Joins in a Database (Almost) for Free. In: VLDB (2001)

[9] Gravano, L., Ipeirotis, P.G., Koudas, N., Srivastava, D.: Text Joins for Data Cleansing and Integration in an RDBMS. In: ICDE (2003)

[10] Hey, J.D.: Do Rational People Make Mistakes? Foundations of Social Sciences, Economics and Ethics (1998)

[11] Hwang, C.L., Yoon, K.: Lecture Notes in Economics and Mathematical Systems: Multiple Attribute Decision Making: Methods and Appllication. Springer, Heidelberg (1981)

[12] Khatri, H., Fan, J., Chen, Y., Kambhampati, S.: Qpiad: Query processing over incomplete autonomous databases. In: ICDE (2007)

[13] Kießling, W.: Foundations of preferences in database systems. In: VLDB (2002)

[14] Lacroix, M., Lavency, P.: Preferences: Putting More Knowledge into Queries. In: VLDB (1987)

[15] Lakshmanan, L.V.S., Leone, N., Ross, R., Subrahmanian, V.S.: ProbView: a flexible probabilistic database system. ACM TODS (1997)

[16] Naumann, F.: Quality-Driven Query Answering for Integrated Information Systems. LNCS, vol. 2261. Springer, Heidelberg (2002)

[17] Naumann, F., Leser, U., Freytag, J.C.: Quality-driven integration of heterogenous information systems. In: VLDB (1999)

[18] Nie, Z., Kambhampati, S.: Joint optimization of cost and coverage of query plans in data integration. In: CIKM (2001)

[19] Qu, H., Labrinidis, A.: Preference-aware query and update scheduling in web-databases. In: ICDE (2007)

[20] Saaty, T.L.: How to Make a Decision: The Analytic Hierarchy Process. European Journal of Operational Research (1990)

[21] Saaty, T.L.: Multicriteria Decision Making: The Analytic Hierarchy Process: Planning, Priority Setting, Resource Allocation. RWS Publications (1996)

[22] Scannapieco, M., Missier, P., Batini, C.: Data quality at a glance. Datenbank-Spektrum (2005)

[23] Simmhan, Y.L., Plale, B., Gannon, D.: A Survey of Data Provenance in e-Science. SIGMOD RECORD (2005)

[24] Stonebraker, M., Devine, R., Kornacker, M., Litwin, W., Pfeffer, A., Sah, A., Staelin, C.: An economic paradigm for query processing and data migration in Mariposa. In: Proceedings of the Third International Conference on Parallel and Distributed Information Systems 1994 (2002)

[25] Wang, R.Y., Storey, V.C., Firth, C.P.: A framework for analysis of data quality research. IEEE Transactions on Knowledge and Data Engineering(1995)

[26] Wang, R.Y., Strong, D.M.: Beyond accuracy: what data quality means to data consumers. Journal of Management Information Systems (1996)

[27] Yeganeh, N.K., Sadiq, S.: Avoiding Inconsistency in User Preferences for Data Quality Aware Queries. In: Abramowicz, W., Tolksdorf, R. (eds.) BIS 2010. LNBIP, vol. 47, pp. 59–70. Springer, Heidelberg (2010)

[28] Yeganeh, N., Sadiq, S., Deng, K., Zhou, X.: Data Quality Aware Queries in Collaborative Information Systems. In: Li, Q., Feng, L., Pei, J., Wang, S.X., Zhou, X., Zhu, Q.-M. (eds.) APWeb/WAIM 2009. LNCS, vol. 5446, pp. 39–50. Springer, Heidelberg (2009)

SemGen—Towards a Semantic Data Generator for Benchmarking Duplicate Detectors*

Wolfgang Gottesheim[1], Stefan Mitsch[1], Werner Retschitzegger[1],
Wieland Schwinger[1], and Norbert Baumgartner[2]

[1] Johannes Kepler University Linz, Altenbergerstr. 69, 4040 Linz, Austria
[2] team Communication Technology Mgt. Ltd., Goethegasse 3, 1010 Vienna, Austria

Abstract. Benchmarking the quality of duplicate detection methods requires comprehensive knowledge on duplicate pairs in addition to sufficient size and variability of test data sets. While extending real-world data sets with artificially created data is promising, current approaches to such *synthetic data generation*, however, work solely on a quantitative level, which entails that duplicate semantics are only implicitly represented, leading to only insufficiently configurable variability.

In this paper we propose SemGen, a semantics-driven approach to synthetic data generation. SemGen first diversifies real-world objects on a *qualitative level*, before in a second step quantitative values are generated. To demonstrate the applicability of SemGen, we propose how to define duplicate semantics for the domain of road traffic management. A discussion of lessons learned concludes the paper.

1 Introduction

Duplicate detection is an elementary part in data cleansing processes and addresses the identification of multiple different representations of one and the same real-world object within a data set [16]. Such cleansing processes are vital components in information systems that integrate multiple data sources, as it is the case in systems that support situation awareness. We are currently developing a framework for realizing ontology-driven situation awareness techniques [2], including duplicate detection techniques [3], in the sample domain of road traffic management. Real-world objects are described by *object representations characterized by attributes* that specify their spatial and temporal extent, for example in the form of a region on a highway defined by a start and end point. From such attributes, qualitative relations between objects can be derived that characterize various aspects of objects. For example, from a spatial perspective such aspects could be size, distance, or mereotopology of objects. In situation awareness, spatio-temporal data on objects is incrementally reported in streams, describing real-world evolution courses. Within these data, duplicates may occur in multiple forms (see [15] for a taxonomy on the subject). Most relevant

---

* This work has been funded by the Austrian Federal Ministry of Transport, Innovation and Technology (BMVIT) under grant FIT-IT 819577.

J. Xu et al. (Eds.): DASFAA Workshops 2011, LNCS 6637, pp. 490–501, 2011.
© Springer-Verlag Berlin Heidelberg 2011

for our domain are those arising from *identical attribute values* (e. g., two representations of the same traffic jam with the same regions), *contradictory values* (e. g., two representations of the same traffic jam differ in terms of spatial extent, which may be caused by measuring or entry errors), or *missing values* (e. g., only the start values of the region of a traffic jam are given). Duplicate detection is therefore typically performed by computing similarity measures for pairs of representations on a per-attribute basis, which are aggregated into an overall duplicate decision [16].

A semantics-driven approach to synthetic data set generation. We have defined the following three requirements for a test data generator that provides synthetic data sets for testing duplicate detection methods: (i) *Variability* within the generated data set has to be configurable with regard to multiple aspects to support testing effectiveness. This entails providing accurate numbers on generated duplicates to allow the computation of measures such as precision and recall. (ii) *Distributions* within an aspect in the generated data sets have to be configurable, enabling testing duplicate detection methods for robustness. (iii) *Different quantitative representations* should be realizable so that multiple duplicate detection methods can be tested. For instance, in the domain of road traffic management duplicate detection methods might be required to interpret regions with their spatial extent specified either in kilometers or with a distance measures basing on nodes in a graph describing highway exits. Therefore, quantitative representation for both cases have to be generated.

In this paper we propose SemGen, a semantics-driven approach to synthetic data generation. It is based on a qualitative definition of duplicate semantics and requires a set of data with pairs of objects marked as duplicates of each other—in the following called *labelled duplicates*—and non-duplicates, which are both first diversified on a qualitative level according to duplicate semantics of a domain, before in a second step quantitative values are generated, thereby enabling the creation of data sets with high variability and in different sizes.

Structure of the paper. In Section 2 we detail on qualitative descriptions of duplicate semantics, before we describe our approach in Section 3. Section 4 discusses relevant related work on synthetic test data generation, and finally Section 5 concludes the paper with a discussion of its findings and an outlook on further work.

2   Qualitative Description of Duplicate Semantics

Describing duplicate semantics using spatio-temporal relations on a qualitative level has been proposed as a basis for duplicate detection in our previous work [3]. In the following, we provide an overview on these qualitative descriptions and show how to use them for controlling variability and distribution in a generated data set.

Qualitative relations between two objects are expressed by employing *relation calculi*, each of them focusing on a certain spatio-temporal aspect, such as mereotopology [17], orientation [9], or temporality [1]. These calculi are often

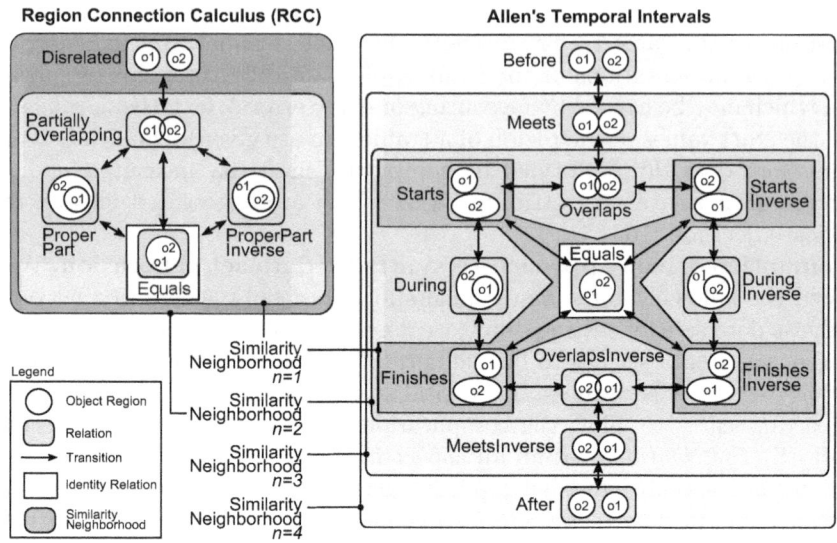

Fig. 1. Conceptual neighborhood graphs of RCC and Allen's Temporal Intervals

formalized by means of *Conceptual Neighborhood Graphs* (CNGs, [10]), which originate in the field of spatio-temporal reasoning. Sample graphs for the Region Connection Calculus (RCC, [17]) and Allen's Temporal Intervals algebra [1] are shown in Fig. 1. In addition, CNGs define similarity between relations since, according to [11], relations are "*conceptual neighbors* if a direct transition from one relation to the other can occur upon an arbitrarily small change in the referenced domain" (e. g., *ProperPart* and *PartiallyOverlapping* are more similar to *Equals* than *Disrelated*). In each such relation calculus, one can define an *identity relation* [3], which states that two objects being in such a relation are most similar according to the particular calculus' aspect of the world (e. g., *rcc:Equals* is the identity relation of RCC, *allen:Equals* the one of Allen's Temporal Intervals). Qualitative relations between objects can be automatically derived from their quantitative attributes using rule-based relation interpretations [2] (e. g., two traffic jams are *PartiallyOverlapping* if their spatial regions overlap). We exploit these relations for describing in which aspects an object and its duplicate should be alike or different. While a number of holding identity relations shows that two objects are *duplicates from identical attribute values* with regard to these aspects, duplicates arising from *contradictory values*, i. e., values describing the same real world object in different ways, can be created by performing *qualitative diversification*. For example, if two objects are in a relation *allen:Equals*, their lifespans are the same, i. e., they "exist" at the same time. Still, they may differ, for instance, in a mereotopological aspect, described by the relation *rcc:PartiallyOverlapping* holding between them. Note that *duplicates arising from missing values* are a special case not reflected on the level of qualitative relations, because, as at this abstraction level we do not know which concrete attributes contribute to a relation, no statement can be made on missing attributes.

For describing duplicate semantics with this definition of CNGs and identity relations, we introduce the concept of *similarity neighborhoods*. A similarity neighborhood is defined by the set of relations reachable within n hops from the calculus' identity relation. Let us denote an instance of a particular object type O_i as a reference object $o_r^{O_i}$, and the similarity neighborhood around $o_r^{O_i}$ as $N_{calculus}(o_r^{O_i}, n)$. Synthetic objects with relations which are part of the similarity neighborhood (i. e., within n hops from the identity relation) are regarded as duplicates to the reference object, whereas objects outside the neighborhood are not labelled as duplicates. In Fig. 1, the similarity neighborhoods for RCC and Allen's Temporal Intervals algebra are shown.

By restricting n for each relation calculus to a particular value, we are able to steer qualitative diversification on a per-calculus basis. In addition, one could define n over multiple relation calculi, with the similarity of the synthetic object being defined by different aspects. For this, a generalization of relation neighborhood from one relation calculus to multiple calculi is necessary, which, however, has already been shown to be straightforward [7] by counting relations in the involved calculi. Finding an appropriate value for n is challenging and requires profound knowledge on the domain's properties. From our experience with duplicate detection in road traffic management we argue that, if using the CNGs of RCC and Allen's temporal intervals as shown above, $n = 2$ is a value yielding reasonable results. Nevertheless, it is desirable to include a larger number of different calculi, which in turn requires an adapted value for n. Using a total of $n = 2$ hops in the CNGs of RCC and Allen's temporal intervals, three different neighborhoods are reachable:

$N_{rcc}(o_r^{O_i}, 2)$: $\{rcc{:}Disrelated\}$
$N_{allen}(o_r^{O_i}, 2)$: $\{allen{:}Overlaps_{Inverse}, allen{:}During_{Inverse}\}$
$N_{rcc \wedge allen}(o_r^{O_i}, 2)$: $\{rcc{:}ProperPart, rcc{:}ProperPart_{Inverse}, rcc{:}PartiallyOver$-
  $lapping\} \times \{allen{:}Starts_{Inverse}, allen{:}Finishes_{Inverse}\}$

Table 1 shows sample duplicates with relations holding between them and the similarity neighborhood they belong to.

Table 1. Sample duplicates with their respective relations

Qualitative Relations	ID	Location		Time	
		begin	end	begin	end
rcc:Equal ∧ allen:Equals: TJ_1' located in $N_{rcc \wedge allen}(TJ_1, 0) \rightarrow$ Duplicates (identical values)	TJ_1	km 6.5	km 8.0	2010-12-01 08:00	2010-12-01 09:00
	TJ_1'	km 6.5	km 8.0	2010-12-01 08:00	2010-12-01 09:00
rcc:PartiallyOverlapping ∧ allen:Finishes$_{Inverse}$: TJ_2' located in $N_{rcc \wedge allen}(TJ_2, 2) \rightarrow$ Duplicates (contradictory values)	TJ_2	km 7.5	km 11.0	2010-12-01 08:40	2010-12-01 00:00
	TJ_2'	km 8.0	km 13.5	2010-12-01 08:20	2010-12-01 09:00

(Legend: TJ = Traffic Jam).

To correlate quantitative values with their qualitative representations, we introduced rule-based relation interpretations that derive relations from object attribute values [2]. As a prerequisite, these relation interpretations assume that attribute values adhere to particular value ranges. These interpretations are domain-dependent, since the definition of such value ranges differs between domains. Again using road traffic management as a demonstration domain and representing mereotopological relations in RCC, let us demonstrate this concept. For deriving such mereotopological relations, using a strictly monotonic, linear space (i. e., road traffic objects, such as traffic jams or roadworks, that occupy a region on a highway) as value range, we can define regions as intervals, whereas given, for example, objects anchored in Euclidian space, we can define a region as a center point with a radius. We can now define the interpretations of relations in RCC ($rcc = \{Disrelated, PartiallyOverlapping, ProperPart, ProperPartInverse, Equals\}$) as functions $f_{rcc} : \mathbb{R} \times \mathbb{R} \to rcc$ mapping object intervals to particular relations (e. g., $PartiallyOverlapping$ may be defined as $o1.start < o2.start \land o1.end > o2.start \land o1.end < o2.end$, as TJ_2 in Table 1 shows). For the purpose of data generation, we use the inverse of these functions, thereby mapping a qualitative relation onto a given value range. For example, for the above specified function f_{rcc} we can use its inverse $f_{rcc}^{-1} : rcc \to \mathbb{R} \times \mathbb{R}$ to map relations between two objects onto the underlying value range. The generation of duplicates arising from missing values can be performed here by providing an inverse function that either maps onto the value range or generates an empty result, such as $f_{rcc}^{-1} : rcc \to \mathbb{R} \lor \emptyset \times \mathbb{R} \lor \emptyset$.

Having laid the foundation for using qualitative data to describe how objects are related, the next section presents our approach to synthetic data generation exploiting the semantics of these relations.

3   Approach

SemGen, our semantics-driven approach to synthetic data generation creates duplicates by taking existing real-world duplicates as the basis for creating additional duplicates that closely resemble real-world characteristics. We control variability in the synthetic data set by using qualitative descriptions as outlined in the previous section. We propose a four-step process as depicted in Fig. 2:

1. *Relation derivation between labelled duplicate pairs* as starting points for the subsequent diversification steps,
2. *Qualitative diversification* to change these relations along the configured aspects,
3. *Quantitative diversification* to map the meaning of each such relation onto the attributes it is derived from, thereby finally characterizing synthetic objects in detail with sample attribute values derived from the attribute values of the labelled duplicate pair, and finally
4. *Export of generated synthetic objects* to an output format suiting the duplicate detection method to be evaluated, such as relational data or an ontology.

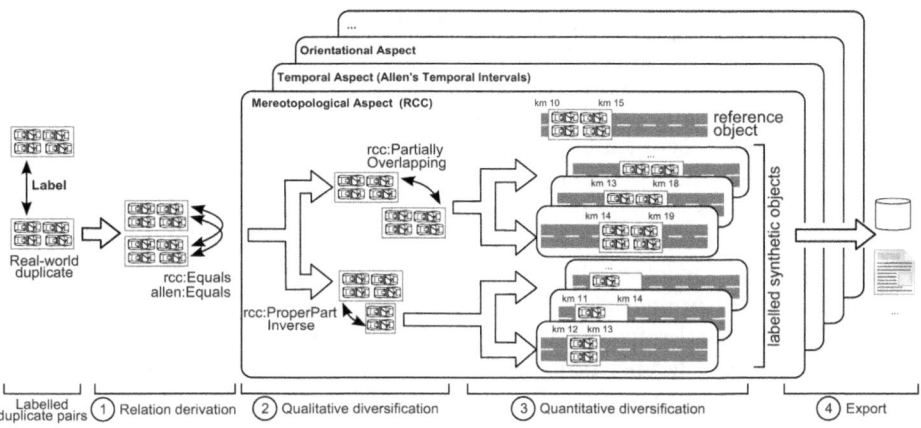

Fig. 2. Overview of synthetic duplicate generation

Input for this process is a reference data set with *pairs of objects labelled as duplicates* as well as *distinct objects*. In road traffic management, such a data set contains information on objects in terms of quantitative attributes, such as spatial extent in kilometers or lifespan given as an interval of timestamps, and in terms of meta-information, such as the data source an objects originates from or an object's type. Common object types are, for example, traffic jams, road works, or lane closures.

Configuration mechanisms allow SemGen's user to control various aspects of the data generation process:

- the *size* of the generated data set,
- the kinds of *relation calculi* that are included in the process,
- the *distribution of relations* within a calculus,
- the *distribution of duplicates* in the whole data set,
- the *distribution of object types* in the resulting data set,
- and the *ratio of duplicates to non-duplicates* per object type.

Given our three causes for duplicates presented above—identical, contradictory, and missing information—it is obvious that the generation of identical duplicates is trivial, since the second step of this process can be omitted, therefore no values other than the original ones can be generated in step 3. Therefore, we focus on the generation of duplicates arising from contradictory and missing information.

(1) Relation derivation. As a prerequisite for the later process steps the relations between objects in a reference data set are derived. These relations serve as starting point for the subsequent qualitative diversification step. Currently, relation calculi relevant to the generation of spatio-temporal data are supported (RCC, spatial distance and size calculi [2], and Allen's Temporal Intervals).

(2) Qualitative diversification with CNGs. On the qualitative level, we employ the conceptual neighborhood of relations and identity relations introduced in Section 2 to define the concept of *similarity neighborhood* for steering qualitative diversification. We can now formulate an algorithm for choosing the relations (of particular aspects) which must hold between a given labelled reference object and its generated synthetic duplicate object, thereby respecting a similarity neighborhood constraint n for each relation calculus. Listing 1.1 shows a pseudocode representation of this algorithm[1]. In principle, the algorithm iterates over the relevant calculi and changes the relations between labelled reference objects and their synthetic duplicates in a random fashion, but constrained by the configured relation distribution. In order to determine the label of the synthetic object, i. e., whether the synthetic object is still a duplicate to its reference object, the algorithm checks whether or not the relation is within the neighborhood of the labelled reference object. If so, the synthetic object is assigned the same label as the reference object, otherwise it is not labelled. During this diversification step, the generation of relations that describe a single synthetic object in a conflicting and, hence, impossible way has to be avoided. An example for such a conflicting configuration occurs, if the *regions of two objects are equal*, while they are at the same time of *different size*.

Listing 1.1. Algorithm to find relations between a reference object and a synthetic object.

```
function select_relations(
  in configuration,
  in ref_object,
  in neighborhood_radii<calculus,n>,
  out relations_to_synth,
  out synth_labels)
  var neighborhood: set<relations>;
      rel: relation;
      relations_to_synth: set<relations>;
      synthetic_labels: set<label>;
  for each (calculus in neighborhood_radii.keys)
    neighbor_relations = N(object, neighborhood_radii[calculus]);
    repeat rel = random_select(neighbor_relations);
    until configuration.is_relation_acceptable(rel);
    if N.contains(rel) then
      // relation is in similarity neighborhood
      synth_label.add(get_label(ref_object));
    end if
    relations_to_synth.add(rel);
  end
end
```

(3) Quantitative diversification. In quantitative diversification, concrete values for object attributes need to be correlated with qualitative relations. For creating sample attribute values, we use the inverse functions to our relation derivation rules as introduced above. Concrete values for a sample region for a synthetic object are chosen randomly from applying the mapped

---

[1] Note that, in this paper, the focus is put on showing its functional principle, thus ignoring possible performance improvements.

Table 2. Exemplary qualitative and quantitative diversification

(a) Qualitative diversification.

		RCC			
		PO	EQ	PP	PP_i
	n	0	1	1	1
Allen's Temporal Intervals	$Finishes_{Inverse}$ 0	⊛			
	$During_{Inverse}$ 1				
	$Overlaps_{Inverse}$ 1				⊙$_{1..n}$
	$Equals$ 1				
	$Starts$ 1				
	$Finishes$ 2		⊖$_{1..n}$		
	$Starts_{Inverse}$ 2				
	$Meets_{Inverse}$ 2			⊙$_{1..n}$	

(b) Quantitative diversification.

ID		Location		Time	
		begin	end	begin	end
⊙$_1$	TJ_1	km 6.5	km 8.0	08:00	08:50
	TJ_2	km 7.0	km 7.5	08:40	09:00
⊙$_2$	TJ_1	km 6.5	km 8.0	08:00	08:35
	TJ_2	km 6.9	km 7.8	08:22	09:00
⊖$_1$	TJ_1	km 6.5	km 8.0	08:30	09:00
	TJ_2	km 6.5	km 8.0	08:35	09:00
⊖$_2$	TJ_1	km 6.5	km 8.0	08:47	09:00
	TJ_2	km 6.5	km 8.0	08:20	09:00
⊛$_1$	TJ_1	km 8.0	km 10.0	08:30	09:00
	TJ_2	km 7.5	km 11.0	08:00	08:30
⊛$_2$	TJ_1	km 8.1	km 9.9	08:45	09:00
	TJ_2	km 7.5	km 11.0	08:40	08:45

interval to the region of the labelled reference object (e. g., a synthetic duplicate that is *ProperPart* of a labelled reference object has randomly chosen interval boundaries that lie within the boundaries of this labelled reference object). Since the relation calculi used for qualitative diversification are designed for reuse, interdependencies between them are not explicitly modeled. For example, the relation *rcc:Disrelated* does not specify in which order and at which distance objects are placed on the highway, leaving many options for quantitative diversification in a strictly monotonic, ordinal value space representing regions on a highway. In case several relation calculi, which describe the same real-world aspect, steer qualitative diversification, interdependencies between them put constraints on quantitative diversification. For instance, consider *rcc:Disrelated* and *spatdist:VeryClose* as a result of qualitative diversification. Then, sample attribute values created during quantitative diversification must satisfy both relation interpretations. To generate duplicates arising from missing values, random null values replacing sample attribute values can be generated during quantitative diversification to better mimic real-world data.

Example. To further illustrate the process described above, we will use two exemplary traffic objects TJ_1 and TJ_2 as shown in Table 1. As a minimal sample configuration based on our experience from road traffic management systems, we choose to use RCC as well as Allen's Temporal Intervals as relation calculi and configure a similarity neighborhood constraint of $n_{RCC} = 1$ and $n_{Allen} = 2$. Table 2a shows the resulting similarity neighborhood as a matrix.

In step (1), we derive relations of the configured calculi for our reference data set consisting, in this case, of two objects TJ_1 and TJ_2, which results in the relations $\{rcc:PartiallyOverlapping \wedge allen:Finishes_{Inverse}\}$ holding between TJ_1 and TJ_2 (denoted in Table 2a as ⊛). This means that their spatial regions overlap and, while the lifespan of TJ_2 begins after the one of TJ_1, both end at the same time. In step (2), these currently holding relations are diversified within the configured relation neighborhoods. This results in 31 possible additional configurations. Finally, in step (3), quantitative representations for these relations are generated based on the original attribute values, with some examples (denoted in Tab. 2a as $\odot$, $\ominus$, $\circledcirc$) shown in Table 2b. Note that attribute values affected in this process are highlighted.

4 Related Work

Automated generation of test data sets is an approach followed in a variety of fields. In the following, we will present domains where data generation approaches are used in order to show commonalities and differences to using data generators as a prerequisite for evaluating duplicate detection methods.

For database systems, Weis [19] distinguishes between data generators that facilitate tasks such as evaluating duplicate detection methods, and those that support the task of testing and improving the performance of a database system. We first cover closely related work from generators that belong to the first category, before we continue with more widely related approaches that fall into the second one. Among those, judging from literature the most well known for generating duplicates in relational data is *DBGen*, also known as *UIS Database Generator*[2], which manipulates records consisting of personal information such as name, address, and social security number by introducing typographical errors or completely changing them in a random fashion [12]. This approach has been refined in [4] to overcome some original limitations, such as poor variability in the set of possible values. Since both approaches are using implicit semantics for domains relying on string-based information, they do not allow to configure variability with regard to multiple aspects, and also lack support for multiple quantitative representations. thus suffering from the limitations described such as the lack of a semantically rich configuration mechanism allowing in-depth control of the generation process. Another approach from the first category is proposed in [8], where synthetic test data also containing duplicates is used to test applications using a relational database. Their goal was to create data sets that allow to verify the correct function of applications that access the database, and to that end, a comprehensive data set covering all relevant cases is required, which also includes the correct handling of duplicates. However, only identical duplicates are regarded, and furthermore configuration mechanisms as proposed here are missing. Thus, they are unable to configure variability with regard to multiple aspects, and do not support more than one quantitative representation.

Other data generators in the database field support the syntactical task of performance improvements by providing a large data set with known statistical properties in an efficient and reproducible way. In the last years, numerous approaches have been presented for generating data, such as [13], which provides efficient generation of large data sets in parallel and is flexibly configurable, or [6], which provides a "Data Generation Language" for specifying the generated data, or [14], using a graph model to control the generation process. But their goal is to efficiently generate large data sets for performance testing which,

---

[2] http://www.cs.utexas.edu/users/ml/riddle/data.html

therefore, have to be consistent and must not contain duplicates. Thus, using these generators for generating test data sets for duplicate detection methods is not feasible. In the area of spatio-temporal databases, frameworks to generate data on moving objects in a quantitative manner have been proposed [5], [18], which focus on generating data to represent the evolution of objects in terms of motion. While they are operating in a similar domain, again the focus is on the generation of consistent data, not of duplicates with known properties.

In summary, although synthetic data generation is an issue in many domains, qualitative approaches have not yet been the focus. Besides, many of these quantitative approaches are heavily domain-specific, limiting their applicability outside their original domain. To date, no data generator for the evaluation of duplicate detection methods in spatio-temporal data has been proposed.

5 Discussion and Further Work

In this section, we discuss several lessons learned during the ongoing implementation of the presented approach, which at the same time represent the directions followed by our further work.

Duplicate variability in real-world data configures qualitative diversification. By deriving relations between labelled duplicates in a small set of real-world data, distribution characteristics per relation calculus can be controlled on a qualitative level (e. g., in RCC, most duplicates may be `rcc:PartiallyOver-`
`lapping`, a smaller portion might be `rcc:Equals`, and some `rcc:ProperPart`). Such distribution characteristics can be used for steering qualitative diversification (thereby promoting CNGs to Bayesian networks), in order to generate synthetic test data sets exhibiting near real-world characteristics.

Qualitative diversification should be aware of error models. In our approach, CNGs steer the qualitative diversification of synthetic duplicates. Although such CNGs can be defined domain-independently, fitting them to the errors encountered in a particular domain is possible. Depending on real-world system factors, such as the type of user interface, various errors may occur: for example, values may be simply outdated (differ from the real value by some offset), or be entered with transposed digits. In road traffic management, for example, traffic jams are either detected by sensors, which may fail arbitrarily (e. g., a large traffic jam may be detected as two smaller, disrelated ones), or entered by humans, which may enter wrong data. Such errors should be represented by adapting the respective CNGs and adding or removing edges, so that errors are in the correct similarity neighborhood.

Characteristics of value ranges bound quantitative diversification. Value ranges provide a model of the world, and may be instantiated to represent different real-world spaces. For example, consider our value range for objects on highways being defined as intervals on a strictly monotonic, linear space. Each concrete highway is an instance of such a linear space that may differ in length from other instances. For quantitative diversification, we may use these additional characteristics as constraints, or deliberately ignore them to also create inconsistent synthetic duplicates.

Generalizing qualitative and quantitative diversification to other domains. Numerous causes for contradictory values in duplicates are known in other domains [16], for instance errors in strings and numbers such as typographical errors and synonyms. Since a representation of these causes as a CNG is possible, extending SemGen towards domains that rely on string representations will begin by defining an appropriate CNG together with relation derivation functions and their inverse functions.

References

1. Allen, J.F.: Maintaining knowledge about temporal intervals. Communications of the ACM 26(11), 832–843 (1983)
2. Baumgartner, N., Gottesheim, W., Mitsch, S., Retschitzegger, W., Schwinger, W.: BeAware!—situation awareness, the ontology-driven way. International Journal of Data and Knowledge Engineering 69(11), 1181–1193 (2010)
3. Baumgartner, N., Gottesheim, W., Mitsch, S., Retschitzegger, W., Schwinger, W.: Towards duplicate detection for situation awareness based on spatio-temporal relations. In: Proceedings of the 9th International Conference on Ontologies, DataBases and Applications of Semantics, Crete, Greece (October 2010)
4. Bertolazzi, P., Santisy, L.D., Scannapieco, M.: Automatic record matching in cooperative information systems. In: Proceedings of the ICDT 2003 International Workshop on Data Quality in Cooperative Information Systems, DQCIS 2003 (2003)
5. Brinkhoff, T.: A framework for generating network-based moving objects. GeoInformatica 6(2), 153–180 (2002)
6. Bruno, N., Chaudhuri, S.: Flexible database generators. In: Proceedings of the 31st International Conference on Very Large DataBases, pp. 1097–1107 (2005)
7. Bruns, H.T., Egenhofer, M.J.: Similarity of spatial scenes. In: Kraak, M.-J., Molenaar, M. (eds.) Proceedings of the 7th International Symposium on Spatial Data Handling (SDH), Delft, The Netherlands, August 1996, pp. 31–42 (1996)
8. Chays, D., Dan, S., Frankl, P.G., Vokolos, F.I., Weber, E.J.: A framework for testing database applications. SIGSOFT Software Engineering Notes 25, 147–157 (2000)
9. Dylla, F., Wallgrün, J.O.: On generalizing orientation information in OPRA$_m$. In: Freksa, C., Kohlhase, M., Schill, K. (eds.) KI 2006. LNCS (LNAI), vol. 4314, pp. 274–288. Springer, Heidelberg (2007)
10. Freksa, C.: Conceptual neighborhood and its role in temporal and spatial reasoning. In: Proceedings of the IMACS International Workshop on Decision Support Systems and Qualitative Reasoning, Toulouse, France, March 1991, pp. 181–187 (1991)
11. Freksa, C.: Temporal reasoning based on semi-intervals. Artificial Intelligence 54(1), 199–227 (1992)
12. Hernández, M.A., Stolfo, S.J.: The merge/purge problem for large databases. In: Proceedings of the 1995 ACM SIGMOD International Conference on Management of Data, SIGMOD Rec., New York, NY, USA, pp. 127–138 (1995)

13. Hoag, J.E., Thompson, C.W.: A parallel general-purpose synthetic data generator. SIGMOD Rec. 36, 19–24 (2007)
14. Houkjær, K., Torp, K., Wind, R.: Simple and realistic data generation. In: Proceedings of the 32nd International Conference on Very Large Data Bases, pp. 1243–1246 (2006)
15. Kim, W., Choi, B.-J., Hong, E.-K., Kim, S.-K., Lee, D.: A taxonomy of dirty data. Data Mining and Knowledge Discovery 7, 81–99 (2003)
16. Naumann, F., Herschel, M.: An Introduction to Duplicate Detection. Morgan & Claypool (2010)
17. Randell, D.A., Cui, Z., Cohn, A.G.: A spatial logic based on regions and connection. In: Proceedings of the 3rd International Conference on Knowledge Representation and Reasoning (October 1992)
18. Tzouramanis, T., Vassilakopoulos, M., Manolopoulos, Y.: On the generation of time-evolving regional data. GeoInformatica 6, 207–231 (2002)
19. Weis, M., Naumann, F., Brosy, F.: A duplicate detection benchmark for xml (and relational) data. In: SIGMOD 2006 Workshop on Information Quality for Information Systems (IQIS), Chicago, IL, USA (June 2006)

Estimating a Transit Passenger Trip Origin-Destination Matrix Using Automatic Fare Collection System

Daming Li[1], Yongjie Lin[2], Xinliang Zhao[1], Hongjun Song[2], and Nan Zou[2]

[1] School of Business Administration, Northeastern University,
No.3 Wenhua Road, Shenyang, Liaoning, P.R. China
damingl@163.com
[2] School of Control Science and Engineering, Shandong University,
Jingshi Road No.17923, Jinan, Shandong, China
yjlin@mail.sdu.edu.cn, nanzou@mail.sdu.edu.cn

Abstract. Automatic fare collection system (AFC) has been widely used for public transport all over the world. However, in China, the most important information, the Origin-Destination (OD) matrix of each bus route, cannot be directly obtained from AFC since alighting information is not recorded at each bus stop. This paper presents an OD estimation model, which applies trajectory search algorithms to track passengers' daily trip trajectory using pre-processed smart card data from all the passengers in one city of China. The results of a rigorous validation with on/off data from a real bus route reveal that the proposed model is quite effective and reliable in estimating the OD matrix for identification of the underlying demand pattern of a transit route. The algorithm is validated using one-day smart card data in Jinan city. The results have shown that the OD estimation from the proposed algorithm match more than 75% with the actual OD pairs. During the peak hours, the matching rate goes up to 85%. Hence, the proposed algorithm significantly improves the utilization of the smart card data. It is valuable to evaluate route network and optimize bus scheduling basing on estimated passenger trip OD matrix.

1 Introduction

Over the past several decades, contending with traffic congestion has emerged as one of the imperative issues during the process of urbanization in developing countries such as China. Development of a transit-oriented urban transport system has been realized by an increasing number of researchers as one of the most effective strategies for mitigating congestion [1]. In recent years, many big cities in China are dedicated to proposing policies and measures for developing efficient public transport systems from both planning and operation perspectives [2]. With the advance of evolving computer and automation technologies, automatic fare collection system (AFC) has been widely used in improving the efficiency of urban public transportation system. In 2010, the utilization ratio of smart cards in Jinan city will reach up to 70%, which could definitely provide reliable data support for the establishment of advanced public transportation system.

J. Xu et al. (Eds.): DASFAA Workshops 2011, LNCS 6637, pp. 502–513, 2011.

Nowadays, most AFC systems around the world only record passengers' boarding information and alighting data is lost, such as bus route, boarding stop and boarding time. To ensure AFC systems utilized efficiently, lots of research issues are proposed, including the stability of data communication, data mining and OD matrix estimation of each bus route. This study will present a new approach focused on estimating OD matrix of transit routes while AFC system does not record passenger alighting data.

In review of the literature, this subject has attracted great attention of researchers in recent years. Barry et al. presented a methodology that estimates station-to-station origin and destination (O-D) trip tables using MetroCard information. A set of straightforward algorithms is applied to each set of MetroCard trips to infer a destination station for each origin station [3]. Trépanier presented a model for the estimation of the destination (deboarding) locations. Some critical issues related to smart card data such as privacy concerns, data structure, data errors, and missing information are discussed [4]. In 2007, Trépanier also presented a model to estimate the destination location for each individual boarding a bus with a smart card. The first application of the model provided a success rate of 66% for destination estimation, reaching about 80% at peak hours [5]. Zhao presented two case studies both in the context of the Chicago Transit Authority [6]. One study proposes an enhanced method of inferring the rail trip OD matrix from an origin-only AFC system to replace the routine passenger survey. The other study is of rail path choice, which employs the Logit and Mixed Logit models to examine revealed public transit riders' travel behavior based on the inferred OD matrix and the transit network attributes. In 2007, Zhao further developed an enhanced method for inferring rail passenger trip origin-destination (OD) matrices from an origin-only AFC system to replace expensive passenger OD surveys [7]. Cui et al. documented the development of an algorithm to estimate a Bus Passenger Trip Origin-Destination Matrix based on ADC system archived data including Automated Fare Collection, Automatic Passenger Count, and Automatic Vehicle Location data [8]. Farzin outlined the process used to create an origin-destination matrix in São Paulo, Brazil, with data available from automated data collection systems [9]. The approach used in this paper addresses a more complex bus network than has been approached before. A very recent study by Lin et al. focused on providing an improved approach for estimating the passenger demand OD matrix for each bus route [10].

Along the line of previous research, this paper presents an enhanced algorithm to estimate passenger trajectory with various transfer distances within satisfied computation time. In addition, algorithms are developed to estimate trip demand matrix of bus stops or traffic zones. Estimated OD matrix represents passenger trip demand so that it can be applied to evaluate or optimize route network and bus scheduling. Especially, in China, this information may be obtained by numerous surveyors and investigators on site, but it would waste much time and fund. Hence, it is greatly significant to present these algorithms to automatically estimate passenger demand OD matrix by using AFC system.

- A *quality attribute* is a collective term that refers to both quality dimensions and quality indicators as shown in Fig. 1. A quality attribute can be regarded as the characteristic of a resource, or a resource attribute.

These terms define a data model to facilitate the resource quality assessment. The assessment consists of subjective resource quality dimensions that are based on pre-defined corresponding quality indicators of an online resource. These indicators can be collected from the resource's original site, or can be harvested from available metadata descriptions of the resource. Values of the collective resource quality attributes can then be derived from these quality indicators enabled by certain techno-logical means. These quality attributes can assist domain experts to make informed value judgments on whether or not to include the resource in a portal. This can be used to facilitate semi-automated quality assessment process since it explicitly defines the attributes which should be taken into consideration when making a decision on the quality of the information resource.

3.1   An Attribute-Based Data Model for the Healthcare Domain

In this section, an attribute-based data model is instantiated for the healthcare domain (see Fig. 1). The model consists of the subjective quality dimensions, corresponding objective quality indicators, and finally the quality attributes in order to automate the resource quality assessment with the use of intelligent techniques.

Resource quality in the context of health information portals is defined as the extent to which information contained in a web-based resource meets a user's information needs and quality perceptions [3]. In this paper, we differentiate resource quality (RQ) from information quality (IQ). While IQ is considered as a resource attribute, RQ is regarded as a relationship between a resource and its actual user. This view requires resource quality to be assessed in a way to reflect user information needs and preferences. Informed by the literature analysis of user quality perceptions and user information needs in the healthcare domain [11-12], we define RQ as a composition of *Reliability* and *Relevancy*, as denoted in the following formula:

$$RQ = Ry \ (Ac, Cr, Cu) \cup Re \ (Up, Ip, Si) \tag{1}$$

Where: RQ refers to *Resource Quality*, Ry denotes *Reliability*, Ac corresponds to *Accuracy*, Cr - *Credibility*, Cu – *Currency*; whereas Re denotes *Relevancy*, and Up relates to User profile, Ip - *Information preference*, and Si - *Subject interest*.

In this study, *Reliability* is defined as the extent to which a resource and its source are regarded as true, credible, and up-to-date. An initial set of quality dimensions was selected, including *Accuracy*, *Credibility* and *Currency*, which were perceived as best describing health information consumers' quality concerns [3]. *Relevancy* in this study is defined as the extent to which a resource is applicable and useful for a user's information needs. It needs to be emphasized that in this paper, we only focus on the measurement of *Reliability* as a multi-dimensional concept.

Having quality dimensions defined, the next step is to define quality indicators, which are regarded as evidences contributing to the evaluation of certain quality di-mensions. In the literature quality indicators have been categorized according to their use in constructing quality metrics. For instance, quality indicators are classified as direct quality indicators versus indirect ones [13], or technical quality indicators

a) The lack of information on route. China's public vehicles are allocated to constant routes. Therefore, the routes are related to vehicle IDs.

b) The lack of information on time and date. The raw data of the same route is obtained according to time order, so missed time and date can be estimated through passenger information taking same bus.

c) The lack of information on stop. By comparing to historical information on route and time, the paper maybe identifies boarding stop that carries the most identical information.

d) The lack of vehicle information. This paper compares the information on route, stop and time from smart card data stored in database, and then selects most similar historical data record to estimate missing vehicle information.

2) False recording information. It mainly refers to a passenger swipe her or his card several times at the same time, but it is impossible for a passenger to take the bus several times at a same time. This problem is mainly caused by the instability of the card-reading equipment, which generates repeated information. One of solutions is to keep the first record.

3) Redundant recording information, such as card style, consumption amount, the balance. This type of information is mainly used to manage finance and information. It is useless to analyze travel demand. To cut the workload of data search, this paper omits this type of information.

2.2 Trajectory Search Algorithms

Assumptions

Suppose passenger's trajectory starts from the origin stop and ends with the destination of each trip, this paper attempts to estimate passenger's daily trajectory from boarding information recorded by AFC at each bus stop. To ensure that the proposed algorithms can be tackled and also realistically reflect the real-world constraints, this study has employed the following two assumptions in the model formulation:

1) Passenger's previous trip destination is nearby the bus stop of next trip origin.
2) Clock of all buses is synchronous.
3) At the end of the day, passengers return to the first boarding stop of the day.

Figure 1 shows a typical passenger's daily trajectory. Here, "P" stands for one bus stop. During the entire day, there are 3 bus trips for this passenger, with b1, b2 and b3 as passenger boarding stops. According to Assumption 1, one can conclude that each trip destination is located around the origin of the next trip. For example, if the starting stop of first trip is b1 and the starting stop of the second trip is b2, the destination of the first trip should be close to b2. The detailed searching algorithms are described as follows.

Notations

To facilitate the following illustration, all definitions and notations used hereafter are summarized below.

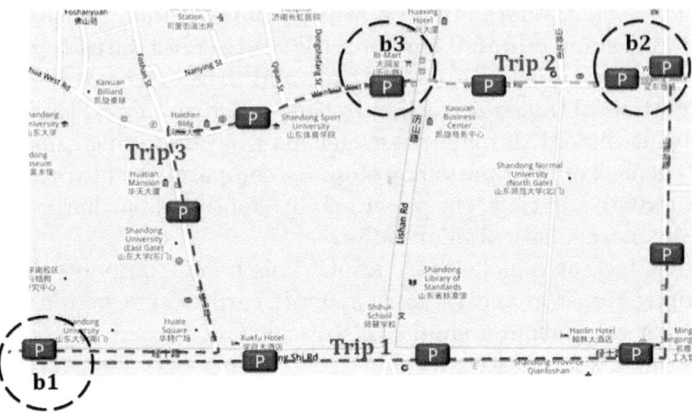

Fig. 1. A passenger's trips distribution

r: The bus route of a city, $r = 1, 2, \ldots, R$,

s_r: The stop of the rth bus route, $s_r = 1, 2, \ldots, S_r$,

sn_{s_r}: The name of the s_rth bus stop from the rth bus route, which is same for overlapped stops from different routes,

$SN = \{(sn_{s_r})\}$: The set of bus stops,

x_{s_r}: The longitude of the s_rth bus stop,

y_{s_r} : The latitude of the s_rth bus stop,

p : The serial number of passengers, $p = 1, 2, \ldots, P$,

or_{p_i} : The bus route from which the pth passenger boards his/her ith trip $i = 1, 2, \ldots, I_p$,

os_{p_i}: The bus stop at which the pth passenger boards his/her ith trip,

ds_{p_i}: The estimated stop at which the pth passenger alights his/her ith trip,

$PT = \{(os_{p_i}, ds_{p_i})\}$: The set of passenger trajectories,

$ta_j = \{sn_{s_{r1}}, sn_{s_{r2}}, \ldots, sn_{s_{rk}}\}$: The set of bus stops that belong to the jth traffic zone,

$m(sn_{s_a}, sn_{s_b})$: The number of passengers who travel from Stop sn_{s_a} to sn_{s_b}. It represents trip demand between sn_{s_a} and sn_{s_b},

$m'(ta_a, ta_b)$: The number of passengers who travel from zone ta_a to zone to zone ta_b. It represents trip demand between ta_a and ta_b ,

D: The standard maximum transfer distance, which is determined by the city's urban economy status, coverage of public transport network and other related factors

θ: The variation of maximum transfer distance, which is determined by one passenger's individual factors, such as age, gender, education, and income level. Usually, θ complies with normal distribution. The distance between two bus stops is defined as follows:

$$d(s_a, s_b) = \sqrt{(x_{s_a} - x_{s_b})^2 + (y_{s_a} - y_{s_b})^2} \tag{1}$$

The distance between one stop and one route refers to minimum distance between the stop and any stop from the route. This distance can be determined by the following equation.

$$d(s_a, r) = min(\sqrt{(x_{s_a} - x_{s_r})^2 + (y_{s_a} - y_{s_r})^2}), \forall s_r \leq S_r \qquad (2)$$

With Equation 1, the area of traffic zone can be figured determined. If the distance between one pair of bus stops does not exceed the scope, these stops can be treated as in the same traffic zone, i.e.

$$d(sn_{s_i}, sn_{s_j}) \leq \gamma_j, \forall sn_{s_i} \in ta_j, sn_{s_j} \in ta_j \qquad (3)$$

γ_j represents the scope of the traffic zone in Equation 3Furthermore, there does not exist any bus stop that belongs to two different traffic zones, i.e.

$$ta_i \cap ta_j \equiv \Phi, \forall i \neq j \qquad (4)$$

Passenger's alighting stop is determined by the next boarding route and stop. If the distance between next boarding stop and current occupied route satisfies the following condition

$$d(os_{p_{i+1}}, or_{p_i}) \leq D + \theta \qquad (5)$$

Then, the alighting stop for this trip can be estimated. Moreover, the alighting stop not only belongs to the or_{p_i} route, but also satisfies the following condition

$$d(s_r, s_a) \leq d(s_c, s_a), \forall s_c \leq S_r \qquad (6)$$

The alighting stop of the current passenger can be estimated with the following equation.

$$ds_{p_i} = s_r \qquad (7)$$

In order to calculate all passengers' trip OD information for one entire city, this paper applies the following algorithms on smart card data, eliminates data errors and estimates the missing data. And then, it developed the improved algorithms with accelerated searching strategies to estimate passengers' trajectory.

Algorithm

The feasibility of transferring in this paper is determined by the transfer distance, which varies among passengers. Once the search criteria are met, passenger trajectory and destination of each trip can be estimated. This algorithm analyzes the data collected by AFC during the whole day. China is a country with dense population. Particularly in large and medium-sized cities, the number of passengers' daily trips is much more than average values worldwide. Taking Jinan city, the capital of Shandong province, as an example, by 2010, the number of daily bus passengers in the city is up to 2.5 million and the amount of daily card-swiping is above 1.5 million. Theoretically, it requires around 2.25 trillion cycles to estimate the trajectory of all the passengers. Moreover, each loop operation also includes massive data comparison and matching calculation. It is very hard to reliably estimate OD matrices for all passengers with traditional algorithm within reasonable time. Therefore, it is necessary to investigate searching

---

Algorithm 1.

Input: RP, the set of smart card data from all the passengers in one city;
n_{s_r}, the set of bus stop number for one city.

Output: Alighting stop of each passenger's every trip. Both boarding stops and estimated alighting ones are combined into the set of trip trajectory PT.

1: Descend RP according to card number and corresponding time;
2: **for** $(P = 1, RP_p = \emptyset; p \leq P; p + +)$ **do**
3:     Query or_p, os_p into RP_p from RP limit 50, and descend RP_p according to card number and corresponding time;
4:     **for** $(i = 1; i \leq I_p; i + +)$ **do**
5:         Let $d = 100000$, $ds_{p_i} = 0$;
6:         **for** $(s_{or_{p_i}} = 1; s_{or_{p_i}} \leq S_{or_{p_i}}; s_{or_{p_i}} = s_{or_{p_i}} + 1)$ **do**
7:             **if** $(i = I_p)$ **then**
8:                 {If $(d \leq d(os_{p_{i+1}}, s_{or_{p_i}}))$ $d = d(os_{p_{i+1}}, s_{or_{p_i}})$};
9:             **else**
10:                 {If $(d \leq d(os_{p_1}, s_{or_{p_i}}))$ $d = d(os_{p_1}, s_{or_{p_i}})$};
11:             **end if**
12:             **if** $(d \leq D + \theta)$ **then**
13:                 $ds_{p_i} = s_{or_{p_i}}$; // Estimated alighting stop
14:             **end if**
15:         **end for**
16:     **end for**
17:     Delete or_p os_p from RP in descending order according to card number and corresponding time, $RP_p =$ empty passenger data set;
18: **end for**

---

algorithms to efficiently analyze massive amount of data. The search algorithms developed in this paper are described below.

By using the above algorithm for estimating passengers' travel trajectory and searching smart card data, it is possible to reliably estimate all passengers' daily trajectories within one specified day.

3   Travel Demand Matrix

Trip demand matrix describes travel demand from each stop to all other stops. It represents the travel demand of every stop of a route, and supports route network optimization and bus scheduling. Passenger trajectory is the basis of travel demand matrix about traffic zone or bus stop. Each traffic zone has included different bus stops. By using the above algorithm, this paper estimates all passengers' trajectories of one city, and the trajectories. After the passengers' trajectories are obtained, one can then calculate travel demands in different traffic zone or bus stop. The estimation algorithm developed in this paper is as follows:

With the above algorithm, it is efficient and reliable to estimate passengers' travel demand matrix. The demand matrix can be used to evaluate or optimize both bus route network and bus stops, and support advanced bus scheduling system by providing real-time data on passenger flow.

Algorithm 2.

Input: PT, the set of trip trajectory and SN, the set of bus stop name for one city.
Output: M, passengers' travel demand matrix for each bus stop; its dimension is
    equal to $sn \times sn$, while sn represents the number of bus stop name for one city.
  1: Descend PT according to boarding stop number and alighting stop;
  2: **for** $(a = 1, M =$ zero matrix, $p = 1; p \leq P; p + +)$ **do**
  3:    **for** $(i = 1; i \leq I_p; i + +)$ **do**
  4:      $M(sn_{os_{p_i}}, sn_{ds_{p_i}}) = M(sn_{os_{p_i}}, sn_{ds_{p_i}}) + 1;$
  5:    **end for**
  6:    Delete the p passenger's trajectory or_p, os_p from PT;
  7: **end for**

Further, this paper presented a new definition of traffic zones. The number of passengers for traffic zones represents trip demand. The travel demand matrix in different traffic zones takes the zone as the starting and end points. This paper mainly studies stop-determined traffic zone, i.e. traffic zone is the set of stops with relatively close physical distance. The travel demand matrix of passengers between any two traffic zones refers to the number of passengers belong to the two traffic communities. Here is the specific calculation method:

Algorithm 3.

Input:
    Passengers' travel demand matrix M and the set of all traffic zones $TA = ta_1 \cup$
    $ta_2 \cup \cdots \cup ta_K;$
Output:
    The travel demand Matrix M of the passengers grouped by traffic zones, and its
    dimension is equal to $K \times K$, with K representing the number of different bus stop
    name for one city.
  1: Let $M' =$ zero matrix;
  2: **for** $(a = 1; a \leq sn; a = a + 1)$ **do**
  3:    **for** $(b = 1; b \leq sn; b = b + 1)$ **do**
  4:      find the cell $M'(ta_i, ta_j)$ where $sn_{os_{p_i}}$ is found in ta_i and $sn_{ds_{p_i}}$ is in $ta_j;$
  5:      $M'(ta_i, ta_j) = M'(ta_i, ta_j) + 1;$
  6:    **end for**
  7: **end for**

With this algorithm, it is feasible to reliably calculate the travel demand matrix of all the passengers in any two different traffic zones. The matrix can be used to evaluate, design or optimize bus route network, bus stops, and traffic hubs. Furthermore, it is significant to support regional advanced bus scheduling, improve bus operation efficiency, give a full play to bus carrying capacity, and mitigate traffic congestion.

4   Case Studies

Taking Jinan as an example, this paper analyzes passengers' information collected by the automatic fare collection system, and validated the accuracy and efficiency of the algorithm. At the end of Year 2010, the total number of bus route is 186, and the total operation length is 3,383 kilometers. The number of passengers served is 2.2 million and the number of card swiping is more than 1.5 million.

This paper takes Route 115 of Jinan city as an example to estimate passenger trajectory and travel demand matrix of every stop by using proposed algorithm and search strategies. The operation time of the inbound route starts from 5:00 and lasts to 20:30 while the outbound service starts from 6:00 and lasts to 21:10. This route passes through both the urban and suburb areas of Jinan. The number of card-swiping activity on this route is about 10,000 times per day. This paper estimates passengers trajectory and travel demand matrix by using a program written in C++ programing language and MYSQL database environment based on the proposed algorithms. With search strategies proposed in this paper, the computation time is 5 times faster than regular calculation algorithm (on a personal computer with Intel Pentium Dual E3200 at 2.4 GHz CPU). Therefore, the calculation time of all the bus routes in the city will be largely reduced. By analyzing historical data, there are morning peak hours and evening peak hours that last from 7:00 to 9:00 and from 17:30 to 19:30 respectively. With the proposed algorithms, the successful rate of daily passenger trajectory estimation is 70% and the rate in peak hours is over 80%. Some trips cannot be reliably estimated due to the following reasons:

1) There is only one trip for one passenger in a day, so it is impossible to estimate the alighting stop;
2) Passengers travel more than once, but the distance between the first alighting stop and the starting point of next stop is too large, so it is not reliable to estimate the destination of last travel. It can also be explained by mixed usage of transportation modes (such as taxi, subway, etc.), which breaks the trip chain of buses.
3) Passengers travel more than once, but they did not return to the starting point at the last travel. The trajectory cannot be estimated in such case.

In the case study, the paper first analyzes the estimated passenger travel demand matrix on the basis of bus stops, and studies the passenger demand of both inbound and outbound routes respectively, which is shown in Figure 2(a), 2(b), 2(c) and 2(d), which are 3-D graphs, with horizontal axis standing for boarding stop while the first vertical axis standing for alighting stop and the second vertical axis standing for the number of passengers, i.e. the number of passengers between the boarding stop and alighting stop. From the figure, it also shows that the travel demands between some stop pairs are quite huge.

Then, it further analyzes the passenger flow in both the going-to-work peak hours and off-work peak hours. By comparing travel demand distribution trend of the going-to-work peak hours and off-work peak hours, the paper has validated the accuracy of this algorithm. In inbound direction, if the first vertical axis is summed according to boarding stop of horizontal axis, boarding passenger travel demand of each stop can be obtained as shown in Figure 3(a). For the same reason, if the horizontal axis is summed according to alighting stop of first

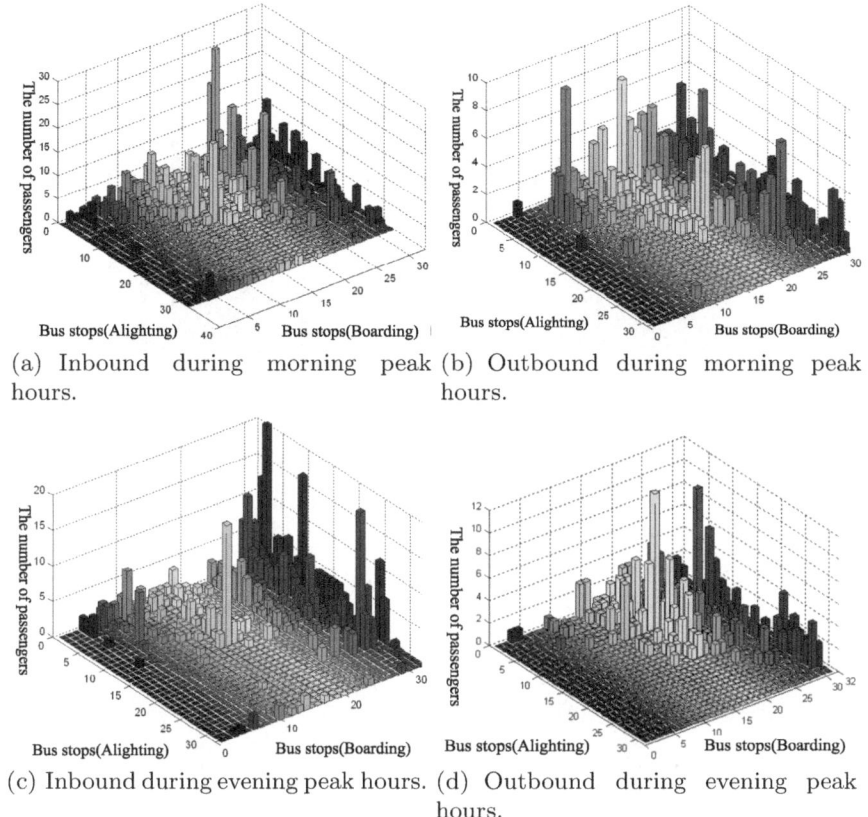

(a) Inbound during morning peak hours.

(b) Outbound during morning peak hours.

(c) Inbound during evening peak hours.

(d) Outbound during evening peak hours.

Fig. 2. The number of passengers' distribution of route 115 in one day

vertical axis, alighting passenger travel demand of each stop can be calculated as shown in Figure 3(b). Meanwhile, the boarding and alighting travel demand in outbound can easily figured out as shown in Figure 4(a) and 4(b).

Figure 3 shows the number of passengers in inbound route of each stop and there are 32 stops, while Figure 4 shows the number of passengers in outbound route of each stop and there are also 32 stops. The difference between them is that there is not only different series number for the same stop in inbound and outbound route, but also the sum of any series number is 33 for the same stop in both inbound and outbound route.

We can be seen there are a lot of boarding passengers at the $1^{st} - 4^{th}, 6^{th}, 11^{th}$, and 14th stops in the inbound while the number of alighting passengers at the 24th and 32nd stops is also quite larger than others. This keeps in touch with reality because these stops are near resident zones, commercial zones, companies, and schools. In Figure 4, it is off-work peak hours, and the number of passengers at the $1^{st}, 9^{th}, 10^{th} - 20^{th}, 26^{th} - 27^{th}, 29^{th} - 32^{nd}$ stops are quite larger than

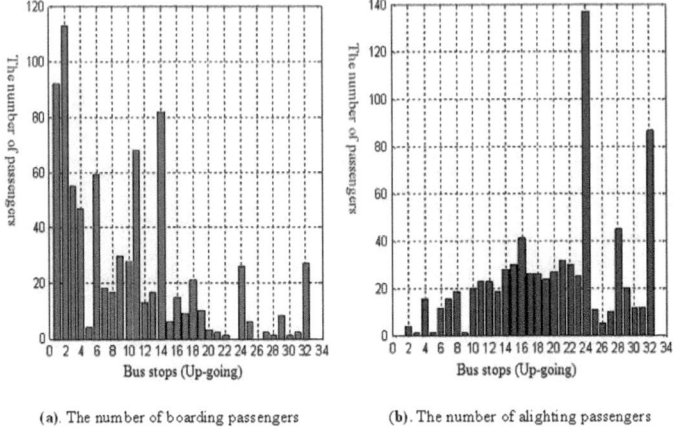

(a). The number of boarding passengers   (b). The number of alighting passengers

Fig. 3. The inbound passengers' distribution during morning peak hours

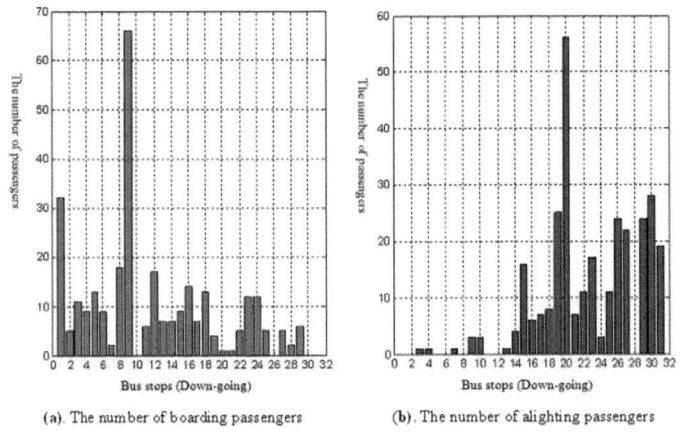

(a). The number of boarding passengers   (b). The number of alighting passengers

Fig. 4. The outbound passengers' distribution during evening peak hours

others, and it is because many workers return to home after work. As some off-work workers go shopping or dinning after work, the passenger flow is a bit scattered.

5   Conclusions

Based on the analysis of China's bus operation management and AFC, this paper summarizes existing algorithms about analyzing smart card and proposes an algorithm that can estimate passenger trajectory based on variant transfer distance. The proposed acceleration algorithm improves the searching efficiency from massive data sets. The developed algorithms can mine smart card data to improve the use of automatic fare collection system. Furthermore, estimated passenger trajectory and travel demand are validated by numerical tests. The matrix in this paper is significant in supporting trip demand estimation, bus dispatching and route optimization for planners and decision-makers. The searching strategies are presented to improve the operation efficiency, which provide the capabilities of estimating OD matrix of each bus routes in large and medium-sized cities in China and around the world.

Note that this paper has presented preliminary for the proposed model through a case study. More extensive tests or evaluations will be essential to assess the effectiveness of the proposed model with more data samples and to account for additional critical issues such as raw data mining, passengers' records recovery, and the calibration of asynchronous clock for each bus.

References

1. Yu, J., Liu, Y., Chang, G.L., Ma, W., Yang, X.: A Cluster-Based Hierarchical Model for Urban Transit Hub Location Planning: Formulation, Solution, and Case Study. Transportation Research Record 2112, 8–16 (2009)
2. Yu, J., Liu, Y., Yang, X.: Cluster-based Optimization of Urban Transit Hub Locations: Methodology and Case Study in China. Transportation Research Record 2042, 109–116 (2008)
3. Barry, J.J., Newhouser, R., Rahbee, A., Sayeda, S.: Origin and Destination Estimation in New York City with Automated Fare System Data. Transportation Research Record: Journal of the Transportation Research Board No. 1817, 183–187 (2002)
4. Trépanier, M., Chapleau, R.: Destination estimation from public transport smart card data. In: The 12th IFAC symposium on Information Control Problems in Manufacturing, Saint-Etienne, France (2006)
5. Trépanier, M., Tranchant, N., Chapleau, R.: Individual Trip Destination Estimation in a Transit Smart Card Automated Fare Collection System. Journal of Intelligent Transportation Systems: Technology, Planning and Operations 11(1), 1–14 (2007)
6. Zhao, J.: The Planning and Analysis Implications of Automated Data Collection Systems: Rail Transit OD Matrix Inference and Path Choice Modeling Examples. Master's Thesis, Massachusetts Institute of Technology, Cambridge, MA (2004)
7. Zhao, J., Rahbee, A., Wilson, N.: Estimating a Rail Passenger Trip Origin-Destination Matrix Using Automatic Data Collection Systems. Computer-Aided Civil and Infrastructure Engineering 22, 376–387 (2007)
8. Cui, A.: Bus Passenger Origin-Destination Matrix Estimation Using Automatic Data Collection Systems. M.S. Thesis, Massachusetts Institute of Technology, MA (2006)
9. Farzin, J.: Constructing an Automated Bus Origin-Destination Matrix Using Fare card and GPS Data in São Paulo, Brazil. Presented at the 87th TRB Annual Conference, Washington, DC (2008)
10. Lin, Y.J., Jia, L., Zou, N.: Estimating Passenger Origin-Destination Matrix of Fixed-fare Bus Smart Card Usage Information. Presented at the 17th World Congress on Intelligent Transportation System, Busan, Korea (2010)

An Approach to Assess the Quality of Web Pages in the Deep Web

Tiezheng Nie, Ge Yu, Derong Shen, Yue Kou, and Dejun Yue

College of Information Science and Engineering, Northeastern University,
110819 Shenyang, China
{nietiezheng,yuge,shenderong,kouyue,yuedejun}@ise.neu.edu.cn

Abstract. Web pages contain a large number of structured data, which are useful for many advanced applications. Existing works mainly focused on extracting structured data from web pages by individual wrappers but ignored the
quality for these underlying web pages, which in fact impact the extracting results seriously. Thus, we define the quality of a web page by the data quality a
wrapper can achieve in extraction. This paper proposes a novel approach to assess the quality of web pages in the deep web. In our approach, we first define
the schema of web data with a hierarchical model. Then web pages are dealt
with as XML documents and parsed into a DOM tree. The data units and attribute values in the web page are annotated with the schema semantics and the
XPATH of position in the DOM tree. Based on the annotation, we build an assessment model for the quality of web pages with two dimensions: the structure
complexity and the text complexity of node in the DOM tree. The quality is partitioned into three quality levels in our model, and the quality of web pages in
the same quality level is compared by the proposed formulas. Moreover, we design an XQuery-based wrapper to extract the web page and validate our quality
model since most of existing wrappers can not handle the data with hierarchical
structure. The wrapper generates XQuery statements to extract web data with
the annotation information. The experimental results demonstrated our approach is accurate for assessing the data quality of web pages. It is very helpful
for data quality control in the deep web related applications.

1 Introduction

In the surface web and the deep web [1], there are a large number of structured and
semi-structured data contained by web pages. These data can be used in many advanced applications and provide high quality data. However, these structured data are
represented in the form of HTML, which is designed for human browsing instead of
data processing. So, using wrapper to extract structured and semi-structured data from
web pages of data sources has become an important work for making full use of web
data. There are many existing works that focus on generating wrappers for extracting
data of the deep web. Early approaches [2, 3, 4] generate wrapper based on manual
techniques, in which programmers find some patterns from Web data pages of sites
by analyzing page code and define rules of extraction in wrapper. Then, some
automatic approaches [5, 6, 7] were proposed for data extraction, which learn patterns

J. Xu et al. (Eds.): DASFAA Workshops 2011, LNCS 6637, pp. 514–525, 2011.
© Springer-Verlag Berlin Heidelberg 2011

from multiple similar pages by parsing Web pages into a tag tree. Moreover, vision-based approaches [8, 9] that analyze the layout of data pages and use the visual representation of pages to extract data were proved effectively too. To extract web data from a large scale of web pages, most wrappers choose a set of web pages from a data source as sample pages to train a template, and use the template to extract data from other web pages in the data source.

In current researches, the quality of web data becomes more important for data selection, customization and integration in advanced applications [19]. However, most of existing works devote to improving the accuracy of extracting data records from source web pages, and they ignore the importance of quality assessment for web data sources. In the practice of extracting the web data, the accuracy of data extraction depends on not only algorithms used in wrappers but also the quality of source web pages which is a part of data quality on the web. The quality of source web pages is a kind of representational data quality in data quality categories proposed in [20] and indicates the data quality a wrapper can achieve in extraction. In web pages of the deep web, the data of hidden databases are reorganized by programmers with HTML to display in browsers. Values of different attributes may be connected in the same elements, and additional elements might be inserted for highlighting some contents of search results. So some web data pages have the structures that are difficult to be accurately analyzed for all wrappers. Therefore, for source web pages, it requires methodologies to assess their structure quality, which is important for applying web data in the deep web related applications.

In this paper, we focus on propose a novel approach to assess the quality of web pages for the extraction of web data. For extract data from a large scale of web pages, existing wrappers are conditioned to analyze the web pages with a tree structure and identify values from text content of nodes. Therefore, we assess the quality of web pages with two dimensions: the structure complexity and the text complexity. Since web pages can be easily formatted into XML documents, we use the DOM tree of XML document to represent the structure of a web page. In the DOM tree, values of attributes of data records are contained by text of nodes. These nodes should regularly appear in the layout structure of the DOM tree. However, additional elements for visualization break the regular structure of data records, which is the structure complexity and leads to the mistakes in the data extraction. On the other hand, text of nodes may contain not only values of a given attribute but semantic tags or values of other attributes, which is the text complexity of web pages. In this case, wrappers must design very complex algorithms to align values for the annotation of web data.

In our approach, with features of the structure complexity and the text complexity, we construct multiple quality levels to describe the quality of web pages. And for a given web page, we transform it into a DOM tree and extract data with manually method that is required for amending mistakes. Then we annotate each data values with the attribute name of predefined schema and the XPATH in the DOM Tree. We propose the algorithms to extract patterns for locating and aligning data values based on the information of annotation. Based on patterns, we classify a web page into an proper quality level. Furthermore, the method that computes scores with patterns is also proposed for comparing web pages in the same quality level.

Moreover, to verify the validity of the quality assessment, a wrapper is required to extract data from web pages. Most of existing wrappers are table-oriented and only deal with the relational structured data. They suppose that the data in web pages are stored in an underlying relational table of databases behind the interfaces. However, not all of data in the deep web are relational data, and some data may have a hierarchical structure in their schema when they are displayed in the result pages. For example, the result data of hotel booking include two levels that are hotel and room types. To address this problem, we develop an XQuery-based wrapper to extract hierarchical data. XQuery[10] as an XML query language has been well defined by the W3C organization, XQuery statements can express queries across all data with XML format. Therefore XQuery can deal with both semi-structured data and structured data in web pages. In our approach, the construction of XQuery statements is based on the annotation of web data. Using XQuery also has other advantages: firstly, XQuery is more easily for human reading and can be re-edited on demand by programmers; secondly, since XQuery is supported by most database systems, it can improve the compatibility of the wrapper.

The rest of this paper is organized as follows. Section 2 introduces the related works of this paper. In section 3, we present the preprocessing on web pages for the assessment. The methodology of quality assessment for web pages is presented in Section 4. Section 5 introduces the construction of XQuery-based wrapper. Section 6 shows the experimental results. Section 7 concludes the work of this paper.

2   Related Works

The assessment [22] is one of fundamental issues for the data quality [19]. Some works [23, 21] have studied this problem in various approaches. However, about how to assess the quality of web pages for data extraction in the deep web, there is few works to research this problem to the best knowledge of us. Methods used by search engines are employed to rank web pages for query results. And methods for web data mainly focus on the intrinsic data quality and the contextual of data quality web pages, e.g. the accuracy and completeness of data. But for data extraction, the representational quality of web pages is very important and determines the accuracy of result data. So in this paper, we evaluate web pages in dimensions of the representational quality.

For the web data, existing works focus on assessing the quality of data with the extraction accuracy of wrappers that has been studied by many researchers. Early wrappers are based on manual techniques, which require human analyzing the feature and structure of pages. Hammer [11] extracts semi-structured information from Web pages by a declarative specification. WIEN [12] uses a combination of empirical and analytical techniques to evaluate the computational tradeoffs among six wrapper classes, in which both expressiveness and efficiency are considered. In WebOQL[13], a web page is abstracted into the structure of syntax tree, and then a user generates the query to locate the position of interested data. XWRAP[14] use a semi-automatic generation of wrapper programs, which use heuristic rules to find interested data.

Recently, many automatic approaches for generating wrappers have been proposed. The structure-based approaches parses web pages into a special structured model, and extract data by analyzing the structure of pages. Most of them use a DOM tree or a tag tree to find patterns in data extraction. Roadrunner [5] uses multiple pages containing similar data records to find patterns or grammars in DOM trees of pages. Works with similar method include [6] and [15]. MDR [16] makes use of the HTML tag tree of the web page to extract data records, but data items does not be aligned and extracted. DEPTA [17] solved shortcomings of MDR. Another kind of approaches is based on the visual information of web pages. They use visual features to locate data regions in web pages. ViPER [18] uses user's visual perception to identify and rank potential repetitive patterns. The visual-based approaches have a common characteristic: they identify and extract data only at record level by visual features.

Though the structure-based wrappers do not provide methods to assess the quality of web pages, the structure patterns of web data they discussed in their algorithms are very helpful for our approach. In data extraction, wrappers map each identified data value to a special attribute of data schema, and group data values into a data record with the patterns they discovered. Existing wrappers have analyzed various layout structures of data records within the DOM tree of web pages and provided the description of the patterns extracted from these layout structures. Moreover, in the alignment [7] for annotating record items, both attribute patterns and the text complexity of node are analyzed. Our XQuery-based wrapper is also partially built based on these patterns.

3   Preprocessing for the Assessment

To assess the quality of web pages, our approach need firstly annotate data values with the underlying schema of web data. And the schema of web data influences the layout structure of data records in the DOM tree of a web page. In this section, we introduce the schema model of web data and the annotation of data items.

3.1   The Schema Model of Web Data

The schema of data is very important for web data extraction. In the deep web, it is called the query result schema of web databases. The schema defines the semantics of data values and the structures of result data. In most existing wrappers, algorithms regard that web data are stored in the form of relational tables. Therefore, web data are always modeled with the schema of relational databases, in which the data of a web page consist of a set of data records. Each record is a tuple in a relational table, and each data value of record is mapped with an attribute of the relational table of the schema. However, an obvious problem is that the data in web pages have a hierarchical structure in its schema. It leads to the difficulty of data extraction and may lead to mistakes for existing wrappers.

Therefore, we propose a schema model for web data which can handle both relational data and hierarchical data. In our approach, the data of a web page is regard as a set of data records which obey the same schema S of entity E in structure and semantics. So in the schema model, we define the schema of web data as $S=<N_E, A>$, where N_E denotes the name of entity E, and $A=\{a_1,...,a_n\}$ is the attributes set of entity E. For each attribute a_l in A, it is defined as $a_i=<N_i, T_i, V_i>$, where N_i denotes the name of attribute, V_i denotes the value of attribute and T_i denotes the type of attribute a_i.

There are three types of attributes: (1) the single-value attribute which is like a pair of key-value; (2) the complex attribute which is a sub entity and consists of multiple detailed attributes; (3) the collection attribute which consists of multiple instances of the same sub entity. If T_i is the complex attribute or the collection attribute, we define it as a sub entity E_i with a sub schema which is denoted as $S_i = <N_i, A_i>$, where N_i is the name of entity E_i, and $A_i=\{a_{i1},...,a_{it}\}$ is the set of attributes in sub schema S_i. For the complex attribute A_i, e.g. date with day, month and year, attributes of its sub-entity can be inserted into attributes of its parent entity E, then $A=\{a_1,..., a_{i1},...,a_{it},...,a_n\}$. But for the collection attribute, we must insert a hierarchical level in the structure of schema.

For an attribute $a_{i,j}$ of sub schema S_i, it may be also a collection attribute with the schema $S_{i,j}$. This structure of schema can be defined in nested way, which makes our model effectively represent the schema of web data with hierarchical structure. For example, a web site of the hotel domain provides the information of booking rooms in its web pages. The data in web pages contains the information of hotels and the details of their empty rooms. So the schema of data has a hierarchical structure in 2 levels. The structure of data is expressed with a tree model. An example is shown in Fig. 1, in which there are three types of nodes: (1) The *collection node* is mapped to the collection of an entity in the schema, such as *hotels* and *rooms*;(2) The *instance node* is mapped to the root node of an instance with the sub schema S_i, such as *hotel*$_i$ and *room*$_i$; (3) The *attribute node* is mapped to data value of an single-value attribute.

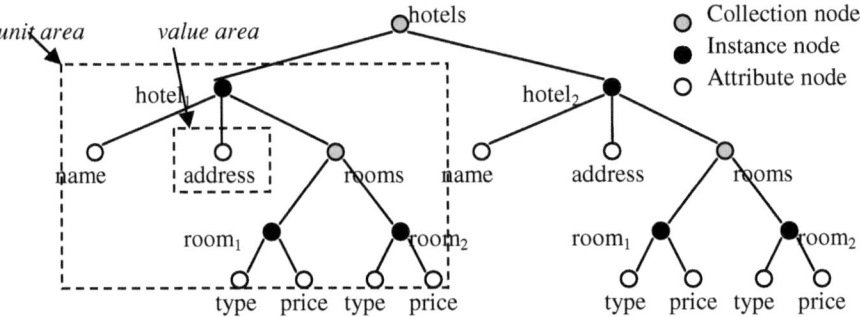

Fig. 1. Tree structure of example web data with hierarchical schema

3.2   The Annotation of Web Data

To analyze patterns of web data in web pages, besides defining the schema of the web data, we should also annotate the web data of web pages with their semantics and position information in their DOM trees. In the annotation of semantics, we map each data value of the web page to an attribute a_x in the schema S defined for the data. The name of the attribute expresses the semantics of the data value.

Since web pages are formatted into XML documents and parsed into the DOM tree structure, we give the definition of *data set area*, *data unit area* and *attribute value area* in the web pages. Their visualized layouts on web page are shown in Fig. 2.

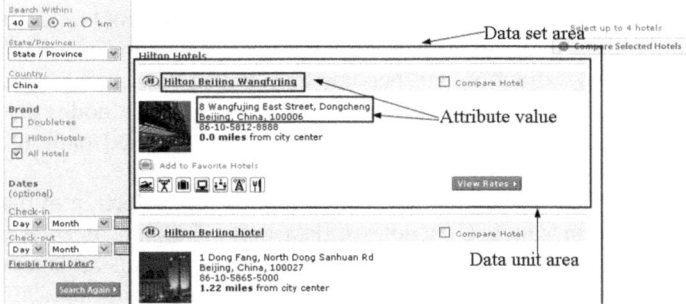

Fig. 2. The data set area, data unit area and attribute value area in a web page

Attribute value area is an atomic content which is mapped to an attribute with single value in the schema. Each attribute value is contained by text content of a node in the DOM tree, e.g. *value area* in Fig.1.

Data unit area is an area of web page which contains a completed instance. A data unit is an instance of entity defined in the schema *S*. In the DOM tree of web page, data unit area is a sub tree that contains all attribute values of an instance, e.g. *unit area* in Fig.1. The root node of the sub tree is the Smallest Lowest Common Ancestor (*SLCA*) [24]of all attribute nodes for an instance and is mapped to the instance node.

Data set area is the area which contains all data units in web page. The data set area is also a sub tree of the DOM tree where its root node is the *SLCA* of all data unit nodes and is mappd a collection node in the result data.

For each above node, we use an exclusive path to annotate its location in the DOM tree. The exclusive path is expressed with the path expression of XPATH. In the path expression, from the root node of document to current node, each node is transferred into a *step* of the path. To separate the node with its siblings, we use its order position as the position predication of the *step*. Therefore, for a given node of DOM tree, its location is expressed with the path *P* as:

$$P = /T_1[p_1]/ T_2[p_2]/......./ T_m[p_m]/$$

Where T_i is the tag of *step$_j$*, and p_i is the position predicate of node of *step*. After the annotation of web data, for each entity *E* defined in the schema, we match it a set of data units $D= \{U_1,...,U_n\}$ in the web page, and for each attribute in schema, we match it a set of attribute values located with their path expressions.

4 The Method for Quality Assessment

We will assess the quality of web pages with two quality dimensions: the structure complexity and the text complexity, which reflect the difficulty of extracting data. In this section, we will introduce patterns in the structure complexity and the text complexity, and discuss the model for quality level to assess the quality of web pages.

4.1 Analyzing the Structure Complexity

The structure complexity is defined to represent whether data units and attribute value are regularly distributed in the structure of DOM tree. If the distribution of values has an obvious regularity, it indicates the web page with a low structure complexity, and wrappers can identify patterns of data units and attributes with high accuracy. Otherwise, the accuracy of extraction will become low. In annotation, data units and attributes are located with the path expressions of their nodes. Since nodes of data units or attributes are distributed in the DOM tree with some specified patterns, we extract patterns from the path expressions.

For a given sample web page, $D_u=\{U_1,U_2,...,U_n\}$ is the collection of data units contained by the page, in which U_i denotes a data unit. The path of U_i is expressed as $Pu_i=/T_{i1}[p_{i1}]/T_{i2}[p_{i2}]/....../T_{im}[p_{im}]/$. For most of web pages, paths of data units have the same tag sequence T_{i1} $T_{i2}...$ T_{im}, but their position sequences $p_{i1}p_{i2}...p_{im}$ are always different. We extract the path patterns of data unit based on paths Pu_i of data units. The path pattern consists of two parts: the common path Puc and the local path Pul. The common path of data units is the Longest Common Path *(LCP)* of all Pu_i, which exactly locates the node of data set area. The algorithm for extracting *LCP* is the similar with the SLCA for XML keyword search. In $Puc =/T_1[p_1]/....../T_k[p_k]/$, $k<m$, for each step $T_i[p_i]$ of Puc, if there is no node $T_i[p_x]$ which is a sibling node of $/T_1[p_1]/.../T_i[p_i]$ with the same tag T_i and has no descendant node whose path is the same with the Puc, we remove its predicate p_i. The local path is used to separate nodes of data units from their sibling nodes which have the same tag and do not contain any data. For each step $T_j[p_j]$ of local path, we extract the pattern by computing the position region from the position sequence ps_j of $p_{1j}p_{2j}...p_{nj}$. If the nodes of current step have the common parent, we classify ps_j into following patterns based on the structure as shown in Fig.3: *continuous, noisy continuous, regular interval, and irregular interval.*

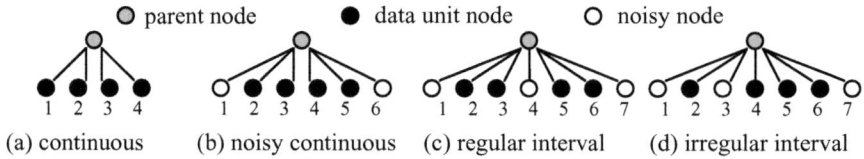

Fig. 3. The patterns of position sequence

In Fig.3, noisy nodes are sibling nodes of data unit nodes with the same tag T_j. For the patterns of *continuous, noisy continuous, regular interval,* they can be easily expressed with regular expression. We define them as the *regular pattern* of the structure complexity. For the *irregular interval* pattern, we additionally check XML attributes of nodes, since XML attributes is wildly used to define styles in web pages. If there exists an XML attribute that just appears in nodes of current step or there is an exclusive value to separate data nodes with noisy nodes, we can use this XML attribute to identify data nodes of current step. We define it as the *attribute pattern* of the structure complexity. Otherwise, the *irregular interval* pattern is hard to be expressed and it requires special operations to identify in wrappers. Therefore, we define it as the *irregular pattern* of the structure complexity. In wrappers, these patterns are used to align data records and data items.

For a given attribute A_i, nodes containing its values can be located with the relative path expression which is relative to the path of data unit. The patterns of a relative path expression are similar with that of data units. Suppose the relative path of attribute A_i is $/TA_{ik+1}[p_{ik+1}]/.../TA_{it}[p_{it}]$, in which $TA_{ix}[p_{ix}]$ is a step of the path. We classify the patterns of steps in relative path by the same method of data unit.

4.2 Analyzing the Text Complexity

Only using structure information is not enough for extracting the web data since detail data values are in the text of node. The text complexity reflects the difficulty of extracting attribute values from the text content of node in the web page.

In web data, many attribute values of data units is not solely contained by a node. Some noise text or values of other attributes may appear in text of the same node with the current attribute value. In this case, it is required to align text in nodes to separate the text of attribute value from other text content. For a given attribute a_i, $na_i=\{na_{i1},...,na_{in}\}$ is the node set that contain values of a_i. Each node na_{ij} in na_i has a text content t_{ij}. We compute the common substrings on all text t_{ij} of a_i. Suppose common strings are $\{cs_1,...,cs_l\}$, the text of node is express as $t_{ij} = t_{ij1}\ cs_1\ t_{ij2},..., cs_l\ t_{ijl}$, in which t_{ijl} is a substring of t_{ij} different each other. Except for the attribute value ta_{ij}, all other text is regard as noisy text. We also classify the text of attribute node into three patterns: *alone value, regular value, irregular value*.

(1) For the attribute with *alone value*, its values are the only text in nodes of it.

(2) For the attribute with *regular value*, its values can be extracted by a regular expression as $cs_k t_{ijk} cs_{k+1}$. For example, many web pages insert the semantic tags to annotate attribute values in the text of node.

(3) For the attribute with *irregular value*, there is no obviously regular way for extracting attribute values. It must exploit complex algorithms and mining in more sample data for the extraction, e.g. the information of reference in papers.

4.3 The Quality Level

In this paper, we propose a model which defines the quality of web pages with three quality levels: *perfect, well* and *acceptable*. There are three quality dimensions in our model for quality assessment.

The first dimension is structure complexity introduced in section 4.1 which includes three types of patterns. For the *regular pattern* and the *attribute pattern*, wrappers are easy to identify the nodes containing data.

The second dimension is the text complexity in extracting attribute values from text content of nodes. We separate it into three types with the complexity of string patterns: (1) the *alone value* is the most simple pattern for extracting attribute values; (2) the *regular value* requires wrappers to align text of nodes and extract the pattern for identify attribute values; (3) the *irregular value* is the most complex condition for wrapper, which may need to learn rules from large–scale data.

The third dimension is the hierarchical structure of data schema. The more layers the schema have, the more difficult for wrappers to extract.

The detail of assessment of quality levels is show in Table 1 in which "Y" denotes available with existing XQuery statements. The web pages in *perfect* level can be extracted by most wrappers with high accuracy. Then the web pages in *well* level provide high quality data, but require wrappers training the pattern from sample data. At last, the web pages in *acceptable* level have irregular patterns on structure or attribute text, and if it has a hierarchical schema which decreases the accuracy of extraction in most wrappers, we classify it into *acceptable* level directly.

Table 1. The levels of the quality model

Quality Levels	structure complexity			text complexity			Hierarchy of Schema
	regular pattern	Attribute pattern	irregular pattern	alone value	regular value	irregular value	
perfect	Y	Y	N	Y	N	N	1
well	Y	Y	Y	Y	Y	N	1
acceptable	Y	Y	Y	Y	Y	Y	>=1

Furthermore, to compare web pages, we propose a novel method to compute a quality score for each web page. We consider the irregular structure pattern and irregular value pattern as the main factors that decrease the accuracy of data extraction. In our model, we build an extraction tree based on the schema tree of web data and the DOM tree of web page, as shown in Fig.4. To extract an attribute value, there are at least four phases of paths from the root node to the attribute value in the extraction tree: *LCP*, *local path*, *relative path* and *text pattern*. In each of path, if there exists an irregular pattern, the possibility of mistakes in extracting an attribute value will increase.

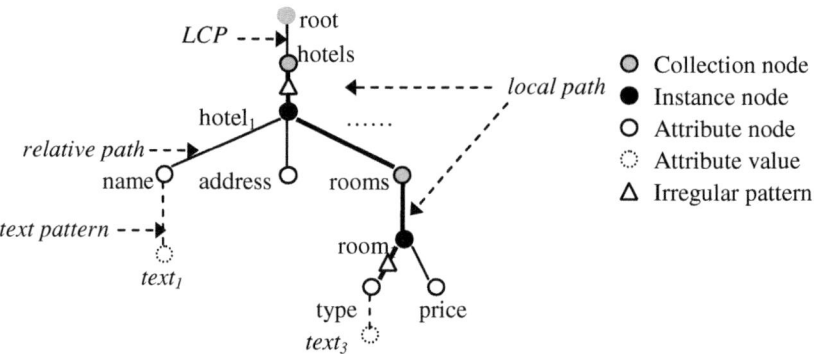

Fig. 4. An example of extraction tree

Therefore, for a given web page, we define the quality for extracting an attribute a of the schema as $Qa=(Nl-Np)/Nl$, where Np is the number of irregular patterns in the extraction path of attribute a and Nl is the phase number of extraction path. For example, in Fig.4, there are 6 phases and 2 irregular patterns in extraction path for values of attribute *room*, and the quality of extracting attribute *room* is 0.667. In extraction tree, the irregular patterns in higher level should have more effect than those in lower level. Therefore, we define the quality score of a web page in formula 1, in which n is the number of attributes of the schema. The score is used for comparing the quality of web pages in the same quality level since it considers both irregular patterns and the structure of data schema.

$$Score = (\sum_{i=1}^{n} Nl_i - \sum_{i=1}^{n} Np_i)/\sum_{i=1}^{n} Nl_i \qquad (1)$$

5 An XQuery-Based Wrapper

To extract web data from web pages and check the validation of our model, we design an XQuery-based wrapper. In the wrapper, we use XQuery statements to operate XML documents of web pages. The structure of XQuery is shown as follows:

FOR *$dataunit* **in** *Path expression*
RETURN *return expression*

The XQuery statement is mainly constructed with the FLWOR expression which is used to query, filter, and return data in xml documents. In our approach, we use FOR clause to query the nodes of data units in the DOM tree of web page. To locate node of data units, we transform patterns of structure into the *position predicates* of path expressions. RETURN clause is used to control the output of result data and extract data values. In RETURN clause, the relative path expression is used to locate nodes of attributes. Patterns of text content are transformed into functions of XQuery, where the *alone value* maps *text()*, and the regular values can be handled with built-in functions of XQuery, and the irregular values must use user-defined functions. In XQuery, the format of result data is determined by the *return expression*. The *return expression* can return both XML format and relational data. The structure of *return expression* for XML format and the relational data are shown as follows:

$<N_E>[< N_i >\{$unit/Pr(A_i)/textFunc(A_i)\}</N_i>][< N_j > [FLWOR(S_j)] </N_j>]</ N_E >$
$(\{$unit/Pr(A_1)/textFunc(A_1)\}, ..., \{$unit/ Pr(A_n)/textFunc(A_n) \})$

Where $textFunc(A_i)$ is the function for extracting values of attribute A_i, and FLWOR (S_j) is a nested FLWOR expression for the sub schema S_j in the schema S.

6 Experimental Results

In experiments, we use the dataset TBDW Ver.1.02[25] and the dataset RBH we crawled from the deep web. The TBDW consist of 51 different web sites, and for each web site there are 5 web data pages. The data in TBDW has complex structures, such as repetitive attributes. The dataset RBH we crawled includes web pages from 60 web sites which belong to three domains: *realty*, *book* and *hotel*. In our dataset, there are some web pages containing hierarchical data.

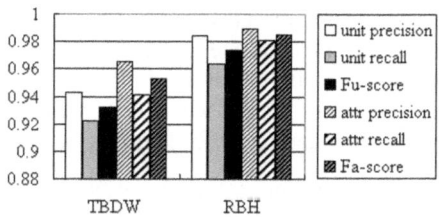

Fig. 5. Experimental results

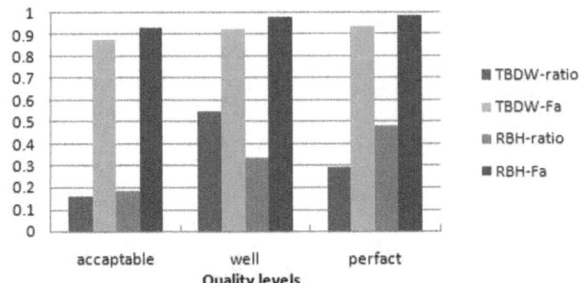

Fig. 6. Experimental results on quality levels

We first evaluate our XQuery-based wrapper according to two metrics, concentrating on the different granularity in data extraction. The metrics are the unit accuracy and the attribute accuracy. The unit accuracy shows how well the data units are identified and extracted. The precision of unit is $P_u=(Nu_i-Nu_e)/Nu_i$, where Nu_i is the number of identified data units and Nu_e is the number of unit identified with errors. The recall of unit is $R_u=(Nu_i-Nu_e)/N_u$, which refers to the number of data units in the dataset. The attribute accuracy shows how well the attributes are extracted. The precision of attribute is $P_a=(Na_i-Na_e)/Na_i$, where Na_i is the number of identified attributes value in all data units and Na_e is the number of attributes identified with errors. Finally, we compute F-score $F = 2PR/(P+R)$. In our experiments, the golden standard of data is specified by manually. The experimental result is shown in Fig.5.

For our data quality model, we classify web pages of dataset into the three quality levels, and we compute the metrics on web pages of each quality level separately. The experimental results shown in Fig.6, in which TBDW-ratio is the ratio of web pages of each quality level and TBDW-Fa is the F-score of attribute metrics on each quality level. It is the same for BRH-ratio and BRH-Fa. We find that, the quality level is consistent with the accuracy of extraction, that is, with the quality level increasing the accuracy of extraction is also increasing. This verifies that the data quality model is correct.

7 Conclusion

In this paper, we proposed an approach to assess the quality of web pages which indicates the data quality a wrapper can achieve in extraction. It is very useful for controlling data quality in further applications. In addition, we proposed an XQuery-based wrapper to extract web data and validate our quality model. The wrapper is very efficient to extract the web data with hierarchical structure which is hardly handled by existing approaches. In our future work, we will improve our model for the quality assessment of web data by considering multiple quality categories.

Acknowledgments. This paper is supported by the National Natural Science Foundation of China (Nos.60973018, 60973021 and 61003060), the 863 High Technology Foundation of China (No. 2009AA01Z131), and the Fundamental Research Funds for the Central Universities (No.N090104001).

References

1. Bergman, M.: The deep web: surfacing hidden value. The Journal of Electronic Publishing 7(1) (2001)
2. Cohen, W., Hurst, M., Jensen, L.: A flexible learning system for wrapping tables and lists in HTML documents. In: WWW (2002)
3. Pinto, D., McCallum, A., Wei, X., Bruce, W.: Table extraction using conditional random fields. In: SIGIR (2003)
4. Wang, Y., Hu, J.: A machine learning based approach for table detection on the Web. In: WWW (2002)
5. Crescenzi, V., Mecca, G., Merialdo, P.: Roadrunner: towards automatic data extraction from large web sites. In: VLDB (2001)
6. Arasu, A., Garcia-Molina, H.: Extracting structured data from web pages. In: SIGMOD (2003)
7. Zhai, Y., Liu, B.: Web data extraction based on partial tree alignment. In: WWW (2005)
8. Liu, W., Meng, X., Meng, W.: Vision-based web data records extraction. In: WebDB (2006)
9. Cai, D., Yu, S., Wen, J., Ma, W.-Y.: Extracting content structure for web pages based on visual representation. In: Zhou, X., Zhang, Y., Orlowska, M.E. (eds.) APWeb 2003. LNCS, vol. 2642, pp. 406–417. Springer, Heidelberg (2003)
10. XQuery 1.0: An XML Query Language, http://www.w3.org/TR/xquery/
11. Hammer, J., Garcia-Molina, H., Cho, J., Aranha, R., Crespo, A.: Extracting semistructured information from the Web. In: Workshop on the Management of Semistructured Data (1997)
12. Kushmerick, N.: Wrapper induction: efficiency and expressiveness. Artificial Intelligence 118, 15–68 (2000)
13. Arocena, G.O., Mendelzon, A.O.: WebOQL: restructuring documents, databases, and webs. In: ICDE (1998)
14. Liu, L., Pu, C., Han, W.: XWRAP: An XML-enabled wrapper construction system for web information sources. In: ICDE (2000)
15. Wang, J.-Y., Lochovsky, F.: Data extraction and label assignment for Web databases. In: WWW (2003)
16. Liu, B., Grossman, R., Zhai, Y.: Mining data records from Web pages. In: KDD (2003)
17. Zhao, H., Meng, W., Yu, C.: Automatic extraction of dynamic record sections from search engine result pages. In: VLDB (2006)
18. Simon, K., Lausen, G.: ViPER: Augmenting automatic information extraction with visual perceptions. In: CIKM (2005)
19. Gertz, M., Ozsu, T., Saake, G., Sattler, K.: Data Quality on the web. Report (2003)
20. Strong, D., Lee, Y., Wang, R.: Data Quality in Context. CACM 40(5) (1997)
21. Even, A., Shankaranarayanan, G.: Utility-driven assessment of data quality. ACM SIGMIS Database 38(2), 75–93 (2007)
22. Pipino, L., Lee, Y., Wang, R.: Data quality assessment. CACM 45(4) (2002)
23. Batini, C., Cappiello, C., Francalanci, C., Maurino, A.: Methodologies for data quality assessment and improvement. ACM Comput. Surv (2009)
24. Xu, Y., Papakonstantinou, Y.: Efficient Keyword Search for Smallest LCAs in XML Database. In: SIGMOD (2005)
25. Yamada, Y., Craswell, N., Nakatoh, T., Hirokawa, S.: Testbed for information extraction from deep web. In: WWW (2004)

Using Machine Learning to Support Resource Quality Assessment: An Adaptive Attribute-Based Approach for Health Information Portals

Jue Xie and Frada Burstein

Center for Organizational and Social Informatics (COSI), Monash University,
900 Dandenong Road, Caulfield East, Australia
{Jue.Xie,Frada.Burstein}@monash.edu

Abstract. Labor-intensity of resource quality assessment is a bottleneck for content management in metadata-driven health information portals. This research proposes an adaptive attribute-based approach to assist informed judgments when assessing the quality of online information resources. It employs intelligent learning techniques to predict values of resource quality attributes based on previous value judgments encoded in resource metadata descriptions. The proposed approach is implemented as an intelligent quality attribute learning component of a portal's content management system. This paper introduces the required machine learning procedures for the implementation of the component. Its prediction performance was evaluated via a series of machine learning experiments, which demonstrated the feasibility and the potential usefulness of the proposed approach.

Keywords: Quality Assessment, Machine Learning, Metadata, Health Information Portals.

1 Introduction

Health information portals act as an effective quality control approach to address both information overload and information quality concerns of online health information provision [1]. Online resources accessible via these portals are assessed for their potential usefulness to information consumers, and described in a way to facilitate indexing, navigation, filtering, and value-added information provision. This requires intensive efforts from domain experts, who have expertise in both medical domain and information management [2]. The need to sustain labor-intensive resource quality assessment processes has been raised as a key concern by domain experts and portal developers [3]. Existing works, which develop generic quality assessment frameworks (e.g. [4-5]) or automatic quality evaluation tools (e.g. [6-7]), do not address the need of measuring multi-faceted resource quality in the context of information needs and preferences of health portal users. The emergence of new solutions, such as supporting the processes with both a user-sensitive quality assessment framework and intelligent quality evaluation tools, is imperative in order to improve the scalability and eventually the sustainability of portal content management.

J. Xu et al. (Eds.): DASFAA Workshops 2011, LNCS 6637, pp. 526–537, 2011.
© Springer-Verlag Berlin Heidelberg 2011

This paper reports a study that developed and evaluated an adaptive attribute-based approach to support resource quality assessment for health information portals. A semi-automated process based on machine learning techniques is proposed in the context of a metadata-driven health information portal. An adaptive attribute-based data model for resource quality assessment will be presented in the next section. The paper then introduces the machine learning procedures implemented in this study to operationalize the model. Evaluation results of prediction performance of conducted machine learning experiments are summarized before the conclusion, and finally the future work is presented.

2   Research Context: The BCKOnline Portal

This study was motivated by and tested in the context of the Breast Cancer Knowledge Online (BCKOnline) portal (www.bckonline.monash.edu.au). The portal provides the breast cancer user community with access to online resources, quality-assured against the portal's selection criteria [8]. The operation of the portal over eight years has highlighted deficiencies with manual processes for content creation and management, in particular the resource quality assessment processes [3].

The BCKOnline portal is based on a comprehensive metadata schema used to describe selected resources from a user-centric point of view. The metadata schema extended the Australian Government Locator Service (AGLS) metadata standard (www.agls.gov.au) and introduced an innovative *Quality* element. This element and its encoding scheme were designed to address portal users' information quality concerns. The specific metadata elements of *Quality* and *Audience*, enable the user-sensitive information retrieval and quality reporting in search results [9].

3   An Adaptive Attribute-Based Approach for Resource Quality Assessment

There is no generic approach to the quality assessment of information on the internet [3]. In this paper we suggest an approach for resource quality assessment, which was developed for application in a healthcare domain. Wang et al. [10] propose the use of an attribute-based approach for data quality management. Their approach defines a *data quality attribute* as a collective term that refers to both subjective data quality dimensions and objective data quality indicators. It is proposed that such an approach can be utilized to facilitate the automation of resource quality assessment in a portal context. Adapted from Wang et al.'s [10, p. 354] definitions, this paper introduces the following terms for the assessment of resource quality:

- *Quality indicators* provide objective information about the characteristics of a resource. A quality indicator is objective if it is generated using a well-defined and widely accepted measure.
- A *quality dimension* describes a qualitative or subjective single aspect of resource quality, the value of which is based on the values of underlying quality indicators.

- A *quality attribute* is a collective term that refers to both quality dimensions and quality indicators as shown in Fig. 1. A quality attribute can be regarded as the characteristic of a resource, or a resource attribute.

These terms define a data model to facilitate the resource quality assessment. The assessment consists of subjective resource quality dimensions that are based on pre-defined corresponding quality indicators of an online resource. These indicators can be collected from the resource's original site, or can be harvested from available metadata descriptions of the resource. Values of the collective resource quality attributes can then be derived from these quality indicators enabled by certain techno-logical means. These quality attributes can assist domain experts to make informed value judgments on whether or not to include the resource in a portal. This can be used to facilitate semi-automated quality assessment process since it explicitly defines the attributes which should be taken into consideration when making a decision on the quality of the information resource.

3.1 An Attribute-Based Data Model for the Healthcare Domain

In this section, an attribute-based data model is instantiated for the healthcare domain (see Fig. 1). The model consists of the subjective quality dimensions, corresponding objective quality indicators, and finally the quality attributes in order to automate the resource quality assessment with the use of intelligent techniques.

Resource quality in the context of health information portals is defined as the ex-tent to which information contained in a web-based resource meets a user's informa-tion needs and quality perceptions [3]. In this paper, we differentiate resource quality (RQ) from information quality (IQ). While IQ is considered as a resource attribute, RQ is regarded as a relationship between a resource and its actual user. This view requires resource quality to be assessed in a way to reflect user information needs and preferences. Informed by the literature analysis of user quality perceptions and user information needs in the healthcare domain [11-12], we define RQ as a composition of *Reliability* and *Relevancy*, as denoted in the following formula:

$$RQ = Ry \ (Ac, Cr, Cu) \ \cup \ Re \ (Up, Ip, Si) \qquad (1)$$

Where: RQ refers to *Resource Quality*, Ry denotes *Reliability*, Ac corresponds to *Accuracy*, Cr - *Credibility*, Cu – *Currency*; whereas Re denotes *Relevancy*, and Up relates to User profile, Ip - *Information preference*, and Si - *Subject interest*.

In this study, *Reliability* is defined as the extent to which a resource and its source are regarded as true, credible, and up-to-date. An initial set of quality dimensions was selected, including *Accuracy*, *Credibility* and *Currency*, which were perceived as best describing health information consumers' quality concerns [3]. *Relevancy* in this study is defined as the extent to which a resource is applicable and useful for a user's information needs. It needs to be emphasized that in this paper, we only focus on the measurement of *Reliability* as a multi-dimensional concept.

Having quality dimensions defined, the next step is to define quality indicators, which are regarded as evidences contributing to the evaluation of certain quality di-mensions. In the literature quality indicators have been categorized according to their use in constructing quality metrics. For instance, quality indicators are classified as direct quality indicators versus indirect ones [13], or technical quality indicators

Table 1. Quality dimensions and indicators

Quality Dimension	Definition (derived from [9, 15])	Quality Indicators	References
Accuracy	Extent to which information contained in a resource is correct, certified as free of error, or conforms to common consensus in the field.	Evidence-based, bias or potential conflicts of interest, third-party labels or seal of certification	[6, 16]
Credibility	Extent to which information contained in a resource is highly regarded in terms of its source or content.	Author, contributor, editor, publisher, HON code of conduct seal, site URL, Google's PageRank	[17-21]
Currency	Extent to which information contained in a resource is representative of up-to-date practice, views and/or wisdom on a particular topic.	Date of creation, date of last update, frequency of update, maintenance of website	[14, 19, 22]

(automatically measurable or detectable) versus non-technical ones [14]. Table 1 below defines the quality dimensions of *Reliability*, and the corresponding quality indicators of each dimension that have been used in previous studies in the literature.

In the context of metadata-driven health information portals, the quality attributes of a resource are considered as value-added metadata that can be inferred by rule-based methods [23]. Informed by an analysis of quality indicators and their relationships to existing metadata models for resource description, including Dublin Core (www.dublincore.org), AGLS and BCKOnline [15], an attribute-based data model was built to facilitate semi-automated resource quality assessment. An instantiation of the model is depicted in Fig. 1 below.

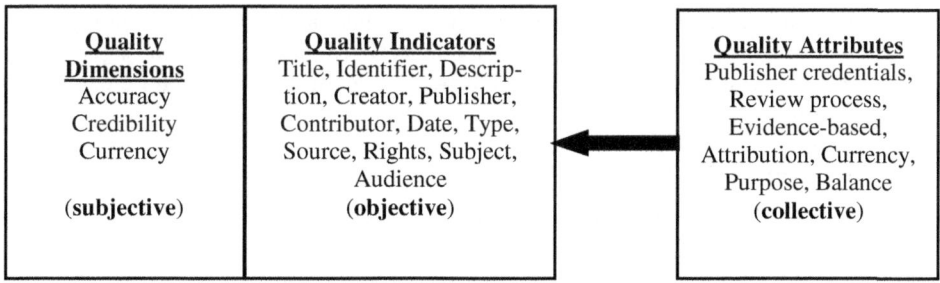

Fig. 1. An attribute-based data model for resource quality assessment

4   Machine Learning for Predicting Quality Attributes

This research proposes the use of machine learning (ML) techniques to detect decision patterns from annotated resources in order to support domain experts making quality judgments on new resources. ML is a technique for the acquisition of structural descriptions underlining data that can be used for prediction, explanation and understanding [24]. The extracted patterns capture the implicit decision structure and explain learning in an explicit way. Decision-making involving human judgment such as quality assessment is a suitable application field for ML techniques.

ML techniques are used to learn concepts from instances and their attributes, while statistical tests are used to validate ML models and to evaluate ML algorithms [24]. Each instance is characterized by the values of attributes that measure different aspects of the instance. A suitable ML method needs to be selected to solve the specific learning problem of this research.

Facilitated by the attribute-based data model defined in the previous section, an intelligent component of a portal's content management system was implemented to help domain experts assess and describe attributes of quality for a resource. The component consists of two modules: a quality attribute learner and a quality attribute predictor. The learner mines domain experts' decision-making patterns from previous value judgments that are encoded in resource metadata records. Based on the learned models, the predictor can suggest values for the quality attributes of new resources, given the other resource attributes have been annotated beforehand. Such an intelligent component can automatically predict values for describing resource quality attributes. Being provided with these suggestions, domain experts can assess and describe resources in a more efficient and consistent manner.

The next section introduces the implementation of a learner component. Open-sourced WEKA data mining workbench [25] was used to perform ML experiments. The WEKA workbench provides the implementation of ML algorithms that can be applied to any given dataset for performing standard data mining tasks, including classification, regression, clustering, association rule mining and attribute selection. It also provides a comprehensive toolkit for data pre-processing and visualization.

5   ML Procedures for Intelligent Quality Assessment

This section introduces the ML procedures for the implementation of an intelligent learning component (the quality attribute learner) that can be used for building classification models for future prediction of quality attributes. The three steps involve 1) learning schema selection 2) data attributes selection, and 3) data cleaning and transforming.

5.1   Selection of ML Scheme

According to Witten and Frank [24], there are four basic learning styles: classification, association, clustering and numeric predication. The first two styles are considered as applicable approaches to solving different learning problems of this study. Classification (also known as supervised learning) can be used to classify a nominal data attribute while association can be used to discover relationships among a number of data attributes.

To illustrate the use of ML techniques for assessing resource quality, only the classification approach was applied, given the attribute was measured in a nominal scale, hence, it has a limited set of discrete values. Resource metadata records from the BCKOnline portal, containing descriptions of resource quality attributes assigned by experts, were used to learn a classification model for each selected quality attribute.

Also, in order to find a classification learning scheme that yields satisfying prediction performance, different classification algorithms and data settings were tested via a series of experiments. The experimental results are presented in section 6.

5.2 Selection of Data Attributes

As mentioned previously, the BCKOnline portal uses a comprehensive metadata schema for describing external online resources suitable for inclusion in the portal [15]. The *Quality* element of the BCKOnline metadata schema is composed of seven quality attributes, which describe different aspects of quality for a resource. Table 2 below outlines the five selected quality attributes that serve as the concepts for classification learning and their corresponding classes.

Table 2. Selected BCKOnline quality attributes for classification learning

Quality attribute (concept for learning)	Definition [9]	Attribute values (classes)
Attribution of sources	Whether or not the actual resource has a quality attribution, which clarifies the source of the information.	Yes, No.
Balance	What kind of issue is the actual resource, and is it noted or not.	Controversial Issue – Noted, Controversial Issue – Not Noted, Non-Controversial Issue.
Purpose	Describes the purpose this article was written for.	Commercial, Discussion Forum, Educational/Informative, Reportage of Results, Review.
Publisher credentials	The authoritativeness and credibility of the individual or organization responsible for the document.	Cancer Organization, Clinician, Commercial Body, Consumer Group, Educational Institution, Government Organization, Lay Author, Researcher, Medical Organization.
Review process	How the actual resource was reviewed.	Editorial Board, Peer Review Process, No Editorial/Peer Review Process.

The ML classification techniques were considered as applicable to build learning models for five quality attributes, which were labeled as Attribution of sources, Balance, Publisher credentials, Purpose, and Review process. Constrained by the available data, the other two quality attributes, which were labeled as *Currency* and *Evidence-based*, were excluded from learning. However, this does not mean that ML classification is not appropriate for these quality attributes in general. For the *Currency* attribute, all existing resources in the BCKOnline metadata repository were tagged as containing "current" information, which made the classification algorithms unable to discriminate "non-current" against "current". On the other hand, the *Evidence-based* attribute allowed multiple values, which brought more complexity in generating and evaluating learned classification models. To simplify the issue, this quality attribute was not selected.

The five selected quality attributes of the *Quality* element served as the class labels in the experiments. Another eight BCKOnline metadata elements were considered as relevant features for building a classification model for each selected quality attribute. These metadata elements were *Title, Description, Creator, Publisher, Type, Rights, Subject* and *Audience* [15].

5.3 Data Cleaning and Transforming

In order to improve the reliability of learned models, the data used for machine learning required cleaning to achieve better accuracy and consistency. The data integrity problems were due to the interface that the domain experts used not supporting lookups and validation. If the value of a quality attribute was misspelled or miscapitalized, an extra possible but unwanted value would be created for that attribute. Therefore, before the data was fed into WEKA for learning, the internal data consistency was checked to eliminate inaccurate, inconsistent or missing values.

Moreover, upon examination of the data format of eight selected metadata elements for ML, three of them were categorical and allowed multiple values, including *Audience, Type* and *Subject*. The other five elements allowed for free text entry, including *Title, Description, Creator, Publisher* and *Right*. Different methods were used to transform these two groups of metadata elements to build classifiers for selected quality attributes. For those categorical elements that allow multiple values, the label of each category was transformed to a data attribute in a binary format. As a result, the applicable eight BCKOnline metadata elements were transformed into 34 data attributes for building classifiers. Later at the time of model building, a WEKA built-in filter was utilized to tokenize the other five string-type elements to word vectors. All transformed word vectors were used as part of the data attributes.

6 Evaluation of Prediction Performance

In order to evaluate the feasibility of the adaptive attribute-based resource quality assessment, a series of ML experiments was conducted. The statistical tests of prediction performance of different classification schemes were conducted by comparing outputs against expert inputs within the WEKA workbench. The resulting accuracy rates are presented in the following discussion, indicating how confident the learned classification models could be to predict new data.

6.1 Statistical Evaluation Method

In order to obtain reliable measurement results of predicting accuracy, the standard *stratified 10-fold cross-validation* evaluation method was utilized by all conducted experiments. According to Witten and Frank [24], when the amount of data for training and testing is limited, the dataset cannot be assumed to have the normal distribution. Therefore, the sample used for training or testing might not be representative enough or on the contrary over-representative. To mitigate any possible bias caused by a particular sample, the statistical techniques of stratification and cross-validation were adopted, which randomly divided data into 10 folds that were approximately of the same size, and then used each data fold in turn for testing and the remainder for training. For each experiment, the training and testing process was repeated 10 times with different random samples, and the overall estimate was averaged on all iterations.

6.2 Datasets for Experiments

All ML experiments conducted by this study used 780 published resources of the BCKOnline portal. That included 780 metadata records that were manually assigned by multiple domain experts. The retrieved resource metadata records were pre-processed and transformed to create datasets that would be suitable for building classification models. For each of the five selected quality attributes, a dataset was specifically compiled for training and testing a classifier for the quality attribute. Each dataset was represented in a flat file, which was a matrix of 780 data instances versus 35 data attributes. These data attributes included the quality attribute itself and the other 34 data attributes that were transformed from the BCKOnline metadata elements (see section 5.3). Therefore, the only difference of these datasets was the last data attribute, which was the quality attribute to be classified.

6.3 Comparison of Prediction Performance

This section presents and summarizes the evaluation results of prediction performance of different ML schemes for classification. The prediction performance of the classifier learned from a training set was assessed by measuring the success rate on a testing set [24].

This study compared the prediction performance of the most common learning algorithms in solving the classification problems of quality attributes. These algorithms included support vector machines, Bayesian networks, decision trees, rules, and lazy algorithms. Although finding the possibly best classification method was not the aim of this study, experimenting different classification algorithms helped to measure how well the concept could be learned and by which means. The WEKA experimenter provided supports to perform the comparison and generated the t-tests outputs as shown in Table 3. Numbers in the above table were an average accuracy over 10 runs with a random 90% train 10% test split +/- the standard deviation. The experiments used paired t-test with the significance level of 0.05. All schemes were compared to the scheme in the first column. In this scenario, SMO was selected as the representative algorithm of the learning method Support Vector Machines (SVMs).

Table 3. T-tests comparison of the most common ML methods for this problem

Dataset	SMO	NaiveBayes	J48	IB1	ZeroR
Attribution of sources	83.20±3.84	85.38±2.82	83.20±3.70	75.26±3.54 •	61.80±0.35 •
Balance	78.59±4.52	77.69±5.17	72.18±4.95 •	78.33±3.39	65.38±0.00 •
Purpose	89.38±3.47	75.70±2.90 •	85.17±1.98	85.03±2.48 •	84.02±0.98 •
Publisher credentials	82.26±5.17	75.58±4.32 •	80.47±4.66	78.06±5.01 •	27.46±0.69 •
Review process	84.61±3.96	71.27±5.02 •	81.28±2.29	81.80±2.34	83.21±0.33

• statistically significant degradation ° statistically significant improvement.

A statistically significant improvement means an improvement over SVMs, while a statistically significant degradation means significantly lower accuracy than SVMs. As can be seen from the above table, the absence of any "statistical improvement" dots means that there were no datasets, where a classifier provided a significant improvement over SVMs (although for some datasets/classifiers the performance was neither significantly better nor worse).

The t-tests show that SVMs were the only classifiers that didn't suffer significantly degraded accuracy on at least one of the datasets. The performance on these datasets may be related to the large number of attribute values, which SVMs tend to handle much better than the other classifiers [26]. The comparison to the ZeroR (zero-attribute-rule) algorithm is important as it indicates the classification success of the other algorithms. The algorithm does not count any attribute and simply classifies everything as belonging to the largest class. On a particular dataset, if no other algorithms is better than ZeroR, this indicates that the dataset is extremely hard to learn and/or highly imbalanced.

6.4 Predicting Accuracy of SVMs

Table 4 below summarizes the learning results on a dataset of 780 published resources of the BCKOnline portal by using the SMO classifier of SVMs. The evaluation results demonstrated that SVMs worked on the learning problem in the way it was designed and that the quality of machine-generated values was satisfactory.

Table 4. Summary of learning accuracy using an SVM

Dataset (35 attributes)	Correctly Classified Instances	Incorrectly Classified Instances
Attribution of sources	658 (84.36%)	122 (15.64%)
Balance	611 (78.33%)	169 (21.67%)
Purpose	701 (89.87%)	79 (10.13%)
Publisher credentials	568 (72.82%)	212 (27.18%)
Review process	660 (84.62%)	120 (15.38%)

For the dataset of *Publisher credentials*, reducing the set of attributes to *Publisher*, *Creator* and *Rights* generated results of much better accuracy for the classification problem. The correctly classified instances were increased to 640 (82.05%) from previously 568 (72.82%). Intuitively, this makes sense since other attributes, such as *Title, Description, Subject, Glossary, and Audience* etc. are probably not relevant to determine a publisher's credentials. The similar cases may be found with the other quality attributes, where using a subset of data attributes will increase the predicting accuracy. However, exploring the relationships between data attributes and a specific quality attribute will be of interest to future work.

Evaluation results in Table 4 also indicate that classifying *Purpose* achieved around 90% overall accuracy using an SVM, which was the highest amongst the five. However, an overall accuracy can sometimes obscure detailed information about the class. If observing at the confusion matrix and the accuracy detail by class, some learning problems can be detected. For instance, Table 5 below shows the confusion matrix of *Purpose*, which has the overall accuracy broken down to recall and precision for each class. For the class of *Educational/Informative*, both recall and precision were above 90%. In contrast, recall and precision values for the other classes were much lower varying between 0% and 80%. These figures imply that the accuracy drops when the class size gets smaller. As almost all of the training data belonged to the *Educational/Informative* class, the classifier favored this value enormously. Problems like this are referred in the ML literature as "imbalanced" that are known to be notoriously difficult to learn [27].

Table 5. Confusion matrix of Purpose

a	b	c	d	e	← classified as	Recall	Precision
639	14	1	3	0	a = Educational/Informative	97.26%	92.07%
26	**53**	0	2	0	b = Review	65.43%	74.65%
12	0	**4**	0	0	c = Discussion Forum	25.00%	80.00%
14	4	0	**5**	0	d = Reportage of Results	21.74%	50.00%
3	0	0	0	**0**	e = Commercial	0.00%	0.00%
					Weighted Average	**89.9%**	**88.4%**

Recall = Number of instances accurately classified as the class / Number of all instances of the class.
Precision = Number of instances accurately classified as the class / Number of all instances classified as the class.

7 Conclusion and Future Work

In this paper, we defined the assessment approach and a set of measures of resource quality on the basis of user information needs and preferences for the healthcare domain. Our research shows how ML techniques could be applied to support resource quality assessment through the proposed adaptive attribute-based approach. Based on the results of ML experiments, SVMs were identified as the classification method suitable for solving the specific learning problems of this study. The achieved prediction performance on a set of online healthcare resources from the BCKOnline portal ranged from 73% to 90%, which proved the feasibility of using ML techniques to generate value suggestions for describing resource quality attributes. As the accuracy of the suggested values is not definitive, the approach can only provide suggestive guidance to domain experts, who are then responsible for making the final judgments on resource quality. Preliminary usability test of this component with four domain experts produced promising results.

In order to achieve higher accuracy or reliability of machine-generated value suggestions, it is necessary to build classification models from incremental data at the time of prediction. This is due to new resources from websites not previously included in the portal, or the available data may be highly imbalanced for classifying certain quality attribute. Therefore, the classification models for prediction need to be fine-tuned continuously by incorporating expert decisions made on new resources.

The future work of this research will include several tasks. Based on the described quality attribute learner, a quality attribute predictor will be implemented as a feature of the BCKOnline portal's content management system. As mentioned in the paper, reducing the attributes set for the quality attribute of *Publisher credentials* increased the predicting accuracy. Similar cases may find with the other quality attributes. The attribute analysis tool provided by WEKA will be used to determine which data attributes (or metadata elements) are dominant factors in determining the classification model for a particular quality attribute. It will also be necessary to explore the relationship between the metadata elements and the different quality attributes outside the healthcare domain.

References

1. Benigeri, M., Pluye, P.: Shortcomings of Health Information on the Internet. Health Promotion International 18, 381–386 (2003)
2. Evans, J., Manaszewicz, R., Xie, J.: The Role of Domain Expertise in Smart, User-Sensitive, Health Information Portal. In: the 42nd Hawaii International Conference on System Sciences, HICSS-42 (2009)
3. Xie, J.: Sustaining Quality Assessment Processes in User-Centred Health Information Portals. In: The 15th Americas Conference on Information Systems, AMCIS 2009 (2009)
4. Stvilia, B., Gasser, L., Twidale, M.B., Smith, L.C.: A Framework for Information Quality Assessment. Journal of the American Society for Information Science and Technology (JASIST) 58, 1720–1733 (2007)
5. Wang, R.Y., Strong, D.M.: Beyond Accuracy: What Data Quality Means to Data Consumers. Journal of Management Information Systems 12, 5–33 (1996)
6. Griffiths, K.M., Tang, T.T., Hawking, D., Christensen, H.: Automated Assessment of the Quality of Depression Websites. Journal of Medical Internet Research 7, e59 (2005)
7. Sessions, V., Valtorta, M.: Towards a Method for Data Accuracy Assessment Utilizing a Bayesian Network Learning Algorithm. Journal of Data and Information Quality 1, 1–34 (2009)
8. Burstein, F., Fisher, J., McKemmish, S., Manaszewicz, R., Malhotra, P.: User Centred Quality Health Information Provision: Benefits and Challenges. In: The 38th Annual Hawaii International Conference on System Sciences, HICSS 2005 (2005)
9. McKemmish, S., Manaszewicz, R., Burstein, F., Fisher, J.: Consumer Empowerment through Metadata-Based Quality Reporting: The Breast Cancer Knowledge Online Portal. Journal of the American Society for Information Science and Technology (JASIST) 60, 1792–1807 (2009)
10. Wang, R.Y., Reddy, M.P., Kon, H.B.: Toward Quality Data: An Attribute-Based Approach. Decision Support Systems 13, 349–372 (1995)
11. Anderson, J., McKemmish, S., Manaszewicz, R.: Quality Criteria Models Used to Evaluate Health Websites. In: The 10th Asia Pacific Special Health and Law Librarians Conference, pp. 337–354 (2003)
12. Williamson, K., Manaszewicz, R.: Breast Cancer Information Needs and Seeking: Towards an Intelligent, User Sensitive Portal to Breast Cancer Knowledge Online. The New Review of Information Behaviour Research 3, 203–219 (2003)
13. Eysenbach, G., Diepgen, T.L.: Towards Quality Management of Medical Information on the Internet: Evaluation, Labelling, and Filtering of Information. British Medical Journal (BMJ) 317, 1496–1502 (1998)
14. Wang, Y., Liu, Z.: Automatic Detecting Indicators for Quality of Health Information on the Web. International Journal of Medical Informatics 76, 575–582 (2007)

15. McKemmish, S., Manaszewicz, R., Cheah, C.: Bckonline Metadata Schema Version 1.0 (2004),
    `http://www.sims.monash.edu.au/research/eirg/BCKO_MetadataSch`
    `ema_Version16.doc`
16. Griffiths, K.M., Christensen, H.: Website Quality Indicators for Consumers. Journal of Medical Internet Research 7, e55 (2005)
17. Price, S.L., Hersh, W.R.: Filtering Web Pages for Quality Indicators: An Empirical Approach to Finding High Quality Consumer Health Information on the World Wide Web. In: AMIA 1999 Annual Symposium, pp. 911–915 (1999)
18. Katerattanakul, P., Siau, K.: Measuring Information Quality of Web Sites: Development of an Instrument. In: The 20th International Conference on Information Systems, pp. 279–285 (1999)
19. Zhu, J., Gauch, S.: Incorporating Quality Metrics in Centralized/Distributed Information Retrieval on the World Wide Web. In: The 23rd Annual International ACM SIGIR Conference on Research and Development in Information Retrieval, pp. 288–295. ACM Press, New York (2000)
20. Conrad, J.G., Leidner, J.L., Schilder, F.: Professional Credibility: Authority on the Web. In: The 2nd ACM Workshop on Information Credibility on the Web. ACM, New York (2008)
21. Freeman, K.S., Spyridakis, J.H.: An Examination of Factors That Affect the Credibility of Online Health Information. Technical Communication 51, 239–263 (2004)
22. Aladwani, A.M., Palvia, P.C.: Developing and Validating an Instrument for Measuring User-Perceived Web Quality. Information and Management 39, 467–476 (2002)
23. Hatala, M., Richards, G.: Value-Added Metatagging: Ontology and Rule Based Methods for Smarter Metadata. In: Rules and Rule Markup Languages for the Semantic Web: Second International Workshop, pp. 65–80. Springer, Heidelberg (2003)
24. Witten, I.H., Frank, E.: Data Mining: Practical Machine Learning Tools and Technologies. Elsevier, Amsterdam (2005)
25. Hall, M., Frank, E., Holmes, G., Pfahringer, B., Reutemann, P., Witten, I.H.: The Weka Data Mining Software: An Update. SIGKDD Explorations 11, 10–18 (2009)
26. Evangelista, P.F., Embrechts, M.J., Szymanski, B.K.: Taming the Curse of Dimensionality in Kernels and Novelty Detection. In: Abraham, A., de Baets, B., Köppen, M., Nickolay, B. (eds.) Applied Soft Computing Technologies: The Challenge of Complexity, pp. 425–438. Springer, Heidelberg (2006)
27. Akbani, R., Kwek, S., Japkowicz, N.: Applying Support Vector Machines to Imbalanced Datasets. In: Boulicaut, J.-F., Esposito, F., Giannotti, F., Pedreschi, D. (eds.) ECML 2004. LNCS (LNAI), vol. 3201, pp. 39–50. Springer, Heidelberg (2004)

Grid-Based Probabilistic Skyline Retrieval on Distributed Uncertain Data

Xiaowei Wang and Yan Jia

School of Computer, National University of Defense Technology, Hunan
Changsha, 410073, China
gfkdwxw@yahoo.com.cn

Abstract. The skyline queries help users make intelligent decisions over complex data. It has been recently extended to the uncertain databases due to the existence of uncertainty in many real-world data. In this paper, we tackle the problem of probabilistic skyline retrieval on physically distributed uncertain data with low bandwidth consumption. The previous work incurs sharply increased communication cost when the underlying dataset is anti-correlated, which is the typical scenario that the skyline is useful. In this paper, we propose a knowledge sharing approach based on a novel grid-based data summary. By sharing the data summary that captures the global data distribution, each local site is able to identify large amounts of unqualified objects early. Extensive experiments on both efficiency and scalability have demonstrated that our approach outperforms the competitor.

1 Introduction

Skyline queries help users make intelligent decisions over multiple dimensional data when different and often conflicting criteria are considered. Specifically, a skyline query returns objects not dominated by any other objects. An object o is said to dominate o', if o is not worse than o' in every single dimension, and better than o' in at least one dimension.

In many real-world applications, massive data is integrated from a large number of data sources, and assembled at query time. Meanwhile, with large scale emerges data uncertainty as a factor that can not be ignored. Consider a web-based hotel recommendation system which integrates from multiple data sources, a common feature of such kind of systems is to allow users to rate the hotels based on their consumer experiences. The favorite rate associated with each hotel can be regarded as the existential probability because it represents the probability that the hotel occurs exactly as claimed in the advertisement. The system is supposed to recommend confidential hotels according to multi-criteria based ranking, such as low price and more bedrooms. Such kind of problems can be modeled as probabilistic skyline query, that is, the system returns the hotels which are not dominated by any other hotels on "price" and "bedrooms" above a confidence level.

This work concentrates on probabilistic skyline query on distributed uncertain data. Our goal is retrieving the global probabilistic skyline objects in a communication

J. Xu et al. (Eds.): DASFAA Workshops 2011, LNCS 6637, pp. 538–547, 2011.

efficient way. The only existing probabilistic skyline algorithm on distributed uncertain data (named e-DSUD) [1] utilizes a priority-based scheme to compute the global probabilistic skyline progressively. However, their approach is limited in communication efficiency because of the lack of global data distribution in local sites. We tackle this problem by a data summary sharing approach which reduces the communication cost by pruning unqualified local skyline objects early. In summary, our contributions are as follows:

- We propose a grid-based data summary which captures the distribution of an uncertain dataset in a data independent manner. Based on the data summary proposed, the general framework of our algorithm is proposed to seemly integrate the knowledge sharing and early pruning process to e-DSUD.
- A big challenge of our algorithm is sharing the data summaries with low communication cost. We tackle this problem by further optimization which significantly reduces the information need to be transferred for a data summary.
- We conduct comprehensive experiments on both real and synthetic datasets. The results have shown that our approach outperforms the existing one in communication efficiency.

The rest of the paper is organized as follows. In Section 2, we formulate the problem. Section 3 addresses the framework of our algorithm. Section 4 describes further optimizations. Section 5 reviews related works. Section 6 reports the experimental results. Section 7 concludes the paper.

2 Problem Definition

Given m local sites $S = \{s_1, \ldots, s_m\}$, each holding a local database $D_i = \{t_{i,1}, \ldots, t_{i,n_i}\}$, which is a horizontal partition of a global uncertain database D in a d dimensional data space U, and a centralized server H which can communicate with any local site via Internet. We consider the tuple-level uncertainty [2] in this paper because it widely used in confidence-aware applications, i.e., each object t in D is associated with a existential probability $p(t)$ and the probabilities of objects are mutually independent. Without loss of generality, we assume smaller values are preferred in all dimensions. And we denote t' dominate t by "$\succ$".

In the uncertain data context, an object takes a probability to be in the skyline. Following the definition in [3], the skyline probability of object t in D (denote by $p_{sky}(t)$) equals the probability that t exists and all objects that dominate t do not exist:

$$p_{sky}(t) = p(t) \times \prod_{t' \in D \wedge t' \succ t} (1 - p(t')) \tag{1}$$

Given a user-defined probability threshold q, we use $q\text{-}SKY(D)$ to denote the answer set of the probabilistic skyline query on the global dataset D. $q\text{-}SKY(D)$ is a set of objects in D each of which takes a skyline probability at least q:

$$q\text{-}SKY(D) = \{ t \in D \mid p_{sky}(t) > q \} \tag{2}$$

In the aforementioned distributed environment, the query delay mainly depends on the amount of data transferred. Therefore, our goal is retrieving the global probabilistic skyline q-$SKY(D)$ from the set of distributed database $\{D_i\}$ with low communication cost.

The recent work [1] is sensitive to the data distribution, and incurs a sharply increase communication cost when the cardinalities of local skylines is relatively large. A feasible approach is pruning unqualified objects early in the local sites, which calls for knowledge of global data distribution sharing among local sites. Nevertheless, this is not trivial as two challenges naturally emerge: (1) A data summary that captures the probabilistic data distribution of the datasets compactly and data-independently, and enable early pruning of unqualified skyline candidates in local sites. (2) A mechanism sharing the data summary across the local sites which introduces minimal additional communication cost.

3   The Grid-Based Probabilistic Skyline Algorithm

3.1   The Framework

Our framework gracefully integrates three pre-processing steps into [1]. We first present the general framework as follows, and then detail each step in the sequel.

The general framework
1. *Loading data* (in local sites): each local site S_i maps local dataset D_i to the local data summary $G(D_i)$ and then transfers $G(D_i)$ to H.
2. *Merge and share* (in centralized server): H merges $G(D_i)$ into a global data summary $G(D)$. Then H broadcast $G(D)$ back to every local sites.
3. *Local pruning* (in local sites): each S_i uses $G(D)$ to prune unqualified local skyline objects.
4. *Skyline computation* (in local sites and centralized server): compute probabilistic skyline using [1] progressively.

3.2   Loading Data

The Grid-based Data Summary. A grid G splits each dimension of the data space U into n consecutive slices. That is, each of the n^d cells in G is a d dimensional hypercube of the same width. A cell $c \in G$ is identified by a d dimensional vector $\vec{c} = <c[0], \ldots, c[d-1]>$. $c[i]$ represents its order in G in the ith dimension from the origin point of U. A cell c is mapped to a one dimensional order $c.id$ to reduce the information representing c. The one dimensional order is computed by Hilbert curve [4], as it can be reconstructed back to coordinates of cells and facilitate computation in local pruning in section 3.4. For sake of brevity, we use c_i to denote the cell owns order i, i.e., $c_i.id = i$. Fig.1 shows an example of cell ordering for a 2 dimensional data space ($n = 4$). For example, $\vec{c_7} = <1, 2>$, its order is 7. An uncertain data set is encoding by a grid-based data summary through mapping each object to the corresponding cell.

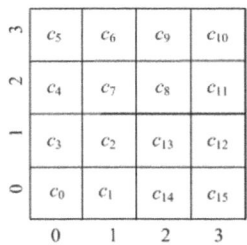

Fig. 1. Cell ordering

Mapping a Dataset to a Grid. When mapping D_i to $G(D_i)$, each object t in D_i is mapped to a cell $c(t)$. We use $c.CS(D)$ to denote the set of objects in D mapped to cell c. Each cell c is associated with a cell probability $c.cp$, which represents the probability that none of the objects in $c.CS(D)$ exist, i.e.,

$$c.cp = \prod\nolimits_{t \in c.CS(D)} (1 - p(t)) \tag{3}$$

After mapping D_i to $G(D_i)$, the local sites S_i send the information of cells of $G(D_i)$ to H. Note that empty cells contribute nothing to data distribution of local database. For cell c ($c.cp < 1$), the information need to be transferred is: $< c.id, c.cp>$.

3.3 Merge and Sharing

In this phase, H merge the received cells to a global data summary G. H first initialize an empty grid G, i.e., for each c in G $c.cp$ is set to 1. And then, once H receives a cell c^L from a local site, it sets the probability of corresponding cell in H (the cell c^H which fulfills $c^H.id = c^L.id$) by the following equation:

$$c^H.cp = c^H.cp \times c^L.cp \tag{4}$$

Lemma 1. *Let $G(D)$ represents the grid-based data summary of D, G is the data summary merged as above, then G is the data summary sketching D, i.e., $G = G(D)$.*

Proof: We need prove that for any c and c' in G, if $c.id = c'.id$, then $c.cp = c'.cp$. By (3) we have $c'.cp = \prod\nolimits_{t \in c'.CS(D)} (1 - p(t))$, note that $D = \bigcup_i D_i$, thus $c'.cp$ can be rewrite to $\prod_i \prod\nolimits_{t \in c'.CS(D_i)} (1 - p(t))$. Then we get $c.cp = \prod_i \prod\nolimits_{t \in c'.CS(D_i)} (1 - p(t))$ $= c'.p$ from (4). Thus we obtain $G = G(D)$.

3.4 Local Pruning

After equipped with the global data summary G, the local sites attempt to prune un-qualified objects utilizing G. We first illustrate the pruning heuristic by the example in fig.2. Suppose a local site S_i holds a grid G received from H. For object t ($c(t) = c_7$) in

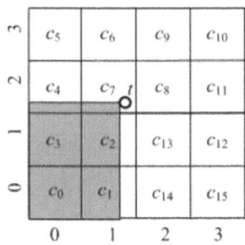

Fig. 2. Upper-bound object t with G

D_i, all the objects that dominate t is in the gray region, which covers cell c_0 and c_3 entirely. From equation (3) we can derive the upper-bound of $p_{sky}(t)$ as $p(t) \times c_0.cp \times c_3.cp$. Then t can be safely pruned if the upper-bound is less than q.

In order to prove the above observation theoretically, we first extend the dominate relationship between objects to cells naturally:

Definition 1. Dominate relationship between cells ("$\succ_c$"): $\forall c, c' \in G$, $c \succ_c c'$ if and only if $\vec{c} \succ \vec{c'}$. i.e., $\forall i \in [1,d], \vec{c}[i] \leq \vec{c'}[i]$, and $\exists i \in [1,d], \vec{c}[i] \neq \vec{c'}[i]$.

Definition 2. Strictly dominate relationship between cells ("$\succ_c^s$"): $\forall c, c' \in G$, $c \succ_c^s c'$ if and only if $c \succ_c c'$ and $\forall i \in [1,d], \vec{c}[i] \neq \vec{c'}[i]$.

Lemma 2. *For an object t in D_i, and a grid G for D, $p_{sky}(t)$ is upper-bounded by the product of $p(t)$ and $\prod_{c \in G \wedge c \succ_c^s c(t)} c.cp$.*

Proof: For any t' mapped to $c(t')$, t' dominates t if $c(t')$ strictly dominates $c(t)$. Thus $\prod_{t'' \in G \wedge t'' \succ t}(1 - p(t''))$ is upper-bounded by $\prod_{c(t') \succ_c^s c(t)} c(t').cp$. From (1) we have $p_{sky}(t) \leq p(t) \times \prod_{c \succ_c^s c(t)} c.cp$.

4 Further Optimization

Besides the empty cells, there are still large portion of cells that are redundant for determining the upper bound of skyline probability. We first present the observation by an example in fig.3. Suppose q equals 0.3, and the cell probability of c_0 and c_1 is 0.6 and 0.4 respectively, i.e., $c_0.cp \times c_1.cp = 0.6 \times 0.4 < q$. Then all the cells strictly dominated by both c_0 and c_1 ($c_{8\sim13}$) are not necessary to transferred to H. This is because, any object that falls into $c_{8\sim13}$ will inevitably strictly dominated by c_0 and c_1, thus its

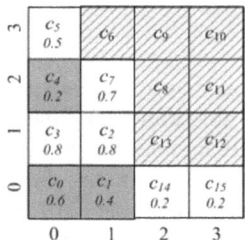

Fig. 3. Redundant cells

skyline probability is always less than q following lemma 2. Then cells in $c_{8\sim13}$ will never contribute to the probabilistic skyline computation and can be safely excluded from the cells that need to be sent. We can exclude c_6 in the same way.

We then formally proof the above heuristic. As depicted by fig.4, we use *DownLeft(c)* to denote the set of cells that dominate cell c as well as c itself, i.e., *DownLeft(c)*={ $c' \in G(D_i)$ | $c' \succ_c c$ } $\cup$ {c}, and use *DownLeft(c).cp* to denote the product of cell probability of cells in *DownLeft(c)*, i.e., *DownLeft(c).cp* $= \prod_{c' \in Downleft(c)} c'.cp$. The cells strictly dominated by c are denoted by *UpRight(c)*, i.e., *UpRight(c)* = { $c'' \in G(D_i)$ | $c \succ_c^s c''$ } holds. Then we obtain lemma 3:

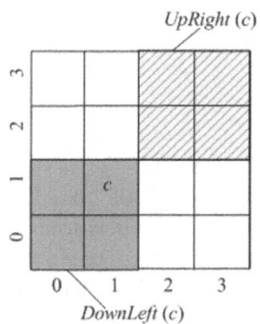

Fig. 4. Illustration of lemma 3

Lemma 3. *For any cell c in $G(D_i)$, UpRight(c) are not necessarily transferred if Downleft(c).cp is less than q.*

Proof: From the definition of *Downleft(c)* and *UpRight(c)* we can conclude that for any cell c in *UpRight(c)*, and any c' in *DownLeft(c)*, it always holds that $c' \succ_c^s c$. If t is dominated by c in *UpRight(c)*, then it must be dominated by any cell in *DownLeft(c)* . In another word, any object that can be prune by *UpRight(c)* will always be pruned by *Downleft(c)* , thus cells in *UpRight(c)* are not necessarily transferred.

5   Related Works

The skyline operator is first introduced into the database community by [5]. There has been considerable works on the distributed skyline query processing. Balke [6] first investigates the skyline computation under the distributed environments, where data is vertically scatted in multiple distributed nodes. Several works deal with distributed skyline retrieval in Peer-to-Peer network where different overlays are considered [7-9]. The recent literature considers distributed skyline retrieval under more general network architectures [10, 11]. However, the above algorithms can not be extended to uncertain data because of the semantic gap.

The skyline query was first extended to uncertain data by Pei [3]. ZHANG [12] studied the probabilistic skyline query in streaming uncertain data. They proposed an in-memory R-tree based approach to maintain the candidate set efficiently. Yiu [13] extended R-tree to facilitate the probabilistic skyline computation under the spatial database context. However, none of the above work considers the distributed prob-abilistic skyline query under the uncertain context.

More recently, DING [1] has proposed the first probabilistic skyline algorithm on distributed uncertain data. The communication efficiency of their algorithm stems from a priority-based scheme, where the central server computes global skyline in the order of local skyline probabilities. However, the communication cost of their algorithm highly depends on the data distribution, and increases sharply when the cardinality of local probabilistic skyline is relatively large.

6   Experimental Evaluations

6.1   Experimental Setup

Both our algorithm (Grid-based Probabilistic Skyline, GBPS) and e-DSUD are im-plemented in C++ and compiled by VC 2005 on a 3.8GHz Dual Core AMD processor with 2G RAM running Windows OS. The performance measures are number of bytes transferred during the query processing. Both real and synthetic datasets are used in our experiments. The real dataset (available from www.zillow.com) contains information about real estate all over the United States. We deal with 2 dimensions namely number of bedrooms and price gap. The price gap of a house is computed as the maximum house price in the dataset minus the price of the house. Intuitively, the real dataset with the dimensions (number of bedrooms, price gap) is more likely anti-correlated as a large room tends to has small price gap. We also generate the synthetic anti-correlated dataset following [5]. The dimensionality d of synthetic dataset ranges from 2 to 4. Both of the real and synthetic datasets contain 1 million objects.

We associate uncertainty to the aforementioned datasets by randomly assigning each object with an occurrence probability following normal distribution with the mean value μ equals 0.8 and the standard deviation σ equals 0.3. The threshold q is set to 0.3.

In each experiment, the dataset of cardinality n is horizontally split to m partitions equally, each simulate a local database. The default setting of d and m is 2 and 100 respectively unless otherwise specified.

6.2 Experimental Results

Optimal value of *n*. We first study the optimal setting of the number of slides on both real and synthetic datasets. As depicted by fig.5, the optimal value of *n* is around 14 and 16 respectively. When *n* is relatively large, transferring the cells consume too much additional communication cost. However, when *n* is relatively small, the pruning power of the grids will be limited. In the sequel, we set *n* to 14 for both datasets.

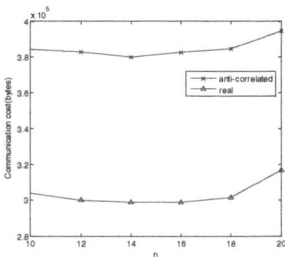

Fig. 5. Optimal value of *n*

Communication cost vs. number of local sites. Both algorithms incur more communication cost on anti-correlated dataset than on real dataset, since the former generates much more skyline candidates. GBPS saves average 12.1% and 19.7% bandwidth against e-DSUD for real dataset(depicted in fig.6 (a)) and anti-correlated dataset (depicted in fig.6 (b)) respectively, which confirms that GBPS perform better than e-DSUD for more skewed data. On both datasets, GBPS scales better with respect to *m*.

Communication cost vs. dimensionality. We test the scalability of GBPS against e-DSUD with respect to dimensionality by varying *d* from 2 to 4 on anti-correlated dataset. As shown by fig.7, GBPS always incurs lower communication cost than e-DSUD. Furthermore, the percentage of bandwidth saved by GBPS increases when *d* gets larger, which indicates that GBPS has better scalability than e-DSUD with respect to dimensionality.

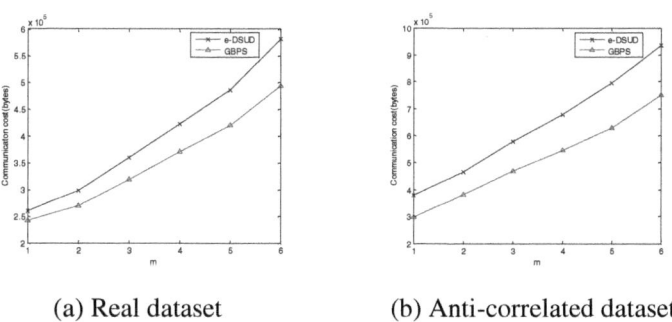

(a) Real dataset   (b) Anti-correlated dataset

Fig. 6. Communication cost vs. Number of local sites

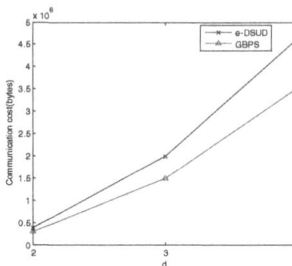

Fig. 7. Communication cost vs. dimensionality

7 Conclusions

In this paper, we propose a grid-based communication-efficient algorithm of probabilistic skyline on distributed uncertain data. By sharing a grid-based data summary, our algorithm improves the communication efficiency against existing approach.

As can be expected, grid-based approach performs not well on correlated and independent datasets. This suggests that, after samples or histograms are used to estimate the distribution of the underlying data, our approach can be an alternative when the underlying dataset is more likely anti-correlated, which is deemed as the challenging problems for skyline computation.

Acknowledgement

This work is supported by the National High-Tech Research and Development Plan of China under Grant No. 2006AA01Z474, No. 2007AA01Z451, and No. 2007AA010502.

References

1. Ding, X., Jin, H.: Efficient and Progressive Algorithms for Distributed Skyline Queries over Uncertain Data. In: 30th International Conference on Distributed Computing Systems. IEEE Computer Society, Genova (2010)
2. Aggarwal, C.C., Yu, P.S.: A Survey of Uncertain Data Algorithms and Applications. IEEE Transactions on Knowledge and Data Engineering 21(5), 609–623 (2009)
3. Pei, J., Jiang, B., Lin, X., Yuan, Y.: Probabilistic Skylines on Uncertain Data. In: 33rd International Conference on Very Large Data Bases. VLDB Endowment, Vienna (2007)
4. Jagadish, H.V.: Linear Clustering of Objects with Multiple Attributes. In: ACM SIGMOD International Conference on Management of Data. ACM, New Jersey (1990)
5. Borzsony, S., Kossmann, D., Stocker, K.: The Skyline Operator. In: 17th International Conference on Data Engineering. IEEE Computer Society, Hannover (2001)
6. Balke, W.-T., Güntzer, U., Zheng, J.X.: Efficient distributed skylining for web information systems. In: Hwang, J., Christodoulakis, S., Plexousakis, D., Christophides, V., Koubarakis, M., Böhm, K. (eds.) EDBT 2004. LNCS, vol. 2992, pp. 256–273. Springer, Heidelberg (2004)

7. Wang, S., Ooi, B.C., Tung, A.K.H., Xu, L.: Efficient Skyline Query Processing on Peer-to-Peer Networks. In: 23rd International Conference on Data Engineering. IEEE Computer Society, Istanbul (2007)
8. Wu, P., Zhang, C., Feng, Y., Zhao, B.Y., Agrawal, D.P., El Abbadi, A.: Parallelizing Skyline Queries for Scalable Distribution. In: Ioannidis, Y., Scholl, M.H., Schmidt, J.W., Matthes, F., Hatzopoulos, M., Böhm, K., Kemper, A., Grust, T., Böhm, C. (eds.) EDBT 2006. LNCS, vol. 3896, pp. 112–130. Springer, Heidelberg (2006)
9. Vlachou, A., Doulkeridis, C., Kotidis, Y., Vazirgiannis, M.: Skypeer: Efficient Subspace Skyline Computation over Distributed Data. In: 23rd International Conference on Data Engineering. IEEE Computer Society, Istanbul (2007)
10. Cui, B., Lu, H., Chen, Q.X.L., Dai, Y., Zhou, Y.: Parallel Distributed Processing of Constrained Skyline Queries by Filtering. In: 24rd International Conference on Data Engineering. IEEE Computer Society, Cancun (2008)
11. Zhu, L., Tao, Y., Zhou, S.: Distributed Skyline Retrieval with Low Bandwidth Consumption. IEEE Transactions on Knowledge and Data Engineering 21(3), 384–400 (2009)
12. Zhang, W., Lin, X., Zhang, Y., Wang, W., Yu, J.X.: Probabilistic Skyline Operator over Sliding Windows. In: 25rd International Conference on Data Engineering. IEEE Computer Society, Shanghai (2009)
13. Yiu, M.L., Mamoulis, N., Dai, X., Tao, Y., Vaitis, M.: Efficient Evaluation of Probabilistic Advanced Spatial Queries on Existentially Uncertain Data. IEEE Transactions on Knowledge and Data Engineering 21(1), 108–122 (2009)

Author Index

GPSR Compliance

The European Union's (EU) General Product Safety Regulation (GPSR)
is a set of rules that requires consumer products to be safe and our
obligations to ensure this.

If you have any concerns about our products, you can contact us on
ProductSafety@springernature.com

In case Publisher is established outside the EU, the EU authorized
representative is:

Springer Nature Customer Service Center GmbH
Europaplatz 3
69115 Heidelberg, Germany

Batch number: 09473985

Printed by Printforce, the Netherlands